Emanuel Swedenborg

AF874161

THE ANIMAL KINGDOM Considered Anatomically, Physically, and Philosophically

Volume 1

Elibron Classics
www.elibron.com

Elibron Classics series.

ISBN 1-4021-7998-7 (paperback)
ISBN 1-4021-2919-X (hardcover)

This Elibron Classics Replica Edition is an unabridged facsimile of the edition published in 1843 by W. Newbery, London.

THE ANIMAL KINGDOM.

LONDON:
PRINTED BY WALTON AND MITCHELL,
Wardour-street, Oxford-street.

4. Hence it is evident, that one truth is never opened without an infinity of other truths being also opened: inasmuch as the one is the conclusion of all the others; and further, that the conclusion—thus derived from the others—is not a truth except relatively to those numberless truths which constitute, that is, form and determine it. Thus a truth, to be such, must not be true in itself simply, or in the conclusion alone, but in those things and their connexion from which the conclusion is derived: whence every circumstance and every different modification varies the thing itself. Howbeit, I admit the existence of absolutely constant and immutable truths, both natural and moral, and pre-eminently, of spiritual truths; indeed, of all those that are in harmony with the perfect order of the universe. But in the animal kingdom—in man considered as a microcosm—there are as many different kinds of order, and as many different states, as there are subjects and human minds variously approximating to, or receding from, the order of the universe; for human minds are the objects into which order and truth flow, and the subjects by which the influx is received. Hence the multitude of dubious, uncertain, and shifting opinions on every subject: whereby knowledge is confused, the will perplexed, and our minds are brought into suspense respecting the plainest truths, and induced to contend with each other respecting them: thus truth is endangered; for it is placed in jeopardy whenever it is made a subject of contention. The consequence is, that the truths in the rational mind do not deserve to be called truths, but principles only, as, indeed, it is usual to call them.

5. I have set out, indeed, by speaking respecting truths, but this is a subject sufficient to fill a volume; nor does it belong to this place to consider what truth is, but rather, how we should investigate it. This must be our present business and labor.

6. There are two usually-received ways or methods for discovering truths; the synthetic and the analytic. The synthetic commences from principles and causes, and passes therefrom to phenomena and effects; thus proceeding from the prior to the posterior, from simple to compounds, from superior to inferior, from interior to exterior; or, what amounts to the same thing, from a universal to singulars, consequently to experience, to

confirm prior things. The analytic method, on the other hand, rises from phenomena and effects to causes, and evolves from them principles; consequently evolves universals from the experience of singulars, interior things from the exterior, simples from compounds, in a word, the prior from the posterior. Thus analysis as a method of proceeding is the inverse of synthesis.

7. Synthesis, which begins its thread of reasons from causes and principles, and evolves and unwinds it until it reaches the effects of the causes, and the phenomena resulting from the principles, assumes some particular principle familiar and favorable to the intellect as formed by previous ideas; and however susceptible this principle may be of doubt or controversy, synthesis seizes it as a truth, and lays it down; and thus presumed, defines and disengages it, and confirms it, first rationally, next empirically. Should any thing adverse appear, synthesis polishes away, represses and removes it, until at length the truth can come upon the stage, naked at first, but afterwards bedecked and ornamented; exactly in imitation of the inverse method of analysis, which is called also the *regula falsi*, or rule of false position. And synthesis, in reality, is nothing but a poor, precocious, and vague analysis; it gives out nothing more than what has crept into the intellect, and among the intellectual ideas, by way of the senses, from a few phenomena of experience, without any general bond to connect them; and for the most part in the first impetuosity of the judgment. The hasty conceptions thus formed at the mind's first glance, are termed opinions, conjectures, hypotheses; whence come systems.

8. This has been the received and established method for ages past, from the very infancy of philosophy, through its later and maturer development; and now also it endures and flourishes by the favor of our contemporaries, being adopted exclusively even at the present day in reasonings on the causes of things, which are naturally abstruse and profound. It is also pleasing, and wonderfully accommodated and in a manner akin to human minds; it enables each mind to indulge its own tastes, to favor its own state, and to assent to an order, whose laws are proclaimed as truths, n. 4. And we are very easily impelled and carried away into ideal games of this kind, inasmuch as they are races of our thoughts from assumed starting-places to the very goals we desire to reach. This also pimps to self-love

and self-glory, for as nothing properly belongs to us but the produce of our own minds, when these have conceived any thing, and supported it by plausibilities, we suppose we have divined the pure reality, opened the true Delphos or heaven itself, unlocked oracles which the genius of our predecessors never penetrated, and, in a word, earned an indisputable palm of victory. But those who commence with this species of scholastic exercitation, that is, who set out relying on mere reasoning, not fortified by the sure patronage of experience, will never, as I think, attain the goal; for they begin from the goal, and hurry to the starting-place; thus they bend their course outwards instead of inwards, contrary to the order which the nature of the human mind prescribes for the discovery of the occult and unknown.

9. But granting, for argument's sake, that any of the chiefs or rulers of the learned world, commencing from synthesis, may have perhaps taken the false for the true, but with the intention of afterwards eliciting from it purer truths, by means of analysis, or the rule of false position, and of correcting, perfecting, and polishing it, like the sculptor working the rude marble :—Tell me then, I pray, which of them has thus followed the rules of analysis? Which of them afterwards has wrought and corrected the visions and appearances he imbibed and predetermined, perhaps in the very dawn of thought, and which were adverse to the truth? Instead of this, have not they all, as experience shews, sought the confirmation of the false, and not of the true? For while the will is directed to the false, it is constantly detained in those things also that confirm it, or are conformable to it. Hence the presumption becomes more and more confirmed by plausible arguments, until at last it has the same power of persuasion as the truth itself. For whenever affirmative reasoning is applied to a preconception, an infinity of particulars, all voting the same way, fly to its assistance,—both the decrees of ratiocinative philosophy, and the phenomena of the world, laid hold of in the fallacious light of the senses. Indeed, there is nothing but may form a constituent part in different series of reasonings, if not directly, at least obliquely; as a single particle of salt may form an ingredient in an infinity of savors, and a single color in an infinity of pictures: and one thing may be engrafted on another, as branch upon branch; thus the legiti-

mate upon the spurious: so that falsehood assumes the form of truth, and the measure of the fiction increases by meditation. At length, when the phantom is led forth upon the theatre of what is called the learned world, multitudes run to it, passionately admire it, favor and applaud it; nay, numerous connoisseurs embellish it with paint and new decorations, so that it looks like a phantom no longer, but like a beautiful Venus, or a Delphian virgin. Whatever is now poured from its mouth, you are to regard as the voice of destiny, or the response of an oracle. But all things have their day; among the rest, the produce of the human faculties,—particularly those misshapen offspring, the monsters of hypothesis. They are conceived, they are born, they grow to maturity, they grow old, at last they die. But from the ashes of each new ones rise; and every hydra-head that is lopped off by the youthful Hercules, produces hundreds of others: whence spectres of similar brood prevail for ages, and, like enchantresses, distract the human mind perennially. Hence errors, mental obscurity, fallacies, and strife; civil wars between the soul and the body; scholastic contentions about straws and trifles; the flight and exile of truths; and stupor and thick darkness in those very things where the light is most brilliant: and this to such an extent, that the very altars and their sacred fire are contaminated; which is the reason why the philosophy of the human mind is solemnly proscribed in the divine records. All these things flow from that single source,—we mean, from the habit and the propensity of reasoning synthetically.

10. The power of divining true principles by the mind alone, and of descending therefrom, in the path of certainty, through their consequences, to posterior things, belongs exclusively to higher beings and powers; to spirits, angels, and the Omniscient Himself, who indeed inhabit the brightest light, and dwell in essential truth and wisdom. They see all things, in one complex, as at once beneath them and within them: they view the last things from the first, the lowest from the highest, the outermost from the innermost; in a word, all the circumferences from the centre; consequently, the very effects of the world, from their causes. Not so human minds, which derive from the senses, or absorb through the senses, all the materials which they have to reason upon. For we are born in dense

ignorance; in process of time organs are opened for us, and ways prepared, and images themselves are sublimated, until they become ideas, and at length reasons; which when connected into series, are brought under the inspection of the reasoning power. Thus by slow degrees only, judgment is developed and reason displayed. This then is man's only way of attaining truths, so long as his soul lives in the body. Can you tell me by synthesis or *a priori*, before seeing the viscera or examining the interior parts, what is contained within the animal body? Can you predict that it contains the liver, the mesentery, the kidneys, the heart, the arteries, and an infinity of other things; still less that they are connected together in one way, and in no other? Must you not rather, like a blind man, afflicted with cataract and suffusion, present to yourself ludicrous imaginations, and dream dreams, at which you yourself, when you shall have looked into them, must ultimately either blush or laugh? But alas! we are so puffed up with self-conceit, that we seem to ourselves to be not in the outmost, but in the inmost; to be standing, not on the earth, but in the sky; and in no faint or uncertain light, but in the brightest radiance; nay, in heaven itself, whence we descend before we ascend, and where we even build our airy palaces; not knowing, that our very height must aggravate the peril of our fall. This, as we before said, is the cause and the source of the insanities of the human mind. So much for synthesis, now for analysis.

11. Analysis commences its web of ratiocination from the facts, effects, and phenomena which have entered through the bodily senses, and mounts to causes, and causes of causes; that is, to the simple principles of the mind; and thus unwinds the thread of the web. In the first place it searches for certain and evident materials, and collecting them from all quarters, heaps them together, and again selecting them from the heap, reduces them skilfully into order. Furthermore, it imbibes all the sciences wherewith nature has assisted the human mind, and not in memory only, but in heart, and learns them for the sake of their application. Enriched with as it were these treasures, and aided thereby, the mind girds herself to her task, and begins to work and to build. If the monument she is essaying to construct may be compared with a palace, a mansion, or a pyramid, she may be said now to lay the foundation first, then to raise the

walls, and surrounding the edifice with ladders and scaffolds, gradually to carry it to the roof or summit. Whatever is now wanting to complete the fitness and coherence of the whole, —as posts, rails, gates, tiles, and the like, is superadded afterwards. Thus the mind, keeping along the path of analysis, founds and rears her palace, not in the air, or in an atmosphere too high for her, which is not her element, and where there is no support, still less foundation; but on the solid ground.

12. This is the only way to principles and truths—to high and almost to heavenly things—and no other appears to be open to us earthborn men; yet truly it is a most toilsome and extensive one, if we enter it with the intention of searching out the truth by all the truths which enter into it, that is, of exploring all the truths which form the one truth, and of connecting them together, or concatenating them, in one general bond. In this case evidently we must lay the broadest foundation—we must compare all things together carefully, and embrace them in one design. We must also make ourselves thoroughly masters of all the sciences, doctrines, and arts, which the work will require for its completion: nay, from those already known, we must generate and discover others: for by these means the work is constructed, and the mind led directly to the summit. In a word, we must cultivate acquaintance with all the muses. To these requisites we must add an innate love of truth, an eager desire of exploring it, a delight in finding it, and a natural gift and faculty of meditating thoughtfully and distinctly, and of connecting reasons together acutely: also of recalling the mind from the senses, from the lusts of the body, the enticements of the world, and its cares,—all which things are distracting forces, —and of keeping it in its own higher sphere, until it has summed up its reasons, and carried its thoughts to their conclusion. In proportion as by these means we ascend to truths, in the same proportion truths descend to us. Above all things it behoves the mind to be pure, and to respect universal ends, as the happiness of the human race, and thereby the glory of God: truth is then infused into our minds from its heaven; whence as from its proper fountain it all emanates. "Plato used frequently to say"— so the philosopher relates—"that when his soul was engaged in contemplation, he seemed to enjoy the supreme good, and incredible delight: that he was in a manner fixed in astonishment,

acknowledging himself as a part of a higher world, and as feeling his own immortality with the greatest assurance and light: at length, that his understanding, wearied with this contemplation, relapsed into fantasy, and that he became sorrowful as the light decreased. That again leaving the body, and returning to the former state, he found the soul abounding with light, and this light now flowing into the body." And again, "The soul, freed as it were from the body, ascends and is enlightened; descending again, it is obscured, but it is afterwards purified and reascends." But this may perhaps appear like a mere fable, to those who have not experienced it.

13. When at length, under the conduct of such an analysis, we have been carried up to the principles of things, we may then properly for the first time commence, or rather return, from principles, and put them forth, as of sufficient authority, by a clear and intelligible definition: for the mind now looks round the whole world as from a mirror, and contemplates all things in a universal manner. Ladders are constructed, and steps interposed, whereby we may equally descend and ascend. These ladders are so many concatenated series of truths, by which we are enabled to steer our course, or to go and return, whithersoever it pleases us. But these very truths, explored by this means, if we must confess it, nevertheless remain still intermixed and entangled with ignorance and twilight shades, being, therefore, after all, only appearances of truths; for the mind is never absolutely purified from the fallacies of the senses, it is never removed from them or placed beyond them, during its conjunction with the body.

14. I have now, therefore, ventured to attempt this method of discovering truths, at present deeply hidden under a veil of hypotheses. And the proper time has arrived; for a rich store of experience is at hand; an accumulated heap sufficient to enable us to build a palace; a luxuriant field where our sickles may reap an abundant harvest; a table where we may enjoy the most sumptuous banquets. Nor do I think we ought to wait any longer, lest haply experimental knowledge should be overtaken by age, night, and oblivion; and the arts and sciences be carried to the tomb; for unless I mistake the signs of the times, the world's destinies are tending thitherwards. The following then is a summary of my intended work.

I intend to examine, physically and philosophically, the whole Anatomy of the Body; of all its Viscera, Abdominal and Thoracic; of the Genital Members of both sexes: and of the Organs of the five Senses. Likewise,

The Anatomy of all parts of the Cerebrum, Cerebellum, Medulla Oblongata, and Medulla Spinalis.

Afterwards, the cortical substance of the two brains; and their medullary fibre; also the nervous fibre of the body; and the muscular fibre; and the causes of the forces and motion of the whole organism: Diseases, moreover; those of the head particularly, or which proceed by defluxion from the Cerebrum.

I purpose afterwards to give an introduction to Rational Psychology, consisting of certain new doctrines, through the assistance of which we may be conducted, from the material organism of the Body, to a knowledge of the Soul, which is immaterial: these are, the Doctrine of Forms: the Doctrine of Order and Degrees: also, the Doctrine of Series and Society: the Doctrine of Influx: the Doctrine of Correspondence and Representation: lastly, the Doctrine of Modification.

From these doctrines I come to the Rational Psychology itself; which will comprise the subjects of action; of external and internal sense; of imagination and memory; also, of the affections of the animus. Of the intellect, that is, of thought and of the will; and of the affections of the rational mind: also, of instinct.

Lastly, of the Soul; and of its state in the Body, its intercourse, affection, and immortality; and of its state when the body dies. The work to conclude with a Concordance of Systems.

15. From this summary or plan, the reader may see, that the end I propose to myself in the work, is a knowledge of the soul; since this knowledge will constitute the crown of my studies. This then my labors intend, and thither they aim. For the soul resides and acts in the principles, not of the body only, but also of the universal world; inasmuch as it is the supreme essence, form, substance, and force of the microcosm; and appoints, establishes, and governs the order thereof, of itself and by its own nature; consequently, it is in the sphere of truths. For these reasons, the soul has engaged the profound attention of nearly all human minds, ever since the infancy of philosophy;

and still holds them in suspense, division, and perplexity. But as yet, her mode of being and her nature are almost absolutely unknown; and such is the general state of doubt and hesitation, as to preclude all distinct thinking. This has given rise to so many obscure guesses on the subject,—it has caused so many clouds to collect round it, that all hope of discovery is nearly at an end. In order, therefore, to follow up the investigation, and to solve the difficulty, I have chosen to approach by the analytic way; and I think I am the first who has taken this course professedly.

16. To accomplish this grand end I enter the circus, designing to consider and examine thoroughly the whole world or microcosm which the soul inhabits; for I think it is in vain to seek her anywhere but in her own kingdom. Tell me, where else can she be found, than in that system to which she is adjoined and injoined, and where she is represented, and momentarily exhibits herself to contemplation? The body is her image, resemblance, and type; she is the model, the idea, the head, that is, the soul of the body. Thus she is represented in the body, as in a mirror. I am, therefore, resolved to examine carefully the whole anatomy of her body, from the heel to the head, and from part to part; and for the sake of a closer approach, to examine her very brain, where she has disposed her first organs; lastly, the fibres also, and the other purer organic forms, and the forces and modes thence resulting.

17. But since it is impossible to climb or leap from the organic, physical, and material world—I mean, the body—immediately to the soul, of which neither matter, nor any of the adjuncts of matter are predicable, (for spirit is above the comprehensible modes of nature, and in that region where the significations of physical things perish); hence it was necessary to lay down new ways by which I might be led to her, and thus gain access to her palace,—in other words, to discover, disengage, and bring forth, by the most intense application and study, certain new doctrines for my guidance, which are (as my plan shews) the doctrines of forms, of order and degrees, of series and society, of communication and influx, of correspondence and representation, and of modification; these it is my intention to present in a single volume, under the title of *An Introduction to Rational Psychology.*

18. When this task is accomplished, I am then admitted as it were by common consent to the soul, who sitting like a queen in her throne of state—the body—dispenses laws, and governs all things by her good pleasure, but yet by order and by truth. This will be the crown of my toils, when I shall have completed my course in this most spacious arena. But in olden time, before any racer could merit the crown, he was commanded to run seven times round the goal, which also I have determined here to do.

19. Not very long since, I published the *Economy of the Animal Kingdom*, a work divided into distinct treatises, but treating only of the blood, the arteries, and the heart, and of the motion of the brain, and the cortical substance thereof; and before traversing the whole field in detail, I made a rapid passage to the soul, and put forth a prodromus respecting it. But on considering the matter more deeply, I found that I had directed my course thither both too hastily and too fast,—after having explored the blood only and its peculiar organs: I took the step, impelled by an ardent desire for knowledge. But as the soul acts in the supreme and innermost things, and does not come forth until all her swathings have been successively unfolded, I am, therefore, determined to allow myself no respite, until I have run through the whole field to the very goal—until I have traversed the universal animal kingdom, to the soul. Thus I hope, that by bending my course inwards continually, I shall open all the doors that lead to her, and at length contemplate the soul herself: by the divine permission.

20. But I know it will be whispered in my ear by many of the most accomplished philosophers of the day, that it is vain and useless to enter the recesses and interiors of the human body, with a view to arriving at the soul, inasmuch as the very things which are far below her, and objects of ocular vision, as the organic parts of the body, their modes, sensations, and actions, are not only obscure to our sight, but are even fallacious under close scrutiny and investigation. Moreover, the human intellect cannot penetrate or know itself even; how then should it penetrate the soul, which inhabits a still higher or superior region? for those things that are superior, inhabit a light inaccessible to the inferior; and if we rashly approach it too nearly, designing to enter it, we shall either cover ourselves with shade,

as when the eye gazes on the sun, or perish outright, like a garment thrown into the flames of an altar. Add to this, that the idea that might apprehend the soul, and the speech that might express her, are both wanting; for nothing that is adequate to body and matter, is adequate to the soul; she is neither corporeal nor material; consequently, she is entirely above that species of intelligence which receives its notions by means of the forms, predicates, and adjuncts of matter—as is the case with the human intellect; and expresses them by the same means—as is the case with human speech. Whence possibly it may be inferred, that it is unprofitable, and absolutely foolish, for any one to attempt ascending thither. But these arguments may properly be met by a few opposite ones. Inasmuch as the soul is the model, the idea, the first form, the substance, the force, and the principle of her organic body, and of all its forces and powers; or, what amounts to the same thing, as the organic body is the image and type of its soul, formed and principled to the whole nature of the soul's efficiency, it follows, that the one is represented in the other, if not exactly, yet quite sufficiently to the life; and that an idea of the soul is suggested to the mind by elevating the forms of singulars, and extracting from them a higher meaning, and by analogies and eminences, as will be seen in our doctrines of forms, of order and degrees, of correspondences and representations, &c. Thus, by the body, we are instructed respecting the soul; by the soul, respecting the body; and by both, respecting the truth of the whole: and in this way we are led to an ample knowledge of the animal kingdom.

21. I shall again suppose an objection made, to the effect, that the human mind appears to be interdicted from prying into those things which transcend or exceed the present state; consequently, into the soul, which guards the threshold of a sacred temple with three recesses; for the way to celestial, spiritual, and divine things, leads to the soul, through the soul, and from the soul. The reason of the interdiction is, that all those things which transcend our present state, are matters for faith and not for intellect: as for instance, that the soul exists, that it is a spiritual essence, that it is intimately united to the body; that it is affected by the rational mind according to the state and influx of the active principles thereof; that when freed from its

chains, it will possess immortal life, and either be happy in the assembly and kingdom of the blessed, or unhappy and accursed in a hellish and demoniac region; not to mention other things of similar import. The province of reason or intellect consists exclusively in considering and inquiring what is reasonable, profitable, and becoming in society, or in the civil and moral world; and what is proper to be done in the kingdoms below it, the animal, the vegetable, and the mineral. Let the intellect be contented with this its lot, and not aspire to higher things, which, inasmuch as they are sanctuaries and matters of revelation, exist to faith only. Furthermore, faith is banished as soon as ever the intellectual power endeavors to open the doors to its mysteries; for the intellect most commonly abolishes all faith in divine things; and what is received by the intellect, is not received by faith, that is to say, not by such a faith as elevates us above ourselves. And those who are inspired by a divine faith, completely despise the assistance of confirmatory arguments; perhaps they will laugh at this very book of mine,—for where there is faith, what need is there of demonstration; as where there is sight, what need is there to talk of light. Thus faith is above all demonstration, because it is above all the philosophy of the human mind.

22. I grant this; nor would I persuade any one who comprehends these high truths by faith, to attempt to comprehend them by his intellect: let him abstain from my books. Whoso believes revelation implicitly, without consulting the intellect, is the happiest of mortals, the nearest to heaven, and at once a native of both worlds. But these pages of mine are written with a view to those only, who never believe any thing but what they can receive with the intellect; consequently, who boldly invalidate, and are fain to deny the existence of all supereminent things, sublimer than themselves, as the soul itself, and what follows therefrom—its life, immortality, heaven, &c., &c. These things, perhaps, since such persons do not perceive them, they reject, classing them among empty phrases, *entia rationis*, phantasms, trifles, fables, conceits, and self-delusions; and consequently they honor and worship nature, the world and themselves; in other respects, they compare themselves to brutes, and think that they shall die in the same manner as brutes, and their souls exhale and evaporate; thus they rush

fearlessly into wickedness. For these persons only I am anxious; and as I said before, for them I indite, and to them I dedicate my work. For when I shall have demonstrated truths themselves by the analytic method, I hope that those debasing shadows, or material clouds, which darken the sacred temple of the mind, will be dispersed: and thus at last, under the favor of God, who is the sun of wisdom, that an access will be opened, and a way laid down, to faith. My ardent desire and zeal for this end, is what urges and animates me.

23. Let us then gird up our loins for the work. Experience is at our side with a full horn of plenty. The nine virgins are present also, adorned with the riches of nearly two thousand years: I mean, all the sciences, by whose abundance, powers, and patronage, the work is constructed. The sciences are indeterminate and of no profit or advantage, unless they be applied and made subservient to uses. What is a knowledge of numbers, ratios, figures, and forms, in arithmetic and geometry, apart from its benefits in civil life? What are the philosophical sciences, with their predicates, qualities, modes, and accidents, without they have reference to reality? All things, at the present day, stand provided and prepared, and await the light. The ship is in the harbor; the sails are swelling; the east wind blows; let us weigh anchor, and put forth to sea.

CHAPTER I.

THE TONGUE.

24. We commence our analytical exploration of the inferior or abdominal viscera, from the tongue, the lips, the mouth, and the fauces; because they are in a manner the thresholds by which we are conducted into the œsophagus, the stomach, the intestines, and the viscera appended to them and enclosed in the peritonæum: and also, indeed, into the trachea and the lungs, which belong to the chest, (although the principal entrances to the latter are the nares, of which we shall treat when we come to speak of the thoracic or pleural viscera.) First, then, we shall premise the observations and experience of the best authorities; of those, who from double superiority in endowment and education, and unwearied diligence in the school of experience, are acknowledged by the scientific world to have possessed an extraordinary insight into nature. Afterwards we shall endeavor to profit by the experience thus afforded, to elicit the connexions of things, their ends and causes; which we shall reduce into a general form, and then confirm and corroborate, either by the same experience, or a resumption of it; for we shall always take especial care not to wander far from the company of experience.

25. Heister. "The situation, singleness and size of the tongue are well known to everybody. In figure it is somewhat pyramidal, but flat. Its anterior, more moveable and sharp portion, is called the apex, and its posterior and thicker portion is called the base or root. It is connected with the os hyoides, the lower jaw, the styloid process, the pharynx, the larynx, and the other adjoining parts, by a number of muscles and membranes. Under its apex is the frænum, (at the root of this, on either side, the salivary ducts of Wharton enter the mouth); and a membranous ligament; by both of which it is connected to the lower jaw, the os hyoides, the larynx, and particularly to the epiglottis.

In the middle of the tongue there is a longitudinal line, called the median line, which divides it equally into two parts. The tongue is principally composed of small muscles, of nervous involucra or membranes, and of fat; also of glands and considerable vessels. Besides the sublingual glands, it has a great number of miliary glands on its superior and posterior parts: in some subjects, I have seen these as large as lentils, and perforated in the centre. The foramen cœcum, as it is called, is often conspicuous among these glands in the superior and posterior part of the tongue; its use was long unknown, but at present it is reckoned a salivary duct. The arteries of the tongue come from the external carotids, and are large and numerous. Its veins are branches of the external jugulars: those which are visible under the apex of the tongue, are called the ranine veins; these are sometimes opened by surgeons for bleeding, particularly in disorders of the fauces. The nerves of the tongue are very large; consisting of two branches from the fifth pair which are usually thought to be gustatory, and of two from the ninth pair which are considered motor. Coschwitz and Duvernoy delineate lymphatics in the tongue. The muscular substance of the tongue is surrounded by three involucra, tunics or membranes. The external, which is continuous with the common membrane of the mouth, forms a number of pyramidal and globular pouches or vaginulæ, which are porous and serve for receiving the nervous papillæ of the third membrane. The middle, or membrana reticularis Malpighi, consists of a beautiful net-work, which transmits the nervous papillæ through its apertures; this is visible only on the upper part; but is more difficult to detect in man than in brutes. The third, or membrana papillaris nervosa, visible only in the same part of the tongue as the last, contains nervous papillæ of different shapes, but chiefly fungiform, (similar to snails' horns or fungi,) full of little holes, and capable of protrusion and retraction; also pyramidal papillæ of different sizes, and sometimes incurvated. Both these kinds of papillæ arise from the internal membrane of the tongue, and from its nerves; they pass through the little foramina in the reticular membrane, and terminate in the vaginulæ of the external membrane. These papillæ are the primary organ of taste. The muscles of the tongue, besides those of the os hyoides, are four pair, in addition to which, however, there are a multitude of inextricable muscular fibres; which contract and expand the tongue,—draw it backwards and thrust it forwards, and perform all its other wonderful motions. *(Comp. Anat.* n. 285.)

26. The tongue has four pairs of muscles. The genio-glossus arises from the chin, above the genio-hyoideus; it enters the middle of the tongue, and moves it forwards. The stylo-glossus arises from the apex

of the styloid process; it descends obliquely to the root and side of the tongue, and moves it laterally, backwards, and upwards. The cerato-glossus arises from the cornu of the os hyoides, and is inserted into the root of the tongue. The basio-glossus arises from the base of the os-hyoides, and runs through the middle of the tongue towards its apex; this muscle with the assistance of the preceding one retracts the tongue, and makes it shorter. The mylo-glossus of some authors is a part of the mylo-hyoideus; hence there is no occasion to consider it as a distinct muscle. Moreover, I have frequently found it inserted not into the tongue, but into the genio-hyoid muscle only. *(Comp. Anat.* n. 323.)

"The foramen cœcum linguæ, delineated by Collins and Morgagni, as well as by myself, was first regarded as a salivary duct by Vater, who proved by a number of experiments, that it communicates with the glandular substance about the base of the tongue; but he did not shew very clearly how this communication is effected. Coschwitz attempted to do this, and to shew that several ducts run through the maxillary and sublingual glands, and have their various places of excretion on the surface of the tongue: these ducts, however, either from some fault of my own, or from the natural differences between different subjects, I have never been able to find. Nevertheless, in a male subject, I lately saw two considerable ducts of another kind, and which terminated in the foramen cœcum; these are figured exactly in my tab. viii. fig. 34, 35. *(Comp. Anat.* n. 278.) In this case, the foramen was so large and deep, that it admitted without difficulty a thick tube, such as I usually employ for inflating the ureters and the bladder. When I blew into the foramen through the tube, in order to discover the duct mentioned by Vater, the posterior part of the tongue became remarkably distended. But as this was not sufficient to shew its true position, constitution, and direction, I made use of a red wax-injection, and this succeeded so well, that afterwards, on opening the foramen cœcum, the beginnings of two remarkable and hitherto undescribed ducts presented themselves, passing for about the depth of a line under the teguments of the tongue. After cautiously removing the teguments, these ducts appeared; one on each side; similar in figure, size, situation, and course, to those in the delineation. The left duct was full of a clear liquor very like saliva, all the way to a tubercle, which was a pellucid vesicle. At the end of the right duct there was no vesicle. As, however, in the foramen cœcum of this subject, there were also three little orifices whose cavity ran downwards, but which did not admit a fine point more than a line or two, I therefore consider this foramen or canal to be the excretory duct of the lingual glands, or a place of concourse of various ducts, which discharge the saliva into the mouth from the glandular substance of the tongue; but these ducts, I suspect, if the large foramen happens to be wanting, (as is, indeed, frequently the case,) discharge themselves by a number of lesser, distinct orifices; and it appears that nature varies greatly in this particular. (*Comp. Anat.* tom. ii. not. 56.) The structure of the orifices themselves was singular; they had the appearance of valves or carunculæ, which had collapsed; but they opened if they were inflated with the blow-pipe directed downwards. (*Comp. Anat.* tom. ii. expl. tab. viii.)

"It is usual to class among the bones of the head the bone which adheres to the base of the tongue, and which has hence been called by some the lingual bone, and by others, from the Greek letter, *υ*, the os upsiloides or hyoides; others also have called it the os bicorne. It is in contact with no other bone. In young subjects it consists of three moveable pieces, a base, occupying its middle portion, and two lateral pieces called the cornua, with which the tongue is connected. In adults, at the junction of the cornua with the base, I have often found two other bones, very small, and thence overlooked by most writers. These are nearly of the shape of a grain of wheat, and may thence be called ossa triticea; there are ligaments affixed to them, by which they are attached to the styloid processes. Sometimes, but very rarely, other little bones are found in these ligaments; and this caused Vesalius, who had met with six of these, and other authors, to reckon eleven pieces to the os hyoides; but without good reason. This bone serves as a firm basis to the tongue; and various muscles of the tongue and of the larynx, which serve for the necessary motions of both, are inserted into it. (*Comp. Anat.* n. 106.) The os hyoides coheres with the base or root of the tongue, so that it is placed between it and the larynx, and moves with the tongue. It is connected by ligaments with the tongue, the larynx, and the styloid processes; by muscles, both with the parts just mentioned, and with the jaw, the scapulæ, the clavicles, and the sternum. It has five pairs of muscles. (*Comp. Anat.* n. 279). 1. The mylo-hyoideus arises broad but thin from the base of the lower jaw, and terminates at the base of the os hyoides. 2. The coraco-hyoideus arises near the coracoid process, from the superior costa of the scapula, and is inserted into the base and the cornu of the os hyoides. 3. The genio-hyoideus arises from the middle of the chin, above the mylo-hyoideus, close to the symphysis of the under jaw, and terminates in the base of the os hyoides. 4. The sterno-hyoideus sometimes arises from the sternum and clavicle, sometimes from the sternum alone, sometimes from the clavicle alone; and is inserted into the base of the os hyoides; this muscle is the antagonist to the preceding one. 5. The stylo-hyoideus takes its rise from the styloid process, and terminates in the cornu and base of the os hyoides; it is often perforated by the digastricus." *(Compendium Anatomicum,* n. 322.)

27. WINSLOW. "We all know the situation and space which the tongue occupies: its figure and size are accommodated to that space. Its posterior and thicker portion is called the base; the anterior and thinner, the apex. Its upper surface is plano-convex, and divided into two lateral halves by a shallow depressed line, the linea linguæ mediana. The margins are thinner than the other parts, and a little rounded. The lower surface reaches only from the middle of the length of the tongue to the apex. The tongue is principally composed of softish fleshy fibres, intermixed with a peculiar medullary substance, and disposed in various ways; some of these fibres are confined to the tongue; others form muscles which go out from it in different directions, and are inserted in other parts. The upper surface of the tongue is entirely covered with a thick membrane or tunic of a papillary texture, upon which lies a fine epidermoid membrane, which is likewise continued over the lower surface, but without papillæ. Three sorts of papillæ are distributed over the upper surface of the tongue: capitatæ, semilenticulares, and villosæ. The papillæ of the first kind are the largest, resembling little mushrooms with short stalks. They lie on the base of the tongue in superficial crypts or fossulæ. They resemble small conglomerate glands, seated on a narrow basis, and a little hollowed in the middle of their convex side. They occupy the whole surface of the base of the tongue, and are arranged in such a manner, that the anterior ones form an angle. These are glandular mammillæ or papillæ, or salivary or mucilaginous glands. We often observe about the middle of this part of the tongue, a peculiar foramen, more or less deep, the internal surface of which is entirely glandular, and filled with small papillæ like those of the first kind. It is called foramen cœcum Morgagnii, from its discoverer. M. Vater examined it closely, and discovered some ducts, apparently salivary, belonging to it. The papillæ of the second kind, or lenticulares, are small orbicular eminences, only a little convex, the circular border of which is contiguous to the surface of the tongue. When we examine them with the microscope in a fresh tongue, we find their convex sides full of small holes or pores, like the end of the spout of a watering-pot. They lie on the middle and anterior parts of the tongue, in greater or less numbers, and are sometimes most visible on its edges. They soon lose their consistence after death, so that by rubbing them several times, they may be drawn out in the form of small soft pyramids, and laid down on one side. The papillæ of the third kind, or villosæ, are the smallest and most numerous, and occupy the

whole of the upper surface of the tongue, and even the interstices between the other papillæ. They would be more properly named papillæ conicæ, than villosæ, from the figure which they appear to have when examined in clear water through a microscope. They are naturally softish, but they become so flaccid after death, that by handling them they may be made short and thick, whereas they are naturally long and small. The internal muscles of the tongue, or rather the fleshy or muscular fibres of which the mass of the tongue is composed, and which go no further than the tongue, called by Spigelius, musculi linguales, are of three general kinds,—longitudinal, transverse, and vertical; and each with different degrees of obliquity. The longitudinal fibres point to the base and apex of the tongue, and seem partly to be expansions of the stylo-glossi, hyo-glossi, and genio-glossi muscles. The vertical fibres seem also to be prolongations from the genio-glossi, and the transverse from the mylo-glossi. Besides these mixed productions, there is a distinct plane of longitudinal fibres which run near the upper surface of the tongue, and a distinct transverse plane under them. All these fibres are interwoven with each other, and terminate about either the borders, the base, or the apex of the tongue, without passing beyond its substance. They lie immediately above those that belong to the genio-glossi. To discover them all, and their different degrees of direction, we need only dissect the tongue longitudinally and transversely, after it has been boiled, or well macerated in strong vinegar. The external muscles of the tongue, are those which enter it by one extremity, and by the other are fixed in some part without the tongue. Of these there are four pairs,—the mylo-glossi, stylo-glossi, hyo-glossi, and genio-glossi. The muscles which move the os hyoides, belong likewise to the tongue, and are the principal directors of its motions: they are, the mylo-hyoidei, genio-hyoidei, stylo-hyoidei, omo-hyoidei, and sterno-hyoidei." We now come to the muscles which are proper to the tongue. "The mylo-glossi are small fleshy places situated transversely, one on each side, between the ramus of the lower jaw and the base of the tongue. They are connected to the lower jaw immediately above the posterior half of the mylo-hyoid muscle, between the prominent oblique line on the inner surface of the maxilla and the molar teeth; from thence they run towards the side of the base of the tongue, and are lost on one side of the glosso-pharyngei. The stylo-glossi are two long, slender muscles, which descend from the styloid processes, and form a portion of the sides of the tongue. Each muscle is attached to the outside of the styloid process by a long tendon; being the uppermost of the three muscles which are attached to this process of the temporal bone the stylo-hyoideus is the lowest, and the stylopharyngeus is in the middle. As it runs down almost opposite to the inside of the angle of the lower jaw, it sends off laterally a pretty broad but short aponeurotic ligament, which, being fixed in that angle, serves for a frænum or suspensory ligament to the muscle in this part of its course. From thence it passes to the side of the base of the tongue, where it first adheres closely to the lateral portion of the hyo-glossus, and then constitutes a large part of the side of the tongue. The hyo-glossi are attached to three parts of the os hyoides that lie near each other; to the base, to the root of the great cornu, and to the symphysis between these two; and on this account, the hyo-glossus has been divided by some into two or three distinct muscles, called basio-glossus, cerato-glossus, and chondro-glossus. In some subjects they are distinct enough, being simply contiguous to each other. The genio-glossi are situated close to each other on the lower side of the tongue. Each muscle is attached to the internal or posterior part of the symphysis of the lower jaw; thence it runs backward towards the os hyoides, to which its lowest fibres are connected by a ligamentary membrane. In this course, its fibres are dispersed through the substance of the tongue in a singular manner. Some run directly towards the os hyoides, all the way to the base of the tongue; some are inflected forward, and go to the apex; and some radiate forward, upward, and backward in the substance of the tongue: the middle fibres expand laterally toward the sides of the tongue. The two genio-glossi run close to each other, as if they formed but one mass; but they are evidently divided by a thin cellular membrane, or middle septum, which also penetrates between the two lateral halves of the tongue; keeping the direction of the median line. When the ends of these muscles are detached from the chin, they presently contract so much, that their anterior extremities, which are naturally under the point of the tongue, are as far back as the middle of it. It is in this disordered and preternatural situation that these muscles are represented in plates. These two muscles, by their posterior straight fibres which go to the base, can draw the tongue out of the mouth, and bring it back again by their anterior bent fibres which go to the apex. They can either successively or all at once make the tongue hollow, and they can at the same time contract it by their lateral expansion; to pass over many other motions which these muscles are capable of performing. The stylo-glossi, by the contraction of their fibres, turn the tongue towards the cheeks, and force the aliment between the upper and lower molares. When they act jointly with the lateral portions of the superior fleshy plane of the tongue, they turn it obliquely upward to the teeth of the upper jaw, and near the cheeks, and thus bring away any food which may remain there after mastication. When they act jointly with the lateral portions of the hyo-glossi, they turn the tongue downward between the lower teeth and the cheek. When all the parts of the hyo-glossi act together, they contract the tongue, and turn its apex between the teeth and the under lip. The superior fleshy plane of the body of the tongue, bends it upward towards the palate, and makes it lick the upper lip. The mylo-glossi serve as a frænum to one side of the base, when the apex is turned towards the other side. The suspensory ligaments of the stylo-glossi may answer the same purpose, and even supply the want of the mylo-glossi. Besides the membranes of the tongue already described, it is customary to mention another,—the membrana reticularis; which is commonly demonstrated from the boiled tongues of oxen or sheep, and some assert that it exists in the human tongue, but I confess I have not been able to find it there. Not long since I shewed that it is not a true membrane, but a species of clear mucilaginous substance, diffused between the papillary and external or epidermal membrane, and which by boiling becomes white, and acquires consistency enough to be taken out in large portions; and that the foramina found in it are caused by the pyramidal papillæ. The tongue is fixed in the mouth not only by muscles, but also by ligaments, which are chiefly membranous. The principal ligament is that called the frænum, which is the fold that appears first under the tongue when we raise it, and which is a continuation or loose duplicature of the membrane which covers the inferior cavity of the mouth. This fold covers the curvature of the anterior portion of the genio-glossi, from the apex of the tongue, almost to the middle interstice between the lower incisor teeth. The other ligaments of the tongue are, first, the small membranous fold which runs along the middle of the convex part of the epiglottis to the base of the tongue; and next, the membranous folds which cover the inferior half-arches of the septum palati. These three folds are continuations of the membrane which covers the neighboring parts. The aponeurotic ligaments of the stylo-glossi may also be looked upon as true lateral ligaments of the tongue, and they adhere a little to the lower part of the internal or anterior pterygoid muscle. The principal blood-vessels of the tongue, are those that appear so plainly on its lower surface on each side of the frænum. They are four; an artery and a vein on each side, named the sublingual or ranine veins and arteries. The veins lie next the frænum, and the arteries on the other side of the veins. The arteries are branches of the second internal or anterior branch of the external carotid, and communicate with branches of the first external or posterior branch of the same carotid, &c. The veins are generally ramifications of a branch of the external anterior jugular vein. The nerves of the tongue: we observe four nervous fasciculi or cords, going distinctly to the base of the tongue, and continuing their course to the apex. Two of these are branches of the inferior maxillary nerves, or of the third branch of the fifth pair from the medulla oblongata. The other two are branches of the ninth pair. The lesser portion or first branch of the eighth pair, sends likewise a nerve to each side of the tongue. The great lingual nerve on each side glides forwards between the mylo-hyoideus and hyo-glossus, under the genio-glossus, and is distributed to the muscular fibres all the way to the apex of the tongue; communicating by several filaments with the small lingual nerve or branch of the fifth pair, and with the nerve from the eighth pair. The small lingual nerve on each side separates from the maxillaris inferior, sometimes at, and sometimes above its passage between the two pterygoid muscles. After separating more and more from the trunk, it passes under the lateral part of the tongue, over the sublingual gland. It gives filaments to the nearest parts of the tongue as it passes, and then entering its substance terminates at the apex, having sent a great number of filaments to the papillary membrane. It communicates, as has been said, with the lingualis major, and with the nerve from the eighth pair. This lingual nerve, a little after it leaves the maxillaris inferior, is accompanied by a small distinct nerve, which runs upward and backward towards the articulation of the lower jaw, in company with the lateral muscle of the malleus; passes through the tympanum between the handle of the malleus and the long leg of the incus, under the name of the chorda tympani: and afterwards penetrating the posterior wall of the tympanum, unites with the portio dura of the auditory nerve. This small nerve has been looked upon by anatomists as a kind of recurrent of the lingual nerve; but as in some subjects it appears to make simply an acute angle with the latter, and as the lingual is somewhat larger after this angle, it should rather be believed to come from the tympanum, and to unite with the lingual nerve, than to arise from this nerve and run up to the tympanum. In some subjects, the union of this nerve with the lingual is in a manner plexiform. The lingual nerve of the eighth pair, which is the first branch of this pair, runs first of all on the inside of the digastric muscle of the lower jaw, and gives filaments to the genio-hyoideus, the neighboring muscles of the base of the tongue, and those of the pharynx. Afterwards it gives out ramifications and forms anastomoses, and lastly goes to the lower part of the tongue, where it communicates with the lingual branches of both the fifth and the ninth pairs." *(Exposition Anatomique de la Structure du Corps Humain; Traité de la Teste,* n. 504—538.)

28. Malpighi. "The following is a short account of what I have observed in the tongue of the ox, and others of the same kind, when prepared by boiling. In the exterior or superior part, extending from the apex to the root of the tongue, an immense number of bodies, disposed in a kind of series, rise from the surface, and slightly curving, exhibit a uniform inclination and position towards the posterior part of the tongue; so as to resemble a carding comb: whence, if we stroke the tongue lightly with the fingers from the root towards the apex, we feel a kind of tearing sensation, from the strain on these bodies. In the ox they are cartilaginous, and seem to have a particular resemblance in figure to the teeth of the boar and other similar teeth; and inferiorly they exhibit a concavity, at their roots especially. They are composed of a dense and tough material, which looks like a collection of little twigs; from which circumstance, on the exterior part of the apex, the constant use and attrition causes them to be of unequal length in some subjects. Near the sides of the tongue they become so small, as to be almost obliterated; and at the base their place is supplied by certain membranous bodies, whereof the anterior resemble cones, but the posterior, obtuse papillæ. All these cornua are invested by the external membrane of the tongue, so that when this membrane is pulled off, the external covering of the cornua is pulled off with it. In the tongues of certain fishes, the same bodies are observed, but osseous, scattered in great numbers over the extremity of the apex; but in the middle of the tongue, grouped all together, and disposed circularly, with their apices pointing outwards; whence they present the appearance of an opening or full-blown flower. Whether or not these cartilaginous bodies in the ox are hollow interiorly, we cannot determine by sight,—they are so exceedingly minute: in some of them the extremities, which are immersed in the mucous substance, are concave, not unlike a common pen. The conical and obtuse bodies, which supply the place of the cornua at the base of the tongue, are evidently hollow, and their subtance becomes so thin and so much dilated, that it not only affords room for the nervous papillæ to enter them from beneath, but also becomes transparent. The cornua themselves proceed from, or rather are firmly implanted in, a tough membrane of the tongue, and a mucous or nervous substance underneath it; particularly in places where we observe singular holes or pits, disposed in similar order; wherefore no small portion of them projects into the interior part of the tongue, being firmly implanted in a kind of mucous substance. The membrane just alluded to, and which completely invests the tongue, has the cornua inserted in its superior area, and covered by it; and in those places where the larger papillæ (which we shall mention presently) protrude, it is very fine and thin, and

hollowed out in the form of small cupping-glasses, with lips projecting internally round their orifices. In the protuberances thus formed, we find a number of most minute pores and orifices, which probably exist also in other parts of the tongue, where the rest of the papillæ terminate, and at the roots of the cornua particularly.

"When this membrane is removed, and pulled off by the nails, we observe a kind of glutinous substance, extending over the superior part of the tongue especially, and of some thickness: this is white where it is connected to the membrane before mentioned; but blackish where it touches the interior part. It extends in the form of a membrane, or thick rete, and has conspicuous openings corresponding to all the cornua; and innumerable little canals between them, which can be seen only by the microscope: these are of different shapes, and open on the surface of the tongue, whence, if it be torn across, or examined by the microscope up against the light, they become visible; and inasmuch as an accurate view shews, that on the inside this membrane represents a dark-colored rete, (such as we see every day in the external circumference of snails, which is of the same color,) I therefore think it ought to be regarded as a cribriform and reticular substance. About the concave sides of all the larger foramina in this nervous and glutinous substance, we observe portions of the same reticular and cribriform substance, shewing that the larger foramina open in it, as well as the canals from the roots of the cornua, and from the surrounding parts of the exterior membrane. The glutinous substance extends completely over the upper area of the tongue; at its sides the dark portion is nearly obliterated; but traces of it are visible in the palate and even in the cheeks.

"After examining the glutinous substance, we next come to a nervous and papillary substance, of a yellowish white color; extending in the form of a membrane over the whole of the superior area particularly, and of considerable depth. This substance, on its inner surface, where it is connected to the muscular parts of the tongue, appears smooth and even; with the exception of certain nervous connexions or processes, scattered here and there between the fleshy fibres, and to which it is connected, or rather grows. On the exterior part, however, it is rendered uneven by certain remarkable nervous papillæ which proceed from it, and which are disposed in a very astonishing order. In the ox, the goat, the sheep, and in the human subject, these papillæ may be divided into three kinds, according to their threefold configuration and magnitude. Some are much larger than the rest; these are situated at the sides of the apex particularly, between certain others to be mentioned presently: they are disposed in a square on the superior area of the tongue; about its middle region, where it becomes white, very few are observed: but there are some, and those of considerable size, at the sides of the base. In substance and shape, these resemble the emissile and retractile cornua which are observed in snails; they stand on a long peduncle, which having risen through the mucous substance, ends in a little round head, that is placed in a sort of cavity in the thin exterior membrane. They originate from the nervous and papillary substance; for they are continuous with it, and the same circumstances, and a very similar structure, are observed in both. They have, however, this peculiarity: in the base of each there is a nervous twig, to which it is appended, or rather grows. The nervous papillæ of the second kind are more numerous than those just mentioned; corresponding on the inside to the cornua on the outside of the tongue. Proceeding from the common papillary substance, they rise to some height, and from their summits put forth nervous offsets, which enter the cavities before spoken of, and meet the roots of the cornua; they are surrounded by almost innumerable papillæ from the same origin, and which rise to the same height, but which are conical and more slender, and entering particular cavities in the mucous substance, at length terminate in the direction of the external membrane. About the base of the tongue, instead of the cornua, the nervous papillæ already described project outwards, and changing their form as they approach the base, they successively become more obtuse, rounded, and depressed: and the largest of them are not very much unlike those observed in the cheeks, at the roots of the teeth. It should, however, be noted, that the same papillary substance, and both the coverings under it, are found also in the palate and the cheeks, although in a very attenuated form; but with this difference, that in these places the larger papillæ stand out in the form of cones, and near them are excretory vessels, implanted in subjacent glands, and among the vessels are scattered a few very minute nervous papillæ. Whence this nervous and papillary substance originates, appears doubtful; for although I have examined it carefully in all parts, I have not come to a satisfactory decision. But since it is closely connected to certain nervous twigs, by which the whole mass of the fibres is connected and held together, and since many nervous twigs proceeding from the trunk terminate in it, hence it will not be absurd to suppose that it originates from these twigs, especially since the nerves designed for sensation in the other sensoria commonly terminate in a membranous expansion. The substance of which the whole mass of the tongue is composed, so far as sight informs us, may probably be considered to be muscular rather than glandular: for immediately underneath the membranes and the papillary substance, we observe straight, fleshy fibres, proceeding from the muscles which retract and shorten the tongue. But the centre of the tongue consists of many kinds of fibres, longitudinal, transverse, and oblique, which are all plaited together into something like matting, as may be seen in my various delineations." *(Exercitatio Epistolica de Lingua.)*

29. SWAMMERDAM. "The tongue of the covered snail (cochlea opercularis) lies under the concave fold of a certain cartilage, and is covered therewith when the snail swallows its food, in the same way as the larynx, in us, is covered by the epiglottis. Its tongue approaches in this respect to the tongues of serpents. Below it is seen a very delicate muscle, which draws the tongue, together with the whole mouth, palate, jaws, and even the brain itself, inwards into the belly, or at least into the neck. On the point of the tongue there is a little horny bone, divided into two or three very tender little teeth, with which, as with a hook, the snail, when it is about to eat, first lays hold of the small herb, and then suddenly draws the piece into its mouth. These creatures are polyphagous, and consume great quantities of vegetables." *(Biblia Naturæ,* p. 109. tab. iv. fig. 6. tab. v. fig. 3. tab. vi. fig. 2.)

"The little tongue of the naked or house snail (cochlea domestica or nuda), in the fore-part is of a bright or transparent red color, and covered on each side with many small parts, like the branchiæ of fish, or like a comb with a double row of teeth. These little parts grow paler towards the back of the tongue, and consist of a substance between horn and bone, such as the little tongue has in its divisions." *(Bib. Nat.* p. 177.)

"The tongue of the aliekruik snail, which the Dutch eat in large quantities, is enclosed in a singular cavity, the whole of which, it seems, may be protruded together out of the body. It is nearly two inches long, and at the same time beautifully convoluted into spiral folds, like a serpent closely coiled; and it is thus placed on the inside, in the body, so that it passes with the gullet under the brain. The part of the tongue inside the body, is cartilaginous, and most elaborate in its construction." *(Bib. Nat.* p. 184. tab. ix. fig. 20.)

"The tongue of the cuttle fish (piscis sepia), which is thought to be a fungous substance, consists, I find, of seven little cartilaginous bones, which lie close to each other, and are besides united by a particular membrane. The upper extremity or apex of the tongue is somewhat curved, but its lower portion is connected with some muscular and fungous flesh, in which it lies as in a hollow tube. This fleshy portion of the tongue is covered with rugæ and elegant folds, which seem to contain a great many salivary ducts. I have found by dissection, that a very considerable salivary duct opens and discharges into this fleshy part. The tongue, and the parts belonging to it, when taken out of the beak, generally bring along with them several of the muscles. When the tongue has been stripped of its membrane, we may see by the microscope that every one of the seven small bones is provided with above sixty curved, dentiform, cartilaginous papillæ, (somewhat resembling the papillæ of the tongue of the ox,) by means of which, the sepia, when feeding, is the better enabled to move its food, and dispose it for an easy swallowing. The fore-part of these papillæ is of a transparent amber color; but the hinder part, which constitutes the base of the tongue, is of a transparent white. If the tongue be inverted, its under side is found to resemble a regular web, produced by the combination of the cartilaginous bones already described." *(Biblia Naturæ, sive Historia Insectorum.* p. 882, 883, 884. tab. l. fig. 4, 5, 6, 7. Leydæ, 1737-8.)

30. Boerhaave. "On the dorsum of the tongue, especially at the apex and edges thereof, under the skin, lie obtuse papillæ, which appear to be of three different kinds; in a tongue that is alive, warm, moist, thrust out, and applied to taste any thing, they project from the surface, particularly when the person is hungry; but in the dead subject they disappear. They originate from a nervous substance which lies upon the muscular part of the tongue, (appearing like a very delicately-cellular, fatty membrane,) from whence they rise through a perforated reticular structure, just as in the skin; and they are then placed in small erect sheaths, or vaginulæ, belonging to the external membrane of the tongue, whereby they are defended against the roughness, acrimony, and warmth of the food. These sheaths, however, are porous, and project so far, that when the aliment is squeezed, it comes upon them with much force, and thus they are enabled to receive it. And it appears not improbable that these papillæ, which are so numerous, arise from the ninth pair of nerves, which is wholly bestowed upon the tongue; but that the branch of the fifth pair serves for the muscular motion of the organ, as in other parts. Laurence Bellini has shewn by careful experiments, that these papillæ are the organ which receives the impressions of taste from sapid objects; but that the other papillæ of the mouth, the tongue, the fauces, and the palate, have not this office: although perhaps those which lie inside the cheeks, near where the molar teeth of the upper and lower jaws meet, may be concerned in sensation. That matter in vegetables and animals from which art extracts salts or oil, mixed or separate, is the true object of taste; as salt, soap, oil and spirit; the same may be said of fossils. Taste, therefore, is occasioned, when the matter to be tasted—attenuated and dissolved to some extent in the saliva, warmed in the mouth, and by its motion applied to the tongue; insinuated into the pores of

the membranous sheaths; and penetrating to the surfaces of the papillæ therein contained—affects and moves the papillæ; whereby the motion impressed is conveyed to the general sensorium, and excites in the mind the idea of salt, acid, alkali, sweet, vinous, spirituous, bitter, aromatic, hot, pungent, acrid, austere, or tastes compounded of these. Hence it easily appears why the same object occasions different tastes, according to the difference of age, temperament, disease, sex, habits, and according to what the person has been tasting previously." *(Inst. Med.* n. 485—490.)

"The genio-glossi muscles arise fleshy from the inside of the chin, and spreading in their course, are inserted into the root of the tongue: they draw forward and contract the tongue. The cerato-glossi arise broad and fleshy from the sides of the os hyoides, and thence ascending, we find them dispersed in large quantities of fibres through the tongue; they draw back, depress and widen it. The stylo-glossi arise sharp and fleshy from the external part of the styloid process of the temporal bones, and descending obliquely forwards, are inserted into the back part of the tongue: this they lift up, draw to one side, and widen. Fleshy fibres proceed from the body of the tongue, to the inner sides of the lower jaw. The body of the tongue is made up of longitudinal muscles, which render it shorter; of transverse muscles, which make it narrower; of perpendicular muscles, which diminish its thickness; of others which act upon its dorsum and sides; of angular muscles which draw the tongue inwards, and of others which depress its sharpened dorsum; and of straight muscles, which compress the base of the tongue. All these being conceived to act either separately, or in various combinations together, easily explain the actions of the tongue, whereby it determines the food between the molar teeth, and fluids and other portions of the food towards the fauces; especially if the joint action of those fibres which pass from the tongue into the external muscles, and with which the tongue acts in concert, be taken into account." *(Institutiones Medicæ,* n. 62.)

31. Let the reader particularly consult the plates of different authors, for one delineation sometimes teaches more than a thousand descriptions: as Eustachius, *Tabul. Anat.* tab. xviii. fig. 2, 5, 8, 11, 18, 19, 20. (Edit. Colon. 1716.) Cowper, *Myotomia Reformata.* Morgagni, *Advers. Anat.* i. tab. i. ii. &c. Heister, *Comp. Anat.* tab. viii. fig. 34, 35. Malpighi, in Mangetus, *Theatr. Anatom.* tab. cix. fig. 15, 16, 17, 18, 19. Ruysch, in *ibid.*, fig. 7. Verheyen, *Corporis humani Anat.* tab. xxix. fig. 6, 7, 8, 9, 10, 11. Bidloo, in Mangetus, *T. A.*, tab. cix. fig. 8, 9, 10, 11, 12, 13, 14, 20; where he exhibits certain dentiform, or hard cartilaginous bodies, which project through the

investing membrane, from the apex to the root of the tongue, and the same bodies of a bicipital form: also, certain pyriform, vesicular, and pellucid globules, placed here and there among these bodies; and the reticular covering placed over them as a sheath, &c.

32. The use or effect which produces the end, must be the first object of analytical inquiry. The nature of a member or organ is known from the use. The use determines what the organ is in itself, or in its own form; what it is, in series with other organs which are contiguous to it or surround it, and which continuously precede and continuously follow it (*a*): and what it is, in order, with those which are above and below, or prior and posterior to it (*b*). All these, and their uses, indicate

(*a*) Those things are *contiguous* to an organ, which are in immediate contact with it, and surround it; properly those to which it is connected: thus the contigua of the tongue are the palate, the fauces, the cheeks, the gums, the teeth, the jaws, the lower jaw particularly, to which the tongue is attached; also the styloid process of the temporal bones, and the os hyoides; respecting which we shall speak presently. But the things which *follow continuously*, are the pharynx, the œsophagus, the larynx, the trachea; lastly, the stomach also, wherewith the intestines are continuous, to which again, by a continuous connexion, the other abdominal viscera are appended.

(*b*) We must clearly distinguish between those things which precede and follow continuously, and those which are prior and posterior: the doctrines of order and degrees, and of series and society, will shew what the distinction is; meanwhile, it is important to explain it briefly in this preliminary stage of our analyses. Those things which *precede and follow continuously*, are in the same series, or in the same degree; members of the same society; and they are homogeneous, for they are referrible to the same, or to similar unities. *Prior* and *posterior* things, on the other hand, are in series and degrees which are above or below; as the cerebrum, the cerebellum, and the medullæ continuous therewith, relatively to the organs of the body; or as the medullary and the nervous fibres, relatively to the moving fibres, which are the unities of the nature of the organ under investigation (*c*). The use and end are the first things that manifest themselves; for the end is in a manner all in all in every stage of the progress, from first to last; the very soul of the thing (*d*). Thus all things that belong to the body, and that act as parts of the body, vividly represent and manifest the soul.

33. Although the tongue is a uniform mass, and appears to be a congeries of muscular fibres, yet it distinctly brings forth a greater number of uses, produces more ends, in a word, com-

the muscles: the former are above, and are not homogeneous with their inferiors, for they are the causes and the principles of posterior things.

(*c*) The anatomy of a single member is by no means sufficient to explain the object under investigation; we must seek for instruction respecting it in a larger field; that is, not by scrutinizing the particular form of that member only, but also by examining thoroughly the other things which are in connection with it, and which succeed it in order. One thing indicates the nature of another; thus the anatomy of the whole body indicates the nature of each individual organ; (see my *Economy of the Animal Kingdom*, treatise i. n. 12, 13, 14, 15, 16:) and ultimately, the nature of the soul; for its superior and inferior operations accord, and by degrees explain what each thing, as a part, provides, contributes, and produces, in its series.

(*d*) This will be better understood, if we observe the operations of the rational mind, and their determination into action by the will. The mind first contemplates ends, and ultimately embraces them in one idea. These are carried into effect by the organs of the body, and unfolded in successive order. All effects conducing to the same end, are means, each of which respects and involves the ultimate end. Thus this individual end is a companion in the progress, in the last things as in the first, consequently, as it were, all in all; and thus in all there is the very mind itself, comprehending the ends. All effects which are without an end, are irrational, destitute as it were of mind and soul. From these considerations we may see clearly how the body is the image and type of the soul, or how the soul is represented in the body. For that which contemplates ends, is the soul; and that which produces ends, is the body. Thus all things of the body contain the soul, because they represent its series of ends. But of these subjects we shall speak in our rational psychology, which is the end of our work; and which, as being the end, must also be the point whence we begin.

prehends and takes part in more divers offices, simultaneous and successive, than any other member of the organic body. This is the cause of its seemingly perplexed and intricate structure (*e*). The whole mechanism of its uses and ends is, indeed, distinctly marked: nevertheless, it appears indistinct, because the eye, which is the mind's informant, sees only the grosser forms and the last compounds of things.

34. The primary, proper, and natural office of the tongue, consists in sucking, sipping, eating and drinking; or to speak more plainly, in receiving food for the nutrition of the body and the blood, in working this food about and forming it into a ball, and in rolling the ball into the œsophagus and swallowing it (*f*).

(*e*) Respecting the tongue's intricacy of texture, Malpighi says, "Why the tongue, in which the secrets of art and nature are displayed, should be so unintelligible and obscure in its composition; and whether this is to be attributed to a real meanness, or rather to an occult majesty,—I cannot tell. On this subject Vesalius remarks, 'Quandoquidem vero universam linguæ fabricam cæterarum corporis partium constructione mihi minus, ut verum fatear, cognitam necessario, ac serie quadam complexi sumus.' Whence it is no wonder, if we find among anatomists various and contradictory sentiments in speaking of the tongue; some regarding the flesh thereof as a loose mass of a spongy nature; some considering it glandular; some as consisting of muscular and glandular substance combined; and others again attempting to reduce it to a peculiar substance, like no other in the body. Nor is there less difference of opinion respecting the precise part in which the sensation of taste resides," &c. *(Exercitatio Epistolica de Lingua.)* In what follows, it will be seen, that the tongue is a compound body, made up of vessels, that is, of arteries and veins; of nervous and muscular fibrillæ; and of glands, papillæ, ducts and membranes, of different kinds; also, that it is full of fat, salivæ, and fluids, which permeate its corpuscules and little tubes, and also surround them.

(*f*) The tongue is a congregate, and, if we may use the expression, a simultaneous body; hence its operations are at first perceived as simultaneous; for instance, its mode of eating and chewing; nevertheless, it is divisible into the successive modes and actions here enumerated: thus each of its uses must be made out, before each of its forms can be extricated from their apparent chaos and confusion. First it *receives* the food and drink, and this on its apex or point, and also

The tongue, considered in general, performs this office; it likewise performs the same office in all its parts; for it foretastes the first fruits, the extracts, the spirits, and the purer essences of the food; that is, it takes them up by little mouths, it nimbly works them about upon little tongues, it drinks them by imperceptible pores, and delivers them immediately to the blood (*g*).

on its edges, which, therefore, are sharper, thinner, freer, and more agile than its body, and capable of being extended beyond the prolabia; for the same reasons, the anterior part of the tongue is loose, and entirely under the government of the muscles; and of an oval shape. Afterwards, it *works about* the materials received, and *makes them into a ball;* presses them closely against the fauces; expresses, macerates, and seethes their juices; and dissolves them in a salivary menstruum supplied for the purpose: on these accounts, its middle portion is connected to the lower jaw, and comparatively thick and immoveable, &c. Lastly, it *rolls down the ball of food,* not only by its own internal and external muscles, but by the assistance of the hyoidei also, of which we shall speak presently. Its modes therefore are in successive order, from the extreme points of the tongue, all the way to its very base, at the gullet, into which they are continued.

(*g*) More will be said on these subjects in the following pages, when we come to speak of the glands of the tongue. The appearance of little tongues all over the surface of the tongue, is even represented in the tabulæ of some authors, as Bidloo; these are so many oval, pointed, or bicipital bodies: in Verheyen's tabulæ, we see certain falcated bodies, and which are hard, cartilaginous or osseous. But Malpighi gives a complete description of them, as follows: "In the exterior or superior part of the tongue, an immense number of bodies, disposed in a kind of series, rise from the surface, and slightly curving, exhibit a uniform inclination towards the posterior part of the tongue; so as to resemble a carding comb. In the cow they are cartilaginous, and have a resemblance in figure to the teeth of the boar; and inferiorly they exhibit a concavity, at their roots especially. At the base of the tongue their place is supplied by certain membranous bodies, whereof the anterior resemble cones, but the posterior obtuse papillæ. These conical and obtuse bodies are evidently hollow. The cornua are implanted in places where we observe singular holes or pits, disposed in similar order. In the protuberances formed in the external membrane, we find a number of most minute pores and orifices, which probably exist also in other parts of the tongue, and at the roots

Thus the lesser parts perform not only the same office on a small scale as the entire tongue, but also the same offices as the tongue performs in conjunction with its associate and contiguous organs,—the lips, the cheeks, the fauces, and the palate; and the least parts, as the tongue performs in conjunction with its continuous organs,—the pharynx, the œsophagus, the stomach, &c.: for the least in every series comprehends an idea of its universe.

35. A second proper office of the tongue consists in feeling and perceiving what is about to be received, with a view to becoming acquainted with its qualities; that is, in tasting. This office makes it necessary that the tongue shall be capable of undergoing all the changes of condition which the sensation of taste presupposes and involves: that it shall have the power of properly expanding and relaxing, extending and retracting its sensoriola or papillæ; of applying them to objects, and touching

of the cornua particularly." (n. 28.) In insects' tongues, which are more simple, similar little cornua or denticles, or cartilaginous and osseous lunulæ, are observed; as in the tongue of the covered snail, and of the naked or house-snail, &c. "The tongue of the house-snail," says Swammerdam, "is covered on each side with many small parts, like the branchiæ of fish, or like a comb with a double row of teeth, &c." (n. 29). Thus, there are as it were denticles, or little teeth, which seize the nutritious particles, and carry them to the little cavities subjacent. But I do not dwell on these circumstances, for in the field of leasts, similitude in mere shape is of no consequence, so long as there be corpuscules that perform a similar use, as we shall shew presently to be the case with the glands. The same thing obtains in all the other members and organs; as in the eyes, where the globular parts in the vitreous humor are so many little eyes; a fact which is best seen in the eyes of the bee and the fly. In the lungs, the least vesicles are so many little lungs. In the cerebrum, the cortical substances are so many primitive cerebellula. In the heart, the lacunæ with their little columns are so many little ventricles of little hearts; see my *Economy of the Animal Kingdom*. The general, therefore, derives its peculiarity of nature from its parts; it must commence from the minutest textures similar to itself: but of these matters we shall speak at greater length in the course of our analysis.

the objects at all points; in a word, of suiting itself to all affections, proximate and remote (*h*).

36. A third office of the tongue (not, however, proper to it) consists in speaking (*i*), or in modifying the sound emitted from the larynx and trachea, in a common or general manner, so as to give it infinite variety; thereby producing words and speech. The office of the tongue is confined to transmitting the sounds properly, and to regulating them in their passage: the rest belongs to the larynx, the palate, the antrum, the throat, the uvula, the lower jaw, the teeth, the mouth, the lips and the nares (*k*). By the office and gift of speaking, the tongue feeds

(*h*) For more on these subjects, we refer the reader to the ingenious Boerhaave, n. 30, above.

(*i*) I do not call it proper, because it only disposes the muscles designed for manducation, at pleasure, in a new way: thus several of the bird tribe, as daws, crows, &c., may be taught to speak, although speech is not a proper faculty of their tongues. It is, however, proper in this respect, that these motions are determined by a previous will; for they all require to be learnt by the young: whereas the motions of manducation are for the most part natural. Not to mention the function of singing, which is an adjunct to that of speaking; or *vice versa*.

(*k*) That all the members here mentioned take part in speech, may be observed by every one who attends to what passes in himself during the act of speaking; the manner in which each contributes, will be shewn in the analysis of those parts. This accounts for the fact, recorded in natural history, that many persons have been able to speak, and to utter words, without the assistance of the tongue. On this subject, Winslow says, "The tongue is likewise one of the principal instruments of speech, and of the articulation of the voice. Riolan, in his Anthropographia, mentions having seen a child five years old, who, though he had lost his tongue by the small-pox, (not, however, the uvula,) continued still to speak almost as distinctly as before. Probably the base of the tongue still remained. M. de Jussieu has published an observation in the Memoirs of the Royal Academy of Sciences, concerning a little girl who could speak, though she was born without a tongue, in the room of which there was only a kind of small tubercle." (*Exp. Anat.*, *Traité de la Teste*, n. 540.) Not to mention other observations on pectoriloquism and loquela pythica.

the higher principles, the very mind itself; by the office of eating, it feeds the lower principles or the body. Thus it may be said to afford food to both the soul and the body; wherefore it guards the meeting of the two ways which lead to the two regions of the body,—to the viscera of the abdomen, through the pharynx and the œsophagus, and to the viscera of the chest through the larynx and the trachea; as well as the crossway which leads to the cerebrum, the hall and palace of the mind. For this reason it is, that the human tongue has a less acute sense of taste than the tongues of the lower animals; for in proportion as we approach the soul, in the same proportion we recede from the body (*l*).

37. The tongue has several other offices, but which are only derivative, proceeding principally from its power of eating and drinking, and of speaking. Such is spitting; clearing the trachea and bronchia; and the pharynx and œsophagus; ejecting things from the mouth; blowing; hissing; blowing horns, pipes, trumpets, &c.; licking the fauces, gums, prolabia, &c. It also assumes different states, corresponding to the different natural affections, as in crying, laughing, &c.

38. It is generally supposed that the tongue has a fourth office, of moistening itself and its organic parts, motorial and sensorial, with a constant stream of fresh saliva. This salivary humor, however, is not collected or derived from any springs or glands in the tongue itself, but from a number of springs or glands situated in the mouth, the gums, the fauces, the palate, and the maxilla; and even more remotely, in the temples, the cheeks, and the orbits; in short, from the sublingual, molar,

(*l*) We may perceive this by attending to the state of our senses, when the rational mind is thinking acutely or very deeply; for in proportion as the external senses are dulled, and lose their acumen, in the same proportion the internal senses are exalted, and the mind's acumen is increased. The mind recedes, as it were, from the senses of the body, whenever it enters itself, and looks above itself. The light of intelligence, and the light of sight, rarely shine together; the one is in a manner obscured and dulled by the other. Therefore man, who enjoys the gift of reason, is not so rich in that of sensation. See the saying of Plato quoted above, n. 12.

buccal, palatine, maxillary, parotid, and orbital glands, &c. (*m*). These streams the tongue brings to it principally by a kind of suction, implanted in it by habit and nature, perhaps in the very womb, and certainly after birth; also, by the agencies of

(*m*) That the lingual or papillary glands—the capitatæ or fungosæ, as well as the lenticulares—do not pour forth saliva, but only absorb the juices moistened with or dissolved in saliva, is a point which will be proved presently from the anatomy of those glands. Furthermore, an immense number of abundant streams are here present, (as we shall shew in the next chapter,) so that the tongue is positively surrounded by them. And the tongue exercises the greatest dexterity in attracting the salivary fluid, and drawing it to itself by suction; for that suction was the first rudiment of its offices or functions, is proved by the infant sucking at the breast; hence proceeds the tongue's natural power of demanding the salivary milk itself through the whole course of life, with the greatest ease, as often as it pleases; and of pouring it over its own mass, and over its ingesta. We will here lay before the reader the words of Malpighi, without further comment: "Whether all the three orders of nervous papillæ mentioned above, or only some of them, contribute to the production of taste, and whether they have any other service to perform, is a question. . . . Angelus Fortius, in time past, taught that the finer particles of the food ascend to the brain through the radicles of the nerves; and this position he proved principally by the following considerations,—that persons wearied by exercise, constantly recover their nervous power by taking food and drink; that when food and drink are taken by some persons, the mind is constantly dulled, and the senses blunted; furthermore, that when food and drink are taken by those who follow the plough, and the animal powers are not quickly recruited thereby, such persons appear not to live many years. To which we may add, that the powers are instantly restored by holding wine and similar things in the mouth only. Wherefore it is no groundless conjecture, that somewhat of the finer and more spirituous food is sucked up by those minute papillæ which we see in abundance on the tongue and palate; and that this produces a sort of tension in the relaxed nerves, and gives sudden strength. . . . I have often experienced in myself, that when the apex of the tongue has been thrust out and wiped free from saliva, it has not again become moist on the upper part, although I kept it out for a long time, until by again compressing the palate it was imbued with fresh saliva." (*Exercit. Epistolic. de Lingua.*).

motion, appetite, relish, and hunger. The saliva is so prepared, as to be singly sufficient for all the offices of the tongue; for restoring and renewing its states, for moistening the food received, and which is to be worked about, for dissolving it when ground by the teeth, for sheathing the juices extracted from it, and afterwards for insinuating them into the lacunæ and canaliculi of the tongue; also for rousing the veins to appetite when languid, and satisfying them when hungry.

39. Such are the numerous uses or effects of this one member, the tongue: and their representation or organization necessarily requires a confluence therein of different organic series, muscular, glandular, and sensile; consequently, of arteries, veins and fibres; and their products, membranes, tendons, ducts, and other structures, to constitute the mass and substance of its body: and all in so befitting an order, and in so exact a form, that each, and the least in each, shall know and perform its part in the clearest manner. All things then must be examined from their uses, in order to shew the construction of the several parts which anatomy displays.

40. The muscles of the tongue pass into its substance, with their moving fibres, little tendons, membranes, and blood-vessels, from all the neighboring bones or cartilages which offer them the nearest fixed points; the stylo-glossi arise on both sides, from the styloid process; the genio-glossi in front, from the chin; the cerato-glossi, obliquely on both sides, from the posterior parts, or cornua, of the os hyoides; and the basio-glossi, (which some anatomists subdivide into mylo-glossi and chondro-glossi,) directly from the base of the same bone. These muscles, entering from all quarters, move the tongue in every direction, each according to the ulterior direction of its fibres—forwards, outwards, backwards—obliquely or directly: they extend, contract, invert, bend, and even twist the tongue, particularly its apex, which is free; and cause its middle part, and its root, (which is fixed by its inferior attachments,) to endeavor to act, and to swell and subside, as far as their limits allow. Some of these muscles have a contrary direction to the rest, and mutually antagonize each other, and so balance the mass, that the tongue rushes into motion at the first intimation from the brain, or on the slightest signal being given. These muscles belong properly

to the first function in the process of eating, or to receiving the food, and also to speech; they are termed the external muscles. In conjunction with other sources peculiar to the tongue itself, they generate a new order of muscles, extending lengthwise, breadthwise, and from above to below; therefore named longitudinal, transverse, and vertical; which do not move the tongue out of its place, but only vary the situation of the parts and the dimension of the whole, and compress, contract, widen, and hollow it in every possible degree and way: thereby producing different states of motion. These latter, in conjunction with the former, belong properly to the second function in the process of eating, or to working about and conglomerating the food; and are termed the internal muscles. We now come to the proper muscles of the os hyoides, which arise from nearly the same regions and points as the proper muscles of the tongue; the stylo-hyoidei from the styloid process; the genio-hyoidei from the middle of the chin, near the synchondrosis of the lower jaw; the mylo-hyoidei from the base of the lower jaw: not to mention others from more remote parts, as the coraco-hyoidei from the superior costa of the scapula; and the sterno-hyoidei from the sternum. Thus, the basilar bone of the tongue, mediate between these forces, as if destitute alike of resistance and gravity, vibrates and carries away the tongue whithersoever the impetus directs it. The latter muscles exert their chief action on the root or superior and least moveable part of the tongue; and in conjunction with the internal and external, or proper muscles of the tongue, complete the third part of the process of eating, that is to say, accomplish deglutition. The determination of the moving fibres in their progress from beginning to end, shews the determination of motion in the body which they describe; and as the former determination is visible, it only requires to be investigated; on this subject therefore our authors may be consulted (*n*).

41. Wherever there are muscles there are also nerves, for nervous fibre is the moving power of muscle. The tongue is supplied by nervous fibres of a threefold origin, and which

(*n*) That is to say, the authors quoted above; Heister, n. 26; Winslow, n. 27; Boerhaave, n. 30; and the others.

consequently are of a threefold nature and use. As respects *origin,* some come from the cerebrum; some come from the cerebrum and the medulla oblongata conjointly; and some come from the cerebellum. The fibres of the fifth pair of nerves arise from the cerebrum; the fibres of the ninth pair, from the cerebrum and the medulla oblongata conjointly; and the fibres of the par vagum, or eighth pair, from the cerebellum (*o*). As respects the threefold *nature and use* of these nerves, the branches of the fifth pair, which arise from the cerebrum, construct the papillæ or sensoria of the tongue, rouse and fashion them to action and passion, and cause in them states corresponding to the forms of objects, to the modes of tangible substances, to the appetency of the blood and the spirits, and to the natural longings and affections; for the same cerebrum has the office of feeling or perceiving what is about to be willed, and of willing what it feels (*p*). The fibres of the ninth pair move the tongue obediently to the cerebrum, or as instruments of the will; and, assisted by a branch of the portio dura, and of the spinal or recurrent nerve, they adapt the volutions of the tongue to the motions of the larynx, the trachea, the bronchia, and the lungs: and these are the fibres which instruct and command the tongue to speak. The fibres of the eighth pair inform the tongue, from the earliest infancy, how it is to receive, work about, and roll down the food, and commit liquids and solids to the stomach; for these fibres proceed from the cerebellum, which neither perceives nor learns from the senses of the body, nor determines

(*o*) That the fifth pair of nerves originates chiefly from the cerebrum, and from the fibres of its medullary portion, will be demonstrated fully in our analysis of the nerves, after we have treated of the two brains; here we have no opportunity of demonstrating it, because no experience has been premised. In the same place we shall shew, that the ninth pair of nerves arises partly from the cerebrum, and partly from the fibres of the medulla oblongata. But that the eighth pair is an offspring or production of the fibres of the cerebellum, see my *Economy of the Animal Kingdom,* treatise i. n. 490, 559.

(*p*) Inasmuch as it arises from the cerebrum, it is both the gustatory nerve and the motory nerve of the sensoria of taste; the one thing being an adjunct of the other.

from any previous will instructed thereby, but simply decrees actions in obedience to the order and laws of nature (*q*).

Moreover, that the ear may be intimately conscious of the manner in which sounds are first articulated, from the very earliest tones and efforts of the infantile tongue, a small branch of the fifth pair, in a manner recurrent from the tongue, enters the membrane and cavity of the tympanum, and unites with the portio dura of the acoustic nerve, as it traverses the Fallopian tube; whereby the ear, forestalling a passage through the fauces, is enabled to know what is going on in the larynx and the tongue: and the voice in its first conception, is rendered in unison with the voice as it issues from the mouth, and is received by the external ear and the membrana tympani. In this way we learn to speak (*r*).

42. The glands of the first class, named *glandulæ fungosæ* or *capitatæ,* scattered and grouped round the edges and over the surface of the tongue, are the organs which take the first taste of the nutrient essences of the food, imbibe them with their little mouths, and transmit them through continuous ducts and channels immediately into the blood. Thus they are minute ante-

(*q*) Nothing better shews the causes and nature of the effects in the organic body, than an analysis of the nerves, or a rational neurology,; for all things are determined by the brains through the nerves; wherefore this analysis must constitute the last part of my anatomy of the body, where I am to treat of the organism of animal motion: from this part the above particulars are taken.

(*r*) On these points we shall speak more fully in our analysis of the ear and of the sense of hearing; where we shall treat also of the recurrent nerve itself, and of its passage through the Fallopian tube; as well as of the use of this tube, which accords with that of the nerve. I have thought it better to insert the above in this place, in order that the connexion of the subject may not be broken. It is well known, that the sound, besides the ordinary passage into the ear, also passes thither along the continuous membranes of the mouth, all the way from the gums; it being a matter of common experience. This latter may be called the *internal* way from the larynx; the former, in which the voice emitted from the mouth is at last received by the ear, is the *external* way.

types, not only of the tongue, but also of the mouth, the fauces, and the œsophagus, in short, of all the members and cavities which belong to the first nutrition; or rather, they represent an idea of the functions of those parts (*s*). The glands of the second class, or *glandulæ semi-lenticulares,* represent the same as the larger glandulæ fungosæ, only in a still more perfect manner. Hence these two sets of organs belong, not to the sensation of taste, but to the libatory and manducatory offices of the tongue; not excreting saliva, as commonly supposed, but drinking the first extracts and occult essences of the food, or the juices that it yields, and satisfying and renovating the needy blood, and breaking its fast. This is abundantly proved by a multitude of effects; by the sudden renovation of the blood; the appeasing of hunger and thirst; the cheerfulness of spirits; the instant intoxication, that follow, when juicy, medicated, or spirituous food, syrups or wines, are only taken and held on the tongue: also, by the spontaneous drying of the tongue even when kept in the mouth, as soon as the saliva ceases to be rapidly supplied by the neighboring sources. And it is additionally confirmed by the minute anatomy of the parts. For the glands of the first class, the glandulæ fungosæ, have an ample and luxuriant head or belly, and a thin peduncle, resembling a neck, stem, or tubule, or a snail's horn; each of these corpuscules is lodged in a small pit, and surrounded on all sides with the liquid food which it is meant to taste: moreover, their surfaces do not bulge out, but are a little depressed and hollowed, and perforated with one or more orifices (*u*); a sign that some fluid matter passes in,

(*s*) On these subjects see what we premised above, n. 34., and n. 38 Schol. *(m.)*

(*u*) All these particulars are strictly conformable to our authors' observations. Winslow has the following, which we quote from the original: "Ces mammelons de la première espèce sont comme de petites glandes conglomerées, posées sur une base fort étroite, et elles ont quelquefois chacune un petit enfoncement au milieu de leur sommité ou convexité. Ils occupent la surface de toute la base de la langue, où ils sont situés ensemble près les une des autres, et de manière, que les plus anterieurs forment un angle par leur arrangement. Les mammelons de la seconde espèce, ou mammelons demi-lenticulaires, sont de petites éminences orbiculaires, d'une convexité applatie, dont le and that none passes out. Another proof is, that their passages decrease in diameter all the way from the orifices, and never increase; and that the canaliculi leading from them, tend from unquiet to more quiet stations; that is, from the sharp extremities of the tongue to its middle, or base (*x*): this is evident

bord circulaire n'est pas séparé de la surface de la langue. Quand on les examine dans une langue saine avec un bon microscope, on en trouve toute la convexité marquée de petits trous ou pores, à peu près comme la convexité d'une dez à coudre, ou le pavillon d'un arrosoir," &c. (*Exp. Anat., Traité de la Teste*, n. 508, 510.) Malpighi says, "In substance and shape the larger papillæ resemble the emissile and retractile cornua observed in snails; they stand on a long peduncle, which having risen through the mucous substance, ends in a little round head, placed in a sort of cavity in the thin exterior membrane," &c. (n. 28.) From all this apparatus—that is to say, from these cartilaginous corpuscules, resembling a claw, a sickle, a half-moon, a boar's tooth, or a little proboscis or epiglottis; also from the little pits and cavities surrounding them; and from their figure being wide above, narrow below; particularly from the hollowing out of their superior surface, and from the little foramina in the hollow part—it is very evident, that these glands are designed for absorbing essential fluids, and not for excreting mucilaginous or salivary ones. Their office is of course imprinted, and stamped universally, on their form. If they poured out any humor, their surface would not be hollowed out, nor perforated in the hollow part. The constant habit of ejecting liquid would inevitably extend them into an elongated oval form. This view is also corroborated and avouched by all the adjacent parts; by the little cornua, recipient, inducent, and appliant; and also by the perforated interspaces surrounding them. But this reasoning alone is not convincing, although when taken in conjunction with innumerable other things, it amounts to proof. We shall bring together further arguments in the next chapter; see n. 72.

(*x*) These two arguments for my proposition, appear at first sight like mere reasonings: the arguments, I mean, drawn from the facts, that the passages of the glands decrease constantly in diameter; and that the little canals go always to successively quieter and safer stations: yet they are matters of experience, derived from the whole anatomy of the body; in short, from a course of observations. Throughout the entire corporeal system, the fluids permeate their canals from the greater to the lesser diameter, *whenever they are going to an outlet;*

from the collection of ducts in the foramen cœcum (y). A similar conclusion is deducible from the intimate structure of

(not so, when they are being carried into their receptacles:) as in the arteries, where the blood passes from the trunks to the branches, from the branches ultimately to the most minute twigs, and then into the glandular receptacles. In the glands, the humor is conveyed from its repositories into constantly narrower cavities, and at last into the finest tubes. So also, when the biliary and pancreatic juices are thrown out through the ductus choledochus into the duodenum. In the ventricles of the brain, the humor is expressed into the infundibulum, and thence through a narrowing tubulus into the pituitary gland. The same thing obtains in large parts, as in the œsophagus, which contracts constantly towards the cardia or superior orifice of the stomach; in the trachea, likewise, which contracts towards the bronchia, and from the bronchia towards the vesicles, and so forth. Whence it appears, that wherever any humor is intended to be driven onwards, it is carried through a narrowing channel. The same is the case in these glands; their capitated part forms a kind of broad belly or vesicle, which contracts afterwards into a slender peduncle. As regards the other point, namely, that the little canals through which any humor received by them permeates, tend naturally to more quiet stations; this also is a deduction from general experience, collected in a course of observations. Thus the chyliferous humors go through the mesentery to the receptaculum chyli, which is placed in the central and most tranquil region of the body. The lymphatics likewise. The urinary matter goes to the kidneys, and through the ureters into the bladder, which is seated in the pelvis, external to the peritonæum. The interior liquids of the cerebrum, go towards either the crista galli, where all the circles meet, or towards the pituitary gland, which sits secure in the tranquil sella turcica. So also in other organs, as will be seen throughout in the progress of the work. Thus the wearied juices perpetually seek some port; they always tend from their circumferences to certain centres; and the little canals go from the extremities of their levers to some hypomochlium or centre of motion. So it is with those proceeding from the glands of the tongue; they pass from its circumferences to the middle of its substance: and the larger canals, which are made up of the smaller, as those discovered by Heister in the foramen cœcum, always tend to the base of the tongue, which is its most quiet region; and this with a constant diminution in calibre.

(y) Respecting this foramen, Winslow says, "We often observe about the middle of this part of the tongue, a peculiar foramen, more or less deep, the glandulæ lenticulares, or glands of the second class (*z*). These glandular forms, then, are not for secreting saliva, but

the internal surface of which is entirely glandular, and filled with small papillæ, like those of the first kind. It is called foramen cœcum Morgagnii, from its discoverer. M. Vater discovered some ducts, apparently salivary, belonging to it. M. Heister found two of these ducts very distinctly, the orifices of which were in the bottom of the foramen cœcum. He observed the ducts to run backward, divaricating a little from each other, and that one of them terminated in an oblong vesicle, situated on the side of the small cornu of the os hyoides." (*Exp. Anat., Tr. de la Teste,* n. 509.) Let the reader here consult Heister, n. 26, above; also his *Comp. Anat.* tab. viii. fig. 34, 35; and not. 56. But with respect to the vesicle mentioned by Winslow, according to Heister it existed only on the right duct, and he appears to have suspected that it was caused by the pressure of the injection; for he says, "The liquid in the vesicle, as it seems to me, was driven thither by the force of the injection, and not being able to yield, dilated the posterior part of the duct into this vesicular form." (*Comp. Anat.* not. 56.) These ducts, in a large form, are analogous to the other smaller ones, of which we have before spoken: for nature for the most part produces a general, in similarity to the particulars. These canals have a wide mouth, they commence from a trunk, they contract as they proceed, and they go to the base and under the basilar bone of the tongue: thus by them, our opinion is strengthened respecting the former ones.

(*z*) We have already cited what Winslow says of these glands; the following is Malpighi's account: "The papillæ of the second kind are more numerous than those just mentioned; corresponding on the inside to the cornua on the outside of the tongue. Proceeding from the common papillary substance, they rise to some height, and from their summits put forth nervous offsets, which enter the cavities before spoken of, and meet the roots of the cornua. About the base of the tongue, instead of the cornua, the nervous papillæ already described project outwards, and changing their form as they approach the base, they successively become more obtuse, rounded, and depressed; and the largest of them are not very much unlike those observed in the cheeks, at the roots of the teeth." (n. 28.) Nature persists constantly in her measures and degrees; thus she appears to have planted on the tongue, glandular papillæ, or recipient and absorbent organs, of two kinds; in order that the purest and more simple halitus, or spirituous dews, may be sipped by these of the second kind; the grosser, by those of the first kind. And it is worthy of observation, that these lenticular for absorbing it, and juices seethed and dissolved in it; and thus for imbibing, carrying away, and directly circulating the purer and more spirituated essences, as well as the oral saliva itself: and so preventing any of the nobler principles of the animal kingdom from being lost. Thus the tongue—the feeder and keeper of the entrance to the stomach and the viscera of the body—not merely prepares the table, but also takes the first taste of the viands, and begins the feast.

43. The papillæ of the third class, *conicæ, pyramidales* or *villosæ*, are the true sensoriola of taste. They are therefore the most numerous of all, and fill the interstices between the rest of the papillæ; they are extensile, contractile, elastic, susceptible of every variety of state and form, and capable of adapting themselves to the minutest points of tangible substances (*aa*);

corpuscules, perforated by numbers of little foramina, like finger-stalls, bud forth and are hidden under the claws of the cornua; so that these simpler dews are determined thither undeviatingly.

(*aa*) From the descriptions given by our authors of these papillæ, we may understand their quality of form, and their faculty of action. "The papillæ of the third kind," says Winslow, "or villosæ, are the smallest and most numerous, and occupy the whole of the upper surface of the tongue, and even the interstices between the other papillæ. They would be more properly named papillæ conicæ than villosæ, from the figure which they appear to have when examined in clear water through a microscope. They are naturally softish, but they become so flaccid after death, that by handling them they may be made short and thick, whereas they are naturally long and small." (n. 27.) "The nervous papillæ of the second kind," says Malpighi, "correspond on the inside to the cornua on the outside of the tongue; they are surrounded by almost innumerable papillæ from the same origin, and which rise to the same height; but which are conical and more slender, and entering particular cavities in the mucous substance, at length terminate in the direction of the external membrane." (n. 28.) Whence it appears that these papillæ are capable of assuming all possible states and forms; for they live by sense; and when life fails, they become flaccid and die; and from their extreme delicacy, they admit of being folded by a touch of the fingers. These papillæ, I say, are properly sensorial: we do not, therefore, deny sensibility to the others, of the first and second classes, but allow it to them in a more obtuse, indistinct, and obscure degree; inasmuch as they are properly libatorial, the first-born offspring of their parent fibres; therefore exquisitely sensitive; most numerous, fine, simple, and modifiable, and most like their parents in nature and genius, on the apex and borders of the organ, at the points which the food first touches, and where it is first received. Analogous forests of papillæ are found in the nasal cavities, on the skin, on the internal membranes of the stomach and intestines, particularly of the small intestines, and of many of the viscera (*bb*). They arise, as it seems, from an unweaving of the fibres of the fifth pair of nerves into their simpler threads, and from the fibrillæ of the eighth pair; for the fifth pair comes chiefly from the cerebrum, (which is both the general sensorium, and the general voluntary motorium,) and descends and is distributed, as general explorer and regulator, to all the sensoria of the body; and also intimately enters, arranges, and weaves them, and associates them with muscular fibres; thus it disposes those textures to the state of sensation, and commands the motions suitable to that state (*cc*). The eighth nerve, on the other hand, arising from

as explained above: that they also enjoy a certain amount of rude sense, may be inferred analytically from the fact, that they, as well as the others, proceed immediately from the fibres and from the nervous membrane: according to Malpighi, "They originate from the nervous and papillary substance; with this peculiarity, that in the base of each there is a nervous twig, to which they are appended, or rather grow." (n. 28.) These observations are confirmed by Heister aud many other authors. But I conjecture that these papillæ originate from the fibres of the eighth pair of nerves, whence their sense does not penetrate to the cerebrum or to the mental consciousness. What inclines me to suspect this is, that the whole work of eating is natural, implanted by the first act of suction in infancy; consequently, these papillæ, which are essential portions and parts of the manducatory function, must have learnt to suck and to milk withont any previous volition.

(*bb*) Respecting all of which we shall speak in the proper places; for a similar villous membrane is found in the stomach and intestines, as well as in other parts.

(*cc*) Since the cerebrum is both the general sensorium and the general voluntary motorium, therefore the fibres arising from its medullary portion, are of a similar nature; insomuch that they not only are sentient, but also dispose and direct their sensoria to receive sensation;

the cerebellum, which has the general cognizance of those affairs in the kingdom of which the mind is unconscious, and which is the general ruler, regulator, and mover of the ordinances of nature, associates with the fifth nerve, and enters the recesses of the papillæ with it: and rules silently and secretly, not from the will, but from nature (*dd*). Thus nature and the will have each their departments, and the administration of the kingdom is divided, lest the will, which follows the order of the mind, should depart from, or even go contrary to the order of nature. These nerves, from these two sources, by means of the organic forms, or papillæ, faithfully announce to the brains the nature of whatever touches and affects them, and the state of the parts and the whole as resulting from internal and external causes (*ee*).

consequently, put them into all kinds of states, move them, and command particular motions suitable to the substances applied to them. For every fibre brings with it its parent's nature, from the natal soil. This is the reason why the branches of the fifth pair communicate so frequently with those of the eighth and ninth pairs: see Winslow, *Exp. Anat, Traité de la Teste*, n. 534; and *Traité des Nerfs*, n. 65—71, 114, 148—153.

(*dd*) That the villous coats of the viscera, as of the stomach and the intestines, are produced by the fibres of the eighth pair of nerves, will be shewn as we proceed. The cerebellum, by this its nerve, as a universal informant, is made aware of every thing that happens or goes on within the boundaries of its kingdom; (the cerebrum being cognizant of whatever happens without them;) in order that it may dispose, recruit, and reform all things, by the constant laws of nature: and as there is more of natural, or, as some persons call it, involuntary, in the tongue, than of voluntary, it may therefore be inferred, that not the fifth pair only, but also the eighth pair, passes into the sensorial papillæ; further, the perpetual anastomoses between these pairs in the tongue, (of which we have spoken above,) give additional countenance to this view.

(*ee*) *External causes* are all those which touch the tongue immediately. The quality of the touch, or tangent object,—its heat, cold, bitterness, sweetness, acidity, or aromatic properties,—instantly change the state of the member; and the papillæ either expand or relax, extend or retract, either become sharper or blunter; which changes of state are not ordained by the mind's prescience, but happen naturally and spontaneously. *Internal causes* are what proceed from any disease,

For every individual particle of the three kingdoms, mineral, vegetable, and animal, has a form of its own, and extrinsically a figure of its own, angular, flat or round, with infinite variety. These elements, or aggregations thereof into corpuscules—covered with liquid and saliva, floating about unimpeded, and insinuating themselves into the pores and interstices of the tongue, and thus striking the sensile villi at all points, impress and produce a likeness of themselves. The cerebrum in its principles in an instant perceives this likeness exactly similar to what it is in the extremes; and the consequence is, sense; in the tongue, taste; in the nose, smell; in the skin, touch. These senses are all related to each other; and similar in origin, form, nature, and modes; but they only apprehend the surfaces of objects (*ff*). Touch apprehends the surfaces of parts; taste and smell, the surfaces of parts of parts; nor do they penetrate deeper in human organs (*gg*).

accident, or sore, or from changes and affections of the animus, or of the mind; for the quality of the animus determines a corresponding quality in the fibre, because in the spirit which permeates the fibre. Hence the variety of sensations according to the varieties of bodily temperament, disease, habit, sex, and age; in a word, according to the varieties of the state of the cerebrum, of the cerebellum, of their interfluent fibres, and of the tongue.

(*ff*) It is our intention to treat of these subjects in Part VI., on the organs of the five senses; where we shall shew also, from experimental premises, that the sense of human organs is not sufficiently acute to perceive the external figures of single particles, but only of many together, as one general amount; that is to say, in the rudest manner: also, that the form, delicacy, nakedness, position, and number of the papillæ particularly and generally, the supply of nerves from different sources, and many other circumstances, produce the diversity of sensation which we find in touch, smell, and taste; for all these senses are of the same genus and family, but not so hearing and sight. Meanwhile, respecting taste, see Boerhaave, n. 30, above.

(*gg*) That the sensual faculties are more obtuse in man than in brute animals, was briefly explained above, n. 36, Schol. (*l*); but I intend to treat of these faculties more fully in my analysis of the organs of the senses: for this obtains not in taste only, but also in smell and in sight.

44. The simples, unities, and particulars of the muscles are the moving fibres: the composites, aggregates, and generals are the muscles themselves (*g*). The more exactly, therefore, the simples are distinguished from each other, the more fitly they are put together into forms, and the more ordinately they are related to their generals, the more perfect is the state of the member: as in this case, of the tongue (*h*). For the muscular fibres in the tongue are perfectly distinct, and most beautifully coordinated; so that every one respects every other, and even whole series of others, which are without it, just as parts of itself; for from their very initiaments they tend to perform their contemplated use (*i*). The consequence of this order is, that every thing simultaneous has in it a kind of successiveness; and also a relation to its universe. The same may be said of the glandulæ fungosæ and lenticulares, which, in their kind, relatively to their use to the tongue, may be considered as so many simples, unities, and particulars, perfectly distinct from each other, yet so admirably disposed, that every one respects every other, as a part of itself; which is brought about by means of the membranes, in which these glands are implanted, and by which they are united together. The same may also be said of the papillæ conicæ or organic villi (*k*): and of all the other parts

(*g*) All things refer themselves to their unities; and the unities themselves to theirs; for every thing is a series and in a series: there is no such thing as absolute simplicity in nature. Hence there is no series that may not be assumed as a unity, if we please: thus also the glands and papillæ, of which we shall speak presently.

(*h*) We shall treat of these subjects in the doctrines introductory to our Rational Psychology.

(*i*) The least or prior forms, as we before pointed out, are the models and the ideas of the larger or posterior forms; each of them comprehends in it all those things that follow in order, and that carry the end to the ultimate effect.

(*k*) The papillæ capitatæ and lenticulares may fairly be called glands, by virtue of their uses, like those which absorb the pituitæ; but the conicæ or pyramidales may properly be called papillæ, and, indeed, organic papillæ, by virtue of their use, or office of sensation. That all these kinds of papillæ rest upon and are in a manner implanted in the nervous membrane, as their natal soil, our authors are unanimous of the tongue, as the canaliculi, excretory ducts, vessels, and fibres; and even of the interstices, foramina, and cavities, which bound the parts. Thus there is a perpetual relation of parts to their generals; and a perpetual determination of all things to their final effect, or use.

45. But in order for these things to produce the effect, that is, the end, in a determinate manner, each, from first to last, must be environed with membranes, very thin in the first instance, and growing gradually denser and more composite, and at last still grosser and more general; the membranes following the order from particulars to generals (*l*): they must also be separated by tendinous and aponeurotic ligaments (*m*): and confined by so many bonds (*n*): and thus bound to the due performance of their uses.

in declaring; at present it is sufficient to adduce the testimony of Malpighi: "Since the nervous and papillary substance" (that is to say, the nervous membrane) "is closely connected to certain nervous twigs, by which the whole mass of the fibres is connected and held together, and since many nervous twigs proceeding from the trunk terminate in it, hence it will not be absurd to suppose that it originates from these twigs; especially since the nerves designed for sensation in the other sensoria commonly terminate in a membranous expansion." (n. 28.) And of Heister: "The papillæ arise from the internal membrane of the tongue, and from its nerves; they pass through the little foramina in the reticular membrane, and terminate in the vaginulæ of the external membrane." (n. 25.) The consequence is, that whenever any thing happens to one papilla, the other is rendered conscious of it: for the common membranous and nervous substratum communicates the circumstances of each to all the others, near and remote; and the common sheath, and the membrana reticularis, when it is present, obliges them all to act and to suffer alike: thus the mode of each part is rendered general, and becomes the mode of all.

(*l*) For all the moving fibres are surrounded severally with their own fine coats or coverings: the larger forms made up *(consertæ)* of these, with theirs; the forms composed of these again, with theirs; and so forth. Likewise the fibres and the nerves; also the glands, the organs, and the viscera.

(*m*) We shall treat of these subjects in our analytical and rational myology and neurology.

(*n*) Respecting the ligaments and fræna, see Winslow above, n. 27.

46. In order that organs with such different functions—namely, sensorial, as the papillæ villosæ; libatorial, as the glandulæ lenticulares and capitatæ; and motorial, as the muscles—may make common cause, end and use, the fibres of three pairs of nerves approach, enter and connect all the parts, and communicate whatever goes on in one, to the others (*o*). Thus the gland is reflectively informed of whatever the papilla feels; and the muscular fibre is conscious of whatever is thus communicated to the gland: consequently, the muscle acts as soon as ever the gland sips, and the papilla tastes. The state of one excites that of the others; the cause calls forth the effect, and the end, the use which the soul intends.

47. In order that nothing may be wanting in ultimates which is intended in principles, perpetual variety is necessary in the tongue; the papillæ, glands and moving fibres, one and all must be various (*p*): perpetual varieties harmonically interwoven pro-

(*o*) That every papilla has its own little nerve or fibril, is declared by Malpighi: "The papillæ," says he, "have this peculiarity; in the base of each there is a nervous twig, to which it is appended, or rather grows." (n. 28.) This twig or fibril comes off from a branch, which branch supplies all the papillæ: consequently, when the papilla is touched, or modified, the fibril instantly takes up the touch or modification, and refers it to the branch; whence all the papillæ arising from the common branch, become partakers of it. This takes place in a still more general manner, when the touch or modification is referred to the underlying nervous membrane, in which all the papillæ have their roots: in this case, the activity is dispersed instantaneously to all those papillæ that are strewn upon the same membrane. But the common or external membrane, which, like a tunica vaginalis, receives and transmits the papillæ, not only communicates between them, but also limits them, and disposes them all to the same effect.

(*p*) According to the foregoing descriptions of our authors, the papillary glands are comparatively large, obtuse, and few in number on the base of the tongue; but smaller, sharper, and more numerous on its apex and edges; for the first sipping, and the preliminary tasting, take place on the borders or circumferences of the tongue. Unity and equality would produce none of the differences of operation and sensation, ranging between greatest and least, as between obtuse and most acute: sense would have none of those degrees of altitude, which cause duce the end and result in the perfection of the whole. The glands, papillæ and muscular wreaths are not absolutely the same in the apex as in the middle, nor in the middle as in the base of the tongue. The use determines the harmony of the varieties.

48. These considerations show not merely what is the tongue's form, active force, and power of action, but also what its substance is: they show that it consists of nervous fibres of a threefold origin, nature and use; of blood-vessels ramifying in all directions; of muscular fibres variously interwoven, separated by interstices, intermingled with fat (*q*), and gently bound down by soft, delicate and beautiful bands: also of tubuli and ducts passing through the middle of the muscular fibres (*tt*), and proceeding from the glands on the surface: and of membranes, ligaments and fine tendinous meshes (*uu*): lastly, of commissures,

all things to be exquisitely distinguished. Wherefore, it must be considered as among the arcana of nature's operations, that no two things are absolutely similar or equal; but nevertheless, that there is a harmony of dissimilars; consequently, a relation and quality in common. This is most conspicuous in the ear, where the semicircular canals and spiral cochleæ increase constantly in diameter and thickness, from the bottom to the top, &c.

(*q*) Our authors make continual mention of the fat of the tongue; this fat is generally found in the cellular membrane that separates the muscular congeries from each other. It serves to grease the joints, and to prevent them from growing together: but it is inserted in those little spaces particularly where the fibres are least active; consequently, more in the dorsum and base of the tongue, than in its point and edges.

(*tt*) Respecting the different kinds of fibres in the tongue, Malpighi says: "The centre of the tongue consists of many kinds of fibres, longitudinal, transverse, and oblique, which are all plaited together into something like matting, as may be seen in my various delineations." (n. 28.) And Leeuwenhoek states, (in his *Arcana Naturæ Detecta*, p. 409, seq. [Lugd. Bat. 1722,] where he delineates and describes the tongue of the ox,) that "he has seen a number of very minute particles cut across, which were too small to admit of being represented," &c.

(*uu*) "We observe," says Malpighi, "a kind of glutinous substance, extending over the superior part of the tongue especially, and of some thickness: this is white where it is connected to the membrane before foramina and lacunæ; containing liquid, saliva and mucus. The more distinctly all these things are placed, the more pliable and soft is the tongue, and the better adapted to its uses.

49. The particulars which remain to be supplied, must be elicited from the neighboring parts, contiguous and continuous; that is to say, from the anatomy of the throat, the palate, the pharynx, the œsophagus, the stomach, and the intestines; and then of the other organs in connection therewith. What is still wanting, must be gathered from the origin of the nervous fibres—from the cerebrum, the cerebellum, and the medulla oblongata and spinalis. Any thing further must come from higher powers and principles, and in the end from the highest (*xx*). All these must contribute a part to the exploration of the use of this one member. The simpler tongues also, of nymphs, chrysallises, caterpillars, butterflies, and the like (*yy*),

mentioned, but blackish where it touches the interior part. It extends in the form of a membrane, or thick rete; and has conspicuous openings, corresponding to all the cornua; and innumerable little canals between them which can be seen only by the microscope: these are of different shapes, and open on the surface of the tongue; whence, if it be torn across, or examined by the microscope up against the light, they become visible." (n. 28.) Such are the terms in which this celebrated author describes his reticular membrane, and which differ scarcely at all from Winslow's observation; the latter confessing that he has not been able to find this membrane in the human tongue, but only a species of clear mucilaginous substance, which becomes white by boiling, and declaring that the foramina in it are caused by the pyramidal papillæ. (n. 27.)

(*xx*) See what we said on this subject at the beginning, n. 32, and the notes.

(*yy*) As the tongues of insects are simpler than other tongues, so they are better adapted to uses, and more perfect, and involve the simultaneous performance of a greater number of actions; (this being the attribute of natural simplicity;) as may be seen from the tongues of both the covered snail and the naked or house-snail (n. 29), which are dentated; so that these creatures can seize their food, and possibly also bruise and devour it with the tongue itself: the same thing is evident from the tongue of the aliekruik snail, which is extensile, pliable, and capable of coiling even quite round, being described and which have scarcely emerged from the prior into the posterior sphere of nature, must complete and crown the analysis.

delineated as gyrating in spiral flexures. Furthermore, those inflected and falcated cartilaginous and osseous corpuscules, or cornicles, seen on the surface of the tongue, and which were first noticed by Malpighi, according to Swammerdam's authority are also found in the tongues of the house-snail and of the cuttle-fish. His words are as follow: "We may see that every one of the seven small bones (in the cuttle-fish) is provided with above sixty curved, dentiform, cartilaginous papillæ, (somewhat resembling the papillæ of the tongue of the ox,) by means of which the fish, when feeding, is the better enabled to move its food, and dispose it for an easy swallowing (n. 29):" thus manifestly for the use assigned above to these papillæ, n. 34, Schol. (*g*). The tongues of irrational creatures, of worms especially, must be incomparably more perfect structures, not merely because they are simpler, but also because such creatures live by sense and nature, and not by reason and will like the human race. Nevertheless, those that excel in mind are intrinsically more perfect than those that excel in body; for the mind looks down upon the organic body from its own high seat, as far below it, and as its minister and servant only; yet values it in proportion to its service.

CHAPTER II.

THE LIPS, THE MOUTH, THE PALATE, AND THE SALIVARY GLANDS.

50. IN the fauces we find the epiglottis, the uvula, and several other organic instruments, besides those enumerated in the heading of this chapter; but we shall treat of these in the second part, among the viscera of the thorax: as we shall there begin from the nasal cavities, and the organs of smell and sound, to which the epiglottis and the uvula properly belong.

51. HEISTER. "The lips are composed principally of muscles. They are covered externally with the common integuments; internally, with the membrane of the mouth; under which, in both the lips, there are miliary and lenticular glands. When the prolabia are deprived of epidermis, and macerated for a time in water, they display a multitude of nervous papillæ: whence they are so extremely sensitive. Each of the lips has its peculiar frænulum, the upper lip under the septum nasi; the lower lip, at the roots of the incisor teeth. The lips are of great assistance to us in speaking, and in eating and drinking. The gums consist of the common lining of the mouth, and of the periosteum of the jaws, to which they adhere very firmly. They are furnished with a vast number of blood-vessels, whence their florid red color. They serve for covering the jaws, and for supporting the teeth. *(Comp. Anat.* n. 280-1.) The muscles of the lips are fourteen in number: consisting of one constrictor, called the orbicularis, which forms the substance of the lips; its fibres are frequently of an arched figure. Two pairs of abductors, which may be called risores; 1. the zygomaticus (on each side) arises from the zygoma, and is inserted into the angle of the lips; it is sometimes a double, sometimes a bicipital muscle, sometimes its termination only is bifid; it is interwoven with the various neighboring muscles: 2. the buccinator (on each side) arises partly from the anterior and lower part of the coronoid process of the

under jaw; partly about the roots of the posterior molar teeth of both jaws; its course, in the erect position of the head, is nearly horizontal; and it terminates, like the former: it submits the food to the teeth, and is called into action in blowing trumpets, &c.: the salivary duct of Steno perforates it about the middle. The elevators are some of them common to both lips, some peculiar to each lip; the common are only one pair, called par caninum: the caninus arises on each side from the cavity under the zygoma, in the maxillary bone; and is inserted into the angle of the lips. Those peculiar to one of the lips are two pairs; the incisorius is proper to the upper lip; it arises from the maxillary bone, just under the orbit, passes close to the ala nasi and gives some fibres to it, and is inserted into the orbicularis, near the incisor teeth; the elevator labri inferioris Cowperi is proper to the lower lip; it arises below the incisor teeth of the lower jaw, under the gums, and descends into the cutis of the chin: whence it may be called the incisorius inferior, as the other is the incisorius superior. The depressors of the lips are three; they are often singularly connected to some of the before-mentioned muscles: these are, the triangulares and quadratus. The triangulares are two; they each arise from the lateral and under part of the lower jaw, about the middle, from whence they ascend obliquely to the angle of the orbicularis. The quadratus is single; it is composed of reticulated fibres, and arises broad from the front of the lower jaw, and terminates all along the bottom of the orbicularis." *(Comp. Anat.* n. 319.)

52. The jaws are two in number. "The upper one is immoveably fixed to the bones of the cranium: the lower is moveable, and by its mobility serves for many purposes. *(Comp. Anat.* n. 77.) The upper jaw is composed of thirteen bones, and when the number of the teeth is complete, of sixteen of the latter. Of these thirteen bones twelve are in pairs: these are, 1. the lachrymal; 2. the nasal; 3. the jugal; 4. the maxillary; 5. the inferior spongy; and 6. the palatine. The thirteenth is a single bone, and is called the vomer. They are united together by a kind of even juncture, called by anatomists, junctura per harmoniam. *(Comp. Anat.* n. 79, 80, 81.) With regard to the jugal, zygomatic or malar bones, we are to note the reason of their name, their situation and circumference; their hard substance, their connexions, their four apophyses, the places whence the masseter and zygomatic muscles arise, and the foramina in these bones, with their uses. With regard to the maxillary bones, we are to note their circumference, their size, their situation, their connexion, and their four apophyses, 1. the jugal; 2. the superior nasal; 3. the inferior nasal, which joins the septum nasi; and 4. the palatine: also the palatine and

nasal excavations, the alveoli of the teeth, the cavity commonly called the antrum of Highmore, its size, and the construction of its orifice or aperture in living subjects: its double use, 1. for assisting the voice, and 2. for the secretion of mucus: also the nasal canal. *(Comp. Anat.* n. 84, 85.) With regard to the ossa palati, we are to note the very irregular figure of each; its extension through the nares to the orbit; its body, and its various apophyses, the pterygoid, the nasal and the orbital; the furrow on its upper surface, where it is joined to the vomer. The use of these bones is, to form the palate, the orbit, and the maxillary sinus: to sustain the palatal membrane and the uvula; and to assist the voice. *(Comp. Anat.* n. 87.) The lower jaw is the large moveable bone which contains the lower series of teeth; on its mobility our power of manducation in a great measure depends. In the lower jaw we have to consider the bone itself, and the teeth. In the bone we are to notice its singleness, its state in infancy, (when it is made up of two bones,) its size and figure; its substance, which is externally firm and internally spongy; its external and internal surfaces, and its external and internal margin; its connexion, and the places where the two bones unite in the young subject; the ridge at the roots of the incisor teeth, for the insertion of various muscles; the chin; the condyloid apophyses, covered with moveable cartilage; their articulation with the temporal bones; also the coronoid processes, and their use; the two angles; the alveoli of the teeth; and the four foramina, forming two canals, and affording a passage for blood-vessels and nerves. *(Comp. Anat.* n. 91-2-3.) The teeth are the hardest bones in the body, and are fixed into the alveoli in the manner of nails; this kind of articulation is named gomphosis. We are to note their situation, in deep sockets; their connexion by means of the periosteum and the gums; their natural whiteness; their number, ranging from 28 to 32; their division into four incisors, two canine, (which in the upper jaw are called the eye teeth,) eight molares, or grinders, in each jaw, and two dentes sapientiæ. Further, we must examine their base, and its two peculiar substances, a stony or vitreous, and a medullary substance; their roots, which are sometimes single, as in the incisor, canine, and anterior molar teeth; sometimes double, triple, or quadruple, as in the hinder molar teeth; their cavities, which are invested with a vasculo-nervous membrane; and the little foramina in their roots, for the ingress of vessels to afford them nutrition and sensation. Finally, we must remark the origin and accretion of the teeth in infants; their change and renewal after the seventh year, and occasionally even in old age; the singular uniting and falling in of the alveoli in old people, after the loss of the teeth. *(Comp. Anat.* n. 94, 95.) A foramen, called the anterior palatine foramen, opens into the nares behind the front incisor teeth: but in both the living and dead subjects, it is so closed by the membrane of the palate, that no aperture can be seen in the mouth, nor, so far as I have been able to observe, can any thing be transmitted through the foramen. As the membrane of the mouth joins the membrane of the nares by means of this canal, it seems to strengthen the connexion of the former membrane with the palate. The posterior palatine foramen, which is common to this bone and the os palati, transmits the nerves to the palate. In the palate-bone, besides the common foramen, there is a proper one, where it joins the pterygoid processes: this serves for the passage of nerves to the palate. In the lower jaw there are two foramina on the inside, serving for the entrance of an artery, vein, and nerve into the substance of the jaw; to nourish it and the teeth: and two on the outside serving for the egress of the same vessels to the gums and the chin." *(Comp. Anat.* n. 103-4.)

53. "The muscles of the lower jaw are six pairs; two pairs of depressors, and four pairs of elevators. The two pairs of depressors are, 1. the platysma myoides, which arises near the clavicle, from the pectoral and deltoid muscles, and also, as Galen and Cowper have observed, from the vertebræ of the neck, and is inserted into the lower jaw; but in such a manner, that it commonly sends fibres to the lips and cheeks, and frequently also to the ears, and thus assists the motions of all these parts. 2. The biventer or digastricus: this muscle arises from the incisure under the mastoid process; its tendon often passes through the stylo-hyoid muscle, and through a membranous ring that is fixed like a pulley to the os hyoides; and it is inserted on the inside of the symphysis of the chin. The mouth is opened in a wonderful manner by means of this pulley. The four pairs of elevators are, 1. the temporal, which arises from the whole region of the temples, and particularly from the frontal, sincipital, temporal, and sphenoid bones; passes under the zygomatic arch, and is inserted into the sharp process of the jaw. 2. The masseter, which arises from the inferior and interior part of the arch, and terminates in the external surface of the angle of the jaw: the duct of Steno passes over this muscle. 3. The pterygoideus internus, which arises from the cavity of the pterygoid process, and terminates in the interior and lower surface of the angle of the jaw. 4. The pterygoideus externus, which arises from the exterior lamina of the process, and the neighboring part of the sphenoid bone, and terminates a little above the preceding muscle. *(Comp. Anat.*, n. 321.)

54. "The palate has numerous glands, which were discovered as early as the time of Fallopius; these are principally situated in its back

part, near the uvula, where it forms a pendulous veil, called velum or claustrum palati. They excrete a mucous liquid for lubricating the fauces, and facilitating deglutition; this liquid they discharge through a great number of little orifices. This membrane also defends the palate bones against corrupting, and by the claustrum it prevents the food from getting up easily into the nares. *(Comp. Anat.* n. 282.)

55. "The glands which secrete the saliva, are the parotids, two very considerable glands, situated one on each side, between the ear and the angle of the lower jaw, and frequently extending over a great part of the masseter muscle: from each of these there runs a remarkable duct, about three fingers breadths long, and of the thickness of a straw, and which has a great number of roots. This duct was discovered by Steno in 1660, and is called the ductus salivalis Stenonianus, or the ductus salivalis superior: it crosses over the masseter muscle, through the middle of the cheek, and perforates the buccinator and the membrane of the mouth near the second or third molar tooth, and there, by its orifice, discharges a large quantity of saliva into the mouth. *(Comp. Anat.*, n. 278.) Helvetius affirms, 'That the quantity of saliva secreted in this gland is so great, that a soldier, having received a wound in his cheek, which severed the salivary duct, wetted several cloths, at every meal, with the fluid which flowed from it while he was eating.' I myself, and others, have seen similar cases. *(Comp. Anat.*, not. 54.*)* The maxillary glands are also of considerable size, placed one at each side, on the inside of the angle of the lower jaw: each has a single duct, arising from various roots, a little slenderer than the duct of Steno, but frequently longer, and called from its supposed discoverer, Wharton's duct, or the ductus salivalis inferior: this opens into the mouth, under the tongue, near the root of the frænum, generally by a single orifice, but sometimes by two or even three orifices. The sublingual are two oblong glands, placed one on each side, under the tongue, and which in human subjects are commonly thought to discharge their saliva into Wharton's ducts. Rivinus was the first to discover a peculiar duct proceeding from each of them, in a calf, and Bartholini afterwards found the same in the lion; hence these ducts are called the Rivinian or Bartholinian ducts. Morgagni was the first who taught me, that in the human subject, these glands open into the mouth, on both sides, between the gums and the sides of the tongue, by very small ducts or apertures; these ducts, I have also found, and figured accordingly, tab. vii. fig. 33. *(Comp. Anat.*, n. 278.*)* Morgagni asserts, 'That in the human subject there are in reality a number of these vessels on each side, which proceed from the exterior surfaces of the sublingual glands, not forwards, but transversely, straight towards the gums; and that at a small dis-

tance from the glands, they each have a peculiar osculum or orifice, which is often large enough to admit a fine bristle.' Since learning this from Morgagni, I have been more careful in my dissections, and have at last found the same ducts myself. These ducts are described at large by Walther, who I believe first delineated them in the human subject; he gives eight of them. In one male subject, I found them much larger and more numerous than I had ever before seen them, and I have been careful to delineate them accurately. *(Comp. Anat.* not. 55, tom. ii.) In the same subject, I observed in the cheeks two glands, one on each side, opening into the mouth by several orifices, near the posterior molar teeth of the upper jaw; these may be called from their situation, the molar glands; see my tab. viii. fig. 36, 37. On the left side, there were also two or three other smaller openings, which pierced the membrane of the mouth and the buccinator muscle, and on pressing the gland slightly, each of them yielded a drop of clear saliva. Fig. 37 shows the external surface of the left gland, which is seen to be composed of several smaller ones, so as in shape to resemble a mulberry. Whether these glands are constantly present, I cannot say: I leave it as a subject for further enquiries. *(Comp. Anat.* not. 57, tom. ii.) In the same subject there was also discovered a new duct, situated under the anterior apophysis of the occipital bone, at the top of the fauces: it was of some size, and on pressure, it discharged a mucous liquid into the fauces: see tab. viii. fig. 38. *(Comp. Anat.* n. 278.) I found it extending from the orifice among a dense glandular substance surrounding it, under the middle of the apophysis of the occipital bone, towards the foramen magnum. In speaking of this region of the palate, Santorinus states, 'That it is covered with a dense and thick membrane, filled with numberless glands, from which a viscid mucus is dischaged: that he has sometimes found this membrane divided into shallow cavities, arranged in a kind of regular order: sometimes divided into irregular loculi, and sometimes so far cavernous as by its conspicuous orifices and deep sinuses, (from which a mucous humor might easily be obtained,) almost to resemble the tonsils.' *(Comp. Anat.* not. 58, tom. ii.) Several miliary and lenticular glands are also found dispersed through the membrane of the mouth, particularly about the lips, the palate, the cheeks, and the tongue: these, according to their situation, may be called the lingual, the buccal, the palatine, the uvular glands, &c. The glandulæ Nuckianæ, found in the dog, are placed near the eyes, on each side; they send down ducts into the mouth, near the last molar teeth but one in the upper jaw: but they do not exist in man." *(Comp. Anat.* n. 278.)

56. WINSLOW. "The cheeks and lips form the walls and entry of

the cavity of the mouth. They are formed in general by the connexion of several fleshy portions of different breadths, attached round the convex sides of the two jaws, covered on the outside with the skin and fat, and lined on the inside by a glandular membrane. Besides all this, the lips seem to have a soft spongy substance in their composition, which swells and subsides on certain occasions, independently of the action of the muscles: and which is mixed with adipose tissue. The substance which forms the red border of the lips, is very different from the adjoining skin; being a congeries of very fine, long, villous papillæ, closely connected together, and covered by a delicate membrane, which appears to be a continuation both of the epidermis and of the pellicle which covers the glandular membrane of the cavity of the mouth: this substance is exquisitely sensitive. The internal membrane of the upper lip forms a small frænum above the first incisor teeth. The gums are that reddish coreaceous substance, which covers the two sides of the whole alveolar border of both jaws, is continued between all the teeth, surrounds the collar of each tooth in particular, and adheres very strongly to them. Therefore, the outer and inner gums are continuous, and together form as many openings as there are teeth. The gums are of a singular structure, resembling the texture of a hat, very compact and elastic. It is not connected to the bones of the jaws immediately, but by the intervention of the periosteum, with which it is perfectly united; and it is covered with a fine, strong, even membrane, which is closely adherent to the substance of the gums, and yet seems to be continuous with the thin membrane which goes to the lips and to the tongue. The arteries which supply the lips, cheeks, and gums, are branches of the external carotid, and chiefly of that branch of it which I term the external maxillary artery. The veins are branches of the external anterior jugular: but concerning the arteries and veins, see the section devoted to the arteries and veins, n. 55, 58, 72, &c. The nerves of these parts come principally from the superior and inferior maxillary nerves, which are branches of the fifth pair from the medulla oblongata; and also from the portio dura of the auditory nerve, or sympatheticus minimus, the ramifications of which are spread in great numbers on all these parts, and communicate freely in a singular manner with the fifth pair, as may be seen in the section on the nerves. So much variety is met with in the muscles of the lips in different subjects, that it is not surprising that anatomists differ in their accounts. In some subjects, portions of these muscles are wanting; in some, it is scarcely possible to distinguish them, by reason of the paleness and thinness of their fibres: in others, there are particular fasciculi, which are not generally to be found. About fifteen years ago I dissected an old female subject, in which I observed an extraordinary number of peculiarities. The muscles of the lips are usually divided into common and proper. The common muscles are those which end at the angles or commissures of the two lips, while the proper are attached to either the upper or under lip only. The common muscles are the semi-orbiculares, the supra-semi-orbiculares, the buccinators, and the zygomatici majores. The proper muscles of the upper lip are the zygomatici minores, the canini, the incisorii laterales and the incisorii medii. The proper muscles of the under lip are the triangulares, the triangularium collaterales, the quadratus, the incisorii inferiores, and the cutanei." *(Exp. Anat., Traité de la Teste,* n. 542—551.) Concerning all which see Winslow himself for further information.

"The palate is the arched chamber or cavity of the mouth. This arch is partly hard and fixed, partly soft and moveable. The solid portion is bounded by the teeth, and formed by the two maxillary and the two palate bones. The soft and moveable portion lies behind the other, inclining backwards, like a veil fixed to the edge of the palate bones: being formed partly by the common membrane of the whole arch, and partly by muscular fasciculi. The membrane which lines the cavity throughout, is like that which lines the great cavity of the pharynx. It is thickly set with minute glands, the orifices of which are not so discernible as in the pharynx and especially in its rugæ. This membrane, together with that of the posterior nares, forms by an uninterrupted expansion, the anterior and posterior surface of the soft portion or septum palati, so that the fleshy tissue of this portion lies in the duplicature of a glandular membrane, excepting which it is made up of muscles." *(Exp. Anat., Traité de la Teste,* n. 486-7-8.) We shall speak of these muscles when we come to treat of the nares, the uvula, and the larynx.

"The salivary glands are the parotid, the maxillary, the sub-lingual, the molar, the buccal, the labial and the lingual glands, the amygdalæ, the palatine, the uvular, the arytænoid and the thyroid glands. The parotids are two whitish glands, irregularly oblong and protuberant, situated on each side, between the external ear and the posterior ramus of the lower jaw, and lying on the neighboring part of the masseter muscle. The superior portion of the gland lies on the cartilaginous meatus of the ear, and touches the zygomatic arch of the temporal bone. From its anterior-superior portion a white membranous canal is produced, by the union of a number of tubuli, resembling so many roots. This canal runs obliquely forward on the outside of the masseter, and then perforates the buccinator from without inward, opposite to the interstice between the second and third molar teeth. It is about a line in diameter, and in some subjects is partly covered by small glandular bodies, united with it in different quantities. The arteria and vena angularis run up over this duct; and the portio dura of the auditory nerve passes through the gland itself, which also receives filaments from the second vertebral pair. The maxillary glands are smaller and rounder than the parotids, and are situated each at the inside of the angle of the lower jaw, near the inferior pterygoid muscle. From the inside or that which is turned towards the musculus hyo-glossus, each of them sends out a duct in the same manner as the parotids, but their duct is smaller and longer than Steno's duct. It advances on the side of the genio-glossus, along the inside, and towards the upper edge of the sublingual gland, to the margin of the frænum, where it terminates by a small orifice in the form of a papilla. The two ducts sometimes open by two distinct orifices, sometimes by only one. The sublingual glands are likewise two in number, of the same kind as the former, only smaller, more oblong, and flatted like a blanched almond. They are situated under the anterior portion of the tongue, one on each side, near the lower jaw, on the lateral portions of the mylo-hyoidei, which sustain them. One extremity of each gland is turned backward, the other forward. The glands are covered on the upper side with a very thin membrane, which is a continuation of the membrane that covers the under side of the tongue. They send out laterally a number of small short ducts, which open near the gums by an equal number of orifices, all in a row, at a small distance from the frænum, and a little more backward. The genio-glossi lie between the two sublingual glands, and also between the two maxillary ducts. The molar glands are two glands nearly of the same kind as those just described, situated, on each side, between the masseter and buccinator muscles; they send out small ducts that perforate the buccinator, and open into the mouth, nearly opposite the last molar teeth." The buccal, labial and lingual glands. "The inside of the cheeks, near the mouth, is full of small glandular bodies, called glandulæ buccales, which open by little orifices on the inner membrane of the mouth. The membrane which covers the inside of the lips is likewise covered by the labial glands. Those of the foramen cœcum of the tongue are called the lingual glands. The palatine and uvular glands are found on the roof and septum of the palate. The arytænoid glands were described with the larynx. We may likewise reckon among the salivary glands those of the top of the pharynx, and of the pituitary membrane and nasal sinuses. The amygdalæ are two glandular bodies of a reddish color, occupying the interstices between the lateral arches of the septum palati, one on each side of the base of the tongue. Their uneven and perforated surface gives them the appearance of an almond shell; being full of holes big enough to admit the head of a large pin. These holes, which resemble a sieve or network, are continued to an irregular sinus or cavity within the gland, filled commonly with a viscid fluid, which as it gathers, is discharged through these holes into the throat. To see the structure of the amygdalæ, they must be examined in clear water." (*ibid.* n. 575—587.) Lymphatic glands. "Near the parotid, towards the mastoid apophysis, there is a small gland of another kind, differing from the former in figure, color, excretory duct, and secretion. It is round, even, and without projections: and it is the uppermost of a number of glands of the same kind, which lie partly below the interstice between the parotid and maxillary glands, and at different distances along the jugular vein, all the way down the neck. Among these glands, and upon this vein, we observe a great number of transparent vessels, with valvular divisions: the fluid they contain is clear, somewhat mucilaginous, and is called lymph. The vessels are likewise termed lymphatic vessels, and the glands, lymphatic glands. These glands are not all equally large or round. The lymphatic vessels pass out by one extremity from one gland, and by the other extremity, enter some neighboring gland. This kind of glands accompany both the parotid and the maxillary glands." (*Exp. Anat., Traité de la Teste,* n. 591-2-3.) Respecting these glands, see also *Traité de la Teste,* n. 591—600.

57. Swammerdam. Of the mouth and fauces of the pediculus, or louse. It appears from the figure, that immediately behind the sucker there is a cavity, lying in front of, and continuous with, the œsophagus; into this cavity the animalcule has the power of transmitting the blood drawn by the sucker. This part is the blind case or pouch of the sucker, which lies concealed in it. When the sucker is extended, retracted, or bent, all these parts coöperate. (See *Bib. Nat.,* p. 74., tab. ii., fig. 3, 4, 5.)

The covered snail, besides having teeth and gums, has also a hiatulus, at the posterior inferior part of which, in the palate, are two minute, narrow openings, through which the saliva is conveyed into the mouth by the salivary ducts. (See *Bib. Nat.,* p. 108., tab. v., fig. 1., *c. d.*

"In the worm from which the musca asilus or gad-fly is produced, there are two blind canals, of a membranous, transparent substance, seated in the thorax, where they make a great many windings and turnings. In color they resemble fresh curds. They unite at last so as to form only one vessel, which terminates at the mouth. Near their termination there are two small particles, which very much resemble little muscles. The same particles appear even in the nymph of the worm, and afterwards in the fly, in which they stretch out straight, and after passing through the thorax, they terminate in the abdomen." *(Bib. Nat.*, p. 661-2., tab. xxxix., fig. 7.)

"In the mite, on the side of the gullet, there is a delicate tube (or salivary duct) which runs towards the fauces. In the breast it divides into two small channels, each of which widens into an oblong globular bladder, and then becomes again contracted into a narrow tube, which once again assumes the form of a bladder, beautifully adorned with a great number of pulmonary tubes running over its whole surface; some particles of fat, very regularly placed, surround one side of these glandular vessels; and underneath are extended into a kind of oblong appendages. I have reason to think that these parts perform the office of salivary glands and ducts." *(Bib. Nat.*, p. 704., tab. xliii., fig. 5.)

Of the salivary glands and ducts in the cuttle-fish; "The two ducts unite into a single canal, and the glands are of the conglobate kind." *(Bib. Nat.*, p. 883, 884, tab. l., fig. 4, 5, 6, 7.)

58. The reader may consult other authors for himself, particularly Boerhaave, on eating, *Inst. Med.* n. 58—64. Winslow, on the muscles of the lips, the lower jaw, and the septum palati, *Exp. Anat.*, *Traité des Muscles*, n. 1220—1235; and on the jaws, *Traité des os secs*, n. 272—286. Palfin, *Anat.*, part ii., chap. x. xi. xii., pl. xxvii., fig. 1, 2. (Paris, 1726.) Drake, *Anthrop.* tab. xviii. fig. 1, 2. Respecting the muscles of these parts, see Vesalius, Cantius, Cheselden, Browne, Santorinus, Verheyen; and particularly Cowper, who has given various drawings of these muscles: also Eustachius, *Tabul. Anat.*, tab. xi. xiii. xiv. xv., and above all, tab. xviii., (where he has expressed the several muscles of the face, the lips, and the mouth, as accurately as any of the later anatomists,) and fig. 5, *k.*, shewing the gland at the side of the chin. Respecting the glands, consult their various discoverers; as Heister, *Comp. Anat.*, tab. vii., fig. 33; tab. viii., fig. 36, 37, 38. Nuck, *Adenog. Cur.*, fig. 9, 10, 11, 12, 14. Wharton, Steno, Bartholin; also Munnicks, who alleges that he found a new salivary gland, whose duct perforated the gums near the first molar tooth. Respecting the membrane of the gums, see Ruysch, *Thes. Anat.* vii., n. 41, not.; also p. 20, and tab. ii., fig. 5., shewing the lips and part of the cheeks, with their epithelia, nervous or villous papillæ, skin, fat, &c.

59. A CONTINUOUS passage extends from the median fissure of the lips all the way to the intestines; first into the mouth and the fauces; next into the pharynx and the œsophagus; next into the stomach; and thence into the small and the large intestines. This long and tortuous passage is partitioned and secured by seven doors and locks. The lips close the first space, or the small cavity in front of the mouth. The bony ridge of the teeth environs the second, or the fauces, which is a larger and deeper cavity than the first. The pharynx guards the third, or the œsophagus, which is an intermediate passage. The cardiac orifice guards the stomach, or the true swallow: the pylorus, the small intestines continuous with the stomach: the valvula coli, the large intestines which succeed them. The sacrum and the os coccygis are placed at the end, which closes the outlet. Each door is locked by a muscular sphincter (*a*). At the first the lips stand sentinel. When separated, they disclose a confined passage, or narrow court, which resembles the trench before a fortification, that is, before the arched barrier of the teeth. This again, when opened, leads into the fauces or palate, a vaulted chamber constructed of bones and membranes (*b*). This

(*a*) Excepting the second, which is closed by the teeth; but this is not properly a door, but a kind of bony fence, formed of the teeth and gums; and thus constituting a barrier between the first and second divisions of the mouth, which form as it were one cavity. The small anterior compartment may be compared to a foss surrounding a rampart, as we shall explain presently.

(*b*) Namely, of the teeth, the two maxillæ, and the ossa palati; which are covered with periosteum, and with vascular and nervous is the first pouch; as it were the cupbearer, the market and repository of the food, whence the succeeding cavities draw their stores. But the use of all these parts shows the reason of their construction.

60. The lips, like two chained janitors,* guard, unlock and open the outer gate; that is, the common mouth of all the doors; whereby they afford a passage into the contiguous fauces and into the palate. Thus they anticipate and commence the offices of the tongue, and anon conspire therewith; namely, in sucking, in sipping, in drinking, in eating, in speaking, and in singing: also in feeling, (for they apprize the tongue, by their sense of touch, of what is proper to be taken and tasted) (*c*): likewise in the other or derivative offices (*d*). The lips, moreover, confer upon the face the power of imaging, in its looks, the affections of the cerebrum.

61. The fauces and the palate are hollow chambers, and resemble a dress, case or form modelled exactly to the tongue: consequently, they are the actual and instrumental causes of which the tongue is the efficient principal cause. Now as the instrumental cause is constantly adjoined to the principal cause, to assist and minister to it, so the palate (*e*) is adjoined to the tongue (*f*) in all the offices which it is about to perform; as when it is about to suck, to drink, to eat, or to speak; and even when it is about to taste the essences and dainties of the food. More-

membranes: and the palate, with a glandular congeries besides, and about the septum, with many muscular fibres.

(*c*) For instance, when any thing rough, prickly, styptic, or caustic, &c., is presented, the lips instantly perceive it, provide against it, and warn the tongue not to receive it: this is the object of their exquisite sense of touch.

(*d*) Respecting the derivative offices of the tongue, see n. 37.

(*e*) The term palate is here used in a wide sense, for the whole of the cavity between the teeth and the pharynx.

(*f*) The tongue is the principal part in this cavity: all the other parts are its instruments. An active, in order to be efficient, must always have a passive conjoined with it; consequently, the two, taken together, produce the effect.

* This appears to be an allusion to a custom of the Romans, of chaining the slaves stationed at their doors: such slaves were styled, *Catenati Janitores.—Tr.*

over, the fauces are ancillary to the tongue, in transmitting and supplying the salivæ which perpetuate its activity; and in absorbing the seethed, extracted and ensalivated juices, spirits and tinctures; in the same manner as the tongue itself absorbs them.

62. But inasmuch as the tongue is active, and moveable upwards, downwards, and to the right and the left, in its deep cavity, therefore the form modelled to it and wrought upon it—that is to say, the palate and the mouth—must necessarily be framed with particular reference to its activity; in a word, this form must be similarly moveable. To accomplish this end, the lower jaw is detached from the upper, and furnished with joints, ligaments and muscles; so that it may serve and adapt itself to the tongue, by imitating and following its movements.

63. As the principal and the instrumental cause, or the active and its passive, act ultimately as one, or amount to one (*g*); so the tongue, the fauces, and the palate perform all things together; and at last constantly advance the ingesta into the pharynx and œsophagus, that is, forwards; and the sonorous egesta of the trachea and the larynx to the door of the lips, that is, backwards: wherefore the palate itself, or the inner gorge, is provided with folded coats, pendulous membranes, septa and columns, and also with muscular fibres, and connected to all these parts. These are the general uses of the palate, the jaw, and the lips.

64. *In order that the lips* (as mentioned above) *may unlock and open the outer gate, afford a passage, and lastly conspire with the tongue and the fauces* (*h*), they require to be supplied largely with muscles, and to have the power of opening and folding their fissure into a mouth or orifice of any shape they please. Hence the numerous muscles composing the lips and epithelia, and the numerous muscles lying against them, and passing in all directions into the circle of the orifice. The prolabia are formed by the orbicularis, which when carefully examined is seen to consist of two muscles, with their fibres decussating

(*g*) According to a common rule in physical and philosophical matters, and which nature attests universally by actual effects.

(*h*) We recur to what we said before respecting the lips (n. 60), with a view of explaining it here in detail.

at the angles of the mouth, and which thus alter its circumference variously in shape and size (*i*). Several muscles pass into the circle formed by the orbicularis; on both sides, but within, there are the buccinators, which, besides their other actions, may also draw the corners of the mouth apart (*k*): descending almost perpendicularly, there are the four elevators, the two common and the two proper muscles of the upper lip; and ascending there are the depressors, the common and the proper muscles of the lower lip. The former draw and revolve the orbicular orifice of the mouth towards the septum and alæ nasi; the latter, towards the chin (*l*). The muscles which enter at more acute angles, are the zygomatici majores and minores: and next to these, the canini, the risores, the incisivi, and many others not yet clearly made out, which when they act jointly beautify the natural arch of the mouth, and when they act separately, frightfully distort it. By these means the aperture of the mouth is folded into all the curves, gyres and flexures that the compass can describe—into all that the tongue desires and use demands: and by these powers, (for the moving fibres are so many powers

(*i*) The orbiculares, which constitute the orifice of the mouth, are delineated by Eustachius and others as uniformly circumflex: but according to Cowper, the fibres of the upper lip have a different direction to those of the lower. Winslow, from his own observations, divides these muscles into semi-orbiculares, and supra-semi-orbiculares; because at the corners of the mouth, where the muscles join, the fibres of the upper lip decussate with those of the lower; and because, in the upper lip, the fibres are separated into two portions by a small interstice.

(*k*) The buccinators, placed transversely between the posterior parts of the maxillæ, form a portion of the cheeks, and when retracted towards the ears, they draw the corners of the mouth apart: but as they are subject to the masseter and the zygomatic muscles, they are, therefore, said to exert their action on the mouth only when the person is eating or blowing some instrument: yet from the connexion of their fibres, it seems impossible to deny their action on their lips.

(*l*) Cowper has delineated all these muscles; he points out, that the proper elevators of the upper lip are the same as the retractores nasi; which latter, together with the two depressors of the lower lip, are also distinctly shewn by Eustachius. With regard to the quadratus, it subtends the whole of the inferior part of the orbicularis.

and active forces embodied,) the lips coöperate with wonderful facility in all the offices of the tongue and the palate. Still more innumerable varieties of configuration are produced in the mouth, from the opening of the buccal cavity by the muscles of the lower jaw: for the labial orifice is widened—perpendicularly, laterally, or obliquely—by the mere action of the jaw, on which the chin and the lower lip are situated; and new circles or orifices are figured or defined, which serve the oral muscles already mentioned as new centres, round which, as diameters, they describe new circumferences, of ever varied forms: whence the diversities increase in multiplied ratio. Were a single variety, or a single function of the mouth to be portrayed in all its details, it would fill volumes; and yet there are particulars besides, which, were they to be set forth with their variations, would inevitably confuse and obscure general notions, which distinctly involve particular ideas (*m*).

65. Besides the offices which *the lips* perform for the tongue and the fauces, they also *confer upon the face the power of imaging, in its looks, the affections of the cerebrum* (*n*). For in order that muscles in plane or convex surfaces may unfold freely, in full and particular compliance to the activity of their fibres, there must be somewhere an elastic fissure, or a cavity expansible into diversiform apertures, whereto the muscles in the plane are related, as diameters to a moveable centre. Thus the muscular part of the face derives its power of motion and mutation, and its conformableness to the various natural affections,

(*m*) For example, were we to describe only the concurrence of the lips with the tongue and the mouth, in the process of eating or speech, or indeed in any single part of either, we should certainly have to run through all the figures of geometry, the forces of mechanics, and the laws of physics; while entering and following the natural fluxions of the organs. But the longer we dwell on particulars, the longer we are kept from that general notion of all things, to procure which must be the first object of analysis; inasmuch as it will afford us the means of proceeding distinctly into particulars. This is the reason why we do not enter further into details.

(*n*) We again recur to our general statements (n. 60), for the purpose of considering them specifically: they constitute points whence we may properly expatiate into particulars.

from the pliability of the lips and the extensility of the mouth. On this account, all the muscles of the face, nasal and palpebral, temporal, frontal and mental, are inserted in a wonderful manner into the muscles of the lips or mouth (*o*); and thus are enabled to portray and represent all the forms and features which are produced by the harmony or discordance of objects in the tongue and the other organs of the senses, and by the likings and dislikings consequent thereon in the cerebrum; and which the cerebrum instantly transcribes into the face. Hence the differences between anatomists in describing the muscles of the lips (*p*); for there are as many countenances as there are minds; as many minds as there are senses (*q*); and as many senses as there are heads.

(*o*) This is very conspicuous in all the descriptions and delineations of these muscles; and also from the natural configurations of human countenances; thus, for instance, the frontal muscle descends in two portions to the eyebrows, and blends with the superciliary and palpebral muscles, and by these means, with the zygomatici, canini, and risores, which are the common and proper muscles of the lips. On the other side, the same frontal muscle communicates by the temporal muscle with the mastoid, and by the latter with these same muscles of the lips. The same holds true of the buccinator and masseter muscles, which form the cheeks. The nasal muscles are almost identified with the labial muscles; so also the mental muscles, with which the digastricus communicates; and this, in hale and young subjects by true muscular fibres, but in old subjects, by worn out or tendinous fibres; and universally, by ligaments and aponeuroses. In this way, one muscle flows into another, and all ultimately into the central or labial muscles. The reader may see these things delineated to the life by Eustachius, tab. xv., fig. 1; and tab. xvii., fig. 1.

(*p*) All myologists differ, in both their plates, descriptions, and enumeration of the labial muscles. Santorinus saw many more than other anatomists. Verheyen added a new risor muscle. Winslow discovered several. "So much variety," says he, "is met with in the muscles of the lips in different subjects, that it is not surprising that anatomists differ in their accounts. In some subjects, portions of these muscles are wanting: in some it is scarcely possible to distinguish them: in others there are particular fasciculi which are not generally to be found." (n. 56.)

(*q*) By the senses generally, we mean all those differences which occur in both the external and internal senses.

66. *The fauces and the palate* (as indicated above) *are cases and forms modelled to the tongue;* for they correspond to it exactly in extent, shape and size (*r*): they are also united to it at the bottom by muscular fibres, ligaments and membranes; but above and at the sides the two are only in contact: nevertheless, a common and continuous membrane lines the convexity of the one, and the concavity of the other: and a ropy saliva flows between them, and in a manner tends to conglutinate the divided parts, and to divide the united parts (*s*). The tongue has two means of enlarging the interval between it and the palate, to give itself the mobility and voluble power required by its various functions: one, proper, by the depression of its body and fundus, by means of the muscles: the other, common, by the lower jaw, also by means of the muscles, and by the joints and ginglymi of the jaw; whereby its proper locomotive powers are, as it were, multiplied in it (*t*). For this end, the lower jaw itself is depressed, elevated and rotated by its muscles, and the tongue, the gums, the lower row of teeth, and the lower lip, which are attached to it, follow its motions. The depressor muscles of the lower jaw, are the platysma-myoides and the digastricus, which latter also exercises a slight rotatory action on the jaw. The elevators are the temporal muscles, and the ptery-

(*r*) As appears when the tongue is raised, and the fauces closed upon it; in which case there is no interstice, except at the most moveable part of the tongue, in the immediate neighborhood of the gums.

(*s*) Wherever in the organic body we find parts divided, which are intended to operate conjointly,—as the passive with its active, or the instrumental with its principal,—some glutinous liquid always flows between them, which were it not discussed by the power and activity of the parts, would tend to unite them together: as we shall prove in Parts III. and IV., when we treat of the genital members. That the saliva is of a glutinous nature, particularly what is brought from the palate, is sensibly evident.

(*t*) The tongue has numberless modes of rolling, the space in which it rolls remaining the same; these are further multiplied if the space itself be varied; as is manifest from its modes during eating, and also during speaking; for according to the manner and degree in which the fauces or the mouth are opened, the sound varies in depth, nature, and character. The variations which the general motion imparts to its particulars, were fully explained above, (n. 64.)

goidei interni and externi, both with fixed and strong origins; namely, from the pectoral and deltoid muscles, the mastoid process, the temporal bones, the pterygoid process, and the neighboring part of the sphenoid bone; where, and in their course, they communicate with all the muscles of the face, the mouth and the cheeks, and with all the muscles of the chest, which can in any way assist the tongue in eating and speaking. Thus the palate is a kind of instrument to the tongue; and yet the two are so united, that they perform a kind of conjugal office for each other.

67. When *the tongue is about to speak or to sing, it conspires in a wonderful manner* with the larynx, the palate, the velum palati, the uvula, the mouth, the gums, the cheeks, and the lips; and indeed, with the whole face; and with the trachea, the lungs, and the whole chest. At one time the tongue enlarges the aperture of the throat, at another time contracts it, and this, in every measure, form and way: likewise the cavity inside the cheeks: at one time the tongue applies itself to the mouth, the gums and the teeth, either high up or low down, either slightly or closely; sometimes it only opens the passage, and simply transmits the sounds formed in the larynx and modified in or below the palate: at other times, it contracts the passage, and drives out the whole, or a part, of the sound, through the nares; at another time it compresses the lips, and suddenly opens them. Thus it articulates sounds into words, with infinite play of variety, either singly or compositely, successively or simultaneously. The tongue has the faculty of guiding the sounds; the larynx and the neighboring parts, of modifying them; and the palate, the mouth and the lips, of carrying them forwards (*u*). All these instruments or organs suffer themselves to be naturally and spontaneously influenced by the innermost laws and principles of the art of music—the

(*u*) These statements must be plain to all who attend to what passes in themselves. Wherefore speaking does not appear to be the proper office of the tongue, as we before remarked in the chapter on the tongue. That the tongue is only competent to the office of regulating, is clear from the circumstance of its application to the parietes of the mouth and of the gums, and to the teeth themselves.

art of combining sounds harmoniously and varying them infinitely. The keys may produce everlasting harmonies from these principles, without ever exhausting the source.

68. When *the tongue is about to eat, it also conspires in a wonderful manner* with the lips, the teeth, the gums, the fornix and the palate, and with the pharynx and the œsophagus; in fact, with the whole sphere of the face, and with the whole region of the chest and the abdomen. It seems as if the tongue, by its fibres, demanded homage of the fibres of other parts, near and distant, as its servants. The lips first give presentiment by touch of what is about to be taken into the mouth, which, if innocuous, they seize eagerly, and submit to the teeth,—to the incisor, the canine, and the molar teeth, and the dentes sapientiæ, in succession. Then the tongue sallies forth and collects the scattered morsels, and, assisted by the cheeks, subjects and applies them again and again to the dental mill: it takes up the parts that are well ground, on its point and edges; gathers them from every hole and corner, and throws them on its dorsum: at the same time it soaks them with saliva, then works and agitates them, compresses them against the parietes of the throat, dashes them together, perhaps bruises the finest of them with its own cartilaginous denticles (*x*), by striking them together expresses their products into the cavity of the throat, then absorbs them, and lastly urges the refuse backwards to the root of the organ, (to be more thoroughly broken up in the stomach,) sheathes it with the thick mucus of the palate; and finally rolls it into the œsophagus. Thus every thing in the tongue, and in the lips, the cheeks, and the fauces, coöperates and conspires from its inmost grounds: one fibre rouses another, one gland another, and one muscle another, in order and series, beginning from the tongue, and extending through the height and depth of the body. All parts sally forth

(*x*) See what we deduced and concluded above, from the observations of Malpighi and Swammerdam, respecting the peculiar denticles and claws of the tongue (n. 34, and n. 49, Schol. *yy*); namely, that the denticles themselves, not only reduce, but also grind, the minuter parts of the food; and that the subjacent glands absorb them in the same manner as the tongue itself does on a large scale.

with the tongue to the prey. It would take ages to follow what the tongue executes in a moment. The sense of taste, produced by the quality of the food, by the appetence of the blood and the spirits, and by the soul's love of living in a healthy body, is what sets the whole machine in motion. The tongue has two senses, touch and taste: touch is the sensation of parts; but taste, of parts of parts. Its organic papillæ, which are the ultimate forms of the nervous fibres, convey these senses, almost instantaneously, through their parent fibres to the cerebrum, and as rapidly through the same to the muscular fibres and the muscles; which thus stimulated on both sides, rush into their natural and habitual actions; which are exactly correspondent to the modes of the senses (*y*).

69. But when *the tongue is about to drink, the palate* particularly assists and *conspires*. The lips first draw in the fluid by their aperture; the tongue takes it up on its apex and edges, pours it in on its dorsum, and rolls it gently into the gorge of the palate; it then raises and wreathes up its base from the root, close under the folds of the palate: the palate also unfolds, and lets down the shaggy velum from above; and in this way the two secure the passage against the return of the fluid; which, thus destitute alike of gravity and levity, glides along the smooth surface of the œsophagus into the stomach. Two singular powers of the tongue and the palate unite in the act of drinking.—*Firstly*, of arresting the fluid at any part of the cavity of the throat, and of pushing it onwards from point to point, either by sips or in streams: this power is owing to the tortility of the tongue, and to the flexibility of the membrana and velum palati.—*Secondly*, of exercising a kind of suction or attraction, at will, on any particular isolated spaces (*z*). The tongue and the

(*y*) I briefly touched on this matter also, in the chapter on the tongue, see n. 46; but I do not think it advisable thus early to expatiate further into its details, since an ample opportunity will be afforded in my analysis of the organs of the senses, and in my psychology, where all these things will be explained.

(*z*) This is wonderfully effected by the tongue, in conjunction with the palate; for wherever the tongue applies itself, whether to the lips, or by folding backward to the little fossa in front of the teeth, or to the palate imbibed these powers with the mother's milk, and momently exercise them so easily, that we ourselves are not aware of their very existence. The consequence is, that liquids and even solids descend from the commencement of the mouth (I do not say from the lips) (*a*) into the gullet, as easily, as if they were absolutely destitute of gravity and levity; and as if, in every position, the lips were upward with respect to them, and the gullet downward; for every point of the tongue and every point of the mouth, acts thither by the two powers already mentioned; thus every point of the fluid is actuated thither by a kind of centripetency. Hence draughts of liquid ascend as easily as they descend; which we see exemplified in jugglers, who will drain bowls while standing on their heads; and in long-necked birds, as the goose, the swan, the crane, &c.; and in quadrupeds, which lap, eat and drink, with their heads hanging down: more plainly still in those insects which suck their food through extensile, flexible and retractile suckers, and carry it thereby into their gullets and stomachs (*b*). Indeed, this mechanism of the throat seems designed to prevent liquid from acting at all of its own gravity; as is clear from the fact, that when water is poured into the mouth of a person lying on his back, it instantly regurgitates from the pharynx: evidently, in order that no intrusion may take place, and that nothing may be carried in without the tongue previously feeling and willing it.

gums, or to the prominent ridge under the molar teeth, or to the teeth themselves, or to any of the uneven portions of the mouth above the molar teeth, or at their sides; in a word, wherever it will, it has the power of exercising a tractile or drawing action; and this it accomplishes momently with such nimbleness and ease, at the time that it is chewing and kneading the food, attracting and drawing forth the saliva from the little ducts, and irrigating and moistening the food therewith, that the action cannot be perceived without the closest attention. Verily, before operations of this nature, human art may stand still in mute astonishment.

(*a*) For the tongue first assists the fluids to gain the inner side of the teeth, either by sipping, as in brutes, or else by pouring them in all the way, as in man.

(*b*) See Swammerdam, on the louse, n. 57, above; and also consult him on bees, flies, and other insects with suckers.

70. *The tongue is also the principal agent in drawing the saliva from the glands,* as from the parotid, the maxillary, the sublingual, the molar and the orbital glands; perhaps also from the mental (*c*), and from the innumerable palatine glands; from which the sallies of the tongue bring down copious showers of saliva. For the ducts, orifices and lips of these glands project and open into the cavity of the mouth; and the tongue is constantly touching, grazing and irritating them with its apex and edges; and subjecting them to its peculiar suction or syringic action. These titillations and irritations in the extremities, instantly creep up the ducts to their sources, the glands, and exact a tribute; almost in the same way as blisters, &c., applied to the pores of the skin, extract the juices from its tubuli. The delicate membranes, fibres and glandular granules of the ducts, directly propagate these impressions and vibrations to their sources: and for this reason, these ducts traverse the gums and the continuous membranes—parts of such exquisite sensibility and extreme irritability (*d*). But this mode of drawing out the saliva is the particular mode of the tongue; there is also a general mode. For these glands and their emissary ducts graze, touch and even perforate the muscles of the tongue, the os hyoides and the lower jaw in many places. Such is the case

(*c*) Eustachius also delineates one gland under the chin, (tab. xviii., fig. 5, *k*,) which is thus alluded to by Lancisi: "We think it worthy of observation, that our author demonstrated a gland at the sides of the chin."

(*d*) Every one is aware that the gums are sensitive, from the toothache and other phenomena: for the membrane of the gums is closely connected to the periosteum itself. "The gums," says Winslow, "are of a singular structure, very compact and elastic. It is not connected to the bones of the jaws immediately, but by the intervention of the periosteum, with which it is perfectly united; and it is covered with a membrane which seems to be continuous with the thin membrane which goes to the tongue." (n. 56.) This membrane, and the continuation of it under the tongue, are perforated by the salivary ducts, the orifices of which, therefore, open when the tongue is in active motion, or brandishing about; and sensibly so when ardent spirits, corrosive powder, or herbs like tobacco, are applied; in which case the mouth is speedily inundated with a deluge of saliva.

with the *parotid glands,* and the masseter and buccinator muscles: with the *maxillary glands* and the inferior pterygoid, hyoglossus and genio-hyoideus muscles: with the *sublingual glands* and the mylo-hyoideus and genio-glossi (*e*): and with the *palatine glands* and the muscular substance of the palate, velum palati and palatine arches, &c. Not to mention, that the more considerable ducts open beside the frænum, the most muscular region of the tongue (*f*); and that the nervous fibres leap from the muscles to the glands, from the glands to the muscles and the tongue, in a wonderful manner; and stimulate each to co-operate with each. So also in the simpler types, as the worm, the fly, the acarus, and the cuttle-fish (*g*). Every time then that the tongue works the food, or every time the os hyoides swallows and the jaw chews it, they excite the muscles, and the muscles the glands and their excretions: and hence there is a perpetual rain of saliva. And furthermore, as soon as ever the tongue prepares for action, the fibres, roused by the cerebrum, apply themselves to the same thing; and hence the ducts foam with desire. This is the reason why the tongue and the mouth sometimes dry during the night; and the stomach is deprived of its portion of saliva. From these considerations it is evident,

(*e*) On these points, see our authors, particularly Winslow, (n. 56,) who describes accurately the friction and contact between these glands and ducts on the one hand, and the muscles of the tongue, of the os hyoides and of the lower jaw, on the other; also the perforation of the muscles by the ducts and glands; and further, that the *molar glands* are placed on both sides, between the masseter and the buccinator muscles.

(*f*) For instance, the *maxillary* ducts, which terminate under the tongue, near the root of the frænum, in the form of a papilla, and open into the mouth. The *sublingual,* which terminate between the sides of the tongue and the gums, not far from the frænum: not to mention the *parotids,* which perforate the membrane of the mouth beside the second or third molar tooth. See what Helvetius relates (as cited by Heister, n. 55) respecting the quantity of saliva which the parotids poured forth in a wounded soldier. And respecting the number of orifices lately discovered under the tongue, see Heister, *Comp. Anat.*, tab. vii., fig. 33.

(*g*) Respecting which, see Swammerdam above, n. 57.

that the tongue does not excrete saliva; for this is the office of the instrumental cause, not of the principal; of the servant, not of the mistress. The tongue simply demands, instigates, and exacts it as a tribute (*h*).

71. The SALIVA, separately considered, is the menstruum and the vehicle of the particles which it dissolves or decomposes. And as every animal has its own natural and peculiar food, hence in order for the saliva to dissolve and decompose its parts, and to serve as their vehicle, it must be different in every genus and species of living creatures, and in every individual (*i*). The saliva of the mouth attacks only the weaker and more general affinities of the food; it also sheaths its solutions, and thus secures them a ready transit through the finer tubuli, which would suffer injury from minute saline spiculæ, or would be irritated by their presence; it smooths and in a manner lubricates the avenues and passages: wherefore, in the anterior part of the mouth it is limpid and, as it were, watery, but more viscid as it approaches the palate (*k*); thus it increases in density; in order that the first, or virgin saliva may carry the simpler extracts immediately into the blood; but the cohesive and uncomminuted portions, through the œsophagus into the stomach. Furthermore, the saliva is not permanent, either in quality, nature, or character, even in the same subject; it sometimes changes many times in an hour; in exact correspondence to the affections of the tongue, and hence of the cerebrum, and hence again of

(*h*) According to a proposition in the preceding chapter; see n. 42.

(*i*) That is to say, it is different in man, who lives on roasted, boiled, and cooked meats, from what it is in beasts, which feed on herbs; and in birds, which eat worms, lizards, frogs, sand, gravel, and even pieces of iron; and in fish, which devour other fish, turf, mud, &c.

(*k*) By a little attention, we may clearly perceive this in our own persons; a more tenacious sputum being spit from the palate, while a comparatively thin spittle trickles, sometimes spontaneously, from the first or anterior springs of the mouth. To produce this effect, the parotid and the maxillary glands, which perforate the ante-lingual region of the mouth, appear to be in association with lymphatic glands. Winslow says, "Near the parotid, there is a small gland of another kind; it is the uppermost of a number of glands: among these we observe a great number of transparent vessels, with valvular divisions:

the mind. It is acrid and acute during good appetite; comparatively weak and obtuse during squeamishness and nausea: like the mind itself it is also affected by a multitude of other causes: it is salacious in the wanton, particularly during the venereal ardor; poisonous in the mad (*l*), breathing, as it were, death and vengeance; rare and subtile in the lively; thick or tough in the sad, the stern, the anxious; mild in the merciful; to say nothing of a number of other modifications, not arising from affections of the spirit, but of the blood, as in diseases; and which make the saliva a medium of contagion. Whatever is displayed in the outermost, flows from a nature which resides in the innermost. External things only represent what internal things contain. Thus whatever takes place in the mind, takes place in the innermost of the blood, and in the innermost of every humor derived from the blood; consequently, in the innermost of the saliva; wherefore the saliva also is a type of an intimate, indwelling power.

72. The hollow parietes of the mouth, and the palate, in imitation of the tongue, also eat and drink the purer extracts and spirituous solutions of the food, and the saliva united therewith, and thus relieve the first hunger of the blood. For these parts have lenticular papillæ (*m*) similar in structure to those of the tongue, and perforated with foramina, which are not con-

the fluid they contain is clear, somewhat mucilaginous, and is called lymph. The vessels are termed lymphatic vessels, and the glands, lymphatic glands. This kind of glands accompany both the parotid and the maxillary glands." (n. 56.)

(*l*) Natural history supplies numerous instances of this kind; shewing that the human bite itself—the bite of an enraged man—like the bite of a rabid dog, has produced dreadful and almost fatal disease in its victim. Thus also the bite of vipers, in which we can discover no poison, causes death; the effect being due to the saliva of the enraged reptile. But of these subjects we shall treat in our pathology, and in many other places.

(*m*) All practical anatomists agree, that the membranes of the mouth and palate are covered over with an infinity of lenticular glands, of the same form as those described above in the tongue. It is sufficient to adduce the testimony of Malpighi. "It should be noted," says he, "that the same papillary substance is found also in the palate and the cheeks; but with this difference, that in these places the larger papillæ nected with any subjacent glands, but from which ducts or tubuli pass inwards; hence the same arguments apply to them as to the papillæ of the tongue. And the mouth, like the tongue, dries at night and sometimes during the day, and requires to be moistened with fluid: and sensation itself assures us, that when pleasant, rich and spirituous fluids are kept in the mouth, they vanish away entirely, without a drop passing into the œsophagus. A supply is in fact continually required and eagerly demanded by the numberless ramifications of the jugular veins which redden the fauces; for the chyle of the body, carried up by the thoracic duct, meets the chyle of the tongue and mouth, carried down by the jugular veins, in the middle of the subclavian vein, not far from the right side of the heart; and the blood of both regions, recruited and exhilarated therewith, ever runs forth anew in its circles. Moreover, the cranial or carotic blood is constantly deprived of its serum in the numerous salivary glands; of its nobler essences in the sensoria of sight, hearing, smell and taste; and of its very spirits in the cerebrum and cerebellum; hence arid, slow, hungry and thirsty in the veins, it burns to be recruited by this fresh and first-born chyle of the mouth (*n*). Here then is the second use which the saliva performs—for the interior organs of the brain, and the external sensoria of the body; which require the purest blood to prevent them from being dulled and defiled; this the salivary glands defecate, and render fit for use.

stand out in the form of cones, and near them are excretory vessels, implanted in subjacent glands, and among the vessels are scattered a few very minute nervous papillæ." (n. 28.)

(*n*) From these causes there arise in all the veins, in the branches of the jugulars particularly, a thirst and desire of imbibing the liquids and juices expressed from the food, and thus, of reabsorbing and as it were ruminating their salivæ: whence arises the general desire of all, which is called hunger and thirst, whereof the latter is sometimes felt to be confined to the mouth and tongue. We shall shew elsewhere with what inherent avidity the blood attracts liquids when it needs them, particularly those which are combined with the saliva. This is the primary source of the general attractive power in the tongue, which prevails also throughout the system; being a kind of principle, at the same time that it originates from a prior cause.

CHAPTER III.

THE PHARYNX, THE ŒSOPHAGUS, AND THEIR GLANDS.

73. Heister. "The œsophagus or gula, called by Cicero and Celsus, the stomachus, is a membranous canal, reaching from the fauces to the stomach, and conveying into it the food and drink taken in at the mouth. It is somewhat infundibular in form; and its upper part is called the pharynx. It is placed in the neck, almost exactly behind the trachea, and along the cervical vertebræ; but in the thorax it turns a little to the right, on account of the aorta; and soon afterwards again, to the left. (*Comp. Anat.* n. 262.) Modern anatomists, however, are not agreed about the situation of the œsophagus. It has been described by most writers as running straight behind the trachea, between it and the cervical vertebræ. Morgagni has figured it as inclining a little to the right of the trachea; while Winslow contends that it is always to the left. Cant accuses Vesalius of an error, in having placed the gullet behind the trachea; and also takes Morgagni to task, for representing the œsophagus at the right side of the trachea, and appeals to the Tabulæ of Eustachius. My own opinion is, that nature varies in the situation of this part: and I can take upon me to assert, that in some subjects the gullet lies immediately behind the trachea; that in others it inclines to the right, as Morgagni has figured it; and that in others again it inclines to the left. Whoever fairly examines the Tabulæ of Eustachius, will find that he has placed the œsophagus behind the trachea. (*Comp. Anat.* not. 46.) The substance of the œsophagus is membranous, consisting of five coats. The exterior coat is membranous, and is continuous with the pleura in the thorax. The second is muscular, and robust; and in the human body it consists of two layers of fibres, a longitudinal, and an annular or circular layer: in the ox it is composed of two decussating spiral lamellæ, and serves for constricting the tube, and for forcing on the food. The third coat is cellular, much like that of the intestines. The fourth is nervous, of

considerable thickness, and divisible into a number of other lamellæ, and furnished with a multitude of vessels and glands, (whence it is divided by Verheyen into two, a vascular and a glandular coat,) and continuous with the interior membrane of the mouth and stomach. The fifth coat is villous, and is usually called the crusta villosa: it is covered with a lubricous humor. There are a number of little glands situated chiefly in the upper part of the pharynx, and excretory openings are frequently discovered with them. The arteries of the œsophagus are from the carotids, the aorta, the intercostals, and the cœliac. (*Comp. Anat.*, n. 262.) They have been noticed and described by but few writers; Nicolai cites only Ruysch and Drake; yet I have myself seen and demonstrated, not once only, but many times, two, or even three little arteries, arising by distinct beginnings from the aorta, and which I have succeeded in injecting with wax. (*Comp. Anat.*, not. 67.) The veins are from the jugulars, the azygos, and the coronary vein of the stomach. The nerves are from the par-vagum. Probably there are also lymphatics. The new excretory ducts of Vercelloni, which convey a saltish liquid into the œsophagus and stomach, arise from three kinds of glands; from the conglomerate gastric glands, which are situated near the left orifice of the stomach; from the dorsal glands, which are situated near the fifth dorsal vertebra; and from the bronchial, tracheal, and thyroid glands. (*Comp. Anat.*, n. 262.) Jac. Vercelloni contends, that the business of digestion has never yet been thoroughly understood by medical writers, because they have never been acquainted with the true digestive fluid; which, he says, is poured into the œsophagus and stomach by very fine ducts; 1. from conglomerate glands situated in the left orifice of the stomach; 2. from the dorsal; and 3. from the bronchial and tracheal glands. But from the thyroid gland, ovula verminosa, (of which, according to him, this gland is a nidus,) are transmitted to the œsophagus and stomach, to give vitality to the chyle. (*Comp. Anat.*, not. 47.) This gland in the human subject is usually single, not double; in shape it resembles a new moon, with the horns running upwards on each side, and adhering to the thyroid and cricoid cartilages, as well as to the œsophagus: its middle part, or isthmus, is joined to the upper cartilages of the trachea. (*Comp. Anat.*, n. 383.) I have seen fibres, indeed, manifest enough, by which this gland adheres to the œsophagus; and I have always found the gland itself hard and firm; but I have never been able to discover either the ovula, or the hollow ducts of Vercelloni.... nor does pressure on it succeed in forcing a drop of fluid into either the trachea or the œsophagus.... On opening the gland, when in a state of turgidity, a quantity of limpid fluid escaped; this fluid was replete with yellowish spherules, which floated in water like very minute drops of oil. Whether these were the ovula of Vercelloni, or something else, it is not easy to determine. *(Comp. Anat.*, n. 385.) Besides this gland, there are also found in the neck a number of other glands, distributed here and there between the muscles and in the fat. They are of various sizes and shapes, and differ in situation and number in different subjects. Those in the anterior part of the neck are called jugulares; those in the posterior part, occipitales and cervicales. Their use is uncertain. . . . The œsophagus, in its upper part especially, has a great number of glands; these have been figured by Valsalva. In some subjects I have found them larger than he has represented them, and a little aperture, resembling the osculum of an excretory duct, in the centre of each: of these glands I have delineations still in my possession. *(Comp. Anat.*, n. 386.) The bronchial glands are of considerable size and of a blackish color; they are situated externally in the larger divisions of the trachea and bronchia. Vercelloni will have it, that they secrete a fluid which assists digestion, and discharge this fluid through certain extremely minute ducts into the œsophagus and stomach: I myself have found considerable fibres, passing from these glands to the œsophagus. In the largest of them I have sometimes found a calculus, equal in size to the last joint of the little finger. In one male subject, I found many black glands of this kind, adhering to the whole posterior surface of the trachea, near the œsophagus, of a size between a grain of wheat and a kidney-bean, and some adhering loosely to the anterior part of the trachea also. About the fifth dorsal vertebra, in the thorax, there is sometimes a remarkable gland, adhering to the posterior part of the œsophagus. It has been figured and described by Vesalius and others under the name of the dorsal gland. It is often of the size of a kidney-bean or of an almond; sometimes larger, and sometimes smaller than either; and not seldom it is either wholly wanting, or extremely minute indeed." *(Comp. Anat.*, n. 388, 389.) [Both our author and Verheyen describe cases in which the sides of the œsophagus were pressed together, from the enlargement of this gland.] "No ducts have hitherto been found, either by other anatomists or myself, from this gland to the œsophagus. Fantonus, and some others before Vercelloni, suspected that these glands discharged a mucous fluid into the cavity of the œsophagus; and Fantonus alleges, that he not only saw the ducts in a dog, but that he found little worms in them. Morgagni also says, that in dogs these glands are sometimes tumid, and inhabited by numbers of oblong and slender red vermiculi. Redi and Leclerc affirm the same; and Morgagni adds, that in these cases there are open passages from the glands into the œsophagus." *(Comp. Anat.*, n. 389, 390.)

74. "The muscles of the pharynx:—the muscles of the pharynx serve to dilate and contract it. Those which dilate it are the six following pairs. 1. The stylo-pharyngæus, which arises from the beginning of the styloid process, and is inserted, on both sides, into the pharynx and the thyroid cartilage. (This and the following pair also elevate the pharynx.) 2. The cephalo-pharyngæus, which arises from the anterior apophysis of the occipital bone, and terminates in the posterior part of the pharynx. 3. The pterygo-pharyngæus, which arises from the pterygoid processes. 4. The salpingo-pharyngæus, which arises from the Eustachian tube. 5. The mylo-pharyngæus, which arises from the lower jaw, behind the last molar tooth. (This muscle in great part surrounds the tonsils.) 6. The glosso-pharyngæus, which arises from the tongue. All these muscles terminate at the posterior part of the pharynx, where a tendinous line is very often conspicuous. The pharynx has only one constrictor, the œsophagæus or sphincter gulæ, which arises on both sides from the os hyoides, and from the thyroid and cricoid cartilages, and surrounds the posterior part of the œsophagus. Valsalva, observing the numerous origins of this muscle, divided it into three pairs: and Douglas, Cant, and Santorinus, into still more, which from their origin and termination, they name hyopharyngæus, thyro-pharyngæus, crico-pharyngæus, &c. But since the fibres of all these are in general so intimately connected, that it is difficult to separate them, and since the divisions which Valsalva represents can seldom be made out clearly, therefore it seems best to agree with Cowper not to multiply parts without reason, but to comprehend all these divisions under the name of sphincter gulæ, or œsophagæus. The vaginalis gulæ of Cowper has been already described under the name of tunica musculosa œsophagi." (*Comp. Anat.*, n. 325.)

75. Winslow. "The pharynx is a muscular and glandular bag, the outer surface of which is closely joined to, and continuous with, the inner surface of all that space, which is at the boundary of the mouth, behind the posterior nares, uvula and larynx; and which reaches from the great or anterior apophysis of the occipital bone, all the way to the œsophagus, which is the continuation of the pharynx. This space is bounded posteriorly, by the muscles which cover the bodies of the first cervical vertebræ; and laterally, by the superior portions of the two internal jugular veins and carotid arteries, by the spinous processes of the sphenoid bone, by the extremities of the petrous portions of the temporal bones, by the sphenoid bone itself immediately above the internal alæ of the pterygoid processes, and by the neighboring portions of both the pterygoid muscles. From these limits and adhesions of

the pharynx, we may determine its figure; which is like the wide part or a funnel, of which the œsophagus is the tube, and in reality, the continuation. The pharynx may be divided into three parts, a superior part, which is the arch of the pharynx; a middle, which is the body or great cavity; and an inferior, which is the fundus, narrow portion, or sphincter. We are to observe in it three openings, that of the arch, towards the nares; that of the body or great cavity, towards the mouth; and that of the fundus, towards the œsophagus. The arch is the broadest part of the pharynx, and ends on each side in an angle towards the fossæ jugulares basis cranii; after this, the great cavity contracts laterally, its other dimensions remaining the same; behind the larynx, it again enlarges on each side, a small space being, however, left between it and the cricoid cartilage. The extremity of the lower portion is very narrow, and joins the base of the cartilage just named. The pharynx is made up partly of several distinct fleshy portions, (which determine its capacity, and are looked on as so many separate muscles,) and partly of a membrane, which lines the inner surface of the whole cavity, and is a continuation of that of the nares and palate. This membrane is wholly glandular, and it is thicker on the arch and middle of the pharynx, than on the lower portion. Immediately above the first vertebra, it forms several longitudinal rugæ, thick, deep, and short, and in the dead subject generally containing a collection of mucus. In the great cavity, there are no rugæ; for the membrane adheres very closely to the muscles, both there and in the upper part. At the lower part where it is thinnest, it covers likewise the posterior part of the larynx, and is very loose, and formed into irregular folds. It dips in from side to side, between the edges of the pharynx. Although the muscular portions, for the most part, form one continued bag or receptacle, they are nevertheless quite distinct from each other, not only by their different insertions, but also by the different directions and crossings of their fibres. They may be looked on as digastric muscles, the middle tendons of which lie backward in one longitudinal line, which in some subjects constitutes a true linea alba. These muscles may be reduced to three general classes, with regard to their insertions. The first class comprises those which are attached to the basis cranii, consisting of the cephalo-pharyngæi, petro-pharyngæi, spheno-pharyngæi, or spheno-salpingo-pharyngæi, pterygo-pharyngæi, and stylo-pharyngæi. The second class comprises those which are inserted towards the mouth; namely, the peristaphilo-pharyngæi, glossi-pharyngæi, hypero-pharyngæi, and genio-pharyngæi. The third class comprehends those attached to the lateral parts of the larynx; namely, the syndesmo-pharyngæi, thyro-pharyngæi, crico-pharyngæi, œsophagæus and adeno-pharyngæus." *(Exp. Anat., Traité de la Teste,* n. 468—474.) For a particular account of these muscles, see Winslow, *ibid.,* n. 475—485.

76. "The œsophagus is partly a muscular and partly a membranous canal, situated behind the trachea, and in front of the vertebræ of the back, from near the middle of the neck to the lower part of the thorax, where it passes into the abdomen through a peculiar aperture in the small or inferior muscle of the diaphragm, and terminates at the upper orifice of the stomach. It is made up of several coats, almost in the same manner as the stomach, of which it is the continuation. The first coat, in the thorax, is formed only by the duplicature of the posterior part of the mediastinum; it is wanting above the thorax, in the neck, where the outer coat of the œsophagus is only a continuation of the cellular tissue of the neighboring parts. The second coat is muscular, consisting of different layers of fleshy fibres; of which the external are mostly longitudinal, yet not all continued from one end of the canal to the other. The next layers are obliquely transverse, and the next to these, still more transverse; the internal incline the contrary way: they cross irregularly in many places, but are neither spiral nor annular. The third is termed the nervous coat, and is like that of the stomach and intestines: it has various longitudinal plaits or folds, being much wider than the muscular coat; and is surrounded by a whitish, soft, fine filamentary tissue, like a kind of cotton; which tissue, when macerated, swells and thickens. The fourth or innermost coat resembles that of the intestines, except that instead of villi it has very small and short papillæ. It is folded lengthwise, like the third coat; so that the œsophagus when cut across, represents one tube within another. A viscid lymph is continually excreted from the pores of this coat. The œsophagus, from its beginning, inclines gradually towards the left, and naturally runs along the left extremities of the cartilages of the trachea." *(Exp. Anat., Traité de la Poitrine,* n. 157—162.)

77. SWAMMERDAM. "The œsophagus of the louse or pediculus is a very small canal, situated a little behind the eyes, where it seems to be carried up above the brain: in the neck it is somewhat enlarged, and afterwards grows small again in the back, until it terminates in the stomach. It lies in the upper part of the back, and also in the head and neck. (*Bib. Nat.,* p. 75.)

"On opening the back of the common day-butterfly, there immediately appear in the thorax some little curling vessels, which lie near the gullet, and are inserted in the forepart of the abdomen. Their beginning is a slender little canal, which divides into two fine tubes, that afterwards dilating, terminate at last about the beginning of the stomach; and they are so firmly united and fastened to it by means of the muscles and fat, that I have not yet by any means been able to trace them further. The gullet, dividing into two little tubes, conveys the food to the stomach. From the lower part of the gullet, near the stomach, there issues a short and small canal, which ends in a slender little bag or air-bladder, into which the air seems to be diverted, whilst the food is making its way to the stomach. This bladder has a remarkable peristaltic motion, and almost always lies in the butterfly's back, above the stomach. In the chrysalis, I found it full of a red liquid. (*Bib. Nat.*, p. 595, 596, tab. xxxvi., fig. 1.)

"The gullet of the cuttle-fish passes under the brain, and descending into the thorax, it lies softly on the two salivary glands, and is connected with them: from thence it runs in a straight line to the abdomen, where it opens into the stomach." (*Bib. Nat.*, p. 889, tab. li., fig. 5.)

78. "On these, and many other points, see the descriptions and plates of anatomical authors; as Eustachius, *Tabul. Anat.*, tab. xviii., fig. 17, shewing the connexion of the pharynx with the larynx, and the separation of the œsophagus from the trachea: fig. 18 and 20, exhibiting the trachea as placed under the œsophagus, and certain muscles terminating in the linea alba; also the œsophagæus muscle, with its external fibres slanting a little upwards; its middle fibres meeting them, and the internal ones becoming longitudinal: fig. 19 represents several glands in the base of the tongue, near the os hyoides.

Respecting the muscles of the pharynx, see Cowper, *Myotomia Reformata*, tab. xxviii., xxix., (fol. Lond. 1724), and *Anat.*, App., tab. ix., fig. 38. (Leyden, 1737.) Cant, *Impet. primi Anat.*, tab. iii., fig. 1. Verheyen, *Corp. Hum. Anat.*, tab. xxiii., fig. 4: the coats of the œsophagus in the ox, *ibid.*, fig. 5, 6, 7: the description, tract. iii., cap. xiv. Bidloo, in Mangetus, *Theatr. Anat.*, tab. xci., fig. 11, 12; where he shews that the œsophagæi muscles are interrupted in many places by tendinous prolongations, and in others by true muscular prolongations, and that thus the œsophagæus becomes divided into several muscles. See also Browne, in Mangetus, *T. A.*, tab. xi., fig. 10. Respecting the offices of the several muscles in deglutition, see Winslow, *Traité de la Teste*, n. 475—485. Boerhaave, *Inst. Med.*, n. 65—74.

Respecting the coats of the œsophagus, see Ruysch: *Thes. Anat.*, vii., n. 40; and *Th. A.* viii., n. 47., on the villous coat: *Th.* viii., n. 47., the internal coat which has nervous papillæ, like the corresponding coat in the gullet of the tortoise: *Th.* iii., n. 77, and *Th.* ii., n. 14., the muscular coat: *Ep.* vi., tab. vii., fig. 3., the œsophageal artery. Schellhammer, *Diss.*[x] viii., n. 54. Santorinus, p. 138. Willis, Steno, Cole, Bellini, who represent the fibres of the muscular coat as transverse.

Respecting the glands of the œsophagus, see Vercelloni, *Dissert. de Glandulis conglomeratis Œsophagi*[x]; of these glands he enumerates ten; namely, the gastric, the dorsal, and the bronchial glands, the glands of the first division of the trachea, three pairs of tracheal glands, and lastly the thyroid gland. Respecting the latter and the dorsal gland, Verheyen, *Corp. Hum. Anat.*, tract iii., cap. xi. Fantonus, *Dissert. Anat.* iii. Morgagni, *Adversaria*, part iii., p. 5; where he states that he observed in a dog, a broad, thick, glandular ring, surrounding the inferior orifice of the pharynx; and two oval tubercles, about the size of small walnuts, on the inside of the œsophagus, at its lowest part; also *ibid.*, part i. and ii. See also Heister, *De Gland. Bronchial.*, in *Ephem.* cent. vii., viii., tab. vi. Valsalva, *De Aure Humana*, tab. v., fig. 2. Cant, *Impet. primi Anat.*, tab. iv.

79. The pharynx is as it were the receiving vessel of the palate; and the alembic or head (*a*) of the œsophagus. It receives, devours and swallows whatever the palate and the tongue commit to it; and in this way it is the gorge and swallow, that is, the gullet (*b*); on the other hand, it forces and precipitates its ingesta and contents into the œsophagus (*c*). Thus the pharynx is at once passive and active: it becomes passive while the tongue and the palate are acting; and it does not begin to act until the tongue and the palate cease (*d*). This twofold nature

(*a*) The pharynx does not resemble an alembic when it is open, but only when it is closed by the tongue and palate.

(*b*) By a little attention to our sensations, we shall be convinced, that during deglutition, the tongue elevates its root and the os hyoides, and wreathes and inflects them upwards towards the gullet; and also that the palate, by means of its lax, membranous velum, assisted by its own muscular structure and that of its arches, likewise inclines the superlingual region inwards, and expands above it. Thus there is a pressure of forces, from below, from above, and at the sides, as well as from every point to every other point, all the way into the interior cavity of the pharynx: for an action similar to that described above in the tongue, (see n. 69, Schol. *z*,) is continued in the pharynx, and in the succeeding parts.

(*c*) There is no doubt that the gullet or pharynx pushes on the ingesta by the action of its muscles; that is to say, by its own proper power. Our own feelings, the effect itself, and the very muscles determined hither, shew this to be the case.

(*d*) The sphere of action of the tongue and palate, appears to extend to the linea alba, the line of incidence and insertion of the muscles of the pharynx; see Eustachius, Winslow, and Heister, above.

it derives from the nerves (*e*) and from the vessels which supply it (*f*); and from the muscles, tendons, and membranes which

For these muscles are in a manner biventral or digastric; and consequently they admit of being driven and folded by the tongue and the palate, towards this line as a kind of boundary. From this line the proper sphere of action of the pharynx itself begins. Thus each organ has its appointed limit. The pharynx likewise has its limit with respect to the œsophagus; and the œsophagus has its limit with respect to the stomach, as we shall shew presently. Every member of the body is so limited and articulated, that it has a sphere of action of its own: being yet so implanted in some other member next to it, that the two are enabled to make common cause together. Thus the pharynx yields when the tongue and the palate are acting, and acts when the latter yield.

(*e*) The inability of the pharynx to act, before the tongue, the os hyoides, and the palate, together with the larynx, and the cricoid and thyroid cartilages are retracted, proceeds primarily from the influx of the nervous fibres. For when the nerves are relaxed—as when the tongue and the palate roll backwards and inwards towards the pharynx—the latter is deprived of all power of acting: the nervous fibre being the sole cause of muscular contraction, as we shewed in our *Economy of the Animal Kingdom,* (Tr. i. n. 512, seq.,) when treating of the muscular mechanism of the heart. This is the case here, as in all other parts of the body, and particularly where the eighth pair of nerves prevails: for the eighth pair, on issuing from the cranium to the jugular veins, near the pharynx, proceeds, after a short inflexion, to the pharyngeal muscles, and runs down along the pharynx; thus its force is relaxed when the introaction of the tongue takes place, and restored when the tongue and its apparatus roll back again and extend. The palate and the larynx similarly contribute to producing this effect. Thus the acting force of one point succeeds and instantaneously follows that of another: so that while the one is active, the other is passive: whence there is a perpetual succession of forces all the way from the beginning, where the motion is first impressed. This is implanted universally in the very organism of the body: to save the cerebrum and the cerebellum from the necessity of putting themselves forth into the singulars of every action: which is the reason why there is a similar succession of motions in muscular members even after death; as in the auricles of the heart, in the stomach, in the diaphragm, and in short, every where. But of these points we shall speak further, when we come to treat of the organism of animal motion.

(*f*) The influent arteries and effluent veins of the muscles, contriconstruct it (*g*). And to prevent it from conspiring unseasonably with the larynx and trachea, with which it is in close connexion,

bute no farther to produce their actions, than by keeping the arterial blood always ready to flow back into the moving fibres, and expand the muscle. In order, therefore, that there may never be any deficiency of blood for these fibres, vessels are derived from all quarters to the pharynx and the œsophagus; that is to say, from the carotids, the intercostals, the cœliac and the aorta; see Heister, Ruysch, Nicolai, and Drake. While the pharynx and the œsophagus are performing deglutition, they act inordinately: at one time with the palate and the tongue; wherefore the carotid is then present, ready to flow in: at another time with the stomach; when the cœliac is called into play: at another time with the heart; wherefore the aorta then supplies them. But when manducation and deglutition are not proceeding, they always concur with the trachea and the lungs; and for this reason they derive their principal branches from the intercostals, and transmit the returning blood into the vena azygos, the jugular vein, and the vena coronaria ventriculi. But of these points we shall speak elsewhere.

(*g*) We might here expatiate at much length on the different actions of the pharyngeal muscles, shewing how they all direct their forces and coöperate, in the act of deglutition; applying our remarks to both those which descend from the cranium, and to those which ascend from the os hyoides and lower jaw, and also from the larynx, the cricoid and thyroid cartilages, and the Eustachian tube. But, to say the truth, in setting out to explain these matters, I should probably only becloud the fact: for how few readers comprehend the names even of these muscles, still less their simultaneous and successive directions and operations. Nor ought we to indulge the vain glory of emptily parading unintelligible details. A single action of the pharynx involves the concurrence of more than a hundred different muscles; of the internal and external muscles of the tongue, of all the hyoidei, of all the palatine muscles, of those of the lower jaw, the lips, the temples, and the cheeks; also, of the muscles of the head and the cervical vertebræ; and by a still more general action, of even the pectoral and abdominal muscles. Thus, by endeavoring to define and explain an action involving so vast a concourse of muscles, we should be more likely to lose all distinct ideas in an abyss of darkness, than to gain a single ray of light. Furthermore, the variety of such actions is infinite, so that in essaying an unsatisfactory account of one variety, we should still plunge the others in darkness. Rational anatomy attains its summit, and may be satisfied with its powers of flight, when it has learnt to portray aright any one little its actions and movements follow theirs, and beautifully reciprocate with them (*h*).

80. The pharynx attracts various excretions, particularly the salivæ of the mouth and fauces: also, the exudations of the cerebrum, and the mucus of the ethmoid and sphenoid bones, and their lining membranes,—the latter alternately with the larynx and the trachea (*i*); and it gently inclines and softly conveys them down the tube, in conjunction with the saliva which oozes or is expressed from its own and the neighboring glands (*k*), into the great cavity of the stomach.

action of the number, and to shew the concurrence of motive forces to any given effect or end.

(*h*) That the pharynx and the larynx are not concurrent in their actions, is a matter of sensible experience; for who eats and speaks at once? who swallows and breathes at once? To secure their non-concurrence, the œsophagæus, which forms a sphincter at the entrance of the œsophagus, arises principally from the cricoid and thyroid cartilages, and proceeding therefrom, encircles the back of the pharynx. Thus, whenever this orifice is relaxed by the pressure of the food and the relaxation of the nervous fibres, the larynx lies compressed and still, and ceases to respire. Not to mention the other powers which act simultaneously, in the glottis, the epiglottis, the uvula, and the septum palati. The reason why at this moment the larynx is obliged to stop the breathing and suppress the voice, is, that the pharynx is then agitated by extraordinary motions, which are by no means synchronous with the pulmonic motion. That the natural motion of the pharynx and œsophagus coincides with the pulmonic motion, will be shewn presently.

(*i*) On all these points we shall treat in the proper places. In our analysis of the parts of the cerebrum, we shall shew how the serous fluid collected within the membranes, and between the fasciculi and lamellæ of the medullary fibres, is pumped out through the foramina of the cribriform plate: of the other particulars we shall speak in our chapter on the nares.

(*k*) Glandular congeries are scattered in great abundance over the internal coat of the pharynx. "The internal membrane of the pharynx," says Winslow, "is wholly glandular." (n. 75.) The portion of it adhering to the tongue, is also so delineated by Eustachius, (*Tabul. Anat.*, tab. xviii., fig. 19.) And Morgagni "observed in a dog, a broad, thick, glandular ring, surrounding the inferior orifice of the pharynx."

81. The œsophagus receives from the pharynx the food that has come from the palate, and drives it with marvellous instinct and art through the cardiac orifice into the stomach (*l*). More-

(n. 78.) Besides the foregoing, the tonsils or amygdalæ are in a manner appended to it on both sides; and at the moments of its contraction, the pharynx expresses a large quantity of mucilaginous humor through the foramina of these glands. We need only attend to what passes in ourselves in these respects, to be convinced that the pharynx is frequently swallowing this humor and the saliva; and that the imperceptible character of the action is caused by its frequency and familiarity.

(*l*) The œsophagus possesses the same power as the tongue, (see n. 69, schol. *z*,) of pushing the ingesta onwards from point to point. The arteries also have a similar power, and likewise the other tubes, as the bronchia, the intestines, and even the very pores; whereby they appear to exercise a species of attraction. For this power, the œsophagus is indebted to its penultimate or muscular membrane, and the membrane to the nervous fibres which flow along it downwards and onwards. The muscle actuates the membrane, the moving fibres actuate the muscle, and the nervous fibres actuate the moving fibres. The nervous fibres, by their determination, cause this membrane to react to the same degree that it is acted upon, (and constantly from above,) on the contents which are to be transmitted; so that a kind of impulsion perpetually urges and advances them forwards, and the tube counterfeits the appearances of attraction and suction. This mode of action commences from the œsophagæus muscle, one part of the fibres of which describes a circle, another part curves obliquely downwards, and a third extends longitudinally, according to the delineation in Eustachius, *Tabul. Anat.*, tab. xviii., fig. 18 and 20. Something similar is also the case with the muscular fibres of the œsophagus itself, as we learn from the careful investigation of Winslow, to whose account (n. 76) we again direct the reader's attention. To accomplish the effect we are treating of, the œsophageal tube appears to be in a manner divisible into two tubes; one consisting of the muscular series, the other, of the three internal membranes; in order that when the inner tubulus is a little receding and depressed, the muscular tube may act upon it freely from above. For the same reason, the inside of the œsophagus is full of longitudinal and oblique folds and grooves; and furthermore, it decreases in diameter all the way to the orifice bordering on the stomach. In short, every thing contributes to give it this power. Thus as soon as ever the œsophagus has taken the food, it deprives it over, it summons salivæ,—intimately endowed with powers of diluting, softening, dissolving, extracting, altering, sheathing, carrying, insinuating; and what is more, of vivifying (*m*),—from every

of all its laws and gravity, and assumes the plenary possession and disposal of it: and this, still more perfectly in those animals which crop their herbaceous food with long and sloping necks, and carry it almost upwards into their bellies; and in which the muscular fibres, therefore, flow in a spiral gyre. This is more remarkably evident in the stomach; see n. 104, (*z*.)

(*m*) The nature of the saliva can be understood only by effects. Chemical analysis throws very little light upon it, and does not shew either what its parts are, or in what form they are connected together. The saliva when decomposed chemically, yields a quantity of a watery liquid, a weak spirit tinctured with a little volatile oil, also a thick fetid oil, lastly, a residuum mixed with a fixed alcaline salt. It lets fall a precipitate and turns milky when treated with sugar of lead: it does not effervesce with either acid or alkali: gentle boiling scarcely produces any change in it. Its particles, when examined in water by the microscope, appear oblong and branching. Leeuwenhoek describes them as like little worms, with heads, tails, and tortuous bodies, and swimming about with great agility. But what do we gain in point of understanding, from this chemical analysis, when we consider that similar liquids, spirits, oils, and residua may be elicited by distillation from all the subjects of both the animal and vegetable kingdoms. The saliva is thus characteristically and finely described by the ingenious Boerhaave: "The saliva is a thin, transparent humor, not coagulating by heat, almost void of taste and smell; when shaken it affords a ropy froth: it is a glandular fluid, secreted from arterial blood. During hunger, it is comparatively abundant, fluid and sharp; after long fasting, it is very acrid, penetrating, detergent and resolvent; it produces and promotes fermentation in farinaceous and succulent vegetable matters, and in syrups," &c. (*Inst. Med.*, n. 66.) But as I before said, the qualities of the saliva are best shewn by effects. Thus we find, that as a watery liquid, it *dilutes* the particles of the food, and makes its way into their recesses, where it can put forth and exercise its other properties distinctly: that thus it *softens* them, by separating them individually, keeping them separate and distinct, and smearing them over: that it *dissolves* them, but like a mild and universal menstruum, which acts from the innermost to the outermost; discussing the saline combinations of the parts, by its aqueous element; and discussing their fatty and oily lumps, by its spirituous element. It *extracts* the occult province of the head, and from every province of the chest (*n*);

quintessences that are adapted either directly or collaterally for creating the blood. It *alters* them into suitable forms: it *sheaths* or *envelops* them, so that they almost fly through the minute smooth pores; through which it *carries*, and into which it *insinuates* the chyle. And in order, in the last place, that in the stomach, it may penetrate the innermost connexions of the parts, set them free, and subdivide them to their unities and principles, and envelop, carry and insinuate these, —it pours a spirit upon them, and thus *vivifies* them. Hence the chyle, already related as it were to every part of the kingdom, is led like a richly-dowered bride into the innermost chambers of the blood. Experience confirms all these propositions by an overwhelming multitude of effects: but at present we are not treating of the saliva, but under the guidance of anatomy, of the œsophagus and its glands.

(*n*) That is to say, all the saliva that is collected by the mouth and the palate from the various scattered glands before mentioned; both that which is poured down by the pharynx, and that which the œsophagus itself supplies from its own bosom and repository, and which it mingles and impregnates with what descends to it: for its internal or nervous coat, according to Heister, "is divisible into a number of other lamellæ, and furnished with a multitude of vessels and glands; whence it is divided by Verheyen into two, a vascular and a glandular coat." (n. 73.) The same author remarks, "The œsophagus, in its upper part especially, has a great number of glands; these have been figured by Valsalva. In some subjects I have found them larger than he has represented them, and a little aperture, resembling the osculum of an excretory duct, in the centre of each: of these glands I have delineations still in my possession." (n. 73.) At the lower part of the œsophagus likewise he saw several "conglomerate glands," which he calls "the gastric glands." (n. 73.) In this situation, in a dog, Morgagni found "two oval tubercles, about the size of small walnuts." (n. 78.) So that according to Winslow, "A viscid lymph is continually excreted from the pores of this coat." (n. 76.) Thus a mucilaginous humor constantly covers the concave parietes of this tube, and lubricates the passage, and just as in the most minute tubuli or pores, dilutes the passing materials, softens, sheathes and carries them, and insinuates them into the stomach; and perhaps, except they fly past too rapidly, even begins to dissolve, extract, and alter them. Thus must respecting the salivæ of the mouth, the pharynx, and the œsophagus. A still larger supply is brought from the towns and villages in the neck and thorax; that is to say, proximately from the thyroid cartilage, (to and it conveys and supplies them to the stomach, in such quantity, and of such quality, as the food, the stomach, the

which the *thyroid* gland, in shape like a new moon, is agglutinated; Heister saw "fibres, manifest enough," proceeding from this gland, "by which it adhered to the œsophagus," n. 73, above:) also, from a certain loculus, placed near the fifth dorsal vertebra, and hence called the dorsal gland, (the communication of which with the œsophagus, Heister, indeed, denies having seen, but Fantonus and Vercelloni affirm it, nor does Morgagni absolutely disavow it, see n. 73, ad fin.:) not to mention the *bronchial* glands, the three pairs of *tracheal* glands, (which Vercelloni also derives hither, although this is not established:) and the *jugular*, *occipital*, and *cervical* glands, mentioned by Heister, n. 73. All these, and perhaps other glands besides, proffer and supply their juices to the trachea and œsophagus: and the œsophagus, as a servant and steward, to the stomach. The very way also in which the œsophagus allures, invites, and receives these juices, instils them into its tube, and pours them into the stomach, will be evident enough to all who examine thoroughly the connexions of the part; to say nothing of the organism of motion, whereby the liquid is expressed from the thyroid gland, (adjacent and connected, as it is, to the œsophagus,) by the alternate expansion and contraction of the pharynx and œsophagus; and from the other glands, by the action of the muscles, ligaments, and membranes: also, whereby the expressed liquid is invited, and presented to the stomach. It is conspicuously apparent, that the œsophagus receives these juices, or creams, first within its *cellular* coat, and thence eructates them through an uninterrupted line of passages and ducts into its common cavity; for this coat, which lies immediately under the muscular coat, is made up of perpetual cells, crypts, and receptacles, and is perfectly capable of receiving the stream of exudations and tricklings from the glands: according to Winslow, "It has various longitudinal plaits or folds, being much wider than the muscular coat; and is surrounded by a whitish, soft, fine, filamentary tissue, like a kind of cotton; which tissue, when macerated, swells and thickens," (n. 76.) Add to this, that the great extensility of the corresponding cellular coats of the other viscera, and particularly of the peritonæum, is shewn by injection, inflation, anatomy, and pathology. This coat of the œsophagus, moreover, is in a manner kept separated from the contiguous muscular membrane; thus presenting the appearance of one tube within another, as remarked by Winslow, Willis, and other anatomists. From these facts we cannot, I think, be mistaken in concluding, that the œsophagus invites and allures the salivary fluids chyle, and the blood require (*o*). But if any thing infest the stomach, gnaw its fibres, contract its folds, corrupt its mem-

from the glands into this coat first, and afterwards eructates them into its cavity, by the alternate expansion and contraction of this and the muscular coat, particularly at the time of deglutition, and of the detrusion of the food through the tube. Hence, when in the state of contraction, "a viscid lymph is continually excreted from the pores of this coat," according to Winslow; nor is there any other source of the transpiration, except from that general cellular receptacle wherein the liquids are mingled first. Whether there be also any passage into the cells of this coat from the cellular tissue of the larynx, the trachea, the bronchia, the mediastinum, and the pleura, and thus mediately from the before-mentioned glands; and whether there be any passage from the same coat of the œsophagus immediately into the stomach, I dare not venture at present either to affirm or deny. Very probably this may be the case, since the cells are continuous and pervious; and according to Winslow, "The outer coat of the œsophagus is only a continuation of the cellular tissue of the neighboring parts" (n. 76); and further, since there is perpetual continuity in the organic body; and thus every member or organ is enabled to imbibe all that it wants, from the common lake: so that every thing is a kind of centre, regarding its universe as made up of circumferences, that lead to it by continuous radii, through which its necessities are supplied. Thus the saliva is supplied in such quantity and of such quality, as the food, the stomach, the chyle, and the blood require; according to our proposition.

(*o*) If we may put in a conjecture, we may here suppose that the stomach, by means of the œsophagus, draws from the thyroid and dorsal glands the spirits or nobler animal essences, whereby the saliva is ultimately vivified, and made capable of investing the innermost parts of the food, and of conveying them through the insensible pores into the blood. Our conjecture is grounded on what Heister says of the thyroid gland; that he found in it "a limpid fluid, replete with yellowish spherules, which floated in water like very minute drops of oil" (n. 73). For throughout the animal microcosm, nature, ever fearful of the loss of her spirits, encloses them safely, either deep in the fibres, or in the blood-globules, or else in spherules or little ova, and never opens these but when the spirits are wanted for use; thus she resolves these spherules or ovula in the stomach itself, while she is resolving the composite parts of the food into the least parts of the chyle. Another ground is, that in this gland Vercelloni found a nidus of ovula verminosa, or branes, or vitiate the chyle, the œsophagus sometimes ejects it by vomiting; particularly the air, always a most unwelcome guest, and which it eructates in the form of flatus.

82. Furthermore, the œsophagus brings together and unites the lowest things and the highest; it connects all things belonging to the tongue and the palate, to the mouth and the lips; consequently to the temples, the cheeks and the forehead; also to the nares and the cranium, and thereby to the cerebrum; in a word, to the head; as well as all things belonging to the chest —in short, all things above the diaphragm—with all things belonging to the abdomen, or with all beneath the diaphragm (*p*):

worm's eggs, which, he says, "are to give vitality to the chyle" (n. 73;) and that Morgagni found "vermiculi" in the dorsal gland (n. 73). Nothing, in fact, is more common than for forms in a state of quick motion, to appear in all those parts where the afflux of spirits is abundant, as in the epididymes, the vasa deferentia, the vesiculæ seminales, and the semen itself: and also, according to Leeuwenhoek, even in the saliva of the mouth. So that it seems as if the substance called animal spirit, were in the constant desire and endeavor, wherever an opportunity is offered, of clothing itself with a body, but which body easily relapses back into its constituent principles or spirits. But I wish this to be considered as purely conjectural.

(*p*) The viscera of the abdomen, as the stomach, &c., or the members of the lowest region, are connected with the members or organs of the head, or of the supreme region, only by vessels, fibres, membranes, and muscles; thus generally by the *carotid arteries* and the *jugular veins;* but these do not preserve the connexion directly; for they are engrafted in the way of large branches on the aorta and vena cava superior, and thus they inflect the thread of continuity in a kind of oblique manner. The *fibres* again belong solely to the par vagum, which descends from the head, its Olympus, to these low regions, and enters them only near the lower end of the œsophagus, just beside its insertion into the stomach. The *membranes* effect the connexion, only through the pleura, the most general membrane of the thorax, and thus by means of the diaphragm with the peritonæum, and by its means again with the viscera enclosed in it. The same may be said of the *muscles*, which superficially cover these two cavities or arks of the body. It is evident, therefore, that there is no specific medium of union between the viscera of the abdomen and the head, but the œsophagus, which absolutely merges in the stomach, and its internal coats in particular, with the stomach, by means of its coats, external and internal (*q*); and by means of its passage down the mediastinum and through the diaphragm (*r*); and also of the eighth pair of nerves, which it collects and transmits (*s*). Now as it

are continuous with the internal coats of that viscus; and as the œsophageal coats are continuous with those of the pharynx, the palate, the tongue, the nares, the cheeks, and the lips, and these again with the coats of the face, and so forth, it therefore follows, that the œsophagus brings together and unites the lowest things and the highest.

(*q*) Not only is the internal coat of the œsophagus continuous with the internal coat of the stomach, but its external coat also is continuous with the mediastinum and the pleura. "Its first coat," says Winslow, "in the thorax, is formed only by the duplicature of the posterior part of the mediastinum; it is wanting above the thorax, in the neck, where the outer coat of the œsophagus is only a continuation of the cellular tissue of the neighboring parts," n. 76. "The exterior coat of the œsophagus," says Heister, "is continuous with the pleura in the thorax," n. 73. That the other viscera of the abdomen are likewise conjoined by means of the stomach and the diaphragm, will be shewn in the course of our analyses.

(*r*) That the mediastinum is continuous with the external coat of the lungs, and also with the pericardium, and thus with the external coat of the heart; and that the diaphragm, in the thorax, is continuous with the pleura, and peculiarly with the pericardium, and thus with the mediastinum, will be seen in our analysis of those septa. Consequently the diaphragm is united to the œsophagus in two ways and by two bonds; so likewise the viscera depending from the lower surface of the diaphragm, and enclosed in the peritonæum; for the œsophagus passes through the fleshy part of the diaphragm, and thus unites with the peritonæum also.

(*s*) The par vagum, or nervi sympathetici medii of Winslow, in their descent and passage through the thorax, embrace the œsophagus, and send filaments across it from side to side; and at length, about the lower end of the œsophagus, at its insertion into the stomach, they unite in a singular manner, and ramify from this point round both sides of the stomach; and then form the great stomachic, mesenteric and other plexuses; see the tabulæ of Vieussens and Willis. The communication of the œsophagus with the members of the abdomen, is immensely strengthened by the par vagum, which intimately penetrates the abdominal viscera, and enters and constructs all their parts, and thus binding them to it, governs them as its own peculiar property.

connects the substances of parts, so it connects their particular and general forces; in order that the nexus of all things may be maintained, their effects properly produced, and that each may regard the others as parts of itself, and *vice versa*.

83. The œsophagus also puts forth certain motions, which pass uninterruptedly through the stomach, and ultimately through all the viscera subjacent and appended to it; in a word, it constantly impresses, not the cardiac, but the alternate respiratory motions of the lungs (*t*): and this, by means of the contiguity of the pharynx to the larynx (*u*); of the continuity of the common coat of the œsophagus with the pleura (*x*); of the passage

(*t*) That all the abdominal viscera, the stomach, the intestines, the liver, the pancreas, the spleen, &c., are subject to the rule of the alternate pulmonic motion, and of this motion only, will be shewn in the anatomy of each viscus. The œsophagus, with its appliances and powers, of which we shall treat in the regular order, is like a running axis provided with thongs and cylinders, whereby it excites, turns and whirls the abdominal wheel into the same reciprocal motions as the lungs.

(*u*) Respecting the great nearness, intimacy and friendship which subsist between the pharynx and the larynx, see n. 85. They are in fact conjoined, head to head and neck to neck, in the closest manner; and what is more, the larynx puts forth, from its thyroid and cricoid cartilages, moving fibres, as it were reins, whereby it completely surrounds and binds the pharynx; this being the origin of the œsophagæus muscle, its only sphincter, in which all its other muscles fix their roots. Thus both are copulated in a kind of conjugal bond, and the pharynx married to the larynx. The consequence is, every time the lungs move the bronchia, these, the trachea, and this, the larynx, that the larynx moves the pharynx, this, the œsophagus, and this, the stomach; and the stomach all the other levers and joints of the entire machine.

(*x*) See what we said above, (n. 82, *q* and *r*,) respecting this contiguity and continuity. The truth is, that the vertebral column, the ribs, the sternum, and the diaphragm itself, with the muscles; in short, the whole cavity of the thorax, is lined by a membrane called the pleura. This also, by a wonderful gyre of continuity, rolls round the lungs, and keeps all the contents of the thorax in their places and general connexion. To this membrane the œsophagus is connected by two bonds; and thus closely fettered, it cannot open, except in the same intervals as the lungs, which constitute the wheel of the machine.

of the œsophageal tube through the fleshy substance of the diaphragm (*y*): of the collection and distribution of the branches of the par vagum, into the stomachic, hepatic, splenic, mesenteric and renal plexuses (*z*); and in particular, by means of the

(*y*) It is very important to recollect that the œsophagus passes through the fleshy part of the diaphragm, and indeed at the very confluence of its muscular fibres, which come thither continuously from the ribs, the vertebræ, and the sternum; or from the whole circumference of the thorax: and that thus again it is inseparably dependent upon the same powers, which carry it away into its destined motions by a most potent natural force. The diaphragm, which in this situation is lined by the pleura, is powerfully impelled into these reciprocal motions, by the superincumbent lobes of the lungs; mediately also by the intervention of ligaments: and this is precisely the case with the œsophagus, which is fixed into the diaphragm, like a nail, and surrounded by the fibres of all the powers which are acting similarly. From these considerations it is evident, that the œsophagus is introduced into these motions from head to foot; at the upper end, as we said before, by the larynx; in the middle, by the pleura and mediastinum; at the bottom, by the diaphragm; consequently, in all points of its cylinder: so that these forces cannot possibly fail of their effect. Thus, I think, there is no reasonable doubt remaining, that the œsophagus is entirely obedient to the lungs in these respects.

(*z*) Respecting this nerve, see what we before said, (n. 82, *s*). It passes through the fleshy part of the diaphragm, with the œsophagus; and then ramifies extensively in the subjacent viscera, and penetrates and constructs their innermost connexions, and regulates all their offices and disposes their actions, like a guardian angel sent down from a higher sphere. Every fibre of this nerve is entirely dependent for its life and activity on the spirit of the cerebrum and cerebellum: thus it conspires to the same vicissitudes of motion, as the lungs; for the animatory motions of the brains are indisputably coincident with the respiratory motions of the lungs. By the power of this nerve, which enters deeply into all the fibres of these viscera, every thing is carried away into the torrent of this motion, *from the innermost;* and also similarly *from the outermost,* as we have before proved: thus both generally and most particularly the same effect is produced by the most constant law. This nerve passes through the same muscular substance of the diaphragm; because thus, during every reciprocal movement of the diaphragm or œsophagus, the same general motions concur with the innermost; for whenever the nerves are affected by any cause in the insertion of the œsophagus into the bosom of the stomach. Hence all the viscera of the abdomen live and subsist entirely under the government of this motion; consequently under the government of the lungs.

84. Nor should we omit to mention the peculiar motions of the œsophagus, which are inconstant or occasional, occurring whenever the tongue eats and the pharynx swallows. These motions are always, indeed, communicated to the stomach; but after a gyre, they soon fall back, (by virtue of the retractive and impulsive causes and forces just indicated,) naturally and spontaneously, (since these causes and forces depend on the par vagum,) into the alternations of the pulmonic motion (*a*).

body, the muscular series dependent upon them are also put into action, see n. 79, (*e*). For this reason, the motions of the lungs coincide with the animatory motions of the brain; and for the same reason, respiration is the essential outermost life of the body, and the animation of all the fibres by the brains is its innermost life: were these two not harmonically coincident, immediate separation would take place between the outermost things of the body and the innermost.

(*a*) It is evidently of great importance to examine carefully the connexions of parts in the organic body, for otherwise the successive fluxions, or the very effects, of forces and motions, can never be explored. But before concluding the present subject, it is requisite to add, that every member, as here the œsophagus, has its determinate sphere of action, so that it cannot put forth the whole impetus of its forces on contiguous and continuous members. This is the case with the œsophagus in regard to the stomach, with the stomach in regard to the intestines, and so forth: as may be clearly seen from the very articulation of the œsophagus with the stomach; for, as Willis says, "The internal or villous coat of the œsophagus is continuous with the palate and the stomach, yet it is very distinct from that of the stomach, the sides of which it overlaps for about three finger-breadths." (*Opera: Pharmaceutice Rationalis*, sec. ii., cap. ii., p. 5., 4to. Genevæ, 1680.) We before observed a similar boundary between the pharynx and the palate, (n. 79, *d*); and we shall again advert to the subject in the next chapter. This law prevails in all parts; yea, even in the most minute, between point and point; constituting the ground of their successiveness. In the present case, this organism is a contrivance to prevent the œsophagus, when in the act of swallowing, from carrying away the stomach into precisely the same inordinate and subsultory motion as itself is subject

85. It is worthy of consideration, that the nearness, intimacy and friendship between the trachea and the œsophagus, are so great, that the two rest their heads together, and as it were mutually embrace and kiss. For the œsophagus lives by the breath of the trachea, which gives it activity; and the trachea, by the food of the œsophagus, which gives it nourishment.

86. The same close friendship unites the two carotid arteries and the two internal jugular veins, to the œsophagus. For the carotids ascend along the œsophagus, and inject into its glands and pouches the salivary lymphs, of which these arteries, as they approach the cerebrum, first purify themselves here (*b*). And the jugular veins, emerging from the fossæ of the skull, close to the pharynx, carry down from the cerebrum the spirit, or the innermost part of the blood, at the same time that the œsophagus carries down the body, or the outermost part of the same blood (*c*). This is the reason of their association and fellowship on the way: identity of end, and community of cause, is the bond which conjoins them.

to: by causing the motions which strike it in the way of continuity, to die away readily into the natural pulmonic motion: as when the œsophagus is affected spasmodically, by various ingesta passing into it, forcibly or otherwise, through the pharynx, and which either stick in its folds, or if pushed deeper into it, excite convulsive motions.

(*b*) That the carotids, as they approach the organs of the external senses, and lastly the cerebrum, or the organs of the internal senses, must purge themselves of impurities by means of the salivæ, see n. 72, ad fin.; first of all, therefore, in the œsophagus, which demands only the grosser impurities; for it takes in undigested substances, and rapidly transmits them, and these it has the office of moistening.

(*c*) These particulars will be explained in our analysis of the blood. The blood derives its spirit, consequently its life, from the brain and its fibres; its body, consequently its material existence, from the stomach and the food: for all parts and forms in the organic system, the greatest as well as the least, have both a soul of their own and a body of their own.

CHAPTER IV.

THE STOMACH AND ITS ORIFICES.

87. HEISTER. "The stomach or ventriculus is a hollow membranous viscus; placed, for the most part, in the left hypochondrium, immediately under the diaphragm, and obliquely between the liver and the spleen. In the stomach we are to observe, 1. Its figure, which is like that of the bag of a pair of bagpipes. 2. Its division, into two orifices and a fundus, which latter is directed forwards. The left orifice, called the cardia, is placed much higher than the right orifice, and is continuous with the œsophagus, and furnished with a great number of nerves: the right orifice, or pylorus, is connected with the intestines, and has a singular valve which serves to close the stomach. *(Comp. Anat.*, n. 209.) The pylorus is not situated horizontally, as some anatomists have represented it, but obliquely; in point of fact, it slants downwards from right to left. Moreover, the pylorus does not consist merely of a series of circular fibres, surrounding the orifice of the stomach, but usually of a considerable membrane, which is sometimes two or three lines broad. We are to observe, however, that this membrane, or valve, is not equally large in all subjects; in some it is scarcely more than one line in breadth; while in some it is also much larger than here represented. [See *Comp. Anat.*, tab. i., fig. 6; and "*Ephemerid. Natur. Curios.*, cent. v., tab. ii., fig. 4."] *(Comp. Anat.*, exp. tab. i.) We are also to observe the ligament by which the pylorus is connected to the upper part of the stomach. The size of the stomach varies; in persons addicted to gluttony it is usually very large; and in men it is, in general, larger than in women. In the human subject it is single; but many animals have several stomachs. Its arteries, called the gastric arteries, arise from the cœliac: the gastric veins run to the vena portæ. Among the veins we must observe the vasa brevia, which go off to the splenic branch; and the coronary vein which surrounds

the stomach. Its nerves principally enter at the left orifice; they come from the par vagum, and are very large; and hence it is that the stomach is so sensile. Its lymphatics go to the receptaculum chyli. The substance of the stomach is membranous; it is composed of five coats. The first coat is membranous, and its fibres are transverse. The second is cellular. The third is muscular; in this coat the arrangement of the fibres is various and apparently inextricable: some of them run circularly from the upper part to the lower; some only on the upper part of the stomach, between its two orifices; others run obliquely from the left side to the right; and others surround the orifices. *(Comp. Anat.*, n. 209.) Helvetius describes and figures a very different situation of the muscular fibres, with their courses and divisions, to that mentioned by other writers; but whether they are always circumstanced as he has laid them down, is a matter of doubt. I incline to think, that the arrangement of these fibres varies in different subjects. Santorinus alleges, that the fleshy fibres of the stomach are not annular, but spiral; and that the spiral fibres are most conspicuous in robust bodies about the pylorus. It is much more easy to distinguish the cellular coat, lately pointed out by Morgagni, under the external or common coat, particularly in that part of the viscus which lies between the two orifices. *(Comp. Anat.*, not. 9.) The fourth, or nervous coat of the stomach, forms a multitude of rugæ, and is furnished with a number of blood-vessels and small glands, which secrete the gastric fluid. These are more readily seen in the pig than in the human subject. The fifth coat is villous, thin, and porous; it adheres very closely to the nervous coat. *(Comp. Anat.*, n. 209.) Malpighi denies the existence of the villous coat, because it cannot be separated from the nervous coat; in consequence, he only reckons three coats: at the same time he admits that there are villi and papillæ in the stomach. There are other anatomists who deny, or at least question the existence of even these; but Ruysch and Santorinus have shewn us, that they may be made visible in the human stomach by macerating it in warm water; and they are still more obvious in the stomachs of sheep, pigs, dogs, and other animals. Morgagni states, that in this coat, in the human subject, he has occasionally noticed ventricular glands, with excretory orifices belonging to them, near the antrum pylori: but Santorinus affirms, that in this situation he saw only certain orifices, which he terms siphunculi." *(Comp. Anat.*, not. 9 * *.)

88. Winslow. "The stomach is a large bag-like receptacle, situated partly in the left hypochondrium, and partly in the epigastric region. Its figure is like that of a bagpipe; being oblong, incurvated, large and capacious at one end, and small and contracted at the other.

Its curvature occasions two distinct arches; one large, which runs along its greatest convexity; and one small, opposite to the former. In what follows, we shall name these two arches the great and small curvatures of the stomach, and the intermediate parts, its sides or surfaces. The stomach has two extremities, one large, and one small, like a crooked funnel. It has two openings, called the orifices of the stomach; one between the great extremity and the small curvature; the other, at the end of the small or contracted extremity. The first is a continuation of the œsophagus; the second leads into the intestinal canal, and is called the pylorus. The stomach is not situated in the left hypochondrium and epigastric region, in the manner represented in most of the figures. It lies transversely, obliquely, and almost laterally; the great extremity and the orifice next to it being to the left, and the small extremity and the pylorus to the right, and lower and more inclined than the former. The great extremity is in the left hypochondrium, for the most part immediately under the diaphragm; yet the neighboring or superior orifice is not in the left hypochondrium, but almost opposite to, and very near the middle of the bodies of the last dorsal vertebræ. One of the lateral surfaces or convexities is turned upward, the other downward; and not forward and backward, as in the dead subject, where the intestines no longer support them in their natural situation. If we divide the stomach along the two curvatures, into two equal parts, we find that the two orifices are not both in the same plane of division; but that the diaphragmatic orifice is in the upper half, and the intestinal orifice in the lower half. Thus the body of the stomach is by no means in the same plane with the œsophagus, but forms an angle or fold immediately at the passage of the œsophagus through the small muscle of the diaphragm; and on account of this angle, the superior orifice is turned a little backwards. The coats of the stomach are commonly reckoned to be four in number, and it is usual to subdivide these into several others. The four coats are, the external or common, the fleshy or muscular, the nervous or aponeurotic, and the villous or internal coat. The first or external coat is simply membranous, being one of the internal prolongations or continuations of the peritonæum. This appears evidently at the connexion of the superior orifice with the diaphragm, where the external coat of the stomach is really continuous with the membrane that lines the inferior surface of the diaphragm. The second or muscular coat is made up of several planes of fibres, which may all be reduced to two, an external and an internal: the external plane is longitudinal, though in different respects, following nearly the direction of the curvatures and convexities of the stomach: the internal plane is transversely circular. The fibres of the external plane slant from space to space, and are often intersected by small, oblique, whitish, and as it were tendinous lines. This plane is strengthened by a particular fasciculus, which runs along the small curvature, and the fibres of which appear to be less oblique than those of the great plane. The fibres of the inner plane are stronger; and rather segments which unite at different distances, than complete circles: for they, likewise, are intersected by great numbers of small white lines, of a tendinous aspect, and very oblique; which all together represent a kind of network, the areolæ or meshes of which are extremely narrow. As these circles or gyres approach the great extremity of the stomach, they become gradually smaller, and form a kind of muscular vortex, the centre of which is in the middle of that extremity. Round the superior orifice, and between the outer and inner planes, there are two distinct planes, about the breadth of a finger, or more, and very oblique, which surround this orifice in opposite directions, and decussate where they meet on the two lateral surfaces, where they are ultimately lost. Along the middle of each lateral surface of the small extremity, there runs a tendinous ligament, three or four lines broad, which terminates at the pylorus. These two ligaments lie between the external and muscular coats, and adhere closely to the former. Between the same coats, there is a cellular tissue, which is closely adherent to the external coat, and which dips in between the muscular fibres of the second, all the way to the third coat, as may be demonstrated by inflating it: it is no more than a cellular portion of the membranous coat, resembling the cellular portion of the peritonæum. The third coat, commonly called the nervous coat, sustains on its convex side, a very large number of capillaries and nerves. On the concave side, it appears to be of a very loose texture, and spongy or filamentary, containing a great quantity of glandular granules, especially near the small curvature, and round the pyloric extremity of the stomach. This spongy tissue resembles fine cotton or wool, as may be seen by macerating it a little in clear water, when it swells and tumefies. It is supported by a kind of web or groundwork of fine aponeurotic filaments, which intersect each other obliquely, much in the same manner as in the third coat of the intestines; and it adheres to the convex side of the villous coat. The fourth coat of the stomach is the villous coat, formerly called the tunica fungosa. We observe in it a number of small holes, answering to the glandular granules already mentioned. These two last coats are of greater extent than the two former, and they form large rugæ on the concave surface of the stomach; the major part of these rugæ are transverse, but irregular and wavy. There are likewise some longitudinal rugæ, which intersect the others; but at the pylorus they all become longitudinal, and terminate there. At the superior orifice these rugæ are in a manner radiated, and appear to be a continuation of the folds of the œsophagus; only they are thicker, and where they meet those folds they form a sort of crown, which bounds the superior orifice of the stomach, and distinguishes it from the extremity of the œsophagus. The interstices of the rugæ often contain an albuminous humor of varying density, with which the whole cavity seems likewise to be moistened. This humor is more fluid in living bodies, and is supplied by the glands of the stomach. It may be termed the gastric or stomachic juice. On the inner surface of the small extremity, near the pylorus, we observe a broad, thin, circular duplicature or border, with a roundish hole in the middle. This border is a fold of the nervous and villous coats. It is formed in part by a fasciculus of fleshy fibres, surrounded by the duplicature of the nervous coat, and distinguished not only from the other fleshy fibres of the extremity of the stomach, but also from those of the intestines, by a thin, whitish circle, which shines through the external or common coat, round the junction of the stomach and intestines. The figure of the pylorus is that of a ring transversely flattened, the inner edge of which, or that next the centre, is turned a little inwards towards the intestines, like the broad portion of a truncated funnel: this inner edge runs naturally into plaits, or gathers, like the mouth of a purse when closed. It is a kind of sphincter, which is capable of contracting the orifice, but not of quite shutting it. The principal arteries of the stomach are the coronaria ventriculi, which runs along the small curvature; and the two gastric arteries, the left or greater, and the right or lesser, both which form one common or gastric vessel, which lies along the great curvature. The coronary artery becomes united in the same manner with the pyloric artery, and both form one common or continuous vessel. These two arterial arches send a number of branches over both sides of the stomach; and these branches gradually divide and subdivide, in all kinds of ways, and many times, and nearly all communicate with each other by anastomoses. From their frequent ramifications and communications, two different networks arise; whereof the first and largest lies between the common and the muscular coats, supported by the cellular tissue; the second, which is smaller, lies on the surface of the nervous coat. The latter is a production of the first, being formed by short offsets from it, which pass through small interstices in the muscular coat. By injection we can shew a third network of fine capillary vessels, which run between the glandular granules or papillæ of the internal or villous coat. These do not seem in the natural state to be genuine blood-vessels, as inflammations and injections might incline us to think. The arteries of the stomach come originally from the cœliac artery, through the hepatic, splenic, and coronaria ventriculi. The pyloric and superior mesenteric arteries likewise contribute to them more or less immediately. They communicate also with the internal mammary and diaphragmatic arteries, and by the left epigastric with the inferior mesenteric artery. The veins, in general, are ramifications of the vena portæ; and, in particular, of the great mesenteric, splenic, and internal hæmorrhoidal veins. They accompany the arteries more or less closely, and form nearly similar arches and reticulations; but with this difference, that they are proportionably larger, their reticular areolæ wider, and their external communications more frequent. Between the common and muscular coats of the stomach we find a great number of nerves of different sizes. Many of them pass, in the form of a broad, flat fasciculus, or band, along the small curvature of the stomach, from the superior to the inferior orifice. The rest are spread in different directions, on the sides and extremities, and towards the great curvature, here and there forming reticular plexuses, from which a great number of filaments are sent to the inner coats. They arise principally from the eighth pair, by the plexus coronarius stomachicus, formed round the superior orifice of the stomach by an expansion of the extremities of two great chords, called the stomachic nerves, which descend along the œsophagus. The great sympathetic, or intercostal nerve, contributes likewise to them, by communicating filaments, which the stomachic plexus receives from the semilunar ganglions, and from the hepatic, and particularly from the splenic plexus." *(Exp. Anat., Tr. du Bas-Ventre,* n. 43—79.)

89. Ruysch. "The principal part of the arterial ramifications of the stomach are distributed to its nervous coat, which is, therefore, much more vascular than its other coats. This is remarkable; and would seem to indicate, that the extremities of the arteries perform some office in this coat, which is not performed in the other coats of the stomach. The glands of the nervous coat, in my opinion, are nothing more than the excreting extremities of the blood-vessels, gathered into as it were rotundo-oblong globules, or else into round globules, and into fasciculi. For when I filled the mesenteric arteries with red injection, it escaped through the so-called glands into the cavity of the intestines; and this was often repeated with the same result. In addition to which, when I placed an inverted duodenum (the arteries of which were injected) in fluid, (after it had been macerated in water,) I found that the so-called miliary glands of the duodenum were disposed in the form of brushes or tufts." *(Thes. Anat.* ii., asser. iii., n. 14, not. 2, p. 19, 20. Amstel. 1741.) "In an injected and in-

verted portion of the stomach of a boy, I observed not only the natural rugæ projecting from the internal surface, but also thousands of granules of the wax-injection, which in endeavoring to escape through the extremities of the little arteries had become fixed to that surface. Moreover, after the arteries of the stomach were filled, the injection came out from their extremities, in drops, into its cavity; and I found this to be the case in the intestines also." *(Thes. Anat.* vi., asser. i., n. 23, not. 1, p. 28. Amstel. 1744.) "Before the stomach terminates in the inferior orifice, it exhibits an immense number of extremely minute, quadrangular, cellular compartments, of different sizes, and which have some analogy with the cells in that portion of the calf's stomach, called the reticulus; excepting that they are much larger in the calf than in the human subject." *(Thes. Anat.* ii., asser. iii., n. 4, p. 15.) Let the reader particularly examine the tables of this author; as *Thes. Anat.* ii., tab. v., fig. 1; where he exhibits the œsophagus and stomach of a child, with the longitudinal fibres and the arteries; also the omentum, furnished with myriads of blood-vessels, but destitute of foramina. *Ibid.*, tab. v., fig. 2; shewing a portion of the adult stomach inverted, with its interior folds and innumerable pores. *Ibid.*, tab. v., fig. 3; an inverted portion of the viscus, taken from above the pylorus; *a*, the place where there are no folds or plicæ, but the internal surface of the stomach is provided with little round prominences or monticuli: *b*, the quadrangular, cellular compartments of different sizes. And *ibid.*, fig. 4, the same magnified. Respecting the villous coat and the papillæ, see *Advers. Anat.*, dec. iii., p. 33, 34. (Amstel. 1737.)

90. LEEUWENHOEK. "I took some chyle (in a glass tube) from the intestines of a lamb, and afterwards some from the lacteals near the liver, in the neighborhood of the pancreas. In somewhat less than four minutes I found it coagulated. Its particles then seemed to form a kind of crystalline humor, in which no distinction of parts was visible, except that certain particles were sticking in it, having been impacted there during the coagulation; some about a sixth part the size of a globule of human blood; many also composed of two, three, four, five, and a few of six such globules [talium globulorum]; and several of them seemed to be at least in superficial contact. The uncoagulated part of the chyle was thin and fluid, and mingled with the particles just mentioned, and with an incredible number of globules besides; so many, indeed, that the whole material appeared to consist of nothing else; although in reality they were only swimming in it." *(Arcana Naturæ Detecta,* p. 11, 12. Lugd. Bat. 1722.)

91. SWAMMERDAM. "Where the stomach of the louse joins the breast above, its figure resembles a fork with two teeth: these are two

blind appendages of the stomach, which go deep into the breast, and on either side near the gullet and spinal marrow, and reach to the first pair of legs. The stomach manifestly consists of two coats: the external coat is furnished with a great number of pulmonary pipes: both coats consist of a vast quantity of globular granules, which are, however, somewhat irregular in form. The motion of the stomach is truly wonderful; insomuch that, by reason of its strong agitations, contractions, dilatations, corrugations and expansions, (which are plainly seen through the body, and strike one with amazement,) one might suppose it an animal within an animal. These are particularly conspicuous, when blood is passing into the stomach at the time of sucking, and when the stomach is full of food; for then the food is observed to be shaken and agitated up and down, and on every side in the stomach. When the louse is hungry, it sucks from a pore, and the blood is observed by the microscope, to ascend to the head in a rapid and as it were frightful stream, in whatever posture the animal is sucking. &c. (*Bib. Nat.*, p. 74—80, tab. ii., fig. 3, 6.) The stomach is in strong and continual agitation, being hardly a moment at rest. (*ibid.*, p. 71.)

"The stomach of the covered snail consists of three coats; when empty, it appears full of oblong grooves; it grows narrow by degrees, aud terminates in the pylorus. Two beautiful vessels run along the stomach and the gullet, and discharge a humor into the upper part of the palate, in the mouth, through two small apertures; they are therefore plainly salivary ducts. These two little vessels arise from two small, clear and snowy parts. A beautiful little vessel runs over their whole surface, and gives a great many branches to each of them. These little parts are laterally connected with the stomach by several vessels, which look like so many delicate filaments. (*Bib. Nat.*, p. 123, 124, tab. v., fig. 6, 7, 8.)

"The stomach of the cancellus or hermit-fish is partly membranous and partly osseous: its little bones are very beautiful: it has many muscles. In the cavity of the stomach, I saw three distinct teeth, each of which was divided into several smaller teeth, &c. (*Bib. Nat.*, p. 202, 203, tab. xi., fig. 3.)

"The stomach of the hemerobios, ephemerus, or as it is sometimes called, the esca or bait, consists of a thin and delicate membrane, corrugated on the inside, and full of reticular plaits; and provided with many tubuli which resemble blood-vessels. (*Bib. Nat.*, p. 248, 249, tab. xiv., fig. 1; tab. xv., fig. 5.)

"The cossus or worm of the scarabæus is almost filled by the stomach; this consists of several coats, and has moving, circular fibres:

the creature lives on the wood of the sumach-tree. The stomach is very narrow where it forms the gullet; a little after it expands, until it makes its upper orifice, answering to the œsophagus. In that part, the stomach is armed with about seventy little tooth-like processes, whereof some are longer than others. These are divided into six rows; of which the two upper ones point inwards, but the other four point partly forwards, partly backwards. All these tooth-like tubuli open into the stomach, in the same manner as the appendages, or the pancreas in fishes, opens into the intestines: each opens separately into an ecphysis, as in the salmon. A little lower down, the stomach is rendered conspicuous by twenty-two whitish, glandular tubes, with their ends directed backwards. Finally, about the lower part of the stomach, a little above the beginning of the pylorus, we again observe thirty such tubes, which are also uneven, and open into the stomach in this situation, with their ends pointing forwards. A sort of suture, like the raphe of the perinæum, divides the stomach: which, moreover, has infinite pulmonary tubes inserted into it. (*Bib. Nat.*, p. 313, 314, 315, tab. xxvii., fig. 11, 12.) In the nymph into which the cossus changes, the gullet keeps its original form; but the stomach is greatly contracted, and its appendages almost totally vanish. The vasa varicosa become more loose and free, though they do not absolutely separate from the stomach. We may easily divide the stomach into its three coats. It is wonderful to observe, how the worm of the hornet, that is to put on the form of a nymph, discharges its excrements, and the internal coat of the intestine, or rather of the stomach, that embraces or contains them, at one and the same time. (*Bib. Nat.*, p. 318, 319, tab. xxviii., fig. 5.

"The stomach of the bee-worm is [like a long bladder, or vesicle,] furnished with numberless pulmonary tubes: certain glandular bodies are seen through it: and it has circular muscular fibres. The gullet of the bee (apis operaria) is exceedingly narrow: the stomach is membranous and thin: it is generally filled with honey, which is easily distinguished by the taste: the pylorus follows the stomach. (*Bib. Nat.*, p. 411, tab. xxiv., fig. 5, 6, and p. 454.)

"The stomach of the caterpillar of the day-butterfly, almost entirely fills the caterpillar's body; being both capacious and always distended. The gullet is a small and slender tube, running towards the mouth and gradually dilating in its progress. A sort of tendinous ligament runs from one end of the stomach to the other; it arises from the tendons of the muscular fibres of this part. The stomach consists of three coats. On removing the stomach, there appear very plainly two little canals, which terminate in two sharp filaments, that are inserted near the intestina cœca. (*Bib. Nat.*, p. 575, 576, tab. xxxiv., fig. 4, 5.) The stomach of the butterfly is strangely altered in shape from the stomach of the caterpillar: it is now sacculated throughout, like an inflated colon: this structure affords a beautiful sight, by reason of its numerous folds and convolutions. On the back part, it much resembles a small intestine, full of most delicate folds. (*Bib. Nat.*, p. 595, 596, tab. xxxvi., fig. 1.

"In the stomach and intestines of the vermiform nymph of the gad-fly a great and wonderful transformation is observed. The upper part of the stomach is found to be contracted, as it were, into five small annular rugæ or folds, through which the pulmonary tubes run very beautifully: a little lower it is divided by longitudinal grooves. It is seen shining through the skin, in this region, like a kind of open tube of a reddish color. The beginning of the small intestine is very beautifully sinuated in six folds, or corrugated convolutions. (*Bib. Nat.*, p. 674, tab. xli., fig. 6.) The inner coats of the intestines come away in the course of the changes. (*ibid.*, p. 686.)

"The stomach of the cuttle-fish, or sepia, resembles a spherical bag, somewhat indented in the middle. It consists of three coats. The internal coat readily separates from the middle, with the ingesta it encloses. Under the rectum lies an appendage of the stomach, and which opens into the stomach by a particular orifice, after making a very remarkable spiral revolution, like a snail-shell. This organ is smooth and slippery on the inside, and contains a matter like the pancreatic juice of other fishes." (*Bib. Nat.*, p. 889, 890., tab. li., fig. 5.)

Respecting the stomach of the acarus or mite, see *ibid.*, p. 702, tab. xliii., fig. 5.

92. Glisson. "Man has only one stomach, whereas some other animals have two, or even four. There is but one in carnivorous quadrupeds, as the lion, the wolf, the fox, and the dog; also in the horse, the ass, and the mule: and among birds, in the eagle, the hawk, the crow, the kite, and the owl. It is to be observed, that animals with incisor teeth in the lower jaw only, have four stomachs; this is the case with sheep, oxen, goats, and, in a word, all horned animals; and also with ruminants that are not horned, as camels, dromedaries, does, and some kinds of sheep. The first stomach, or reticulus, has three orifices: by the superior it receives the food from the mouth, through the gullet; by the second, it sends it into the great cavity; by the third, into the omasus. This third orifice is so constructed, that whatever it admits must be conveyed to it along a soft and very sensitive groove in the reticulus. When all the food is collected together, the animal begins to ruminate; that is to say, the great stomach pours back the

grass properly macerated into the reticulus, and the reticulus pours it back into the mouth; where it is comminuted, and again descends indiscriminately into the reticulus. The third stomach, called by the Latins the omasus, ordinarily, the centipellio, receives the food, now macerated and well manducated, from the reticulus. The inside of the omasus is entirely covered with large, broad, transverse membranes, of a lunated figure, to prevent the food from passing out of it with undue rapidity. The abomasus, or fourth stomach, differs very little from the human stomach, except that its internal coat is furnished with larger and more numerous folds, all of which are straight. This stomach receives none but the food that is made fine, and almost converted into chyle; and completes the process. In young, sucking animals, the other stomachs are scarcely used at all, this fourth one alone being exercised. Granivorous birds have two stomachs; one membranous, the other fleshy: also an intermediate stomach, but which is only a passage from the first into the second. The first is the crop, which is sometimes found double, as in pigeons; in which case the gullet is also double." (*Tractatus de Ventriculo et Intestinis, &c.*, p. 140, 142, 145, 146. 12mo. Lugd. Bat. 1676.)

93. Martin Schurig, among his collectanea, has the following particulars which merit insertion in this place. "Ruminant animals have four stomachs; a large stomach, or penula; a second, or reticulus; a third, or centipellus, called also the cellular chamber, or omasus (maniplus); and a fourth, or abomasus; which last is the true stomach. There are instances on record of human rumination." (*Chylol.*, p. 380-1.) With respect to the stomach, this author records cases of irregular position thereof, and of large, small, thin, double, perforated, and ruptured stomachs, (*ibid.*, p. 367—398): of food remaining in the stomach for a long time, as grapes, pills, pieces of sausage, pickled ginger, suet, lettuce, thorns, oats, plum-stones, and rose-leaves: also, of hydatids, stones, lice, a three-pronged fork, and various other things, being found in the stomach. (*ibid.*, p. 435—445.) "J. Hartmann exhibited the stomach of the piscis silurus, and shewed its wonderful motions: the motion began in particular spots, not far from the pylorus, contracted vortically to three different points, and afterwards extended back from the same points through the same vortices, with a kind of regular systole and diastole. The motions of the reciprocating vortices set in motion the rest of the upper part of the stomach, as well as the intestine adjoining the pylorus; and every time we pressed the spots before mentioned, they seemed to rise into the same kind of rhythmical motions, which continued at short regular intervals: and this not only when the stomach was entire; for even when it was cut in pieces, in

the evening, it still exhibited similar motions. The spots above mentioned were occupied by innumerable nerves, lying like twisted cords, without any membranous covering, on the external surface of the stomach." *(ibid.*, p. 446, 447.) Schurig afterwards enumerates certain substances which have passed accidentally into the œsophagus and stomach: little bones, the prickles of fish, the jaw of a fish, crab's claws, large morsels of food, buck-wheat, pebbles, the stones of fruit, Persian apples, a packing needle, common needles, large and small, pieces of wood, a diamond, bits of cord, of leather, goose-quills, a living crab, a ring, nails, money, pieces of lead, marbles, a pipe, a child's rattle, the prongs of a fork, a spoon, a golden cross, a surgical speculum, the links of a chain, a clasp, a pair of scissors, an iron whetstone, the stalk of an iron pot, the point of a knife. *(ibid.*, p. 278—321.) Schurig also speaks of knife-eaters, and mentions noted cases at Prague, in Prussia, at Basle, Halle, London, and Harburgh; also of glass-eaters, stone-eaters, poison-eaters, and of those who would eat any thing whatever, *(ibid.*, p. 322—343:) of various substances that have got into the trachea, *(ibid.*, p. 349 seq.:) of division, cartilaginous alteration, dilatation, and rupture of the œsophagus, *(ibid.*, p. 361, 2:) of impeded deglutition, arising from defect of the uvula, erosion of the epiglottis, ulceration of the fauces, cartilage, fleshy growths and tumors, growing-together of the parietes of the œsophagus, aneurism, polypus and stricture of the œsophagus, small-pox, paralysis of the orifice of the stomach, the use of mercury and tobacco, the bite of a viper, polypus of the nares and fauces, and from a number of other causes, besides external ones. *(ibid.*, p. 256—277.) He mentions the case of a Maltese drunkard, who, after drinking twenty or thirty glasses of water, would presently vomit them up, for the amusement of spectators, and either spirt the glasses full again, one after another, or else throw the whole body of fluid to a distance of twenty feet, or more. He was accustomed also to repeat the same thing with rose-water, spirits of wine, &c. *(ibid.*, p. 528.) In dissecting the glutton or polyphagus (vielfras), Schurig found the cavity of the stomach full of bones or tendons, all sticking together, and of a sort of digestive juice resembling chyme: the stomach itself was expanded and soft, and the intestines exhibited the natural convolutions. *(ibid.*, p. 29.)

Schurig also records instances of diseased appetite, or of pica and malacia: of persons who have desired and eaten sand, lime, mud, earth, cinders, coals, sulphureous turf, pitch, cart-grease, glue, stones, alabaster, tartar, pieces of pot, loaves of bread, wheat, beans, millet, buck-wheat, leaven, marjoram, pepper, ginger, mace, cinnamon, onions, kitchen salt, almond-emulsion, vinegar, treacle, fungi, cherries, rotten citrons, herrings, crabs, a live eel, a calf's-liver, lizards, frogs, spiders, scorpions, lice, a live chicken, skins, tow, wool, cotton, hair, gloves, linen and thread, neckcloths, blotting paper, hay, dung, the mire of cattle, human flesh, raw meat, and metals. (*ibid.*, p. 42—90.) He also mentions that pica has been noticed in brutes, as in the cat and the dog: and in man, in the forms of depraved appetite for water, milk-whey, vinegar; of longing for peculiar colors and smells, for sucking the wind out of bellows, for throwing eggs into other persons' faces, for receiving slaps, and for thieving. (*ibid.*, p. 92—96.)

He also relates several instances both from sacred and profane history, of prolonged abstinence from food and drink; (*ibid.*, p. 175, seq.;) of loathing and antipathy to certain foods; and to different animals; (*ibid.*, p. 96, seq.;) likewise of various antipathies founded on the senses of sight and hearing, (*ibid.*, p. 163—166:) of which we shall speak elsewhere. (*Chylologia Historico-Medica.* 4to. Dresdæ, 1725.)

94. The stomach, like a large bladder or tubulated retort, receives in its bosom, through what is called the cardiac orifice, every kind of saliva and available food;—commixes, circulates, squeezes, strains, bruises, triturates, macerates, seethes, extracts, in a word, *digests* it;—then carries it outwards, drives it through certain foramina, and evaporates and sublimates it along certain ducts, its appointed passages:—summons and sharpens the menstrua (*a*), and increases the forces, according to the measure, degree and success of the operation; and again repeats the processes; that is to say, reduces, filters, corrects and seethes the materials which have been once digested; and all the time, transmits the rectified portions through foramina into tubular passages; but sends off whatever it has not thoroughly laid open *(reclusit)*, through the pylorus into the intestines (*b*).

(*a*) We shall consider presently what it is that performs the office of menstruum; and shew, that this office is not confined to the salivæ, but extends to all kinds of liquids taken by the mouth, as water, milk, syrups, juices, beer, wine, spirits, &c.; which the stomach therefore desires, sometimes with a certain secret craving. At the same time, it also summons suitable salivæ, and applies all their properties to use; for it decomposes even the saliva itself, and draws from it whatever can contribute to its purposes, throwing it on the food, and mixing it therewith; the aqueous portion of the saliva with saline particles, the volatile spirituous portion, and the subtle phlogiston, with oily particles. Moreover, it imbues and vivifies all things with the very spirit of its fibres, and thereby enables them to flow congenially through the pores into the vessels, and finally to be fitted for the blood.

(*b*) The uses of the stomach are brought together here in a simul-

95. The stomach performs and completes all these things, by its motion, warmth and delay. Thus it gives the salivæ an opportunity of acting, and of making use of their powers and forces, or natural properties, and thereby producing their effects (*c*). Apart from their great chemical vessel, and its activities, the salivæ would be barren, impotent and void of office (*d*). Hence it is evident, that nature, or the soul in which this nature dwells, here disengages, develops and realizes all the latent resources of art and science, chemical, physical and mechanical, from their innermost grounds and principles; which if a man can touch on the surface only, and with the edges of his lips, he is an Apollo.

96. Here nature, in her operations, proceeds by degrees from outmost to inmost; and also from both to both at once: thus nothing is presented, that she does not subdue, dissolve, remodel, convey outwards, and at last into the blood, for which it is adapted.

97. The stomach—convoluted of a manifold series of membranes; these, of a manifold series of muscles, tendons, ligaments, glands and papillæ; and these, of a similar series of fibres, vessels and ducts—winds in a stupendous gyre, and thus generates a cavity or ampulla, which in its folded and corrugated state, is designed to be the very perfection of geometrical, ter-

taneous series, or in a summary; from which I shall draw them forth, one by one, in successive order, in the proper places.

(*c*) The properties of the saliva were discussed above, (n. 81, *m*), where it was shewn to have powers of diluting, softening, dissolving, extracting, altering, sheathing, carrying, insinuating, and vivifying.

(*d*) The stomach is the agent; the food, the patient; the saliva is the active medium belonging to the agent. By coöperation, the agent becomes an efficient sufficient cause, and produces a full effect. This is the case universally. Were the agent without appliances and means adequate and accommodated to its particular activities, and to the passivities of those substances upon which the action is directed, its efforts and readiness to act would be in vain. Here, however, all the requisite means are most complete. Thus the most intimate and hidden things are those exclusively which nature here draws from her bosom, or, what amounts to the same thing, from the principles of our human arts and sciences.

restrial and corporeal forms; and to emulate and aspire to a kind of infinity, and to powers of spontaneous activity and motion. For the stomach has two curvatures or arches, and on its concave surface, respects poles, axes and foci; by these, through their radii, which are so many circular forms, circumferences; and by all points of these again, their poles, axes and foci: and so on, in an everlasting gyre (*e*). A similar form

(*e*) I am speaking of the stomach in its corrugated and folded state, while digestion is proceeding, and of its form in that state; not of the stomach when expanded. Whenever any action is intended, a corresponding motion is induced, and a form imprinted and stamped corresponding to the motion; namely, on substances, as the proper subjects from which motions, forces, powers, and modes result: thus the form, or, as it is commonly called, the structure or frame of substances, coincides with the form of the active forces, consequently with the form of the motion which produces the action. That such a form is imprinted on the stomach, is proved by the plain evidence of the senses; which also is singularly corroborated by the leading facts of anatomical experience. That the first, or most general determination of this motion, goes forth into the vortex, from the cardiac orifice as its centre and focus, is shewn by Winslow, where he says, "At the superior orifice the rugæ are in a manner radiated, and appear to be a continuation of the folds of the œsophagus." (n. 88.) But at the pylorus, the limit or goal of this gyre, and which I term the second or general focus, the same rugæ, lines or radii, again draw in, converge, and become longitudinal; according to the descriptions of all our authors. The stomach has also its polar points or cynosures. "As the circles or gyres," says Winslow, "[of the fibres of the muscular coat] approach the great extremity of the stomach, they become gradually smaller, and form a kind of muscular vortex, the centre of which is in the middle of that extremity." (n. 88.) This spot is in the relation of a kind of pole to the foci or orifices themselves, whereby they, as well as all the radii and circumferences proceeding from them, are brought into relation with each other; and opposite to it, we also observe a kind of antarctic pole, in the extremity of the lesser arch. With respect to the circles or circumferences, they are the two greatest but continuous arches, forming a kind of equinoctial line, wherefrom proceed the lesser arches, which are related to the greatest, and by this means to the poles. Winslow terms "these two arches the great and small curvatures of the stomach, and the intermediate parts, its sides and surfaces;" the two ends, he names its "extreoccurs in the intestines, or in the ultimates of the body; likewise in the brains, or in the principles of the body; and also throughout, in the intermediates. This form must be called the perpetual-circular, or properly, the spiral form; it is the essential form of motion, or of the fluxion of organic substances in the animal world (*f*).

mities." (n. 88.) The lesser circles, which more properly speaking are the diameters of the larger or largest, running from it as circular curves, are delineated in the tabulæ of Ruysch and Willis, which represent the inverted stomach with its rugæ or folds. Something exactly similar to these, both as respects their form and their mode of circumvolution, is observable in the cerebrum; where also there is a similar formation of poles, axes, foci, and circles, great and small, as we shall shew in our analysis of that organ: also, in the intestines, of which we shall treat in the next chapter. Inasmuch as this form of fluxion has never yet been brought within the domain of geometry, therefore its winding and meandering determination baffles and eludes the mind. Wherefore, in order to escape from the labyrinth, many writers, as Boerhaave, Santorinus, and others, sometimes call this motion *spiral*, but more commonly, the *peristaltic* and *vermicular* motion. "The arrangement of the [moving] fibres," says Heister, "is various, and seemingly inextricable," (n. 87.) "The structure" [of the stomach in the butterfly], says Swammerdam, "affords a beautiful sight, by reason of its numerous folds and convolutions." "In the louse," says he, "the motion of the stomach is truly wonderful; insomuch that, by reason of its strong agitations, contractions, dilatations, corrugations and expansions, (which are plainly seen through the body, and strike one with amazement,) one might suppose it an animal within an animal," (n. 91.) Still more vividly does Hartmann describe the motion of the vortex in the stomach of the piscis silurus, or sturgeon, in which he has succeeded in actually demonstrating it to the senses. "The stomach," says he, "displayed wonderful motions: the motion began in particular spots, not far from the pylorus, contracted vortically to three different points, and afterwards extended back from the same points through the same vortices, with a kind of regular systole and diastole. The motions of the reciprocating vortices set in motion the rest of the upper part of the stomach, as well as the intestine adjoining the pylorus," (n. 93.)

(*f*) I intend to explain the nature of the spiral form in an especial doctrine of forms. Meanwhile, for the better understanding of the subjects mentioned in this chapter, I will here state, that forms ascend

98. We learn the nature of this form, or of the determinations of motion in the stomach, principally from its windings, furrows and ridges (*g*): from the fluxion of the fibres in its

from the lowest to the highest, in order and by degrees, as do also the essences and substances of all things. The lowest form is the *angular;* which is also called the *terrestrial* and the merely *corporeal* form, inasmuch as it is peculiar to bodies having angles and rectilinear planes; the measurement of which is the primary object of the present geometry. The second and next-higher form is the *circular,* or *spherical* form; which may also be called the *perpetual-angular,* since the circumference of the circle involves neither angle nor rectilinear plane, because it is a perpetual angle and a perpetual plane: this form is at once the parent and the measure of angular forms; for it is the means of shewing the properties of angles and figures, as trigonometry teaches. The form above this is the *spiral,* which is the parent and the measure of circular forms, as the circular form is the parent and the measure of angular forms. Its very radii or diameters are not rectilinear, nor do they converge to a fixed centre, like those of the circle, but they are variously-circular, and have a spherical surface for a centre; wherefore the spiral is also called the *perpetual-circular.* Our science of geometry rises almost to this form, but dare not enter it, or peruse its spires; for at the first glance it strikes us as inextricable, and seems to sport with our ideas. This form never exists or subsists without poles, an axis, foci, a greatest circle, and lesser circles which are its diameters; and as it again assumes a perpetuity which is wanting in the circular form, namely, in respect of diameters and centres, therefore it emulates and breathes a natural spontaneousness in its motion: as also appears from the stomach and its segments after death, for when its nerves are only touched, it rolls and wreathes as in the living subject, and flows spontaneously into its gyres, as though it were still hungering, and longing to grind the food: there being nothing that can prove an obstacle; inasmuch as there are no angles, and consequently no hindrances to motion. There are other still higher forms, as the *perpetual-spiral,* properly the *vortical:* the *perpetual-vortical,* properly the *celestial:* and a highest, the *perpetual-celestial,* which is *spiritual,* and has in it nothing but what is everlasting and infinite. But these remarks are only by the way; in order to give the reader some general preliminary idea, whereby to comprehend the motions of the stomach and intestines.

(*g*) In the Tabulæ of authors which represent the stomach inverted, coats, particularly in the muscular coat (*h*) : also of the nervous fibres; and of the arteries and veins, which are constantly found on the banks of the stream of motion, or as it were in the ports of the vortex (*i*) : likewise, from the determination of the ligaments and tendons, general and particular (*k*): and from the

as in those already mentioned of Ruysch and Willis, we have a good view of the nature of these windings and convolutions, intersected rugæ and depressions, and of their peculiar gyre of fluxion; from which we find, that they are precisely similar to those in the intestines, and precisely similar to those in the cerebrum. They have no discoverable beginning or end, but the spires interweave with each other by wonderful wreaths and inflexions, wind sinuously in and out near the same spots, and reappear as new spires, although they are perfectly continuous throughout: thus they assume a kind of perpetuity. Their general determinations only, to large circles, poles, axes, and foci, are displayed to the careful and contemplative mind.

(*h*) Winslow describes the fibres of the muscular coat in the following terms: "The fibres of the external plane slant from space to space, and are often intersected by small oblique lines. The fibres of the internal plane are rather segments which unite at different distances, than complete circles; for they likewise are similarly intersected." (n. 88.) See also n. 97 (*e*). It appears from this description, that the fibres describe, not absolutely circular spires, but a kind of everlasting spire, which is freely and easily flexible into every species of circular form: and that all points of the moving fibre deviate from the straight line towards the gyre or circular. On the other hand, the fibres of the external or common membrane, according to Heister and Ruysch, are transverse, consequently simply circular: for the circular form is generated by the spiral, as the angular form by the circular; that is to say, by the spiral evolving itself. But of these subjects we shall treat further in our doctrine of forms.

(*i*) That the nerves pass along the arches, along the lesser arch particularly, may be seen in the neurological Tabulæ of Vieussens and Willis, and in representations of the stomach where those nerves are figured. From this common circle, they throw themselves in all directions into their whirling sea; that is to say, they send distinct branches to the lesser gyres, whereof the largest or common circle, that, namely, of the stomach itself, is composed; which branches similarly pass along their largest circles, and are distributed therefrom; always maintaining the relations of poles, axes, and circumferences.

peculiar conglomeration of the glands, ducts and papillæ: finally, from the position of the orifices, no two of even the most minute being ever in the same plane (*l*).

99. This motion in the greatest and in the least things, is exactly synchronous with the reciprocal respiratory motion of the lungs: it commences and goes forth from the œsophageal orifice, and completing a gyre during each respiration, ceases in the intestinal orifice (*m*). So also in the lesser forms; only that

The gastric and the coronary arteries and veins pass likewise along the arches and curvatures. For all the motions of the stomach run forth over its surfaces and sides, from this largest circle; and run back to the same circle; it being a kind of continued pole from one to another. Wherefore this circle is the general resting-place, port, and equilibrium of the motions, where the greatest tranquillity prevails; from which the arteries may safely pour forth their blood, and where the veins may receive it safely. On this account, the arteries and veins are said to be kept on the banks of the stream of motion.

(*k*) The ligaments and tendons are bonds, and at the same time fræna and limits to the motions; being the parts in which the motions and the moving fibres begin and terminate. Respecting the course of the ligaments, see the authors above quoted.

(*l*) I shall endeavor to demonstrate, in the doctrine of forms, that in the fluxion of parts according to the spiral form, the foci are never found placed in the same plane; nor consequently, the superior and inferior orifices of the stomach. "If we divide the stomach," says Winslow, "along the two curvatures, into two equal parts, we find that the two orifices are not both in the same plane; but that the diaphragmatic orifice is in the upper half, and the intestinal orifice in the lower half." (n. 88.) They cannot be in the same plane, consistently with perpetuating the motion. The case is different in circular forms, where the diameters and ordinates meet either in one centre, or in one plane or line, in which the motion is arrested and ceases.

(*m*) It has, I think, been shewn above, (n. 83,) that the œsophagus conforms exactly to the respiratory motion of the lungs. On this depends the motion of the stomach itself, which the œsophagus enters with its coats and nerves, and by strong and multiplied powers compels to the same alternations. There are even additional proofs of the fact on the part of the stomach; to wit, that it is connected to the under side of the diaphragm, and its external coat is there continuous with the common coat of the abdomen, that is to say, with the peritonæum.

they breathe forth their little gyres of motion, at the same instants, not into the common orifices, but into particular foraminula and spiracles of their own, which are so many most minute pylori (*n*).

100. It is a constant law in the organic body, that large, compound, or visible forms exist and subsist from smaller, simpler, and ultimately from invisible forms, which act similarly to the larger forms, but more perfectly and more universally; and the least forms, so perfectly and universally, as to involve an idea, representative of their entire universe. Thus whatever is

The diaphragm is constantly elevated and depressed by the superincumbent lobes of the lungs; by it the effect is communicated to the stomach, which is subjacent and appended to it; and indeed to the minutest points thereof, by means of the nerves and the fibres which flow into it through the fleshy part of the diaphragm, along with the œsophagus, and pass over its arch, whence they are distributed in all directions. Hence precisely the same effect must exist in the living as in the dead stomach, which latter, when these nerves are stretched, still rushes into peristaltic gyres. The ribs also strengthen this effect, for being variously connected to the border of the diaphragm, they obey exactly the respiratory movements of the lungs, and propagate and concentrate them to the spot where the œsophagus penetrates the septum. The same may be said of the abdominal muscles, which expand and contract the abdomen synchronously with the expansion and contraction of the lungs, and communicate their forces and reciprocations to the peritonæum, and by it, as well as by means of the diaphragm, to the stomach. Thus the stomach is in a manner in the centre of all the motions which the lungs excite and pour forth. The same result is effected by more simple means, in insects, which have no apparatus of ribs, muscles, and diaphragm; their stomachs being furnished with an infinite number of pulmonary pipes, see n. 91; and that of the butterfly, according to Swammerdam, with a complete pneumonic vesicle, "which has a remarkable peristaltic motion, and almost always lies above the stomach." (n. 77.) These considerations may also give us some idea of the rapidity of the vermicular motion in the stomach, inasmuch as during every respiration it runs from the diaphragmatic orifice, through the convolutions of the viscus, that is to say, through an immense extent of windings, all the way to the duodenal orifice.

(*n*) That similar things occur in the lesser forms, is a point which will be better elucidated by what follows.

manifested in compounds and ultimates, arises from those innermost or primal forms; from the law, rule, power and energy of which, every property or action of compounds absolutely depends and results (*o*).

101. Thus the stomach, which is a large, compound and visible form, is generated and fashioned by smaller, simpler and invisible forms, that have their largest circles, poles, axes, foci and orifices, in the same manner as the stomach (*p*); and perform similar offices, only more perfectly and more universally;

(*o*) This is a general law, derived from a number of experimental facts; thus from the magazine and storehouse of experience herself. It is therefore a product of analysis, and one which it is necessary to premise in this place, to give the reader a better understanding of what follows.

(*p*) Microscopic researches have clearly established the existence of similar ventricular forms, or antetypes of the larger form. The contracted stomach exhibits a number of large rugæ, wonderfully convoluted and subdivided, and others which are quadrangular and cellular: see Boerhaave, *Inst. Med.*, n. 77. Ruysch has a figure and the following description of the latter: "Before the stomach terminates in the inferior orifice, it exhibits an immense number of extremely minute, quadrangular, cellular compartments, of different sizes, and which have some analogy with the cells of that portion of the calf's stomach, called the reticulus; [besides which,] the internal surface of the stomach has little round prominences, or monticuli." (n. 89.) The way in which these meshes or areolæ are separated from each other by tendons and ligaments, is thus described by Winslow: "The fibres of both planes of the muscular coat," says he, "are often intersected by small, oblique, whitish, and as it were tendinous lines, which all together represent a kind of network, the meshes of which are extremely narrow." (n. 88.) Swammerdam observes of the butterfly, "That the stomach is sacculated throughout, like an inflated colon," forming "numerous folds and convolutions," (n. 91:) and that the stomach of the ephemeris "is corrugated on the inside and full of reticular plaits, and provided with many tubuli which resemble blood-vessels." (*ibid.*) Similar antetypes were before observed in the tongue, and they are observed every where else, in exact agreement with the law just laid down (n. 100), that these little stomachs are as it were the unities of the larger, or its numerical constituents. We must now, therefore, pursue the thread of the subject, and unwind it analytically, and consider and the action of the larger form results from their power and energy. They are so many, as it were, lesser stomachs, which, connected in a stupendous chain and gyre, generate a large or common form, which is the stomach.

102. The materials that have been rudely and imperfectly digested by the large stomach, these smaller stomachs (*q*) *receive, commix, circulate, squeeze, strain, macerate, seethe, extract* (*r*); *carry outwards, drive through certain foramina, and evaporate along certain ducts, their appointed passages* (*s*): *also summon and*

whether all the properties and ratios of the functions of the larger originate from these its lesser stomachs.

(*q*) In this and the following paragraph we repeat what we before gave in one series, (see n. 94, 95, 96), respecting the offices of the stomach. We are now to develop these particulars, and to raise them as it were to higher powers, that is to say, to their principles; and thus, by the analytic method, to unlock the causes of those effects which exist in the common visible stomach.

(*r*) That is to say, the common stomach receives the crude and imperfectly comminuted food from the œsophagus; but these little stomachs receive it, after one digestion, from the stomach; and the least again receive it from these, and reduce and rectify it still more perfectly.

(*s*) The infinity of little foramina which open on the concave surface of the internal coat of the stomach, is fully displayed in one of Ruysch's Tabulæ: this coat, therefore, is considered by anatomists to be exceedingly porous. Furthermore, "Santorinus affirms, that he has seen certain orifices, which he terms siphunculi," near the pylorus. (n. 87.) Winslow describes the ducts or passages into which the orifices lead. "The concave side of the nervous coat," says he, "is of a spongy texture, resembling fine down or wool; it is supported by a kind of web or groundwork of fine aponeurotic filaments, which intersect each other obliquely, and it adheres to the convex side of the internal or villous coat." (n. 88:) and again, "By injection we can shew a third network, of fine capillary vessels, between the granules or papillæ of the villous coat. These do not seem in the natural state to be genuine blood-vessels, as inflammations and injections might incline us to think." (*ibid.*) Thus, not only are there little foramina or outlets answering to minute pylori; but there are also ducts resembling diminutive intestines, the windings of which construct the nervous coat, and give it the woolly and downy appearance already described. All these are provided with *sharpen the menstrua* (*t*), *and increase the forces, according to the success of the operation; and again repeat the same processes.* The portions that are still not properly digested and prepared, they in turn commit to yet smaller, or to least stomachs; which straightway commence a similar work (*u*), but still more perfectly and more universally; *reduce, filter and rectify* the materials committed to them; and sublimate them determinately, through tubular passages, until they reach the blood. They perform and ultimately complete all these things, by their motion, warmth and reiterated agency, or delay. *Thus nature proceeds at once from outmost to inmost, and from inmost to outmost; and nothing is presented to the stomach, that she does not subdue, dissolve and remodel.*

103. Every operation, then, of the grand stomach, results from the manifold agencies of lesser stomachs, whereof a type or general idea reaches the senses, and by their means comes under the cognizance of the intellect. Thence the *motion* derives its origin, spire, and perpetuity (*x*). Thence the *warmth*

fibres and vessels, and bounded by little ligaments and tendons of their own. If then there are such numerous orifices in the middle of the rugæ and furrows of the stomach; and if the furrows, when folded and compressed, filter and express the purer chyliferous humor; is it reasonable to suppose that at the instants of the compression the fine fluids exhale into the cavity of the stomach, or are sent down into the intestines? Must they not rather be insinuated by the force then in operation, into the orifices close to them; and thus inwards, along the intestinular ducts, as their appointed channels?

(*t*) That is to say, they lay open the very salivæ, and the tinctures and essences suitable for rectifying their essences, and at last set free the spirits. See n. 94 (*a*).

(*u*) It is impossible to deny that nature advances progressively in all her operations; and preëminently in the process of nutrition; where at first she comminutes the food imperfectly, next dissolves its connexions more thoroughly, and a third time, more thoroughly still; until, indeed, she penetrates to its very leasts and simples; by which alone the blood will permit itself to be impregnated and nourished.

(*x*) If we consider the motion of the stomach from its very origin, we shall see its stupendous nature; for it may be varied in infinite ways, that is to say, it may form spires or volumes, either great, broad, deep, derives its prime vital fuel, increase and fire (*y*). Thence the *delay*, for the maceration of the food, derives its times and successive seasons. For as the food is successively transmitted through finer and finer gyres, meanders and strainers; moistened, impregnated and penetrated by purer and purer menstrua; expressed to dryness; and the process repeated again, again, and again,—its intimate essences are set free and extracted by an occult force. These operations, collectively, are termed *digestion*. The humors resulting therefrom are called *chyle* and *chyme;* (whence the term, chymistry): the menstruum that decomposes the food, is the *gastric juice:* the warmth, the *calor digestivus*. The delay causes *maceration*, and this generates a peculiar *fermentation* and incipient *putrefaction;* and hence *rancidity, sourness*, and *fœtor*. The motion is sometimes termed *peristaltic* and sometimes *vermicular*.

104. When the stomach is in its vermicular or peristaltic motion, whatever its rugæ, circles or spires then lay hold of—whether dry or moist, heavy or light—it seizes, enfolds and carries away; and conveys through its mazy convolutions, in a rapid stream, to the pylorus. Thus the food committed to the

and long; or small, narrow, low, and short; also, assume every geometrical curve which the circular form comprehends: wherefore no determinate figure can be assigned to its gyrations; but the figure is adapted to the circumstances. There are infinite articulations and internodial spaces, whereby it is enabled to body forth its spires and roll its waves, in every possible way that is intended or required.

(*y*) That the heat of the body is hidden in the inmost parts, and proceeds thence by degrees to the outmost; that far within there is a gentle warmth, (if warmth it may be called,) kindly fomenting the interiors, and lighting up the fires of all the workshops scattered through the system, see my *Economy of the Animal Kingdom*, Tr. 1, n. 80—85. Hence when the saliva is laid open all the way to the animal spirit, this heat comes forth more vividly, and gently combines with the inmost parts of the food also similarly laid open, and evaporates them rapidly through the apertures and pores towards the veins, or other places of destination. Thus, as there are progressive degrees of motion, and of all the operations mentioned above, so there are also progressive degrees of heat, from the vital warmth of the spirit, to the sensible heat in the blood, in the humors, and in the organic structures themselves.

stomach, when once embraced in its folds, and launched in the whirling gyre, no longer has any law, tendency or gravity of its own, but is at the absolute disposal of the stomach; and falls headlong whither the spire draws it, upwards, downwards, obliquely, forwards or backwards, indifferently, (speaking according to our perceptions in the general sphere;) for the centre of the motion is the centre of gravity, to which the food now tends and advances. This is a property of circular forms, and in even greater perfection of the other higher forms, which assume relations like those of the great sphere of nature. And hence the food in the stomach is conveyed, with scarcely any effort, to the pylorus, although it rises considerably above the fundus of the viscus; and with equal ease to a still higher orifice, and upwards, through the œsophagus, to the fauces, when the spire rolls backwards, and the axis is reversed (*z*). The same thing goes

(*z*) It is well known what perplexity has been occasioned, by the circumstance of the stomach throwing the food continuously through the pylorus, although the orifice of the latter is considerably higher than the fundus of the stomach; and thus in an upward instead of a downward stream. But the truth is, we are deceived in the idea, that the same centripetency which governs bodies in the general sphere, must also extend to the stomach. For when we are examining the organic body, we ought entirely to put aside all ideas taken from without, and to attend solely to the parallel consequences of peculiar causes within. Certain it is, that the food in the stomach rises not only above the horizontal level to the intestinal orifice, but also to the œsophagus, and through it perpendicularly to the fauces, as in vomiting: that in the intestines, likewise, it passes upwards and downwards, and in all kinds of oblique directions; in the ruminants also, from one stomach to another; from the great stomach to the reticulus, and from the latter to the tongue and the teeth, up and down, at the will of the animal. The Maltese drunkard, mentioned by Schurig, could at will reject from his mouth, in the form of a fountain, by successive cupfuls, all the liquid he had before taken. (see n. 93.) It is a law in the animal microcosm, that it simulates and imitates the macrocosm in all its properties, particularly in the direction of corpuscular substances to a centre; not the common centre of the macrocosm, but a peculiar centre or centres of its own. Thus whatever the stomach receives, it withdraws from the powers of the great world, and subjugates and appropriates to itself. In on more perfectly in the parts, with their little gyres and vortices; which, as we observed before, in respect of function, although not of structure, resemble so many lesser stomachs.

105. The stomach sends its spirits, tinctures, milky products, phlegm and refuse, through various doors, into receivers, far and near: the purest and most refined immediately to the cerebrum and cerebellum, and their exquisitely organic substances (*a*); the next purest, directly into the veins and the

the present case, with respect to the food once taken, the stomach so subdues it to itself, that it no longer has any laws, gravity, or tendency of its own; as we before observed of the food, when speaking of the tongue, the throat, and the œsophagus. (n. 69.) In order to produce this effect, the stomach assumes a perpetual-circular motion, and along with it, all the causes of centripetency; in exact imitation of the great sphere of nature: thus all that passes to the centre of this motion, flows downwards; all that passes from that centre, flows upwards; other things respect these two points obliquely. These views are supported and proved by an infinity of experimental phenomena. Thus when any cylinder, wheel, ring, or other circle, is driven round, it carries ponderable bodies with it in its revolution; and if pieces of thread or small feathers be disposed on the axis of a rotated vessel—a glass, for instance—they become erect, and point like spicula from the centre towards the circumferences. Again, saline equally with aqueous particles, flow backwards and forwards through the surface of bubbles, as in one continued horizontal plane, without any tendency to fall to the centre of the sphere. The same property prevails constantly in all gyres, and if in those which describe a circle, much more in those which describe a spiral, or perpetual circle, as in the stomach. The centre of its common motion is the pyloric orifice, the last boundary of the spirals, whither they all incline their arcs and bend their courses. The extreme and unrestrained velocity wherewith they hasten and fly thither, was alluded to above, (n. 99, *m*); also, that the stomach can form rugæ of different sizes, and gyres with different curves and directions. (n. 103, *x*.) Thus whatever the stomach embraces in its spires, broad or narrow, it deprives of all its gravity; and so subdues, that itself is the agent, but the food the patient: by which means all things in the animal microcosm proceed properly, according to the order of nature. (see n. 69, *z*; 81, *l*.) Such I believe to be the true solution of this difficult problem.

(*a*) In my chapters on the skin and the brain, I intend to prove by blood (*b*); the next, into the cellular coat (*c*), and thence either

many arguments, that the most subtile elemental essences, designed for the service of the purer blood and of the animal spirits, ascend to the cortical substances of the brain, through the imperceptible pores and channels of the skin, and enter those substances; that is to say, they pass through what may properly be called the corporeal fibres, which construct the internal coat of the arteries, and ultimately the cortical substances themselves. At present, however, I will not presume to enter further into these considerations. That similar halitus and vapors are also sublimed from the stomach, is rendered probable by a number of phenomena; by what we find both respecting sudden changes of the disposition and the mind consequent upon taking food, and respecting sudden changes of the blood and the body, as in hypochondriasis and similar affections. And it is to be remarked, that certainty in these matters is not to be expected from ocular evidence, or from seeing the channels and tissues themselves, but more clearly from effects, and sudden affections of the interior faculties.

(*b*) Three arterial and venous networks are mentioned by those anatomists who have examined the intimate structure of the stomach with the greatest diligence and success: one network on the external coat of the stomach; a second, communicating with the first, on the nervous coat; and a third, which is said not to carry red blood, but a juice similar to chyle, in the internal coat. "The first and largest network," says Winslow, "lies between the common and the muscular coats, supported by the cellular tissue. The second lies on the surface of the nervous coat, and is a production of the first, being formed by short offsets from it, which pass through small interstices in the muscular coat. By injection we can shew a third network of vessels, between the granules or papillæ of the internal or villous coat. These do not seem in the natural state to be genuine blood-vessels" (n. 88). Ruysch discovered similar networks. External sense, in its utmost acuteness, may indeed be unable to discover any orifices here; but when we consider that reticulations and nooses of vessels are placed to meet the vapor of the chyle, wheresoever it penetrates; that these vessels communicate with each other; above all, when we consider that there undoubtedly are purer essences in the chyle; and that veins are present in all parts, which, agreeably to their habit and nature, may absorb those essences; and that these veins are much more numerous and much larger than the arteries; and that the chyle exists for the sake of the blood, and constitutes its serum, and ultimately forms the blood globules themselves,—when we consider all these circumstances, our into the capillaries (*d*), or the lymphatics (*e*), and so into the receptaculum chyli: it appears to take back the residue from the cellular coat, and to ruminate it (*f*). The less defecated mat-

intimate sense cannot help concluding, that the purer principles of the chyle are insinuated into the blood immediately. Heister therefore suggests that the stomach, among its other uses, besides expelling the food in a dissolved and altered state to the pylorus, "also perhaps absorbs its finer portion" (*Comp. Anat.*, n. 209): and this is confirmed by all those changes which befall the blood suddenly, and of which we shall speak in our pathology.

(*c*) I am led to believe this, by the numberless orifices in the internal coat, (see Ruysch's Tabulæ,) which when the folds of the stomach are compressed, cannot possibly fail to imbibe fluid, and to drive it home into the loculi and cells of this coat. These cells also may be inflated, like the analogous cells in the cellular coats of the intestines, of the peritonæum and of the pleura: and in many diseases, particularly in hydrops ventriculi, it is common for them to be much distended; which seems to shew that they constitute a reservoir for some kind of lymph.

(*d*) "The nervous coat," says Ruysch, "is much more vascular than the other coats; this is remarkable; and would seem to indicate, that the extremities of the arteries perform some office in this coat, which is not performed in the other coats of the stomach," (n. 89.) The office here alluded to appears to consist, in absorbing the fluids which traverse the tubuli; for according to the authority and experience of Ruysch, the ends of the arteries construct these tubuli, and also their sources, the glands; likewise, the vascular network of the cellular coat, which communicates with the network of the nervous coat. See Winslow above, note (*b*). But understand that I am now speaking of the venous network, not of the arterial one; unless indeed the arteries be of the same nature here as in the intestines, and communicate their blood reciprocally.

(*e*) The lymphatics lie on this coat, and have their roots in it; as also in the peritonæum, with which this coat of the stomach is continuous. "The lymphatics of the stomach," says Heister, "go to the receptaculum chyli," n. 87.

(*f*) This follows from the first position, that the cellular coat imbibes the serosities from the stomach; also from the observation, that when the membranes of the stomach are squeezed, a pituitary or gastric humor exudes; and from the circumstance, that during long fasting, ters, which require further purification, it sends through the veins, to the liver (*g*). The refuse, which is not properly exhausted, yet nearly effete, goes through the pylorus into the intestines, there to be again subjected to a motion similar to that of the stomach, and to be laid open and broken up by the cystic and hepatic bile. Thus the stomach conveys supplies to the body, and particularly to the blood, by various passages, appointed and constructed with stupendous art; exerting the most provident care, lest our ultimate or bodily life should be dependent to a dangerous degree, on a single and slender thread, like the thoracic duct.

106. The stomach also possesses most exquisite sense (*h*), which is conveyed by the fibres of the par vagum to the cerebellum, but not to the cerebrum, and consequently not to the mental consciousness. That nerve faithfully announces all the states of the stomach, and all their changes, immediately to the soul:—all changes, I mean, arising from either the quantity or quality of the ingesta, solid or fluid, and of the salivæ;—from the want and the need of bodily renovation, and the consequent appetency of the blood, the spirits, and the humors; in a word, of the whole body;—from inclinations excited by the influences of the blood, of the external senses, or of the mind itself (*i*):

the stomach sometimes consumes the fat of the adipose tissues of the omentum and of the viscera. The adipose and cellular tissues of the viscera, of the œsophagus, the intestines, and the peritonæum particularly, are continuous with this coat of the stomach, and regard it as central to them: see what we said above respecting the motion of the stomach, (n. 99, *m.*) But this proposition must await the result of more diligent enquiry.

(*g*) All the gastric veins go to the liver; also those of the intestines, the omentum, the pancreas and the spleen; for the reason of which, see the chapter on the liver. In order for the blood to convey the chyle thither, it is absolutely necessary for the veins and the arteries to take this course.

(*h*) For the internal, or villous coat, according to Malpighi, Ruysch, Santorinus, Winslow, and others, abounds in the same kind of organic papillæ as the internal coat of the œsophagus, and the external coat of the tongue.

(*i*) This is the origin of malacia, pica, and indiscriminate appetite

—from the condition of the stomach and its coats, consequently of its fibres, vessels, glands, papillæ; texture, form and motion (*k*): likewise, from the condition of the other viscera, and parts of viscera, which either touch or are connected with the stomach; also from period of life, temperament and diseases. The soul is exquisitely conscious of these things; and if they be disordered, she immediately takes the field, and by the instrumentality of the cerebellum, renews and re-establishes all their ties and harmonies, particularly during sleep, when the mind sleeps, or is bereft of the will.

for improper as well as proper food, (see the cases brought together by Schurig, n. 93:) likewise, of nausea, and of antipathy to particular kinds of food: also, of likings and dislikings, from some previous imagination, or from some opinion or knowledge of the wholesomeness or unwholesomeness of certain species of aliments or medicines. For whenever any state or animus is induced on the brains, a corresponding one is also induced on the fibres; therefore on the viscera, which are contextures of fibres.

(*k*) For example, if the stomach be unduly expanded, relaxed, or coated, so that the lesser structures, that is to say, the little stomachs, cannot perform what they have to do, properly, then indigestion, crudity and other bad consequences arise, and ultimately vitiate the mass of the blood. To say nothing of an infinity of similar things.

CHAPTER V.

THE INTESTINES.

107. Heister. "The intestines are large and long membranous canals, reaching from the stomach to the anus. In these we are to observe their length, which is usually six times as much as the height of the man they belong to. Their wonderful circumvolutions. Their connexion, by means of the mesentery, with the lumbar vertebræ. Their number; for though properly one, they are usually regarded as six. Three of these are called the small intestines; the first of which is the duodenum, so called from being about the length of twelve [duodecim] fingers. This intestine commences at the pylorus; it first ascends a little, then descends, and afterwards ascending again, it runs transversely towards the left kidney. At the distance of three or four fingers from the pylorus, it receives by a single prominent opening, the orifices of the ductus cholidochus and of the ductus pancreaticus, for taking up the bile and the pancreatic juice. Its coats are thicker, and its cavity larger, than those of the other small intestines. In its beginning, it has none of the valvulæ conniventes Kerkringii, or rugæ, but in its continuation it has great numbers, called by authors, juga. It has also great numbers of the glandulæ Brunneri, serving for the secretion of a thin fluid. Its arteries are from the cœliac, and its veins, like the rest of the intestinal veins, from the porta. The second of the small intestines is the jejunum; so called, because it is usually found empty, owing to the fluidity of the chyle, the greater stimulus of the bile in it, and the abundance of its lacteals. It is placed in the region above the umbilicus; and has a great many valvulæ conniventes. It begins where the duodenum ends; and it terminates at the part where the valves are obliterated. Hence I have observed that its length varies in different subjects; ranging from thirteen to sixteen spans. It is generally shorter than the ileum. The third of the small intestines is the

ileum; so called from being placed principally below the umbilicus, near the ossa ilei. Its length is various, sometimes scarcely fifteen spans, at others, more than twenty. It begins where the valvulæ conniventes are no longer conspicuous, and ends where the large intestines begin: in which place it is inserted into the left side of the colon, in a remarkable manner, so as to form a valve, which is called the valvula coli, Bauhini. It has no valves of its own; but its glands are in general more numerous towards its termination, than in any other part.

108. "There are three large intestines, called from their size, intestina crassa. 1. The cœcum, which resembles a bag, with a vermiform appendage. It is placed at the right os ileum. It begins at the termination of the ileum, and terminates at the bottom of the bag which it forms. Its length is not more than three or four finger-breadths. In the appendix, which generally opens into the side of the cœcum, there are some glands, which together with its usually erect situation seem to shew that some fluid is secreted there. In fowls, &c., this appendix is double; in a great many kinds of fish, there are multitudes of appendices; but in man, the organ is sometimes wanting. 2. The colon, which is placed at the circumference of the small intestines, and forms singular and various flexures, differing in different subjects. It commences above the termination of the ileum, and ends at the sacrum. It is connected, by means of the peritonæum, with the os ileum, the right kidney, the gall-bladder, the liver, the omentum, the stomach, the spleen, the left kidney, &c. Its length varies from less than five to more than seven spans. It is the most capacious of all the intestines. It has three ligaments, running over it longitudinally, and terminating in the vermiform process. It has certain external, adipose appendiculæ, which serve to lubricate the intestines. The valvulæ conniventes are largest in this intestine, and its coats are stronger than those of the small intestines. 3. The rectum, so called from lying almost straight [rectus] upon the sacrum. It is about three hand-breadths long; about three finger-breadths in diameter. It commences opposite to the last of the lumbar vertebræ. Its extremity, called the anus, is furnished with three muscles; a sphincter, for closing it, and two elevators, for drawing it up *(Comp. Anat.*, n. 210). The sphincter which closes the extremity of the rectum, arises from the bottom of the os coccygis, and from the skin under this bone, and its fibres ascend, separating from each other, and surrounding the anus on all sides, and in the male subject are inserted into the lower part of the bulb of the urethra, or into the accelerator muscles that surround the bulb; in the female, into the lower part of the vagina. Many fibres also descend from the interior and lower part of the os pubis, near the symphysis, on both

sides; and forming a kind of oval loop, about an inch in breadth, they surround the extremity of the rectum, and constrict and close it. There are scarcely any such purely circular or annular fibres in the sphincter, as are usually ascribed to it. The levatores ani arise on each side, with a broad beginning, from the internal part of the os pubis, from the sheath of the obturator internus, from the inside of the os ileum, and from the sharp process of the ischium: from these parts, the fibres radiate, as from circumference to centre, towards the sphincter, and unite in the hinder part of the intestine, which they surround; including, at the same time, the neck of the bladder, the prostate glands, and the vesiculæ seminales, in the male subject; and the vagina, in the female. They are inserted partly in the upper and posterior portions of the sphincter; partly, they blend their fibres with the oval and annular fibres of the same muscle, and so they may serve not only for sustaining and elevating the anus, but also for compressing the vesiculæ seminales and the prostate glands, in the act of coition. Besides these levatores, which Douglas calls levatores majores, the author we have just mentioned, and Riolan, describe a pair of levatores minores; which arise partly tendinous, partly fleshy, from the tuberosity of the ischium; and thence cross transversely towards the anus, to be inserted into the sphincter, not far from the bulb of the urethra. There are also two muscles of the os coccygis, one on each side, called coccygæus, which seem to assist the anus. They arise from the sharp and posterior apophysis of the ischium, and are implanted into the sides of the coccyx. These muscles assist the levatores ani, by pulling the coccyx; and when it has been forced too much back, they bring it forwards again into its natural state, and prevent it from being readily luxated, or pushed too far back. *(Comp. Anat.*, n. 346.) The rectum is connected to the sacrum, the coccyx, and the bladder, in men; and to the vagina, in women. Its coats are the most thick and fleshy of those of any of the intestines. Like the colon it has no valves, but only certain slight rugæ; this is to prevent the passage of the fæces from being retarded. It has lacunæ, which are often large enough to admit a bristle, and which terminate in little sacs. It is surrounded with a quantity of fat, to the end that it may be easily dilated, in the evacuation of the fæces.

109. "The structure of the intestines is membranous, and they consist of five coats. The first or common coat is membranous, derived from the peritonæum. The second is the tunica cellulosa Ruyschii; which is continuous with the mesentery, and is to be discovered by inflation: in fat animals, this coat frequently contains fat. The third is muscular, composed of a double series of fibres, longitudinal, which

are chiefly seen in the part opposite the mesentery, and annular, which assist the motion of the intestines. The fourth is nervous, furnished with an abundance of glands, vessels and cells, and more capacious than the other coats: for the rugæ and valvulæ belong to it. *(Comp. Anat.,* n. 210-11.) Albinus teaches a way in which the nervous coat may be converted into an elegant new cellular coat, distinct from the tunica cellulosa Ruyschii; also, that the latter is situated between the common external membranous and muscular coats, and does not positively surround the intestinal canal, but is deficient in the part against the mesentery, of which it is an offset: but that his new cellular coat is situated between the villous and muscular coats, forming the true nervous coat, and completely surrounding the canal. He also shews, that the fleshy longitudinal fibres exist only in the part opposite the mesentery; and are not like threads passing all along the intestines, but interrupted, as it were, like the musculi recti abdominis. He observes also that the annular fibres are not inserted into the mesentery as a tendon, but that they contract towards the mesentery. Moreover, that by inflation, all the rugæ and valvulæ of the nervous, or new cellular coat, disappear immediately, and that by this method the villous coat may be separated from the nervous. Also, that the villi of the villous coat, by means of a slight maceration in warm water, may be rendered distinctly visible; and that these villi, especially such of them as communicate with the veins, will be rendered yet thicker, if they be previously well distended with wax injection. *(Comp. Anat.,* not. 11.) The fifth coat is the villous; this sustains the terminations of the blood-vessels and the beginnings of the lacteals: and hence, when nicely examined, it has the appearance of a sieve; it is, therefore, the organ that percolates the chyle. *(Comp. Anat.,* n. 211.) It has on it fine thin villi, resembling the pile of velvet; these are particularly evident in the dog, the hog, &c., after macerating a part of the intestines for some time in warm water, and shaking it well from time to time. Ruysch has noticed the same villi; but on the other hand, Helvetius denies their existence, and contends that this coat consists entirely of spongy, flat papillæ, (he gives a delineation of them, as they appear under the microscope,) of an irregular shape, and which absorb the chyle by their small orifices, and convey it to the lacteals; and for this reason he gives this coat the new appellation of membrana papillaris. Helvetius, in the same place, speaking of those fibres of the intestines which writers declare to be annular or spiral, denies that they are either, but describes and figures them as perfectly irregular, and as segments of circles collected into fasciculi; but I have never yet been able to see them as such. *(Comp. Anat.,* not. 12.)

110 "The vessels of the intestines run over their substance in great abundance. They are called the mesenteric arteries. The superior mesenteric artery supplies the small intestines; the inferior, the large intestines. These vessels form wonderful anastomoses. The mesenteric veins go to the vena portæ and the liver. *(Comp. Anat.*, n. 211.) Albinus has observed, that the veins of the intestines may be filled with injection through the arteries, and the arteries through the veins; and that it may be forced through either into the cavity of the intestines: and that in this case, the wax is found adhering to the villi of the inner coat, in the form of little worms: a fact which gives rise to singular speculations. *(Comp. Anat.*, not. 11.) The nerves of the intestines are from the par vagum and the intercostals. They have lacteal vessels, or lymphatics. The rectum also receives vessels from the hypogastric arteries. Besides the Brunnerian glands of the duodenum, there are the glandulæ Peyeri in the small intestines. These glands are usually small, congregate, and miliary; sometimes they are single, or solitariæ. They are larger as they are nearer the duodenum; and smaller as they approach the large intestines. They deposit a liquid in the intestines. In the large intestines, and in the vermiform appendage, there are glandulæ solitariæ, large and lenticular glands, and which are largest of all in the rectum. They have orifices, which excrete a liquid to lubricate the fæces and the intestines." *(Comp. Anat.*, n. 211.)

111. WINSLOW. "The curvatures of the intestinal canal form two arches, a small one, by which it is connected to the mesentery and mesocolon; and a large one on the opposite side, which lies loose. *(Exp. Anat., Tr. du Bas-Vent.*, n. 89.) The duodenum, immediately after it has arisen from the pylorus, is reflected a little backwards, and obliquely downwards; then it passes towards the right kidney, to which it is more or less connected; thence in front of the renal artery and vein, and of the vena cava, ascending insensibly from right to left, till it comes before the aorta, and the last dorsal vertebræ. It continues its course obliquely forwards, making a gentle turn, which may be reckoned the third incurvation and also the extremity of the duodenum. Through this whole course, the duodenum is firmly bound down by folds of the peritonæum, especially by a tranverse duplicature, which gives origin to the mesocolon. The two laminæ of this duplicature, being at first separate, and soon afterwards uniting, naturally leave between them a kind of triangular space, which is lined with cellular tissue. In this space it is that the duodenum adheres, by means of the cellular substance, to the parts already named: and it is so encased therein, that without dissection, we can see nothing but

its two extremities, and even these are hid by the colon, and by the first convolutions of the jejunum. The first coat of the duodenum does not invest the whole circumference of the intestine, because it lies in the triangular space: and for the same reason, there is more cellular tissue on it, than on the external coats of the other intestines. The second or muscular coat of the duodenum is thicker than the same coat in the jejunum and ileum. The tunica nervosa and villosa form conjointly on the inside of this intestine, a great number of small duplicatures, which advance into the cavity more or less directly, like circular ligaments, with one border fixed to the intestine, the other, free: these are the valvulæ conniventes. The floating edge of these valves is formed into small gathers in the natural state; and the entire surface of the valves is villous, as well as that of the interstices between them. The villous coat of this intestine is thicker than that of the stomach; but its substance does not so much appear to be villous, as fungous and granular, composed of an immense number of fine papillæ of different figures, in which the microscope discloses a multitude of depressed points and pores, by which their whole surface is pierced: and also, throughout the villous coat, small villous eminences or tubercles, rising like little warts, at different distances from each other. This substance supports a vast number of capillary vessels; for besides the blood-vessels, we observe a quantity of whitish filaments, which run through it, and end at its inner surface, like so many capillary roots of the vessels called lacteals. The spongy substance, which binds these capillary filaments together, and surrounds them, is very tender; and the capillary extremities of the vessels distributed through it, seem to be turned towards the pores of the papillæ. Through these pores is continually discharged a mucous fluid, which moistens the cavity of the intestine. The internal surface of the duodenum is furnished besides with a great number of flat granular or glandular bodies, raised circularly on the sides, and depressed in the middle: and they are more numerous near the beginning of the intestine, than in any other part of its course. Towards the pylorus, they lie in clusters, and from thence the distance between them increases all the way to the other extremity, where they are single. These glands, when examined carefully, appear to be follicles, with their orifices turned towards the cavity of the intestine, and their bodies sunk in the fungous substance, next the nervous coat. They furnish a peculiar fluid, which is often viscid and glutinous. On the internal surface of the duodenum, almost at the lowest part of its first curvature, there is a longitudinal eminence, which ends like a beak in a peculiar aperture; this aperture is the orifice of the biliary duct; within which, the pancreatic duct likewise opens.

This intestine is invested by a greater quantity of cellular substance than the other intestines, especially while in its triangular case, where it is not entirely surrounded by the membranous coat, like the other intestines.

112. "The jejunum commences from the last curvature of the duodenum, and is there connected to the beginning of the mesocolon: from thence it bends downwards from left to right, away from the dorsal vertebræ, and makes several convolutions, which lie in the upper part of the umbilical region. Throughout this course, it is connected to the mesentery. It is generally distinguished from the duodenum by its red color, and the number of its valvulæ conniventes; and also by the difference of its situation; but the best way is, to divide it and the ileum into five equal parts, and to reckon two parts for the jejunum and three for the ileum. The coats of the jejunum are thinner than those of the duodenum. The common, membranous or external coat, is a continuation of the mesentery: its cellular tissue is less in quantity than in the duodenum; and indeed seems to be altogether wanting along the great curvature of its convolutions, where the longitudinal fibres of the muscular coat adhere closely to the membranous coat. The muscular coat is not so strong as that of the duodenum. The longitudinal plane of fibres is very thin and almost imperceptible, excepting along the great curvature, opposite to its connexion with the mesentery, where we see through the membranous coat, a kind of whitish ligamentary band, about four or five lines in breadth, which is continued along the great curvature of all the convolutions of this intestine, and of the ileum. This band or fascia adheres very firmly to the common coat of the intestines, and to the longitudinal fibres of the muscular coat, which are here more visible and stronger than in any other place. The nervous, or more properly speaking, the reticular coat, with its cellular or lanuginous tissue, is just like that of the rest of the intestines. By artificial inflation, it may be made to swell so much, as to obliterate all the valvulæ conniventes, and to take up the entire extent of the coat into the cavity. These duplicatures or valves in this intestine are very broad, very numerous, and very near each other. On the side of the great curvature, their circumference is continuous and uniform; but next the small curvature there are breaks in them, the extremities of some advancing beyond the rest, and terminating in points. Some of these valves go quite round, others only partially so, and some of them are very small, and form oblique communications between the larger ones. The papillæ of the villous coat are here more raised and wavy than in the duodenum; and each of them seems to be divided into several others, in a peculiar manner.

They have been very accurately delineated by Helvetius. The glandular lacunæ of the jejunum, are of the same structure as the glandulæ Brunneri or duodenales; but they are disposed in a different manner: being partly single or solitariæ, partly dispersed, and partly in clusters, like flat, oblong bunches of grapes; these clusters are called, plexus glandulosi Peyeri. They are in the largest quantity near the great curvature of the intestines; and they cross through several valvulæ conniventes at once.

113. "The convolutions of the ileum surround those of the jejunum, and pass from left to right, terminating beside the right kidney: its lateral convolutions are supported by the os ileum. Its valvulæ conniventes decrease gradually in number and size, and at last, instead of being transverse or circular, they become longitudinal. We observe likewise in this intestine solitary glands or lacunæ, and also reticular glands, or glands in clusters; some of which of large extent are found at the termination of the intestines; but the glands are flatter here than in the jejunum. The cellular substance is in less quantity here than in the preceding intestines, and this intestine is less red than the jejunum.

114. "The intestinum cœcum and valvula coli. The cœcum is no more than the extremity of an intestine, being only a kind of round, short, broad bag, with its fundus beneath, and its mouth or opening above. It lies under the right kidney, and is hid by the last convolution of the ileum. It is about three finger-breadths in length, and its diameter is nearly double that of the small intestines. *(Exp. Anat., Tr. du Bas-Vent.*, n. 105—138.) At the place where the cœcum joins the colon, a portion of the circumference of both is depressed, and forms on the inside a large fold, which advances into the cavity of the intestine. It is open in the middle, and its extremities are very thick, by reason of the mutual duplicature of the cœcum and colon. The extremity of the ileum is as it were grafted in the opening of this fold, and strongly united to its sides by the mutual adhesion of the transverse fibres of the two intestines. The ring, which advances into the common cavity, is formed into gathers on the inside, almost like the lower extremity of the œsophagus, the pylorus, or the inside of the anus. Its circumference is nearly oval, and forms two prolongations, which Morgagni calls the fræna valvulæ coli. The membranous coat of the extremity of the ileum is continued on the cœcum and colon, without sinking into any fold, at the place where the ileum enters the colon. The longitudinal fibres of the muscular coat seem here to be confounded with the nearest circular fibres of the cœcum and colon. The nervous and villous coats of the ileum likewise enter the cavity, and meet those

of the cœcum and colon on the edge of the ring, so that the circular rising, or short muscular tube, is covered both on its concave and convex sides with a nervous and a villous coat; the concave side being supplied by the ileum, the convex, by the large intestines. *(Exp. Anat., Tr. du Bas-Vent.,* sect. viii., n. 156—161.)

115. "The vermicular appendage is seen on one side of the fundus of the cœcum, resembling a miniature intestine. It is about three lines in diameter. By one extremity it opens laterally and a little obliquely into the fundus of the cœcum; the other extremity is closed, and is sometimes narrower, sometimes wider than the rest of the appendix. It has some contortions, like those of a worm when it is touched. Its internal coat is plentifully furnished with follicles, almost like that of the duodenum; and it is likewise reticular, the meshes being the glandular lacunæ, which continually discharge a fluid into its cavity.

"Through the common coat of the cœcum, we see three white, ligamentary bands, which adhere very closely both to the common and muscular coats; one of them is hid by the adhesion of the mesocolon, and all the three divide the cœcum longitudinally into three nearly equal parts; and unite together on the vermiform appendix, covering its whole outer side, immediately under the common coat. They are made up interiorly of fleshy fibres, which strengthen the longitudinal fibres of the muscular coat. The villous substance of the internal coat of the cœcum, is very short, besprinkled throughout with glandulæ solitariæ, larger than those of the small intestines, and flattened and depressed. When we blow through a pipe into these lacunæ without touching them, the folliculi are inflated, and represent little caps, with a hole in the middle of their convex side.

116. "The colon. The whole convex side of the colon is divided longitudinally into three parts, by three bands, which are continuations of those of the cœcum. Two of these bands run on the sides, along the great curvature of the colon; the third, along the small curvature. They perform the office of longitudinal fræna, between which the intestine is compressed into transverse folds and tuberosities. All the folds are duplicatures, which form valvulæ conniventes in the cavity of the intestine, and the tuberosities form so many cells. All the coats concur equally to the formation of these duplicatures and cells, the depth of which decreases gradually towards the extremity of the intestine. The common coat of the colon on one side is a continuation of the mesocolon, and on the other side it contributes by the same continuation to form the epiploon or omentum. The longitudinal fibres of the muscular coat are slender, and the annular ones are mere segments. The glands are larger and more numerous than in the cœcum. The manner in

which the colon is connected to the duodenum, to the right kidney, and to the gall bladder, is very remarkable. Along the great arch and the two last incurvations, there are a kind of fatty fringes, called appendices coli adiposæ. *(Exp. Anat., Tr. du Bas-Vent.*, n. 139—155.)

117. "The rectum appears to descend from the lumbar vertebræ, in front of the sacrum, to the end of the coccyx, where it ends in the anus. Properly speaking, it is a continuation of the colon, though it takes a different course from that intestine. When empty, it is irregularly cylindrical, and sinks into transverse folds. It is capable of much extension, even to the size of a large bladder, so as to represent a kind of stomach. The ligamentary bands increase in breadth on this intestine, and approach each other. The muscular, and nervous or filamentary coats, are much more capacious here than in the other intestines; and hence when the rectum is empty, they fall into folds and rugæ. The internal coat exhibits a plentiful sprinkling of glandulæ solitariæ, and it is always moistened by an abundant mucus. Near the extremity of the intestine, the rugæ become in a manner longitudinal, and at last, towards the circumference of the inner margin of the anus, they form semilunar lacunæ, the openings of which turn upwards, towards the cavity of the gut. These lacunæ are something like those at the lower extremity of the œsophagus, or upper orifice of the stomach. *(Exp. Anat., Tr. du Bas-Vent.*, n. 169—177.)

118. "Vessels. The artery of the duodenum, called the duodenalis, comes off sometimes from the coronary artery of the stomach, sometimes from the pyloric, and sometimes from the great gastric and hepatic arteries; and the superior mesenteric and splenic arteries also supply several small branches. The duodenal artery, and the other accessory branches, form a network round the muscular coat of this intestine, and send out a great number of capillary branches towards both the outer and inner sides. The veins of the duodenum are branches of the vena portæ, and are distributed in almost the same way as the arteries, except that they anastomose more frequently with each other, and with the hæmorrhoidal vein. The arteries of the jejunum come principally from the superior mesenteric artery; the veins are for the most part branches of the great mesaraic vein. The small mesaraic or hæmorrhoidalis interna, and the splenic, also supply some branches. The principal trunks of these arteries and veins accompany the cellular tissue between the laminæ of the mesentery, and there ramify and form belts and arches. Those which are next to the intestines form two vascular planes, which separate from each other distinctly, and surround the intestinal canal in a reticular manner. It is also worthy of note, that as the arteries and veins pass through the mesentery, they supply

branches to its glands, laminæ, and cellular tissue; and also that there is a kind of communication and meeting between many of the small mesaraic veins and the lumbar and spermatic veins. Of the vessels of the ileum the same may be said as of those of the jejunum. The arteries of the cœcum and vermiform appendix are ramifications of the last branch from the convex side of the arch of the superior mesenteric artery; the second and third branches also supply some twigs. The same may be said of their veins. The colon is supplied with arteries by the second branch from the concavity of the same arch, and sometimes by the third, and also by the first: and thus the singular common arch of the two mesenteric arteries is formed. The sigmoid flexure of the colon is supplied by the other branches of the inferior mesenteric artery. The veins of the colon are all ramifications of the vena portæ, and principally of the subordinate trunks, the great and small mesaraic, or the internal hæmorrhoidal veins. The arteries of the rectum come from the internal hæmorrhoidal artery, the last branch of the inferior mesenteric, which communicates with the hypogastric, and, in a peculiar manner, with the external hæmorrhoidal artery. The veins communicate with the external hæmorrhoidal veins, as well as with the hypogastric veins, which are distributed to the internal parts of generation in both sexes. There is a successive continuation between all the arteries of the intestinal canal, and likewise between all the veins; which latter are, as usual, thinner and more capacious than the arteries, and more notably so than in other parts of the body.

119. "The nerves of the duodenum proceed from the middle plexus of the semilunar ganglia, and some filaments come from the stomachic and hepatic plexuses. Those of the jejunum, ileum, and mesenteric glands, come from the superior mesenteric plexus, from the post-mesenteric fasciculi, and from the inferior mesenteric plexus. Those of the cœcum, from the inferior mesenteric plexus, and from the post-mesenteric fasciculi. Those of the colon and its arches, from the same fasciculi, and from the inferior, superior, and sub-mesenteric plexuses. The nerves of the rectum come from the inferior mesenteric, sub-mesenteric or hypogastric plexus, and from the ganglia of the same plexus. Those of the anus and its muscles, from the ganglia of the sub-mesenteric or hypogastric plexus, from the lower cords of both the intercostal nerves, and from the common arch at the extremity of those cords." (*Exp. Anat., Tr. du Bas-Vent.*, n. 224—247.) Respecting the cutaneous ligament of the coccyx, the interosseous ligament of the pubes, and the muscles of the anus, see *ibid.*, n. 181—193.

120. LEEUWENHOEK. "In dissecting a certain kind of fly, I was struck with admiration at the intestine, with the chyle passing along it

backwards and forwards by turns; and whenever the chyle was protruded from one part of the canal towards another, the intestine was rapidly closed. This protrusion of the chyle, and closing of the emptied intestine, took place three or four times in a very short period; and the wonder was, that I saw no agitation, or next to none, in the intestine itself. *(Cont. Arcan. Nat. Detect.,* ep. 94, p. 12. Lugd. Bat. 1722.)

121. "I have frequently examined the moth of the silk-worm, as it came from the chrysalis, and more than once I have taken out the intestine, and always remarked with wonder its uncommon motions. At one time this intestine would contract in some particular part, and protrude the small quantity of matter it contained towards its outlet; whither the matter had scarcely arrived, before the intestine would again contract and drive it back. And this reciprocal motion, pushing the matter contained in the intestine backwards and forwards, was repeated so often and so long, in a space not more than the sixteenth of an inch, that my eyes were fatigued with watching it. *(ibid.,* ep. 146, p. 421-2.) I have seen the motions of the intestines in brutes, which motions their intestines retain for a long time after they are removed from the body. *(Arcan. Nat.,* tom. i., p. 61.)

"I have always found, that the supposed pituita in the intestines, is not phlegm, but consists of parts and organs necessary to the intestines. While examining this imaginary phlegm, or as some persons call it, mucous substance adhering to the inside of the intestines, I was astonished at the vast number of fine blood-vessels that I saw, the minute branches of which lay so near to each other, that there was not the twentieth part of a hair-breadth between them; in addition to which there were other vessels, seemingly colorless, and of which it was hard to say whether they were lymphatics or lacteals. To these adhered a kind of viscid and pellucid phlegm or mucus, covered with many most minute globules: and where I endeavored to detach this phlegm gently from the intestine, with a delicate instrument, I frequently found that I had wounded the fine blood-vessels, and also that I had actually detached many of them, together with other vessels, from the intestine." *(Arcan. Nat. Detect.,* p. 54, 55.) Leeuwenhoek also delineates "the extremities of a fibrous substance, which some call pituita, others, mucus of the intestines, but which," says he, "I shall now name, the interior muscle of the intestines." In the figure it exactly resembles the cortex cerebri with the fibres hanging from it; inasmuch as a kind of fibres proceed from all its parts. *(ibid.,* fig. 7, p. 61.)

121. SWAMMERDAM. "In the sepia, or cuttle-fish, the straight-gut, or rectum, issues immediately from the stomach, and is the only

intestine I could discover in the animal: so that the veins must take the aliment immediately from the stomach. Under the rectum lies an appendage of the stomach, and which opens into the stomach by a particular orifice, after making a very remarkable spiral revolution, like a snail-shell. I take this to be the pancreas, which appears of the same construction, except its not being convoluted, in a great variety of fishes. (*Bib. Nat.*, p. 890, 891, tab. li., fig. 5.)

"In the worm of the acarus, the stomach succeeds the œsophagus; it is very long, and supplied with a great number of ramifications from the windpipe: and in appearance, it is like an intestine. Towards the termination of this small canal, there rose from it two little slender intestines or cæca, like vasa varicosa: these two then divided, each into two others; two of these contained a greenish-yellow substance; and the two others, a matter that was partly green and partly white, and that looked as if it was coagulated. The motion of this substance through the intestines, was so quick, that my eyes could not keep pace with it: from this motion, we may reasonably infer, that these intestines are furnished with spiral muscular fibres. The pylorus is seen below the insertion of these four intestines, and close under it is the colon, which is followed by the rectum. (*Bib. Nat.*, p. 702, 703, tab. xliii., fig. 5.)

"In the day-butterfly, after the stomach, or under the pylorus, there appear six intestinula cæca, or vasa varicosa, which are much more slender in this state than they were in the caterpillar, and of a perfectly different form. Under these lie the small intestines, which are transparent, and full of a globular substance. A little lower, the gut widens considerably to form the cloaca; then it contracts again, to dilate a second time into a lesser sinus, in which it terminates. Next follows the rectum, ending in a ring, of a substance between bone and horn, which forms the anus, is covered with hair, and drawn up within the abdomen. At the sides of the anus appear its proper muscles, &c. (*Bib. Nat.*, p. 595, 596, tab. xxxvi., fig. 1.) Respecting the wonderful disposition of the intestines in the caterpillar, see *Bib. Nat.*, tab. xxxiv. fig. 4: the description of them, p. 575, 576: two tubuli near the stomach, tab. xxxiv. fig. 5.

"In the common or working bee the pylorus follows the stomach; and after this comes another little part, somewhat more protuberant, and inclining to a color between red and yellow. Next follows an intestine, which is perhaps the colon, and which is much thicker than the stomach, especially when it is full. It has, moreover, strong muscular fibres, which, when they act, convolute it into many wrinkles and folds. Further down, this gut is considerably contracted, and at this spot an infinite number of white filaments are seen; these are closely fastened to the gut; their connexion with which is extremely strong, and is effected by means of the air tubes. The gut, after its contraction, dilates again on a sudden, though here it seems to be altogether membranous and transparent; some little whitish, oblong parts being seen through it from the outside. These are six in number, and are glandular, and stand out very considerably on the inner surface of the gut. Where the colon ends, it is once more contracted, but again dilates, and at length forms the rectum. By laying these parts on a plate of glass, you may see not only their circular fibres, but even their valvulæ conniventes. (*Bib. Nat.*, p. 454, 455.) In the worm of the bee there are four vessels, or intestina cæca, which are inserted just below the pylorus." (*Bib. Nat.*, p. 411, tab. xxiv., fig. 6.)

Respecting the intestines of the cossus, during the metamorphosis of the worm into a nymph, see *Bib. Nat.*, p. 318, 319, tab. xxviii., fig. 5. Respecting those of the cossus itself, or vermis scarabæi, *ibid.*, p. 313, 314, 315, tab. xxvii., fig. 11, 12. Respecting the intestines of the hemerobius, ephemerus or diaria, *ibid.*, p. 248, 249, tab. xiv., fig. 1; tab. xv., fig. 5. Of the cancellus, or hermit-fish, *ibid.*, p. 202, 203, tab. xi., fig. 3. In the viviparous crystalline snail, the branchiæ or gills are placed at the sides of the rectum, which they accompany for a long way into the curvature of the body: if viewed with the microscope, they resemble a comb with broad teeth, though in softness and delicacy they are like little membranes. (*ibid.*, p. 172, 173, tab. ix., fig. 6.) Respecting the intestines of the covered snail, see *ibid.*, p. 123, 124, tab. v., fig. 6, 7, 8. Of the louse, *ibid.*, tab. ii., fig. 3.

122. MARTIN SCHURIG brings together cases from different authors, of the intestines being too narrow; of their having grown together; of their being unduly expanded; of their position being changed, in hernias, and by wounds; of the colon being found in the thorax; of adhesions of the intestines to different parts; of warts, excrescences, schirrus, inflammation, gangrene and sphacelus, of the intestines; of their perforation by lumbrici; of fissures, ruptures, and atrophy of the intestines: of foreign bodies in the intestines, as lumbrici, calculi (*Chylol.*, p. 537—563): and preternatural substances, as stony scybala, a glandular substance weighing seven ounces, a fleshy mass, a membrane, a leaden ball, a piece of a brazen shell, the head of an arrow, money, needles, the point of a sword, a pair of scissors, a diamond, bones, pieces of whalebone, a hog's tooth, the stones of cherries and plums, straw, seeds, hair, leather, worms, a serpent, an eel, a snail, a slug, a caterpillar, beetles, a mouse, a mole, a toad,

frogs, the fœtus of a dog, a human fœtus. *(ibid.,* p. 709—736.) Furthermore, he relates, that the intestines have been found double; the cœcum double; the ductus cholidochus double: that cases have occurred of a diverticulum, sacculi, and knots in the ileum; also, of reduplication of the intestines, in man and brutes: of strangulation of the ileum, wounds of the same; of the cœcum being in the left side; of two or more cæca; of a pistol ball in the cœcum. John Peter Albrecht relates the case of an inhabitant of Frankfort, in 1667, in whom there was a passage from the cœcum into the bladder. (*ibid.*, p. 564—583.) Schurig also enumerates the causes of passio iliaca; among which he mentions inflammation, gangrene and sphacelus, reduplication and folding of the gut, ulceration, &c. He gives cases of hernia arising from hardened excrements; and from the bite of the scorpion *(ibid.,* p. 632—649); cases of excrement passed through the penis in boys, through the vulva in girls; through the mouth, umbilicus, groin, and other parts of the abdomen; through ulcers, wounds, &c. Of purging produced by imagination, smell, sight, effluvia, affections of the mind, rage, terror, fear, thunder, loss of sleep, the smell of purgative drugs, beef-broth, chicken-broth, &c. *(ibid.,* p. 662—702.) Respecting the peristaltic motion, he observes, that according to Mich. Albert, it is contractile, acting on both the length and breadth of the intestine, and momently intermitted and repeated, beginning from the œsophagus, where it is obscure: that anti-peristaltic motions have been observed after death: that Mery saw a motion of this kind, in a female who died in difficult parturition: that others have seen the same in wounded persons; &c. *(Chylologia,* p. 592—596.)

123. Respecting the intestines, the reader may also consult other authors. Respecting the glands of the duodenum, see Conr. Brunn. On other matters, see Santorinus; Grew, *Compar. Anat.;* Albinus, *De Intestinis;* Helvetius; Boerhaave, the action of the intestines upon the ingesta, *Inst. Med.*, n. 90—98; the matter of the fæces, *ibid*,. n. 108—113. See particularly the plates of different authors; as Eustachius, *Tabul. Anat.*, tab. ii., fig. 3, where he shews the intestines and their convolutions; the colon with its ligaments and cells; also the stomach, and a portion of the omentum: fig. 5, 6, representing the wonderful inflexions of the colon, with the cœcum and the vermiform appendage: tab. iii., the intestines, and the vessels of the mesentery. Ruysch *Th. Anat.* vi., tab. v.; *Ep.* xi., *(Exercit. de Gland. Int.,)* tab. xii., fig. 1, 2, 3, 4; *Catal. R.*, p. 149, 154; *Th.* i., Arc. 1, 2; iv., n. 33, 37, 67, 76, 79, 92; v., n. 101, 102; ix., n. 62; x., n. 50—55; *Adv.* ii., Tab. iii., fig. 4. Morgagni, *Advers.*, part iii., fig. 2, 3; shewing the valve of the colon, &c. Heister, *Comp. Anat.*,

tab. ii., fig. 7; vii., 29, 30, where he exhibits a portion of the ileum, cœcum, and colon, with the valvula Bauhini, the vermiform process, the frænum cœci, and the valvulæ conniventes coli, in a new-born infant. Verheyen, *Corp. Hum. Anat.*, tract ii., cap. xi., tab. vi., fig. 2, 3. The several plates brought together in Mangetus, *Theat. Anat.;* tab. lviii., fig. 1—6, from Willis and Verheyen: fig. 9—15, from Ruysch: tab. lix., fig. 1, 2, 3, from Peyer: fig. 4, 5, 6, from Kerkring: fig. 7, from Swammerdam: tab. ii. extra ord., fig. 1, 2, from Conr. Brunn: fig. 3, from J. B. Bianchi. Respecting the muscles of the anus, see Santorinus, *Obs. Anat.*, p. 174, seq. Morgagni, *Adv.*, iii., p. 89. Riolan, *Anthrop.*, lib. 5, cap. xxxviii. Also Douglas, *Myograph.* Cowper, *Myotom. Ref.* Heister, *Comp. Anat.*, n. 346, not. *(a);* where he affirms, that of no muscles of the body are authors less agreed in their descriptions than of the muscles of the anus.

124. The viscera or integral parts of the body may be compared to lakes in a kingdom, which arise from running streams, and empty themselves into the sea by one or several rivers; or to fountains, produced by the confluence of subterranean veins, which go forth in single channels, and afterwards spread in all directions in numerous rivulets. Thus the heart, which is the channel and whirlpool of the blood of the whole kingdom, grows from numberless invisible veins, uniting into larger ones, and presently whirls back its blood into the body, and distributes it universally by the arterial streams (*a*). Thus the lungs, which are composed of the bronchia and the vesicles, pour themselves forth incessantly on every side, either by various currents of force and motion, or else, as in worms, by an infinite number of aëriferous tubes (*b*). Thus the brain, which is a globular

(*a*) That the heart is not only produced by the confluence of the small veins, the living springs of the cardiac lake, but that it is also excited by them into its peculiar systaltic motions, may be seen in my *Economy of the Animal Kingdom,* treatise i., n. 201, 205, 514, 516; and will be farther shewn in Part III. of this work. Thus the heart arises from the veins, as the trunk of a large tree from its roots; and expands into the arteries, as the tree into branches and boughs. Not to trace its origin back more remotely, to the causes of the veins themselves, and their last ramifications and productions to those causes; that is to say, to the fibres of the brain, whence come the fibres of the body, which return in a circle to the brain, and weave its substances.

(*b*) Respecting the origin and composition of the lungs, see Part II., On the Viscera of the Thorax. That their motions pour forth into all points of the body, see my *Economy of the Animal Kingdom,* treatise i., n. 367, 368, 369. But that in insects the lungs pour composition of the minutest arterial threads and cortical granules, puts itself forth universally by the medullary and the nervous fibres (*c*). Thus the stomach, which is made up of a peculiar internal coat of its own, of the tunics of the œsophagus, and of membranes consisting of vessels, fibres, ducts, and papillæ, discharges itself into the intestinal tube, as a lake into a river, or a fountain into a large stream; and the tube, so formed, straightway pursues a course full of winding turns and meanders. This origin of the intestines causes them to be a kind of new, continued, or consecutive stomach (*d*); and to have the same nature as the stomach; that is to say, similar substances, powers and forces, motions and actions, and similar modes of action; for they make common cause with the stomach, and perform a consecutive and proximate part in the series of one and the same function. Since the stomach is thus continued in the intestines, we may derive instruction on nearly all points from the one concerning the other, both with regard to *structure*, with regard to *motion*, and with regard to *chemical action* (*e*).

themselves forth by aëriferous or pulmonary tubes, see Swammerdam's descriptions of the anatomy of insects.

(*c*) These points will be proved in detail, in the parts on the cerebrum, and on the organism of the animal body.

(*d*) What is here said of the intestines is applicable to all the other viscera in the body. Thus the arteries, collectively, may be said to be an elongated, continued, or consecutive heart. The fibres, or their compound, that is, the whole body, may be said to be a continued brain; for the brain is universally present with the fibres. The pulmonary pipes in nymphs, aurelias, chrysalises, and worms, are their continued lungs.

(*e*) Continuity simulates identity, producing apparent presence between the continuous parts. If only we explore the beginning and the progress of any viscus, together with the use, and take anatomy for a guide, and reason for a minister, we shall be enabled to disengage and interpret all that belongs to it, how intricate and secret soever it may be: that is, provided we follow the thread of continuity from one thing to another, and do not blindly turn aside to disconnected and merely collateral matters, nor wildly jump at things which are either above or below the subject of enquiry. In a word, the discernment of universal connexion and continuity, amounts to the discovery of truth.

125. With regard to STRUCTURE: the membranes of the intestines are of the same number and of the same nature as those of the stomach (*f*); and the blood-vessels, muscular and nervous fibres, ducts, siphunculi, glands, papillæ, &c., have a similar origin, nature, and determination, in both (*g*). The convolu-

(*f*) There are five membranes in both the stomach and intestines. The *common*, or external coat of the stomach, is continuous with the peritonæum; so likewise the external coat of the intestines is connected to the peritonæum in many places. The *cellular* coat of the stomach corresponds to the tunica cellulosa Ruyschii of the intestines. The *muscular* coat of the stomach has layers of longitudinal and transverse fibres; so has the muscular coat of the intestines. See Albinus, as cited by Heister, n. 109. In both cases there are also interrupted fibres, describing as it were segments of circles, and which are connected by seemingly tendinous lines. "The fibres of the external and internal planes of the muscular coat of the stomach," says Winslow, "are rather segments which unite at different distances, than complete circles; for they are intersected by great numbers of small white lines, of a tendinous aspect" (n. 88). The same is true of the muscular coat of the intestines. "The [muscular] fibres," says Helvetius, as cited by Heister, "are neither annular nor spiral, but like segments of circles collected into fasciculi" (n. 109). The *tunica cellulosa Albini* of the intestines (respecting which, see Heister, n. 109), is not indeed mentioned as having any coat corresponding to it in the stomach; yet, from an observation of Winslow, it appears very probable that the stomach has such a coat: "The concave side [of the nervous coat of the stomach]," says he, "resembles fine cotton or wool, as may be seen by macerating it a little in clear water, when it swells and tumefies. It is supported by a kind of web or groundwork of fine aponeurotic filaments, which intersect each other obliquely, much in the same manner as in the third [or corresponding] coat of the intestines" (n. 88). The *internal*, or villous coat, in both the stomach and intestines, is covered with papillary villi, (as Ruysch has demonstrated by injection,) but which are somewhat larger in the intestines than in the stomach, and seemingly of a fungoid structure (see Winslow, n. 111); for a reason which will be seen presently. In the stomach, this coat is likewise capable of being expanded into elevated juga or rugæ. "The two [internal] coats," says Winslow, "form large rugæ on the concave surface of the stomach; the major part of these rugæ is transverse, but irregular and wavy" (n. 88).

(*g*) The *arteries* of the intestines are continuous with the coronaria tions also, and erratic windings of both—the rugæ, furrows, and grooves—have nearly the same structure, spire and fluxion; but with this marked difference, that the spires in the concave surface of the stomach, are developed into equal solids in the intestines, and, forming canals and closed in by coats, perform the same gyres in their volume as the stomach performs on its surface; so that it seems as if the stomach merely unfolded itself continuously below the pylorus (*h*). This similarity in structure

ventriculi and pylori; speaking in a general manner, those of both the stomach and intestines are continuous with the cœliac artery, as a common stock; by their relation to which they all belong to the same family. The *veins* of both the stomach and intestines go to the liver. Their *nervous fibres* come from the par vagum and mesenteric plexuses; thus from one and the same trunk, which passing down along the œsophagus, mounts over the stomach; so that the intestines, in all their functions, communicate with fibres which are continued from and through the stomach: besides which there are other accessory fibres, derived from the intercostal nerve, which like the par vagum is a product of the fibres of the cerebellum, and may be called the twin of the former nerve. See my *Economy of the Animal Kingdom*, treatise i., n. 485, 559. Similar *glands* also, miliares, agminatæ and solitariæ, are found in both. Respecting those in the intestines, see Brunn and Peyer; and respecting those in the stomach, see Heister and others. And similar foramina are seen in the villous coats of both. Of those in the intestines, Boerhaave says, "The villous coat is perforated with the orifices of the lacteals, and with large pores distinct from these." (*Inst. Med.*, n. 91.) And Brunn says, "The internal coat of the duodenum is perforated with an infinity of little pores." (*Gland. Duoden.* p. 23. Francof. 1715). See also Winslow, n. 111.

(*h*) The winding folds and perpetual convolutions of the stomach were treated of above (n. 97): they are very conspicuous in plates which represent stomachs inverted and corrugated, as in the Tabulæ of Ruysch, Willis, &c. We shall obtain a clearer idea of the form of this motion, if we compare it with the analogous forms of the cerebrum, and of the intestines. For the same serpentine convolutions are seen in the stomach, as in the cerebrum; and so wonderfully folded, as to elude the gaze that would follow their intricacies, and to confound its distinctness. Thus, while in one place they hide their heads, as though they had terminated, or were stopped in their course, in some other place, close to the point of disappearance, they emerge again, and produces a similarity in determination of MOTION and in form of fluxion, which, in both the stomach and intestines, is perpetual-circular, or spiral, and continually maintains an intrinsic relation of poles, foci, axes, large circles, and lesser diametral circles (*i*).

126. With regard to CHEMICAL ACTION; all things in this continuous duct, which is the grand alimentary canal, or nutrient duct of the body, from the threshold of the lips and mouth to the end of the intestines, proceed in series, and are performed in order: its operations being, in fact, successive, in the same manner as its articulations. The lips receive the crude, undivided, and undigested food; the teeth comminute it; the tongue

continually fold and open into new gyres, in fresh directions. A similar play is seen in the intestines: the jejunum and the ileum throw themselves into the centres and nodes of the convolutions of the other intestines, and fold together so often, and dip down so deeply, that they become lost to sight altogether: nevertheless, after having finished their gyres, they come up, and reappear in some other place, and as it were meet themselves again, and their next convolutions; just like the analogous forms in the stomach, and in the cerebrum. This similarity proves, that nature produces the spires of both the stomach and intestines according to the same preëstablished laws; but with this distinction, that in forming the intestines, an evolution takes place from superficial spires into similar solids; in order that the food, after having been previously embraced, compressed and triturated in the spires of the stomach, may be wrung similarly, but longer, more fully, and more violently, in the succeeding spires of the intestines.

(*i*) The structure which demonstrates the particular form of fluxion, is generally termed mechanical; but it ought rather to be termed organic: inasmuch as it transcends the ordinary mechanics, founded upon the forms, figures, measures, and proportions of geometry. The truth is, that in this *spiral form*, a new and distinct perpetuity is assumed, which geometrical curves and circles do not possess; for its very diameters describe some circular line; and hence there arises a superior force, and a corresponding structure, which in the living frame especially, takes the new name of organic. Therefore, whenever nature would so far exalt her powers, as to return to her own spontaneousness, she betakes herself to spirals, and ultimately to perpetual spirals. Respecting the relation of the spiral form, to poles, axes, and greater and lesser circles, see above, n. 97 (*f*).

and the fauces further reduce it, and transmit whatever is not properly reduced and dissolved, to the pharynx and œsophagus. These, in their turn, transmit it to the stomach, which dissolves and digests it still more perfectly. The stomach sends the portions that require further dissolving, digesting, seething, and extracting, through the pylorus into the small intestines; and these, in like manner, reject the undivided and unexpressed portion of their contents through the valvula coli into the large intestines. Thus each articulation constantly delivers its impure, exhausted, and recrementitious materials to the articulation which succeeds it: and thus the operations are in the same successive order as the articulations (*k*).

Hence the small intestines receive from the stomach nothing

(*k*) It is commonly thought, that all the chyle made in the stomach, is conveyed, promiscuously with the refuse, through the pylorus into the intestines, and from these alone, through the lacteals, into the receptaculum of the thoracic duct. But from the anatomical analysis of all the members, it is perfectly evident that the stomach does not send away anything through the pylorus into the intestines, excepting what it has not itself sufficiently triturated, percolated and extracted, and evaporated, either towards the liver by the veins, or towards the receptaculum chyli by the lymphatics. That it throws out the refuse alone by the pylorus, is plain from the rancid, vapid, and acrid nature of the fæces of the stomach, and from many other indications; as well as from all the points which have been demonstrated in the foregoing Analyses. For the pores and ducts of the stomach drink and devour the properly chylified essences, as greedily as the hungry belly itself, or as the craving and impoverished blood and spirits. It is incontestably true, that the lips receive only the solid food, in an uncomminuted state, (unless, indeed, it has been previously submitted to culinary processes; for otherwise they receive it raw and undigested, as in the case of all animals but man;) and that the denticular mills grind it first, but only rudely; and deliver it, when ground, to the tongue. And that the tongue, after having imbibed with its little veins and other vessels the purer juices which escape, commits the remainder, or impure portion, to the pharynx and œsophagus; and the œsophagus to the stomach; and that the stomach likewise commits it to the intestines: as follows, if the stomach digests what it receives, and throws out its extracts through determinate ducts; according to what was shewn in the preceding chapter. That the refuse in the intestines themselves, is but its effete, vapid, rancid, and putrescent food, or rather refuse, which they, like the stomach, mix up, circulate, agitate, squeeze, strain, grind, macerate, seethe and extract, also expel, evaporate, and sublimate through certain foramina and ducts, its appointed channels. But as these materials are more refractory and obstinate, the intestines chastise them with a more vehement motion, with a grosser heat, with a longer delay, and with a more acrid salivary menstruum. The intractable portions they eject through the valvula coli into the large intestines, where they are chastised with still ruder violence, and subjected, if we may use the expression, to the lash and the rack. Finally, the residue that no process can cleanse—foul, filthy, worthless—a useless burden—the caput mortuum and fæces of the belly—is cast, by the large intestines, into the sink of the rectum; and after the extremest infliction, it is thrown out thence, by the common effort of the whole duct, and the specific effort of each articulation, and discharged from the body. In this manner the intestines do no more than continue and gradually complete the work, already begun and prosecuted in the antecedent laboratories (*l*).

Thus each articulation of the alimentary canal (*m*) has its

drained and rejected onwards in a similar order, will be shewn presently in our exposition of the several intestines.

(*l*) This follows from the proposition (n. 124), that the intestines are a new, continued, consecutive, or elongated stomach. Nearly the same process takes place here as in distillation; where the heat first sends off phlegm, impregnated with both pure and impure tinctures; as in the mouth and fauces: then the weaker, and afterwards the more rectified spirits; as in the stomach: next oils, gradually more and more dense, and at length fetid; as in the intestines: the residue being a caput mortuum, or fæces. In the animal laboratory, however, particular organs are supplied for the different parts of the process, to absorb each essential juice and spirit separately, and to transmit and dispense it for the uses and purposes of life.

(*m*) The alimentary canal begins from the orifice of the lips, and is continued through the mouth, the fauces, the œsophagus, the stomach, and the intestines. See n. 59. Respecting its several articulations, namely, of the palate, the pharynx, the œsophagus, &c., see also the foregoing pages. Respecting the articulation of the œsophagus own peculiar department to perform, in the series of chemical or chylopoietic operations, agreeably to the law, that each, in effecting and producing something particular for itself, also effects and produces something general for what comes after it, and something most general for all in the series. For this reason the alimentary canal is not only distinctly articulated, but even subarticulated; the intestines into the small and the large; the small intestines into the duodenum, the jejunum, and the ileum; the large intestines into the cæcum, the colon and the rectum; and perhaps the subdivision may take place again, even many times: always with the same relation of successiveness, and with the same unbroken continuity of effect (*n*).

with the stomach, see n. 84 (*a*). Winslow describes it as follows: "At the superior orifice, the rugæ are in a manner radiated, and appear to be a continuation of the folds of the œsophagus; only they are thicker, and where they meet those folds they form a sort of crown, which bounds the superior orifice of the stomach, and distinguishes it from the extremity of the pylorus" (n. 88). Respecting the articulation of the pylorus with the duodenum, see Heister, *Comp. Anat.*, tab. i., fig. 6; and respecting the articulation of the ileum with the colon, see the following pages. Thus each joint or articulation has its peculiar sphere of activity, within which it is enabled to roll to and fro, and to effect somewhat particular, somewhat general, and somewhat most general.

(*n*) These particulars are indeed true, as here stated, in the normal, healthy, and natural condition of the stomach and intestines; but they are infinitely diversified in different subjects, according to the state of the stomach, of the intestines, of the abdominal viscera, of the blood, of the humors, and of the disposition and the mind; according to the quantity and quality of the food; also, according to habits of intemperance; all of which are frequent causes of an aberration from the settled laws of nature; and lay the foundation of indigestion, adulteration of the humors, and complaints, diseases and defects of the stomach and intestines: respecting which, see Schurig, n. 122. The order then becomes perverted, or even absolutely inverted. Digestion in its last and lowest forms precedes digestion in its middle and first forms; and the stomach transfers its offices to the intestines, or *vice versa:* hence we have crudity, after the proper time allowed for digestion has elapsed; or untimely wasting of the food, and consequent nausea and loathing; or undue expansion of the membranes, and obliteration, displacement,

The saliva also increases in digestive power and sharpness, in a similar successive series, proportionately to the acting and resisting forces. It is mild and limpid in the anterior part of the mouth, comparatively thick in the palate and the pharynx, fruitful and potent in the stomach, but acrid in the intestines, where it is no longer called saliva, but bile; being, however, tempered by the more mild and homogeneous pancreatic juice, to meet every requirement of nature, and to correspond to the forces and modes of the intestines as active, and of the food as passive (*o*). The last species of saliva, adapted for macerating the fæces, distils and runs continually from the vermiform appendix into the cæcum and the colon (*q*).

127. The vermicular movements of the intestines perform alternate and reciprocal gyres, synchronously with the œsophagus, the stomach, the lungs and the brains; consequently, the stomach and the intestines digest and rise into action every time the brains animate and the lungs respire. The brains are

or widening of their sinuosities and rugæ, ending in gluttony and polyphagism, &c., &c.

(*o*) That the bile is a species of saliva, but of a very acrid nature, will be explained below in the chapter on the liver. It is worthy of remark in this place, that each articulation of the alimentary tube has its own proper salivary spring, and its own fresh supply; and in insects, certain vessels, called the vasa cœca and varicosa, always pass into the intestines, close under the pylorus; there being sometimes two such vessels, sometimes three, sometimes four; (see Swammerdam, on the worm of the acarus, on the day-butterfly, on the common or working bee, and other insects, n. 121;) that these vessels are channels for saliva, and carry a stream accommodated to the condition of the food which is to undergo solution in this part of the intestines, cannot, I think, be doubted for a moment.

(*q*) Respecting this last source of saliva, in the cavity of the vermiform appendix, we shall speak presently. It is now evident, that each articulation is of use to itself, to the articulation next to it, and to all in the series. For when the saliva has ceased to be of use to its own member, it is instantly derived into the cavity of the next member, and so on, through all to the last. Thus the salivæ also, according to the same law, perform their particular, their general, and their most general uses.

perpetually renewing these reciprocal movements of the intestines from within, by means of the fibres and the spirit of the par vagum and of the intercostal nerves; the lungs, from without, by means of the vertebræ, the ribs, the sternum, the diaphragm, the peritonæum, and their muscles (*r*); the stomach also, by means of the pylorus, and by continuity and influx.

128. The vermicular motion of the intestines, like the peristaltic motion of the stomach, is in constant flux and reflux from one orifice to another, as it were from stage to stage (*s*); the stomach from the cardia to the pylorus; the small intestines from the pylorus to the valvula coli; and the large intestines from the valvula coli to the sphincter ani; to and fro. But the order is not the same in the large as in the small intestines. For while the duodenum, the jejunum and the ileum are in their onward career, wreathing their volumes forwards to the valvula coli, the cæcum and colon wreath theirs backwards

(*r*) The peritonæum, which is the common coat or covering of the abdomen, is surrounded with a kind of muscular mail, formed principally of the abdominal muscles. These, and others of the same region, which are connected to the ribs and vertebræ, very sensibly expand and contract during every respiration of the lungs, and elevate and compress the cavity; and consequently the intestines which lie under them, and are connected to the peritonæum in many parts; and which must coincide with their momenta, unless the motion should be extinguished by the meeting of as it were two currents thereof.

(*s*) The motion of the intestines is more truly vermicular and serpentine than that of the stomach; wherefore, with the reader's leave, I shall call the motion of the former, *vermicular*, that of the latter, *peristaltic*. This motion is too palpable to require further proof. It has been witnessed in the living subject, in cases where the abdomen was opened: and in worms, through their transparent integuments. "The reciprocal motion [in the moth of the silk-worm,]" says Leeuwenhoek, "was repeated so often and so long, that my eyes were fatigued with watching it. In brutes," continues he, "I have seen the motions of the intestines; which motions the intestines retain for a long time after they are removed from the body" (n. 120). Respecting their motions in the worm of the acarus, the working bee, and other insects, see Swammerdam, n. 121. And that anti-peristaltic motions also have been observed by Mery, see Schurig, n. 122.

to the same point. Consequently the small and the large intestines both meet with their extremities in this valve, that is to say, in the cœcum, as a centre (*t*); and at the same time, or during diastole; but during systole, they mutually recede from the valve. Hence the singular structure of this valve, the bulging form and great capacity of the cœcum, the evident articulation of the colon and ileum therein, and the meeting of the extremities of both intestines in the focus of the gyrations.

The simple expansion and contraction of the muscular coat, and thence of the intestinal tube, produces of itself, not only the vermicular gyration, but also the reciprocation and concurrence of the intestinal movements. For the small intestines are so exceedingly sinuous, that they cannot expand without a change of place, and a spiral evolution following the course of

(*t*) Heister has given exact delineations of the valvula coli, in his *Comp. Anat.*, tab. ii., fig. 7, and tab. vii., fig. 29, 30; in the former figure shewing the perpetual hinges and inter-articulations between the colon and the ileum; in the latter, the particular contortion of the two intestines in an infant just after birth. Peyer also has represented the same articulations, by drawings of several sections of the cœcal cavity and valve *(Exerc. de Glandul. Intestin.)*; and many authors have likewise given descriptions of them. Of these I select that of Winslow: "The cœcum," says he, "forms on the inside a large fold, which advances into the cavity of the intestine; its extremities are very thick, by reason of the mutual duplicature of the cœcum and colon. The extremity of the ileum is as it were grafted in the opening of this fold, and strongly united to its sides by the mutual adhesion of the transverse fibres of the two intestines. The ring which advances into the common cavity, is formed into gathers on the inside. The membranous coat of the extremity of the ileum is continued on the cœcum and colon. The longitudinal fibres of the muscular coat seem here to be confounded with the nearest circular fibres of the cœcum and colon. The nervous and villous coats of the ileum likewise enter the cavity, and meet those of the cœcum and colon on the edge of the ring," &c. (n. 114). That both the ileum and the colon wreathe in hither, as it were meeting at the end of their course, and again invert their hinge of motion, is very evident from the knot which they form in the cœcum, and from the cavity where they fold and unfold: so much so, that there is no need of any great penetration to discern it; for nearly every point in the parts indicates and attests the fact.

their sudden turns and reflexions (*u*): and the colon likewise, inasmuch as it passes in a large circle round the convolutions of the other intestines, and is attached throughout its course to the neighboring members (*x*), can undergo extension only where some free space is afforded, or like the ileum, towards the cœcum. The determination of the extension is the sole cause

(*u*) The plates of anatomical authors, particularly Eustachius, *Tabul. Anat.*, tab. ii., fig. 2, shew the manner in which the jejunum and the ileum form the central, and consequently the shorter gyres of the intestines, making very sudden wreathings and reflexions: so that when they extend, and their diastole is taking place, their smooth surfaces necessarily unfold, and suffer evolution; which could not happen, unless the extremity of the ileum were wreathed into the cavity of the cœcum. The jejunum and the ileum conspire to this motion, not by their curvatures alone, but also by their connexion with the mesentery and the neighboring members. This may be easily represented mechanically, by inflating a leathern tube, convoluted into similar circles, and bound down with similar connexions, to those of the intestines; or even by inflating the intestines themselves, as they lie in their natural position: in which case, the contrarious passage and the meeting of the two volumes will become evident. This is the cause of the contorsion and strangulation of these intestines; whence passio iliaca, and several other affections, of which we shall speak in our pathology.

(*x*) See the circumgyration of the colon figured by Eustachius, *Tabul. Anat.*, tab. ii., fig. 3, 5, 6. In this circumgyration the colon occupies the circumference of the abdomen; so that while the gyres of the small intestines, (which are internal to it,) unfold, the colon necessarily swells and elongates; and *vice versa*. Moreover, "the colon is connected, by means of the peritonæum, with the os ileum, the right kidney, the gall-bladder, the liver, the omentum, the stomach, the spleen, the left kidney, &c." (see Heister, n. 108;) and also with the duodenum; and throughout, with the mesocolon. Thus, inasmuch as the motion resulting from its extension, cannot be determined towards the rectum, except during the evacuation of the fæces, hence it is necessarily determined towards the other extremity; in short, towards the cœcum. This arises from the simple mechanical law, whereby a tube, in the act of extension, may be extruded at either extremity, indifferently: nor can it be said, on this account, that the determination of the two motions depends upon two different causes; but only that one and the same cause is followed by a reciprocal effect.

of the arrangement and concentration spoken of above. Were the case otherwise, the reciprocating gyres of the intestinal tube could never exist, much less be perpetuated.

129. The intestines alternately dilate and contract, and when they dilate, they also extend, and when they contract they also retract. This is the primary source of their motions, and the wheel of the machine. The rest of their actions harmonize therewith, and result mechanically, as in an automaton, from organic form and connexion of parts, precisely in such wise as the grand principle of use determines.

130. The circumgyration of the intestines is apparently so erratic and confused, that it seems to be destitute of all form, or at least to be perfectly inordinate and lawless; for the intestines are convoluted and complicated together, and in attempting to disentangle them, our ideas and analyses themselves are apt to be lost in their labyrinthine mazes (*z*). And yet it is important that the maze should be threaded, for it is essential to an understanding of causes. We must regard the gyration as analogous to the folds of the stomach, and to the convolutions of the brain; in which there is nothing angular and rectilinear—nothing but what is perpetual-circular or spiral (*a*). This spiral convolution, or organic form, unlike the circle or sphere, does not respect a fixed, immoveable, single centre, with convergent radii; but instead thereof, a circle or orb, with spires for radii: consequently, the centre of this sphere is a circular

(*z*) In the labyrinthine fluxion of the intestines, viewed cursorily, or at first sight, no regularity is perceptible. Nevertheless, it is certain, that nature never takes the most trifling step, except in order, according to laws, and for the sake of an end; consequently, that this apparently inordinate and irregular gyre, in itself, and in respect of uses, is perfectly ordinate; wherefore a similar gyre is constantly met with in every individual of the same species in the animal kingdom. Thus this circumstance is not in the slightest degree accidental; although we, in our ignorance of the nature and properties of the intestinal form, view it much in the same manner as mules view a water-wheel.

(*a*) Respecting the peristaltic gyration of the stomach, see above, n. 97 (*f*); where some preliminary description was also given of the spiral form.

gyre, whereby, and as it were afar off, it regards a fixed centre, which is that of its central circle. Thus the intestinal gyre respects the fimbriated border of the mesentery, (which is of a circular form,) and mediately thereby the receptaculum chyli, which constitutes the innermost centre of the sphere, and is fixed and immoveable, because it is the centre of a circle. Moreover, the spiral form has the following properties. Its radii or semi-diameters, which are directed to the central circle, are all of them circumferences, and *vice versa*, its circumferences are all of them diameters; and with sinuous flexure they are perpetually rolling into and again revolving from that moveable centre (*b*). And in its volutions, this form is always describing some everlasting curve, related to the circle; either an elliptical, parabolic, hyperbolic, or some other geometrical curve (*c*), but ever evolved, after the manner of the cycloid (*d*). Hereby, the intestinal gyre draws forth every property and peculiar virtue

(*b*) This idea may be illustrated by figures whose properties are known. Thus the circle, in all its points, which are infinite, by infinite radii, respects only a single centre; which is absolutely fixed by the perpetual and as it were infinite concentration and meeting of the radii; being thus rendered immoveable and most inert. But let us suppose that the circle itself occupies the place of the centre; and that its semi-diameters, so many spirals, are similarly directed to it; in this case, there will necessarily be no fixity or concentration of any of the radii, but a perpetual circumvolution, emulating a kind of infinity in its fluxion. That the intestines respect, not merely the mesentery as their included orbit, whereto their spires are related, as to a moveable centre, but also the mesocolon, will be shewn in the following chapter, on the Mesentery.

(*c*) The circular form is of very vast comprehensiveness, insomuch that it includes all kinds of geometrical curves; in the same way as the angular form includes universally all possible figures bounded by angles and lines, or rectilinear planes.

(*d*) The spiral, in its development, is perpetually departing from the straight line, and even from the circle. It never returns to the point it started from, as the circle does; consequently, it describes none of the curves of a conic section, but a different kind of curve, which it evolves; thus a curve similar to what is called a cycloid. There are, however, as many species of cycloids as there are species of curves which are capable of evolution.

that resides in the circle or curve, and constantly converts it to the use intended by nature. And hence this form derives the power of infinite variation; consequently, of accommodating itself not only to every possible space, under every volume, but also to every possible use and end (*e*). Thus it is not surprising, that the idea of the intestinal circumgyration, with its almost infinite varieties, at first sight appears inextricable, and seems to elude analyses founded on the measures, methods and figures of geometry; for even in its lowest form it utterly transcends them.

131. From what we have stated, it follows, that the intestines, in the exercise of this their gyre, induce on their contents a kind of endeavor and gravitation from their centres to their

(*e*) We may easily conclude what an immense number of variations, and consequent changes of state, the spiral form can assume, if we consider, that all possible curves may serve it as centres, and that it may evolve its spires into all possible species of cycloids. Its perfection principally consists in its power of varying itself infinitely, and yet, in every change, of still remaining constantly in its own essence; for thereby it is capable of accommodating itself to all circumstances and to all kinds of use. Thus it differs in every genus and species of living creatures; and in the minutest animalculæ; according to the nature of the food, to the mode of digestion, and the condition of life, in each. It is also capable of adapting itself to every space; as exemplified in the same instances; and in cases of obesity, and hernial protusions of the intestines. Inasmuch as this form, like a Proteus, is capable of transformation into infinite forms, (all, however, of the same family,) it therefore appears so intricate to the superficial observer. But in order to obtain a clear idea of it, we must not attempt to study all its gyres at once; for such a course would be more apt to stupify than to enlighten us: but we must first consider it as perfectly regular, and without variety; with a simply-circular circumference for a centre, and evolved circles for spirals, (as it is exactly in the fluxion of the intestines in snails;) afterwards, we may expatiate into its variations. Precisely so also in the other forms. In the angular form, for instance, we are not to begin from parallelograms, trapezia, or polygons, but from triangles; which are the bases of all the rest; and first, indeed, from the right-angled triangle. So likewise in the circular form, we must begin from the circle, but not from curves, still less from variously-inflected curves.

circumferences, and next from those circumferences to the centre of the circle, or the common innermost centre. For the intestines first throw all their contents, from their axis, which is their continued centre, towards their concave surface or parietes (*f*); this produces a perpetual mingling of both alimentary and excrementitious materials, and a moistening of the whole with saliva; accompanied by a tendency in the materials themselves, not only to change their places, but also to apply themselves closely to the circumferences of the canals, and to pass out through the pores and foramina which there meet and open for them, to their ulterior destination (*g*). But as soon as the chyliferous essences, by this headlong career, have arrived at the bounds of the intestines, or in the cellular coat, the former tendency is supplanted by another, from circumference to centre, in a word, from the borders of the mesentery, by recti-

(*f*) Almost the same thing occurs here as in the arteries, where the blood extrudes its serum towards the circumferences, and separates the light from the heavy portions by perpetual processes of excretion; or as in a barrel or cylinder when rotated, where the contents of the sphere are thrown from the centre, and precipitated as it were downwards. This power (which is exerted particularly in the gyre of the stomach and intestines,) see n. 97 (*f*), is imparted to the motion by the first or primary forms. But when once the materials have been driven and as it were detruded to the proper surface of the intestines, they are then translated into the sphere of a different motion, where there is a tendency driving them from the circumference to the common centre of the whole gyre. All the operations of the intestines mentioned above, are produced by these circumstances as causes; and that this is a true statement of the case is evidenced not by effects only, but also by the laws of motion, as proved from physical phenomena.

(*g*) Since all the parts, and preëminently the fluids, tend towards the circumferences, it follows, that one part is continually substituted in the place of another; and that the parts mutually contend for priority of passage; whence results a perpetual commixtion. Also, that the orifices which do not imbibe juices, but pour them forth, as the orifices of the cystic and pancreatic ducts, cannot pass safely into the intestinal tube, without performing various windings and tortuous flexures between the coats: for otherwise they also would be forced, by the same tendency, to admit the chyliferous shower.

linear radii, to the receptaculum chyli, which is the centre alluded to. A two-fold tendency and impetus thus results from this gyration of the intestines.

132. Now, since the dominant tendency of the vermicular rotation of the intestines is centrifugal, it may seem as if no force was present to urge their contents onwards from the pylorus to the rectum. But let it be remembered, that the intestines are continually expanding and contracting. By these movements, as the stomach, by its corrugations, they send their contents forwards, from place to place, to the intended goal (*h*); whence they are driven down and discharged, by the forcible compression of the particular muscles of the intestines, of the common muscles of the abdomen, and by the joint effort of all the viscera of the body.

133. Hence it appears, that neither the esculent nor the feculent materials enclosed in the intestines, follow their own peculiar laws, but that they are now subject to the laws of the intestines; consequently that they no longer act by their own gravity, as in the general sphere of nature, but passively obey the motion of the containing and surrounding parts; and that their tendencies are directed, and their forces determined, (whether such forces result from gravity, or from any other cause,) to the centres of those viscera (*i*). The animal microcosm

(*h*) When the intestines contract, the points of their surfaces change places, and the anterior come up into the place of those behind them; hence, when they again expand or unfold, they are gradually moved onwards and forwards; and this, moreover, by the folds all passing in the same direction, and as it seems, in a spiral manner, judging from the juga or valvulæ conniventes of the jejunum especially. To produce this effect, the colon, which has to act upon the fæces or more renitent and inert materials, describes a very large arch, and contracts into rugæ and even lacunæ, and expands again alternately, more powerfully than the other intestines; circumstances which shew with sufficient clearness, the manner in which the intestinal tube acts throughout. In insects, and in those particularly which are destitute of feet, a similar locomotive power is displayed; for they go forwards by the rapid elevation of their spines, or by acts of contraction and extension.

(*i*) See what we said above on these subjects in the chapter on the stomach, n. 97 (*f*). This law prevails in all the worlds within the imitates the macrocosm in all its properties; and whatever it receives from the visible and circumambient world, it withdraws from the powers of that world, and subjugates, and appropriates to itself. Consequently, to avoid being misled in investigating the body, it is necessary to put aside ideas derived through the senses from without, touching the extension of the government of the general sphere of nature to the interiors of the animal world; and to confine ourselves to the consequences of similar causes within.

134. The DUODENUM commences from the pylorus, and makes a short course with several curvatures; leaving an open field for the rest of the intestines to produce and perform those gyres which are best adapted to nature's uses (*k*); thus it inaugu-

world; nay, and in all things within our little world; for every thing is a being and a substance by itself, and in its first powers or principles represents the universe: hence the universe is only the aggregate and sum of infinite similar leasts. Wherefore one fruitful source of errors is, that we judge of those things which act within us, from those things which pass without us. The effects, consequences, and conclusions are indeed alike in both cases, for the microcosm operates similarly to the macrocosm; but the causes themselves, by which the effects are produced, are peculiar to the microcosm. Thus we are instructed by the one respecting the other; by the organic animal body respecting the phenomena of the world, and *vice versa.* If each were not modelled to the other, there would be no agreement between them, nor would the animal world live in harmony with the order and laws of the universe.

(*k*) The flexures of the duodenum are represented in anatomical tables, particularly in Eustachius, *Tabul. Anat.*, tab. ii., fig. 4; and they are thus described by Winslow: "The duodenum, immediately after it has arisen from the pylorus, is reflected a little backwards and obliquely downwards; then it passes towards the right kidney; thence in front of the renal artery and vein, and of the vena cava, ascending insensibly from right to left, till it comes before the aorta and the last dorsal vertebræ. It continues its course obliquely forwards, making a third incurvation" (n. 111). The uses intended and appointed by nature are, that the process of digestion and nutrition may be performed in order and series, according to her first and deepest laws; for upon this depends the life of the body, nay, even of the senses, and that supreme life appertaining to the mind, which constitutes our most essential life.

rates the intestines into their spiral form and motion. It receives the exhausted and undigested contents of the stomach, and vigorously detrudes them into the jejunum: it besprinkles them with mixed bile and pancreatic juice, corresponding in quantity and quality to its own constitution and nature, and to that of the viscera, the blood, the spirits, and the food itself. It also absorbs their first products; one part of which it sends to the veins, (making use even of the arteries for this purpose) (*l*), and thence to the vena portæ: another part into the tunica cellulosa Albini, and thence, through the lacteals of the mesentery, to the receptaculum chyli (*m*); in both cases by means of certain glands, papillæ, foramina, and siphunculi (*n*): a third portion,

(*l*) According to Heister: "Albinus has observed, that the veins of the intestines may be filled with injection through the arteries, and the arteries through the veins; a fact which, [as he says,] gives rise to singular speculations" (n. 110). Thus the blood-vessels in these parts seem to assume a kind of double and ambiguous nature, and the arteries themselves to be in a manner veins; so as to lick up and imbibe the juices and chyles, with their little lips and orifices, similarly to the veins: which strengthens my conviction, that the foramina here do not pour forth any serosity, mucus, or phlegm, but on the contrary, absorb them; likewise, that none of the vapor which is taken up by the internal coat of the arteries (but not by the corresponding, or external coat of the veins) rises immediately from the intestines to the cortical substance of the brain, but only from the stomach. See above, n. 105.

(*m*) Albinus describes the coat which goes by his name—the tunica cellulosa Albini—as "situated between the villous and muscular coats," and as "forming the true nervous coat, and completely surrounding the canal" (n. 109). That the lacteals imbibe their chyle from this coat especially, as a kind of perpetual reservoir, is a point which I intend to shew in the following chapter, on the Mesentery; consequently, that the conglomerate ducts, and the downy villous structure, convey hither their fine fluids, and present them to the lacteals. Thus according to nature and nature's order. The manner in which the tubuli converge or radiate hither, is excellently shewn by Swammerdam in a figure inserted in Mangetus, *Theatr. Anat.*, tab. lix., fig. 7.

(*n*) The glands, papillæ, foramina, and tubes of the duodenum and the other intestines, are described in detail by various authors, whom it would be useless to quote in this place. See Peyer, Pechlin, &c., Winslow, n. 111. "In the upper part of the duodenum," says Conr.

less pure, into its second coat, the tunica Ruyschii cellulosa, and thence into the peritonæum and to the external circum-

Brunn, "above the ductus cholidochus, there are great numbers of spherical, glistening granules, set close together; these are made up of other lesser and least acini, much in the same way as the substance of the pancreas. The coat of this intestine is perforated with an infinity of little pores. In the horse, the larger glands are composed of vast numbers of other tubulated glands, so put together, that their tops, (which are perforated with exceedingly minute foramina or pores,) project into the cavity of the intestines; they excrete a matter which thickens by boiling, like the white of egg." *(Glandulæ Duoden.* p. 22, 23, 32.) Moreover, that both the glands and the papillæ consist of minute networks of vessels, organically disposed into a glandular and papillary form, has been shewn by Ruysch, and by other anatomists who have used his injections. From these circumstances, I have no hesitation in concluding, that the glandular clusters and the papillary villi, are so many fine organs for absorbing passing fluids, but not for excreting mucus. The quantity of liquid which passes by them, and the quantity which necessarily enters them, is plain from the quantity taken by the mouth, from the quantity of juice expressed from the food, and of saliva constantly poured forth by the extraneous salivary glands; and also from these little organs themselves being so full of perforations, not to mention the other foramina which open in the interstices between them: shewing that there is no want of fluids wherewith to bathe the intestines; and that they do not stand in need of further supplies from sources of their own. Furthermore, the minute arteries also assume the nature of absorbents, becoming in a manner veins. The same things occur in less perfect animals, where the mesentery is commonly wanting. Thus, in the cuttle-fish, according to Swammerdam, "the veins take the aliment immediately from the stomach" (n. 121). Consequently, the intestines do not give, but receive; agreeably to what seems to be their office. And besides, a force is perpetually operating towards the concave surface of the intestines, and into it, but not outwards; and the glands themselves and their ducts are siphunculi and organs of attraction, not of expulsion. At first sight, it certainly appears as if the glands when stimulated, as by purgatives, did pour forth a certain quantity of mucus into the cavity of the intestines; but the fact seems rather to be, that when the sharp particles strike the glands, they suddenly contract and close their orifices, in order to imbibe no more; consequently, the liquid that is drank, and the saliva, being unabsorbed and kept out, so drenches and deluges the recrementitious materials, ference of the viscera (*o*). It rolls the remainder into the jejunum and the ileum, and involves it in their serpentine convolutions.

and so lubricates the passage, as readily to escape from the intestines: meanwhile, the humor, now pressed into the coats, and in a manner stagnating there, renders them loose and flaccid. For these and other reasons, I am convinced, that all the glandular and papillary forms of which we are speaking, are the beginnings or principles of the intestines, and absorb the digested portions of food, and possibly contribute to the more perfect preparation of the chyle. If the glands and papillæ do absorb the chyliferous streams, then they also present a part of the current to the veins, inasmuch as they themselves, and their very orifices, (according to observations,) are made up of vessels. The existence of still purer, and indeed of absolutely invisible organs of a similar nature, which adhere to the intestines under the appearance of mucus, is rendered probable by Leeuwenhoek's remark, that "the supposed pituita in the intestines, is not phlegm, but consists of parts and organs necessary to the intestines. While examining the mucous substance," he continues, "I was astonished at the vast number of fine blood-vessels," &c. (n. 120). And by Swammerdam's constant assertion, that a similar mucous coat is detached from the intestines at stated times, and replaced by a fresh and more suitable coat. On this subject, see also n. 149 (*h*).

(*o*) Respecting the true cellular coat, as being exterior to the tunica cellulosa Albini, and situated between the external and muscular coats, see the authors cited above. Winslow describes this coat as connected throughout to the peritonæum. "The duodenum," says he, "is firmly bound down by folds of peritonæum, especially by a transverse duplicature: the two laminæ of this duplicature naturally leave between them a kind of triangular space, which is lined with cellular tissue," &c. (n. 111.) So likewise the other intestines, particularly the colon, which by its external and cellular coats, communicates with all the viscera of the abdomen. That this coat of the intestines, and by continuity, the corresponding coat of the peritonæum, and by further continuity, the corresponding coat of the abdominal viscera, is covered with a kind of oily, aqueous, and sometimes comparatively acrid liquid, will be shewn in the chapter on the peritonæum. Now, since the intestines, by means of the foramina, communicate therewith, it follows from the continuity of the passages, consequently, from the chain of causes and reasons, that an abundance of liquid escapes from them into the cells of this and similar coats, and is thus diffused in all directions. But as this humor is thrown out into a kind of sea, it would seem not

Sometimes, when this intestine is irritated, it spews out the gifts of the stomach, and even returns upon itself, and vomits the cystic bile and the pancreatic juice into the pylorus (*p*). Thus, like the stomach, the duodenum is capable of an inverse movement, contrary to the ordinary course of its spires.

135. The JEJUNUM, busy in the centre of the intestinal gyrations, performs the part of a secondary stomach. It receives the materials brought by the duodenum, with their mixture of mild and acrid salivæ; it rolls them through its spirals, upwards and downwards, with equal ease (*q*); it tumbles them over with apt facility, and mingles and entangles them, and applies them

to be of the same purity as that which passes directly to the thoracic duct and to the liver.

(*p*) From the almost rectilinear descent of the duodenum, from the pylorus as far as the ductus cholidochus, and from the consequent power it possesses of rolling in on itself from either end; also, from the villosity and consequent sensibility of the duodenum in this interval; still more, from the fact of large quantities of bile being occasionally vomited from the stomach, although there is no apparent communication between it and the gall-bladder; and from the parallel fact, still more difficult of accomplishment, of the stomach throwing its entire fundus towards the œsophagus,—from all these circumstances, it may be inferred, I think, with absolute certainty, that the duodenum, when carried away by a contrarious motion, rolls into the pylorus, or rolls the pylorus into the stomach; and eructates the bile in large quantity at such times, and the bile alone; for the duodenum is always empty in this interval. Many circumstances prove, that any one of the articulations, by the stimulus of peculiar causes, may be the subject of this inverse movement.

(*q*) That the direction of gravitating substances within the sphere of the body, differs from their direction without, in the general sphere of the macrocosm, see above, n. 131. This is particularly evident in the jejunum and ileum, which wind in everlasting circumvolution, after a kind of spiral manner. Were the food to gravitate within this sphere as it gravitates without, it would lie at the bottom of every hollow in these intestines, and never admit of being raised. But the truth is, all their contents are carried about, as if they were destitute of weight; and this, whatever may be the position of the intestines, or of the body generally.

to its innumerable syringic pores (*r*), and greedily and gladly imbibes them; selecting, however, only what suits its purpose (*s*). Like the duodenum, it sends the pure portion to the mesenteric glands; the portion intended for purification, to the liver. It gives no respite, but thrusts down whatever it does not imbibe, with rapid rolling and redoubled forces (*t*); and delivers it for

(*r*) It may be shewn by indubitable proofs, that the least tubes exercise a certain syringical force, or power of suction, more perfectly than the larger cavities. Thus two forces are at work;—1. The gravitation of the contents of the intestines towards their parietes, and the close application of those contents to the pores and foramina: 2. The bibulous power of the pores themselves. Consequently, there is a tendency of all things towards the surfaces. That they tend first into the cellular coat, before they tend into the area of the mesentery, is a constant phenomenon, and perfectly manifest in the initiaments of the viscera; for the first-formed members are immediately surrounded with a sort of vesicle, into which the juices are attracted, before they are determined towards the interiors. In these circumferences, the juices are as it were in a natural state, and out of the force of the common motion; and in a condition to be determined afterwards to whatever ulterior destination the force may carry them.

(*s*) The glands, pores, and foramina which absorb the various species of liquids in the intestines, are always placed among a crowd of papillæ. These papillæ, by virtue of their sensibility, perceive the nature of the liquids driven against them, and whether these liquids are angular, rough, and sharp, or properly purified, and convoluted or rolled into globules. According to Leeuwenhoek, the chyle consists of perpetual globules. By its globular form, it is enabled to pass with safety into the little openings; which at the touch of any sharp or irregular figure, would instantly contract, and completely prevent all passage and approach: as indeed is the case universally with the nervous fibre, (particularly when divided into its simpler threads,) at the least contact with any pungent or injurious object. That all the pores and tubuli are closely surrounded with guards of this kind, to serve as harbingers of coming good or evil, is a circumstance fully proved by our authors; (see Winslow, n. 111, where he speaks of it at considerable length;) and furthermore, that the papillæ, springing as they do immediately from the nervous fibrillæ, possess exquisite sensibility on their own account, and thus apprize themselves of all contingencies.

(*t*) In the jejunum, as well as in the duodenum and the ileum, purification to the ileum, which is at once its partner and successor in function.

136. The ILEUM, folded in wreathing gyres, receives, detains, macerates and repeatedly agitates these jejune materials; and after seething and squeezing them, it imbibes their liquid products, and their chyloid drainings and juices (*u*). The sterile

there is the same kind of organic power of transmitting along the gyre the materials which come in contact with, and become affixed to, the concave surfaces, as in the stomach itself. For there are prominent folds and rugæ, called also juga or valvulæ conniventes, continued without interruption obliquely across and around the intestine, and which at last become longitudinal. "The valvulæ conniventes in this intestine," says Winslow, "are very broad, very numerous, and very near each other. On the side of the great curvature, their circumference is continuous and uniform; but next the small curvature, there are breaks in them, the extremities of some advancing beyond the rest, and terminating in points. In the ileum, instead of being transverse or circular, they become longitudinal" (n. 112, 113). These valvulæ are also delineated in the tables of many authors. In the jejunum, they seem to act with increased power, like the sharper turns in a screw. Thus a similar power of detrusion to that of the stomach continues in the intestines; besides the contraction and expansion, which is attended by a continual substitution of one portion in the place of another. See above, n. 132. Add to this, that the muscular fibres of this intestine are "more visible and stronger" (n. 112); likewise that here the digestive menstrua, as the bile and the pancreatic juice, enter into union with the menstrua which descend from the stomach; moreover, that the jejunum is furnished with an infinity of absorbing orifices, and even the folds and valvulæ conniventes themselves, with oscula; also, that the border of the mesentery lies close against it, eager for its chyle; (see Heister, *Comp. Anat.*, tab. ii., fig. 8; Verheyen, *Corp. Human. Anat.*, tab. vii., fig. 4;) and that thus it is situated close to the very orbit or circle of the gyre. Hence it is that this intestine drives out the food, and detrudes it into the ileum with such rapidity; whence it is generally found empty, and therefore named, jejunum. The fact then of this intestine evacuating its contents more rapidly than the others, is a purely organic result of causes intrinsic and extrinsic to it; as well as of its gyre, and of its connexion with the neighboring parts.

(*u*) The ileum imbibes none but the less pure chyle, which has remainder it pours into the *cœcum* through the valve, which the colon now opens; or should it fail to do so, then the ileum attempts to open it, sometimes without effect; and meanwhile, it wrings the ingesta so long as they remain in it. The convolutions and spires of this intestine sometimes become knotted together, (from causes either remote or proximate,) producing the affection called, passio iliaca (*x*).

137. The CŒCUM is connected with the ileum above, and with the colon beneath: it is the common hall, the place of meeting and the hinge of these two intestines; the goal of the gyratory course of the small intestines, and the starting point of the outward career of the large intestines: therefore, a short, capacious and dilatable sac (*y*). Every time the ileum expands, it forces its extremity, and its contents, into this cuplike intestine; and the colon moving up to meet it, receives and swallows its egesta. The *valvula coli* opens and shuts the door (*z*); and also guards against, and forcibly prevents, any reflux of materials from the lower into the higher intestines.

been left unabsorbed by the jejunum: this it expresses by means of maceration, and by friction and attrition; its spirals therefore are more frequent, and their gyres shorter.

(*x*) The *proximate* causes which hinder or stop the continuance and vermicular fluxion of the gyration of the ileum, are wounds, imposthumes, constipation, strictures, and contusions; the *remote* causes are violent anger, pain of mind, fright, and many other things whose influence is communicated through the fibres, consequently, by the brains; and which produce either a revolution and contrarious rolling of its gyres, or a cessation of its motion; for in this case the contents gravitate and subside by the action of the general sphere of nature, and force the intestine into knots.

(*y*) The cœcum is very little else than a bag, box, and receptacle for the extremities of the two intestines. Its very figure and form shew that they meet each other therein. See above, n. 128 (*t*).

(*z*) See above, n. 128, and all the descriptions of this valve. We shall find the whole of their particulars clearly applicable to, and coincident with, the use here assigned to the valve; also that it opens at the time when the intestines meet, which takes place during their diastole or expansion; but at the time of their contraction, that is to say, when they mutually recede from each other, the cœcum itself seems to lose its bellying form, and to be drawn out into a tube or intestine.

138. The APPENDIX CŒCI VERMIFORMIS, the tongue of this balance of motion (*a*), opens its pores, and expands its cavity, synchronously with the cœcum; and pours a new liquid, adapted for anointing and lubricating the wavy folds of the colon, and particularly for macerating the fæces, into the fundus of the cœcum and the gorge of the colon (*b*). This liquid is proxi-

(*a*) I term the vermiform appendix, the tongue of the balance, because it is placed in the middle, and at the meeting-point of the motions. The directing fasciæ or ligaments of both intestines also come together upon it. "The white ligamentary bands [of the colon and cœcum,]" says Winslow, "unite together on the vermiform appendix, covering its whole outer side, immediately under the common coat" (n. 115). See also Heister, n. 108. So that it necessarily adapts itself to the motion of both; consequently opens during the expansion of the cœcum; that is, during the contraction of the colon and ileum.

(*b*) From all we have premised it is impossible to avoid the conclusion, that this little vermicular intestine is appended here, for the purpose of pouring some fluid into the large intestines; for each articulation of the alimentary tube has its own salivary springs. The lips receive their liquids from without—as water, beer, milk, juices and wines, of all kinds—which perform an office of the most general kind in the digestion of the food. The mouth and the fauces derive their liquids from numerous glands extraneous to them, as the parotids and the maxillaries. The pharynx, from the amygdalæ or tonsils. The œsophagus, from the thyroid and other glands, external and internal to it. All these streams flow together into the stomach, to be made use of by it. The small intestines, immediately after leaving the stomach, are supplied with the bile of the liver and with the juice of the pancreas. The large intestines cannot but be furnished with similar springs. This conclusion results from the series of effects; and also indeed from the series of causes. For if the intestines, according to the proposition (see n. 134, *n*), do not pour forth fluid at all, but only absorb it, then in this case, without a fresh spring, they would soon dry up; and the food, which in these intestines is a heavy, tenacious, and fetid mass, would collect and harden, to the entire destruction of life. To avoid these consequences is apparently the purpose of the vermiform appendix, which continually pours a humor, adapted for performing this last office, into the beginning of the large intestines. The structure of the appendix, and many experimental observations made upon it, bear out this view. "Its internal coat," says Winslow, "is plentifully furmately obtained from the cellular coat of the intestines; remotely, from the cellular coats of the peritonæum and of the

nished with follicles, and it is likewise reticular, the meshes being the glandular lacunæ, which continually discharge a fluid into its cavity" (n. 115). And Wepfer says, "In the appendix I found a white, turbid mucus: when this was wiped off, the whole of the internal surface appeared to be covered with little foramina, from which, on squeezing the appendix, a white, turbid liquid issued; and it was plain that the whole appendix was of a tubular composition[x]." Morgagni also holds the same opinion, because, when the cœcum and ileum were distended with flatus, no air was observed in the appendix; nor (according to an observation of Zambeccari) when the appendix was cut off, were any fæces found to pass into the cavity of the abdomen; and further, because the appendix is inserted in the end of the cœcum, or the beginning of the colon, and in two cases, a very remarkable ridge was observed in this place, which rendered it impossible for any thing, either flatus or food, to pass into the appendix. *(Advers. Anat.* iii.) And what would be the use of so small a cavity, even supposing it were filled with food, when the intestine itself is so capacious? The appendix also is furnished with numerous glands, which, according to Heister (n. 108), are comparatively erect, and not flat as in the intestines, (?) and which consequently are emissary or excretory. Similar little appendices or cœca, also called vasa varicosa, are found in all worms and insects, being inserted into both the colon and the duodenum; plainly for a similar purpose. From the meeting of the ligaments on the appendix, it may be inferred, that its excretion is more abundant during the full action of the colon. And for this reason, animals which live on hard foods, as fowls, birds in general, and fishes, commonly have several appendices: deficiency in this respect is always attended with a corresponding deficiency in the natural functions. Moreover, Wepfer affirms, that similar springs of humor are found in the colon. "I have observed," says he, "another osculum, large enough to admit the tip of the little finger; it resembled a round stomach, of the bigness of a walnut, and was full of a white, turbid mucus[x]," &c. But this was perhaps preternatural; for the glands of the cœcum are formed on the same type as the glands of the other intestines, and absorb the fluids which come to them, and even the air itself, when they are blown upon without being touched. "When we blow," says Winslow, "through a pipe into these lacunæ without touching them, the folliculi are inflated, and represent little caps, with a hole in the middle of their convex side" (n. 115). Meanwhile, it appears probable, that the colon, in its passage, where it is connected abdominal viscera: the appendix cœci draws off and discharges the useless and harmful portion of it, just as the cœcum itself draws off and discharges the alvine fæces (*c*).

139. The COLON, much thicker and stronger than the preceding intestines, commences from the termination of the gyre of those intestines, at once superior and internal to it, and presently passes to the circumference; and after forming gyres, or making a circular revolution, at two different stages of its course (*d*), and attaching itself to the several viscera of the abdomen, and giving them as it were the final salute, it passes

to the viscera of the abdomen, extracts a pituitary somewhat from their cellular coats; for the colon is a general emunctory. I remember once having seen certain ducts passing between its membranes, in the same oblique manner as the ductus cholidochus passes between those of the duodenum. These may perhaps be the sources of some of the liquid part of the evacuations in cases of purging. But the retention of clysters clearly proves that the colon does absorb liquids in abundance.

(*c*) That the appendix discharges the superfluous lymph of the peritonæum, may also be concluded from the continuation of the cellular coat from all the intestines to the peritonæum, and from the peritonæum round the abdominal viscera; also from its concentration and as it were termination on this appendix. The quantity of liquid which is sometimes found collected in the cellular coats of the peritonæum and of the viscera, as in dropsy and its varieties, anasarca, hyposarca, leucophlegmasia, tympanitis, &c., is well known; nor can it be denied that there is a perpetual circulation of liquid through these coats and cells. And surely, to prevent it from stopping too long, and inundating the membranes, and thus depriving the enclosed viscera of their power of action, it must necessarily be discharged somewhere; and no place appears more opportune, than the end of the intestines, where the other fæces also are evacuated; particularly as this fluid likewise performs the office of anointing and lubricating the parietes of the colon, and at the same time serves for macerating the fæces. In the body, the recrementitious humors themselves always perform some use before they are thrown out: as the cystic bile, which is the antiquated and feculent part of the blood and chyle; the more pituitary and tenacious portion of the saliva, which is the rejected serosity, &c.

(*d*) That the colon wreathes into two gyres, see Eustachius, *Tabul. Anat.*, tab. ii., fig. 6. By this arrangement it is enabled to act more readily in other parts of its extent.

down to the last and lowest parts of the body, and terminates in the rectum. Unlike the preceding intestines, it bestows no mild treatment, no gentle correction, no tempered digestion, on its now putrid and filthy contents; but it dashes them together and chastises them, and wears them down by prolonged and loathsome maceration (*e*); and thus loosens and wrings from them a last residuum, to prevent them from carrying away a single particle of the chyle. To fit the colon for these operations, it is fortified with robust fibres and ligaments; connected to numerous members, and grooved and channelled with large furrows, corrugations and lacunæ (*f*); that thus it may con-

(*e*) The principal function of the colon seems to consist, in macerating the refuse delivered to it by the ileum, not by delay only, but also by the humor of the vermiform appendix. By this means, the connexions of all the parts are dissolved, and fermentation produced, (in some instances even flatulence and colic, from the escape of gases,) together with the fæcal odor. The slightly circular curve of this intestine, its sinuosities, lacunæ, and other circumstances, produce this effect. In those animals which live on hard food, a large cavity is formed, where the ingesta are detained for a long period, and more fully macerated. The same is the case in certain insects. "In the day-butterfly the gut widens considerably to form the cloaca; then it contracts again, to dilate a second time into a lesser sinus, in which it terminates;" (Swammerdam, n. 121.) In the cossus, the colon is divided into two parts, and very capacious. (*Bib. Nat.*, tab. xxvii., fig. 11, 12.) In the cuttle-fish, the intestines are so separated, that there are those which receive the excrementitious materials immediately from the stomach, as well as others which receive the finer juices that are to be digested into chyle. (*Ibid.* tab. l.)

(*f*) Three ligaments run over the convexity of the colon. "The whole convex side of the colon," says Winslow, "is divided longitudinally into three parts, by three bands, which are continuations of those of the cœcum. They perform the office of longitudinal fræna, between which the intestine is compressed into transverse folds and tuberosities. All the folds are duplicatures, which form valvulæ conniventes in the cavity of the intestine, and the tuberosities form so many cells" (n. 116). In insects likewise, as in the working bee, "the colon," according to Swammerdam, "has strong muscular fibres, which, when they act, convolute it into many wrinkles and folds" (n. 121). With regard to their particular appearance, see Eustachius, tinually punish its outlawed contents with stripes and imprisonment.

140. The RECTUM, commencing at the termination of the colon, lictor-like inflicts the last punishment on the outcast, stinking and worthless fæces, and consigns them to their doom in the jakes. It is, therefore, strengthened with muscles, divided by grooves, and perforated with lacunæ; and united in the bonds of fellowship with all the other outlets, which, like itself, proscribe and eject either superfluous or effete materials from the system (*g*).

Tab. Anat., tab. ii., fig. 3, 5, 6. These provisions are for the sole end of giving the colon increased powers of trituration and concussion, and especially, of maceration; such as the fæces require, now that they have been deprived of their finer essences. Thus the operations successively increase in violence; and the instruments and organs also are accommodated for successively producing the same effect; and they so correspond to the operations, that we may pass with unfailing induction from the continents to the contents, and *vice versa*.

(*g*) The rectum is properly the continuation of the colon. See n. 117. It is worthy of remark, that this intestine communicates, by means of membranes, ligaments, muscular fibres, nerves, arteries and veins, with the bladder, the bulb of the urethra, the prostate gland and the vesiculæ seminales in males; in short, with all the extremes of the body, which like itself, discharge superfluities. Of the causes and effects of this consociation, I shall treat in the part on the genital members.

CHAPTER VI.

THE MESENTERY AND THE LACTEALS.

141. Heister. "The mesentery is a thick, fat membrane, placed in the midst of the intestines, particularly of the smaller ones, its name being derived from its situation. We are to observe its substance, which consists of membranes, fat, all kinds of vessels; and in the human body, of a number of glands. Its connexion, in the upper part, with the three superior lumbar vertebræ; in the lower part, with the intestines, and particularly with the jejunum and the ileum; to which it also gives an outer coat. Its division, into the mesocolon, comprising that part which joins the colon; and the mesentery or mesaræum, comprising that part which joins the other intestines. Its circumference; in which, when separated from the intestines, there are several folds, giving it a resemblance to gloves. Its length is about three ells; but the intestines which are joined to it are at least four times that length. It has two coats or membranes, an anterior and a posterior; between which lie the cellular tissue containing the fat, and the glands and mesaraic vessels. Many persons not improperly reckon this a third coat of the mesentery, and call it the cellular coat. The blood-vessels are the same with those of the intestines, and form wonderful arches and anastomoses. The nerves come from the par vagum and the intercostals. Of the lacteals and the lymphatics we shall speak presently. Many glands are dispersed throughout the mesentery. These vary greatly in different subjects, in number, size and situation. In dogs there is only one gland, but that very large, and called, pancreas Asellii. These glands almost disappear in decrepit subjects. Their office is, to secrete a fluid, wherewith to dilute the chyle; for the lacteals pass through them. The uses of the mesentery are, to suspend and connect together the intestines, and to retain them *in situ;*

to support the blood-vessels of the intestines, and the lacteals; and to shorten the passage for the latter to the receptaculum chyli.

"The vasa chylifera comprise the lacteals and the thoracic duct. They are fine canals in the intestines and the mesentery, and which convey the chyle. They were discovered by Aselli in 1622; though in fact they had been observed long before by Erisistratus and Galen, who took them for arteries containing milk. The best time for demonstrating them in animals, is three or four hours after a full meal, when the animal should be strangled; in this case, the lacteals are turgid with chyle; but when digestion is not going on, they carry only lymph, and are lymphatics. The best method of demonstrating them is, by tying the thoracic duct of the animal with a thread, although they are often sufficiently apparent without using a ligature. The difficulty of demonstrating them in the human subject, arises from our not making our dissections soon enough after death, while the body is yet warm; for when it grows cold, they almost entirely disappear. Nevertheless, there are often opportunities of observing them, in the bodies of those who have been suffocated or strangled, or have died of particular diseases. They originate from the intestines, and principally from the small ones; in all of which, their roots are extremely numerous, and much more so, indeed, in the human subject, than in the dog. These roots exist also, but in very small numbers, in the large intestines. *(Comp. Anat.*, n. 212, 213.) In a robust youth, who had died a violent death, I demonstrated to a numerous auditory, so vast a number of lacteals under the membrane of the mesentery, and upon its fat, that the intestines and mesentery seemed covered as it were with these vessels, all full of a white milky fluid. Their quantity was vastly greater than it ever is in dogs, however well they may have been fed previously: for in dogs, the lacteals only follow the course of the mesenteric vessels; but in this subject, they extended promiscuously over the whole mesentery, and had a multitude of inosculations or anastomoses; and some of them immediately entered the glands about the intestines, while others were buried in the fat of the mesentery, so that their further course could not be traced; and a part went to the more remote glands, as in tab. ii., fig. 8, in which, however, the painter and engraver have not been able to represent adequately either their quantity or fineness. After this, in 1718, in another subject, I observed the lacteals very numerous, and disposed as in the former case; and evidently proceeding from the duodenum, which latter fact, some authors venture to deny. In the former case, I also saw the motion of the chyle in the well-filled vessels. *(Comp. Anat.*, not. 14.) These vessels are distinguished into lacteals of the first kind, which run from the intestines to the mesenteric glands; and lacteals of the second kind, which run from the glands to the receptaculum chyli, where they terminate. The latter are larger but less numerous, than the former. The lacteals have semilunar valves: these are double, and placed opposite to each other, to prevent the reflux of the chyle; but they are not so frequent as in the lymphatics. The use of the lacteals is, to carry the chyle and the lymph from the intestines, through the mesentery, to the receptaculum chyli. Birds have no lacteals, but their chyle enters the mesenteric veins." *(Comp. Anat.,* n. 213.)

142. Winslow. "The great bundle of intestines is bound down by a membranous web, which prevents the convolutions from being promiscuously folded and entangled. This web is the mesentery, so named, from being, in some sense, in the middle of the intestines. It is divided into two portions, one of which is very broad and much plaited, and is connected to the small intestines; the other is long and convoluted, and is surrounded by the large intestines. These two portions are in reality only one and the same continuation of the membranous lamina of the peritonæum, doubled back upon itself, and they are distinguished by nothing but their breadth. Taken together, they form a kind of spiral roll, more or less plaited in its circumference. The first of these parts is called the mesentery, the other, the mesocolon. The mesentery begins at the last incurvation of the duodenum, and descends obliquely from left to right along the lumbar vertebræ. In this space, the membranous portion of the peritonæum is detached on both sides, and produces a duplicature by two elongations, and thus forms the mesentery. It is narrow at its lower and upper parts, but particularly at the upper. The middle portion is very broad, and the edge towards the intestines is everywhere much plaited. The plaits are only waving inflections, such as may be observed on chamois leather; they make this edge of the mesentery very long, and they run through about a third of its breadth. The two laminæ are connected by cellular substance, which contains glands, vessels, and nerves; and in some subjects is replete with fat, which keeps the two laminæ at a distance from each other. Along the whole circumference of the mesentery, the two laminæ are naturally separated, and applied to the two sides of the small intestines, which they invest by their union, or rather reciprocal continuation, on the great curvature of that canal. This forms the external or membranous coat of the intestines. The mesocolon is only the continuation of the mesentery, which at the extremity of the ileum contracts and changes its name. At this place, the particular lamina at the right side forms a small transverse fold, called ligamentum coli dextrum. Afterwards the mesocolon ascends towards the right kidney,

where it seems to be lost by the immediate adhesion of the colon to that kidney. Then it appears again, and increasing in breadth, runs almost transversely under the liver, stomach and spleen, where it begins to turn downwards, under the left hypochondrium, towards the left kidney. Through this whole course, the mesocolon extends in breadth, and forms nearly a transverse semicircular plane, very little plaited at its great circumference; by which it is connected to the whole length of the arch of the colon, and hides one of its ligaments. By its small edge, it forms the triangular case of the duodenum, and by its great edge, the external coat of the colon; in the same way as the mesentery forms the corresponding coat of the small intestines. As it passes under the large extremity of the stomach, it adheres a little to the lower portion of that extremity, as the diaphragm does to the upper. Near the left kidney it contracts, and forms another transverse fold, the ligamentum coli sinistrum. Afterwards it expands again, but not so much as in the upper part, and runs down on the left psoas muscle, towards the last lumbar vertebræ. This descending portion is fixed to the convolutions of the colon, in the same manner as the superior or transverse portion is, to the arch of that intestine. The rectum is likewise invested by a particular production of the peritonæum, which is commonly called by the barbarous name of mesorectum. This prolongation is very narrow, and about the middle of the foreside of the rectum it forms a transverse semicircular fold, which is visible only when the intestine is empty.

143. "Between the laminæ of the mesentery, a great number of glands lie in the cellular substance. In the natural state, these glands are something of the figure of lentils or beans; some of them being orbicular, others oval, and differing in number in different subjects, but all of them a little flattened, and occasionally surrounded with fat. These glands are reckoned of the conglobate kind, the structure of which is not yet sufficiently known; they appear to be cellular, surrounded by a thin coat, that under the microscope seems to consist of an interlacement of particular filaments, which Malpighi believed to be fleshy fibres. Injections have not hitherto given us any satisfaction about these particulars, for though they be made with all possible care, they always fill up the follicular structure of these glands; and though by this means we may discover certain blood-vessels, yet we cannot distinguish the secretory, excretory, and blood-vessels from each other.

144. "Besides the blood-vessels which are distributed in a reticular manner in the mesenteric glands, and besides the nervous filaments, there are also a number of particular vessels of another kind, communicating freely with each other, and running from gland to gland.

These vessels are extremely thin and transparent, and furnished on the inside with numerous valves, which give them a nodular appearance externally. The vessels go out from each gland by ramifications, as by so many roots, and after having formed a trunk, they divide, and again ramifying, enter some neighboring gland. They are termed lymphatic vessels, because they usually contain a clear limpid lymph. But as they have likewise been observed to be filled with a white, milky fluid, called chyle, hence they have been called vasa chylifera, or lacteals. The lacteals may be divided into three classes. Those of the first class arise from the villous surface of the intestines, and particularly of the small intestines, by a great number of capillary roots. These roots form a kind of rete mirabile between the coats of the intestines, which rete surrounds almost the whole circumference of the intestinal canal, between the muscular and external coats. This network of lacteals keeps close to the external coat, and leaves the intestines and it along the mesentery, where it forms two planes of ramifications, clearly distinguished from each other by cellular tissue, and adhering closely to the inside of the two membranes of the mesentery. These two planes run separately on the respective laminæ of the mesentery, as far as the first mesenteric glands, where they unite again into one plane. We shall now speak of those of the second class. After this union, the lacteals are distributed almost uniformly through the whole extent of the mesentery, from its circumference as far as its connexion with the dorsal vertebræ, and between the mesenteric glands, which they pass through, forming frequent anastomoses. As for the third class, the lacteals, proceeding towards the spinal column, decrease in number and increase in size, and having passed the last mesenteric glands, they terminate about the middle of the adhesion of the mesocolon, in small common trunks, which receive also a great number of lymphatics from the lumbar and several other glands. A fourth class may be made of the lacteals of the large intestines; of which I demonstrated several to the Royal Academy, in a human colon. The lacteals are visible after death, when the mesenteric glands have become schirrous, particularly in children.

145. "The lacteals of the third class run down along the descending aorta, between the extremities of the small muscle of the diaphragm, and terminate in a kind of cistern, called receptaculum chyli. The greatest part of the receptaculum chyli lies behind the right crus of the inferior muscle of the diaphragm, on the right side of the aorta, at the junction of the last dorsal with the first lumbar vertebra. It is a kind of membranous vesicle, and its form varies in different subjects. Sometimes it is of an oblong and uniform oval, almost like the gall-

bladder; sometimes it is divided into several small and irregularly rounded bags, more or less flattened; and sometimes it surrounds the aorta like a collar. It is composed of very thin coats, and its cavity is divided by fine membranous septa, the disposition of which is irregular. It is chiefly round the lower part of this receptacle, that the last lacteals are inserted, some on the sides of the aorta, and some behind it; and they are accompanied by numerous lymphatics. The upper portion is contracted between the aorta and vena azygos, and forms a canal, which runs up through the thorax, and is called the thoracic duct." *(Exp. Anat., Tr. du Bas-Vent.,* n. 195—223.)

146. See Boerhaave, on the connexion of the mesentery with the intestines, *Inst. Med.,* n. 95: on the action of the mesentery upon the chyle, *ibid.,* n. 113 to 116: on the action of the mesenteric glands, *ibid.,* n. 117 to 123. Ruysch, "*Dilucid. valvularum in vasis lymphaticis et lacteis,*" and *Thes. Anat.* vi., n. 118, where these words occur; "It is well known that the mesentery consists of two membranes: between the folds of these, I find a third, which I denominate the cellular membrane, because it consists of myriads of cells, but which are not seen, until they are inflated, or filled with water. In their natural condition, they are not visible at all, although in obese persons they are replete with fat. I have a preparation in liquid of the cells of this coat, in a morbid state, distended by flatus, and retaining their spherical figure. The filling of the blood-vessels is beautifully seen." Also *Thes. Anat.* x., n. 53. Eustachius, *Tabul. Anat.,* tab. iii., fig. 1, where there is an excellent representation of the mesenteric vessels, and numerous glands, and of the manner in which some of the vessels pass into the glands: fig. 2, represents the convolutions of the mesentery and the mesocolon, and their glands: tab. xi., fig. 2, shews the conjunction of the cœliac artery and the vena portæ, and their branching out and subdividing through the mesentery and the mesocolon, all the way to the anus; with the wonderful semicircular convolutions formed by the vessels. Cowper, in Mangetus, *Theatr. Anat.,* tab. lix., fig. 8, where he delineates the receptaculum chyli as formed of three trunks, one of which is very large, the other two much less: and in the same tab., various lymphatics passing into the receptaculum. In the Appendix to his *Anat.,* he says, that mercury passes from the arteries into the lacteals[x]. Heister, *Comp. Anat.,* tab. ii., fig. 8, where he shews the lacteals of the jejunum and mesentery, and their perpetual communications: also in *Ephim. Nat. Cur.,* cent. v., p. 234, the origin of the lacteals. Schelhammer, *Anal. An. Præf.* Ray, *Phil. Trans.,* n. 307, p. 2289, (London), the lacteals seen in the mesocolon. Nuck, *Adenogr.,* fig. 9, 10, 11, 12, 13, 14. Malpighi,

Op. Posth., *De Struct. Gland. Conglob.* Verheyen, *Corp. Hum. Anat.*, cap. xii., tab. vii., fig. 4, 5; tab. viii., fig. 1, 2, 5. Musgrave, in *Phil. Trans.*, n. 275, p. 996, where he shews how the lacteals may be filled with a blue liquid. Respecting the membranes of the mesentery, see also Wharton, *Adenogr.*, cap. viii. Cheselden, *Anat.*, tab. xv. xvii. Drake, *Anthrop.*, tab. xx., &c.

147. When bodies are set in motion according to their determinations, or sphere of circumgyration, they form for themselves directions, efforts, and tendencies; consequently relations from circumferences, through diameters, to centres, and *vice versa.* These determinations or spheres, one beside another, and one within another, in wonderful coördination and subordination, pervade the entire body. There are as many spheres as there are viscera; as many spheres within them as there are integral parts in the viscera; and as many spherules within the spheres of the general sphere as there are parts in the parts (*a*). Thus there are innumerable antetypes of the large in the small, and innumerable microcosms in the macrocosm or universe; and each with a form, an essence, a nature, an order, and a state of its own (*b*).

(*a*) The body itself is the complex of all the viscera, consequently of all the spheres, being a kind of universe. Thus each individual animal is a little universe in the universe, and therefore a real microcosm; and this microcosm is so perfectly distinct from the grand macrocosm wherein it lives, that it has the choice of either following the order thereof, or of departing from it in favor of any other order that has been formed in itself. Thus it is exempt from the laws and power of the macrocosm; and this is peculiarly the case with man, who is endowed with an understanding and a will; see n. 133.

(*b*) Our assertion, that every the least particle has an order and a nature of its own, must necessarily be regarded as a paradox, at least by those who have no insight into causes, but rest satisfied with vague generalities, and dwell in externals. And yet nothing is more true. For every particle has its own determination, and its own sphere of action; consequently its own form; if its own form, then also its own

148. When the intestines are set in motion according to their sphere of circumgyration, they form for themselves circumferences, diameters, and centres of this kind; and perpetual relations conformably thereto (*c*). By all their gyres they

essence and nature; consequently its own order; consequently again its own laws; and its own states. Its form makes it what it is, and determines its particular mode of action: thus all things which belong to it—attributes, accidents and qualities—are consequences of its form. Every general derives its nature from parts, and every universal, from singulars. (I am not now speaking of external accidents, which also result, accordantly with internal accidents, from the determination of singulars, as accompaniments.) Thus effects, which are very general as perceived by the senses, derive every thing from their causes; and these, from their principles. Were there not ideas in the innermost corresponding to their visible type in the outermost, we should then have effects without efficients; which would be tantamount to conclusions without reasons, analogies without terms, and terms without their determination, whereby they may be brought, as genuine and homogeneous parts, into particular analogies. But the truth is, we stop in the lowest things, without exercising any reflexion, and therefore are carried away by a torrent of ignorance into all kinds of fantasies, until at last we believe that there is nothing in what actually contains every thing.

(*c*) The relations and determinations arising from spheres of circumgyration, were treated of in the chapter on the intestines; see n. 128 to 134. I shall now shew, or rather demonstrate, by a single circumstance, that the intestines, in their fluxion, describe a spiral form. In the snail, the intestines wind circularly to the apex of the shell, closely following its spiral volutions, and after a circumgyration visible even externally, they return to the border, where they discharge the fæces. Here it is perfectly plain, not only that the gyration is regularly spiral, or perpetual-circular, but also that the momenta of the circumgyrations are exactly synchronous with the respiratory movements: for the air cannot enter excepting when the intestines contract. Furthermore, it is evident, that the inspiration itself, when increased, and determined to the colon or rectum, expels the fæces, just as in the human subject; for the place of evacuation is in the border, near the orifice for respiration: so that the one hole is necessarily shut when the other is opened. But I entreat the reader to verify these statements by consulting Swammerdam, *Bib. Nat.*, tab. v., fig. 6, 7, 8, and his respect a particular, distinctly-convoluted orbit or plane (*d*), which they surround and enclose; that is, the MESENTERY and MESOCOLON (*e*). Again, by all points of the circumference of the mesentery, they respect, but only remotely and extrinsically, the centre of a particular circle, and which is fixed, immoveable, innermost, and single; that is, the RECEPTACULUM CHYLI (*f*). The relations, therefore, determine corresponding directions, and

description of the intestines of the snail; for without a clear idea be obtained by means of figures, words alone will prove indeterminate and inconclusive.

(*d*) The mesentery together with the mesocolon, is not an orbital plane, as commonly represented in plates, but a sort of membrane convoluted into the form of a truncated cone; so that by its border it comes in contact with, and is applied to, all the extremes of the gyre of the intestines. The convolution itself is shewn in Eustachius, *Tab. Anat.*, tab. iii., fig. 2. The mesentery and mesocolon, according to Winslow, "taken together form a kind of spiral roll, more or less plaited in its circumference" (n. 142). Hence, when spread out, it is of considerable length; three ells, according to Heister (see above, n. 141); and four, exclusive of the mesocolon, according to Verheyen, *Corp. Hum. Anat.*, tract. ii., cap. xii.

(*e*) The mesocolon forms a membranous roll continuous with the mesentery; being connected to the large intestines all the way from the cœcum, just as the mesentery is connected to the small intestines. But the one is very distinct from the other; because the motions of the large and small intestines are reciprocal, and meet in the cœcum. From the reciprocal motion we may conclude respecting the causes of the distinction between the mesentery and the mesocolon; and *vice versa*, from the causes of the distinction we may conclude respecting the reciprocal motion. "The mesocolon," says Winslow, "is only the continuation of the mesentery, which at the extremity of the ileum contracts and changes its name. At this place, the particular lamina at the right side forms a small transverse fold, called ligamentum coli dextrum" (n. 142). Another proof of distinctness exists in the fluxion of the third class of lacteals. "Having passed the last mesenteric glands," says Winslow, "they terminate about the middle of the adhesion of the mesocolon, in small common trunks" (n. 144).

(*f*) That the receptaculum chyli is placed in the centre, not only of the mesentery, whose sphere is circular, but in one sense in the centre of the universal body, will be seen presently.

the directions, corresponding efforts, and consequently corresponding tendencies.

149. The alimentary particles contained in the intestines, in the first place, tend and actually advance on all sides, from their axis and diameters, towards their concave circumference (*g*); the spirituous portions, round, polished, and slippery with saliva, prevent and fly past the corporeal and heavy portions, which are besmeared with a grosser humor, and impeded and rendered inert by their angles and projections. As soon as the nutritious particles approach, the internal, villous, or fungous coat, and the glandular, vascular, or nervous coat, receive them joyfully, and throw open their manifold doors, avenues, and receiving-rooms, and hailing them as welcome guests, transmit them through innumerable channels and passages into the chambers appointed for them (*h*), that is, into the cellular coat, which

(*g*) See the preceding chapter, n. 131.

(*h*) From all the phenomena occurring in the anatomy of the intestines, I am fully convinced, that the papillary and fungous substance, and the glandulæ conglomeratæ and solitariæ, absorb chyle and liquids, and do not discharge mucus; see above, n. 134 (*n*). For the fungous substance consists of an immense multitude of fine papillæ, full of pores. The glands also are flat, with a raised border, like a turban; and at the bottom they are planted in a fungous tissue: moreover, they have open mouths which terminate in passages; see Winslow, n. 111; also Peyer, Brunn, and other authors. In fact, the very substance of the intestinal mucus, according to Leeuwenhoek, and as figured by him, resembles the cortical substance of the cerebrum, with the fibrillæ appended to it; see above, n. 120. And not from structure only, but also from the quantity of liquids and salivæ which come together here, and are abundantly absorbed, and carried into the lacteals and the blood, it may be concluded, that all these little mouths are absolutely necessary, to perform the copious imbibition which goes on. And since the intestines are inundated by liquid from so many reservoirs, therefore they require no springs of their own; being supplied by the stomach with the liquids that are drank; that are expressed from the food; that are poured out by the salivary glands; and that afterwards fill the arteries and all the passages of the body; and evaporate and exude in large quantities from the skin, or are thrown out by the bladder. And if we sum up the origins of the lacteals, we shall find that they so occupy every point, as to leave no room for excretory vessels. Besides, is the first receptacle, asylum, and dock of the purer chyle and lymph (*i*); whither, when the chyle arrives, it has reached the

what disturbances, and what a contrariety of currents a flux and reflux of humors would create. For these reasons, I consider the intestines to be set apart for imbibing various fluids, but not for excreting fluids at all.

(*i*) Many circumstances conspire to prove, that the first effort carries the chyle no farther than the tunica cellulosa Albini, between the nervous and muscular coats; (this tunic, according to Albinus, completely surrounds the canal, and forms the true nervous coat, which may be changed into it, see n. 109.) Consequently, that the emissary ducts of the glands and papillæ, terminate here in a cellular web, from which they rise again by new roots; just as is commonly the case in the conglobate glands, where the ramifications converge either to a follicle, or to considerable cells, whence they afterwards emerge again by corresponding ramifications. Wherefore, the entire cellular coat appears to resemble a conglobate gland thrown into a plane. The vast number of lacteals proceeding from this coat, is well known, from what are seen in it during life, and in representations drawn from life. "Some thousands of branches," says Verheyen, "may be easily counted in the external coat of the intestines, but no industry of man can reckon those in their interior substance." *(Corp. Hum. Anat.*, tract. ii., cap. xiii.) There is also, at stated times, a direct passage from this coat into the borders of the mesentery. The lacteals of the other viscera, as the stomach, the lungs, the spleen, the liver, &c., all derive their origin from the cellular coat; most evidently so, the lacteals of the peritonæum, which rise from it to the surface; as will be shewn in the course of our analyses. And why should not the lacteals of the intestines, which are lymphatics in an eminent degree, also originate in their native soil and country? Why should they have a dissimilar origin to the other lacteals, when their nature is not dissimilar? Besides, I am not aware that any one has ever found a single lacteal running immediately from the cavity of the intestinal tube, through all the coats, into the mesentery: unless possibly the case were preternatural, or happened in some decrepit subject. For the sphere of the motion of the intestines there terminates, and that of the mesentery, whose activities are produced by absolutely different causes, begins. But effects themselves, of which we shall speak as we proceed, will greatly swell the number of proofs of this position.

extreme limit of the gyre of the intestines, and is beyond their sphere of motion (*k*).

150. This is the ultimate goal and station which the novitiate chyle attempts to reach, and where it lays its efforts aside; yet it does not stop here; for a fresh effort, communicated to it by another set of forces, instantly supervening, impels and possesses it; and like a torrent carries it away through the mesentery and mesocolon, often by winding ways and through many diverticula, to the receptaculum chyli. Thus the chyle, yielding to the power of the active forces, is translated from the sphere of one motion into that of another, and floats down the stream (*l*).

151. The forces or powers which call forth and convey the chyle from its natural place and horizontal reservoir in the circumference of the intestines, are peculiar to the mesentery, inherent in its plane, and results of its organization. They may be divided into general, specific, particular, and individual. The *general* forces are the alternate expansion and contraction of this membranous plane; and likewise of its minutest parts; exactly synchronous with the respiratory movements of the lungs; or what amounts to the same thing, with the alternate movements

(*k*) We must seek the confirmation of this problem, from the phenomena of motion and modification. From these it appears, that corpuscules, when gyrated, tend through the diameters of the gyre, to the circumferences. If the particles contained in the cavities of the intestines, are carried through a similar spiral contorsion to the intestines themselves, they must necessarily be at last determined, according to the nature of that spire, into a similar perpetual circle or everlasting spiral; which when they have attained, they are then at the goal, beyond the limits and the sphere of the general motion, and by no means endeavor to project themselves further. But the miracles of nature in the animal body cannot possibly be explored, without a previous doctrine of forms, to shew the nature of these determinations and tendencies. Without such a doctrine they must sound like paradoxes.

(*l*) That the chyle does not leave either the cellular, or the external coat of the intestines, of its own accord, but by virtue of an extrinsic force, is evident from its fluxion over the jejunum and the ileum, (see Heister, *Comp. Anat.*, tab. ii., fig. 8,) for it is determined from a circle into a straight line.

of the stomach and intestines (*m*). The *specific* forces are as many in number as the glands, which rise and fall, like bellows or lungs, keeping the same alternations; and thus allure, attract and suck up the lacteal or chyliferous stream, like so many pumps or syphons (*n*). The *particular* forces are the myriads of analogous glandular particles of which the glands are com-

(*m*) There are even more proofs of the alternate contraction and dilatation of the mesenteric expanse, than of the contraction and dilatation of the œsophagus, the stomach, or the intestines: respecting the coincidence of whose motions with those of the lungs, we have treated already. For the mesentery is entirely covered by a production of the peritonæum; it is connected to three of the dorsal vertebræ; also to the lowest part of the diaphragm itself, where the alternate motion is concentrated; it is likewise adherent to the intestines in the whole of its circumference, and in one part to the stomach; and the grand plexuses of the par vagum and the intercostal nerve approach and enter it. Granting, therefore, its alternate swelling and subsidence, the inevitable consequence is, a physical attraction, similar to that of bellows; or of the lungs themselves, which rising alternately, receive and give forth the air; or of pumps, which suck up fluids by the action of their pistons.

(*n*) If we grant the alternate expansion and contraction of the glands, the alternate attraction of the lymph by them is a necessary consequence. The number of glands which are situated not far from the intestines, to communicate the first effort to the chyle, is evident from the description of the first class of lacteals. "Their roots," says Winslow, "form a kind of rete mirabile between the coats of the intestines, which rete surrounds almost the whole circumference of the intestinal canal, between the muscular and external coats. This network of lacteals keeps close to the external coat, and leaves the intestines and it along the mesentery, where it forms two planes of ramifications, which run separately on the respective laminæ of the mesentery, as far as the first mesenteric glands, where they unite again into one plane" (n. 144.) Moreover, the second kind of glands similarly absorbs the streams proceeding from the first, as also ultimately does the receptaculum chyli, which is the great gland of all. The whole of the glands necessarily have the same power; for "they appear to be cellular, surrounded by a thin coat, that under the microscope seems to consist of an interlacement of particular filaments, which Malpighi believed to be fleshy fibres" (Winslow, n. 143); which shews that they are expansile and contractile.

posed. The *individual* forces are the channels or lacteals themselves, the whole of which are siphunculi and spiracles of a similar nature. And in order that the whole and the parts may perform this office together in unanimity, both the vessels and the glands are inserted, and as it were entangled, in the meshes of a peculiar cellular tissue; which meshes, while the mesentery expands and contracts in the general inspiration and expiration, coöperate, and not only keep the lacteals and blood-vessels, and the glands and nerves, within their proper bounds, but also constantly dispose and excite them to similar reciprocations (*o*). Such are the powers and forces by which the mesenteric plane acts; and allures and steals the chyle from the extensive and uninterrupted periphery of the intestines, into its fimbriated border, and thence, from place to place, to the receptaculum chyli, and to the fountain of the thoracic duct.

152. To promote and effect this end, the great intercostal and vagus nerves are brought hither, and intromitted in crowds of eager fibres through the central gate of the circus, whence they go forth in all directions, and form stupendous plexuses, reticulations, and interlacements (*p*); and the fibres, being the essential powers and the vital forces of the body, by their insertion into the minutest parts, communicate their own inspirations thereto, and weave corresponding organic forms; that thus all things may proceed uninterruptedly, according to the order, laws, and conditions of nature. But in order that the blood-

(*o*) Ruysch, speaking of this coat, has these words: "The mesentery consists of two membranes: between the folds of these, I find a third, which I denominate the cellular membrane, because it consists of myriads of cells" (n. 146). That this membrane contains lacteals, vessels, and nerves, see n. 142, 144; and glands, n. 142, 143.

(*p*) Respecting the mesenteric plexuses, see Willis, Vieussens, and Winslow. "The great mesenteric plexus," says Verheyen, "is chiefly formed by the meeting of branches from the different other plexuses. It distributes its fibres through the whole of the mesentery; these accompany the mesenteric vessels, and wind around them in various ways, afterwards proceeding to the intestines" (*Corp. Hum. Anat.*, tract vii., cap. v.) It is very evident that the plexuses of the other viscera respect that of the mesentery as their principal and central plexus, inasmuch as they all either arise or are reflected from it.

vessels may obey the determinations of these forces, they enter the sphere in an opposite direction, or from the circumference, by way of the intestines, and meet the essentially-active nervous fibres. And because they pass from the sphere of one motion into that of another, and are mediate, and as it were ambiguous, between two forces, therefore they wreathe backwards and forwards, and form wonderful gyres, meanders, anastomoses, and islands, to decorate the borders and fringes of the mesentery (*q*).

153. But in the organic body, attraction never exists unaccompanied by propulsion: as we see exemplified in the lungs; and in bellows, syphons, and other pneumatic and hydraulic machines which act by suction or attraction. Expiration always accompanies inspiration. The forces in the mesentery—the glands and the vessels—both draw the chyle from the parts preceding them, by attraction, and extrude it into those succeeding them, by impulsion: acting alternately, not only from gland to gland, but also from cell to cell, and from point to point—from the first little follicles or stations in the cellular coat of the intestines, by a motive force in the fibres (*r*), and thence, by a continual series of appliances, to the receptaculum chyli.

(*q*) See the beautiful delineation of the vessels in Eustachius, *Tabul. Anat.*, tab. xi., fig. 2; and in Ruysch, *Thes. An. Max.*, tab. iii., &c.

(*r*) The cellular coat of Albinus, (into which, as a kind of public market, we suppose the chyle to be brought by the ducts, and thence to be driven into the channels or lymphatics of the mesentery,) lies so completely under the muscular coat, that when the latter opens out its fibres, it must likewise open all the little cavities and pores of the former; and when it again contracts, it must firmly close them. For, according to Albinus, "The longitudinal fibres [of the muscular coat] exist only in the part opposite the mesentery; and the annular fibres are not inserted into the mesentery, as a tendon, but they contract towards the mesentery" (n. 109). The forces appear to be so organized, that they are at once propulsive and attractive, and coincide in their times or momenta. Hence arises a kind of spontaneousness in the determination and in the passage. Yet that there is nevertheless a certain amount of resistance, is evident from the numerous valves in even these lymphatics, which preclude and hinder the advancing stream from falling back again, at the times of the contraction of the glands.

These two forces, united in one and the same member, produce a kind of spontaneous passage of the chyle from the beginning to the end of the course.

154. The receptaculum chyli resides in the very centre, not of the mesenteric plane alone (*s*), but also of the abdominal viscera, and in fact, in the centre of the whole body or trunk. All other things are in a manner its circumferences and diameters. This centre, occupied by the receptaculum, is situated opposite the first lumbar vertebra, where the vertebral column is no longer articulated or connected to the ribs, and where it ceases to follow the alternate respirations of the lungs (*t*)—where the diaphragm and its muscular fibres terminate, and are attached to the vertebræ (*u*)—where the peritonæum, after having finished its gyre round the abdominal viscera, at last forms a knot; and whence, saluting the pleura, it runs forth in a fresh career (*x*)—where the great sympathetic, intercostal and vagus

(*s*) Inasmuch as the receptaculum chyli is the centre at once of so many viscera, and of the whole body, therefore it does not communicate by any means with the middle of the mesenteric plane, nor is this placed in the middle of the gyre of the intestines; for the intestines pass along its border only. That the receptaculum occupies a kind of central position in the dog, is shewn by the pancreas Asellii, which is a single organ, and has vessels radiating to it direct, without passing through any other glandular diverticula. The diverticula in the human subject are very numerous, for reasons which we shall mention presently. The succenturiate and auxiliary glands in the human mesentery, all serve as diverticula, collecting the lacteal streams, and drawing out such a quantity and quality of chyle as the states of the body and the mind demand, and whenever they demand it. Moreover, there appears to be a passage from these glands, and an expiration by them, of the chyle into the veins, and of the humor extracted from the blood, from the veins into the glands.

(*t*) The receptaculum is situated at the junction of the last dorsal with the first lumbar vertebra (see n. 145); thus at the place where the costal vertebræ, or the levers of the pulmonary cavity, terminate.

(*u*) "The greatest part of the receptaculum chyli," says Winslow, "lies behind the right crus of the inferior muscle of the diaphragm" (n. 145).

(*x*) "The membranous portion of the peritonæum," says Winslow, nerves meet, as in a circus, with troops of fibres, and run their race; and whence they go forth in all directions into the viscera of this region (*y*)—whither also the cerebrum and cerebellum, by the medulla oblongata and spinalis, but within the vertebral theca, descend with their principles or cortical forms, and terminate (*z*)—whither the arteries from the cœliac, and the veins from the whole field of the intestines and of many of the viscera, sail home as it were into port, and unite, and pour their blood into the vena portæ, as a secondary cardiac chamber (*a*)—where the aorta and vena azygos, the grand channels of the blood, rest after their labors; the aorta sometimes passing through this very centre or receptaculum, and allowing itself to be clasped and surrounded by it as by a collar (*b*)—lastly, where the muscles which cover and protect the two cavities of the body, meet, and whither a great muscular hood descends from the occiput, and concentrates the sphere of forces of the other muscles of the chest (*c*).

"is detached on both sides, and produces a duplicature by two elongations, and thus forms the mesentery" (n. 142). Consequently, it runs from the place where the receptaculum is situated, close beside the first lumbar vertebræ.

(*y*) See above, n. 152 (*p*).

(*z*) The cortical or cineritious substance is continued from the circumference of the cerebrum and cerebellum, first of all in striæ through the medulla oblongata; afterwards from the foramen magnum through the whole axis of the medulla spinalis; and terminates beside the first lumbar vertebræ, nearly over against the receptaculum, where the fibrous substance, or cauda equina begins. Hence it is manifest, that in this region there is a kind of ultimate station of the active forces of the two brains.

(*a*) See the Tabulæ of Eustachius, Ruysch, and Verheyen. The ramifications of the cœliac artery, distributed through the intestines, in the mesentery again unite into considerable branches, and not far from the receptaculum, into two trunks.

(*b*) "The receptaculum," says Winslow, "lies on the right side of the aorta, and sometimes surrounds it like a collar" (n. 145). Here also the azygos, the venous channel which collects nearly all the blood of the intercostal vessels, and of the whole respiratory field of the lungs, passes through the diaphragm.

(*c*) That the common muscles of the abdomen, of the thorax, the

155. It is essential to the well-being and interests of the kingdom that the receptaculum be tranquil and perfectly secure; for the safety and ultimate life of the body, that is, the life of the blood, depends upon it and proceeds from it (*d*). From this position, the receptaculum looks round it upon all things as external to it, just as the universal centre of the macrocosm looks round upon the planets and the stars; proximately, upon the lacteals, the veins, the fibres and the glands of its ground, the mesentery; not remotely, upon the jejunum and the ileum, and upon the duodenum, the colon and the rectum, with their lacteal currents hastening to it (*e*): a little more remotely, upon the

vertebræ, and the loins, as it were concentrate their forces and actions hither, will be shewn in the Part on the Organism of Animal Motion. Indeed there is evidently a concentration of forces to this spot; for the whole strength of the body depends upon it. And it is from a knowledge of this fact, that athletes are enabled to resist the united efforts of four horses, and to raise heavy weights, by tying a rope round them in the situation alluded to.

(*d*) Life in its most essential form is not predicable of the body and the blood, but only of the soul and the spirit. This essential life, however, makes use of the blood, in order to associate itself to the corporeal life. Consequently, the life of the body, and the blood-life, are synonymous terms. The receptaculum, by the chyle, provides for the blood alone; for the chyle furnishes the serum, from which the blood derives its corporeal essence.

(*e*) We are bound in reason to believe, that the colon also transmits a chyliferous fluid into the receptaculum. "A fourth class of lacteals," says Winslow, "may be made of those of the large intestines, of which I demonstrated several, very full of chyle, to the Royal Academy, in a human colon. The late M. Mery having seen that with the end of my finger I could push the white liquor uniformly into the colon, in many places, seemed at first to be satisfied; but for his further conviction, he desired me to open one of these vessels before him, and to take out a drop of the liquor, which having laid upon the nail of my thumb, he was entirely convinced." (*Exp. Anat.*, *Traité du Bas-Vent.*, n. 218.) The rectum likewise transmits a chyle, by what is called the meso-rectum. "In the dog," says Morgagni, "Needham observed a little vessel, full, not of limpid humor, but of chyle, passing from the rectum to the receptaculum. Lacteals were also seen issuing from the duodenum, immediately beneath the pylorus." (*Advers. Anat.* iii., *An.* xvii.)

stomach, the liver, the spleen, and the other abdominal viscera: in the last place, upon the heart and the lungs, in the chest: wherefore it attracts and derives lymphatic rills from all, according to their respective relations, distances and connexions. With respect to the cerebrum and the cerebellum, the mesentery also largely invites to it the vessels, or, in other words, the fibres of the latter; but not a fibre is allowed to enter it from the former: for the cerebral fibre, determined and actuated as it is by the conscious will, here—at the fountain-head of the body's life—would soon defile the crystal spring, and divert and perhaps arrest the natural and temperate streams (*f*). The prevailing dominance of the will in the kingdom of the body, and the disorderly impulses and actions which proceed from it, occasion a great number of guard-houses or glands to be stationed in the

What would be the use of the mesocolon, unless it fulfilled the same office for the large intestines, as the mesentery fulfils for the small intestines? In the large intestines we find similar structures,—glands, papillæ and tubuli. The materials are not absolutely exhausted when they come from the ileum.

(*f*) That the intercostal nerve and the par vagum arise, not from the cerebrum, but from the cerebellum, was shewn in my *Economy of the Animal Kingdom,* and will be further proved in my analysis of the nerves. The ramifications of these nerves are the only branches which enter the area of the mesentery. In the animal kingdom, the empire is divided between nature and the intellect. The former acts constantly from her own order, and the laws thereof; but the latter, from innumerable other principles, as it were edicts or decrees, sometimes contrary to nature; and determines them into action by means of the will. If therefore the mesentery and the receptaculum were subject to the guidance and government of the will, the kingdom would go to pieces and utterly perish in an hour. For the will acts for the most part, not from the decisions of the soul, but from the decrees of the senses, of the blood, or of the body; in short, from mere ignorance and conjectures; generally, therefore, from the insane concupiscences and desires by which it is constantly inflamed and actuated. But on the contrary, nature, or the soul wherein this nature dwells, is inflamed by nothing of the kind, being influenced by no other desire than that of being enabled to act with a sound mind in a sound body, and thus to break the lusts which threaten either the body or itself with danger.

human mesentery; many being required to prevent the little streams which are continually running into this wheel, from taking a wrong direction; or from supplying any other kind of lymph or chyle, than what conduces, under the auspices, not only of nature herself, but also of a sound mind (*g*), to the general good.

156. Every fibre carries with it, wherever it goes, the animus or affections of its parent cerebrum or cerebellum (*h*); particularly, into this central plane; consequently, different kinds of love, desire, hatred, and loathing—longings and antipathies, and all their ever various states. In this manner the fibre brings from its natal soil into the receptaculum, mesentery and intestines, malacia, or diseased craving, sometimes for incongruous substances (*i*); many species of antipathy and nausea;

(*g*) I say, under the auspices of a sound or rational mind; for man has a rational mind given to him, that he may check or repress the lusts of the body, and moderate and assuage their fires. For if the chyle, (agreeably to the proposition in the next paragraph,) be tempered according to all the affections of the animus, and if those affections be kept under the government of the rational mind, then it follows, that the chyle, in this manner, is also kept under its auspices; thus not indeed directly, but obliquely.

(*h*) The cerebrum and the cerebellum produce, and as it were continue and unfold themselves into the body, by the fibres; consequently, both the brains are universally present in the body by means of the fibres. This is very conspicuous from all our sensations and actions. When any thing touches the fibres, the thing itself, and its mode of contact, are felt and perceived by the cerebrum in the most exquisite manner. And when the cerebrum wills to execute any action in the body, it performs it by its fibres in the muscles, in exact accordance to the form or idea preconceived in the intellect. Consequently, each particular mode of affection in the cerebrum is attended with a perfectly similar affection of the fibre, wherever in the body it be distributed. Nothing is more certain, than that the brains are the organs which wish or desire, and that the affections arising in them flow into the universal body.

(*i*) As in pregnant women, who long and crave for incongruous substances, from a kind of intimate principle; likewise in others not pregnant, who sometimes have desires for which they cannot account. In the same manner, the sick also sometimes desire healing food; and

and also of bulimia, or insatiable appetite. The mesenteric fibres, then, thus animated by the brains, are what command and cause the delicate mouths of the lacteals to seize with avidity whatever things are desired, and to reject with loathing whatever are disliked; and to open their little mouths, apply their little lips, and drink with willingness, in proportion to the intensity of the desire; or to corrugate, retract and close their orifices, in proportion to the intensity of the loathing; for at the first approach, contact, or salute of any chyle or juice, whether desired or abhorred, they suddenly change their state (*k*). Thus the fibre, in all its little organs, and least forms, as in the papillæ, the glands and the vessels, accomplishes nimbly and distinctly, what in generals or compounds, as in the tongue, the lips, the palate and the pharynx, it accomplishes only sluggishly and obtusely.

157. Thus from the stores of the stomach and intestines, likewise from the stores of the blood and serum, and from the other repositories, the innermost mind evokes and extracts whatever essences and chyles it intends or wants, and feeds

all species of animals seek aliment homogeneous to their nature. These effects are commonly attributed to instinct; in other words, to an entirely obscure and unknown cause; which is much the same as referring them to some nature of which one has no conception.

(*k*) We perceive the changes of state themselves in compounds; as in the tongue, the mouth, and nearly the whole of the alimentary canal. The tongue rejects what the animus dislikes, and imbibes abundantly what the animus desires; often indeed against the perceptions of touch and taste. But the objects of sense or consciousness are only general effects, made up of the affections of all the similarly-affected parts; consequently of the little affections of absolute singulars, which first become perceptible in this general manner. Such is the case with hunger and thirst: the singulars and individual elements of the body are what really thirst, hunger, and feel appetite; but these states are not cognizable in their particulars separately, but only all together, in their sum. How obscure, perplexed, and indistinct, then, must all our first perceptions be; nor can they be made distinct, excepting by rational analysis. The senses discover nothing. If we adopt synthesis, we then exchange perplexity for absolute dizziness, and our reasonings run mad.

thereon, and nourishes the flame of its desires (*l*). Hence the numerous glands planted over the human mesentery; the numerous lacteal torrents determined to each; and the frequent anastomoses between the lacteals, to enable each gland to imbibe its milk, if it pleases, not from its own streams, but in a manner from the common lake, and from all the streams at once; and this, both in the beginning, or in the cellular coat of the intestines, in the progress, and in the mesenteric ground itself (*m*). Hence, also, the little arteries and veins are inserted into the glands, and mix with the lymphatics (*n*); and just before

(*l*) These particulars lead legitimately to the conclusion, that the nature of the animus determines the nature of the chyle; consequently, of the blood; and ultimately, the state of the body. In fact, by the quality of the chyle, that is to say, of the humor which flows into the vessels, the animus is not only fed, but encouraged. For instance, in anger the blood suddenly attracts bile, and assumes a certain roughness, whereby it is excited to unseasonable motions. In envy, it attracts particles which give it a livid hue: in grief, it contracts a sluggish viscidity, which produces torpor. In mania and canine rabies it contracts a poisonous spume, so that the very salivæ are infected; and so in all other instances. Consequently, similar things are attracted into the chyle itself, exactly corresponding to the intimate affections of the animus. The extremities of the fibres put on the state of the animus; and reject everything which disagrees with it, and select and seize everything which agrees with it.

(*m*) From descriptions and plates it is evident, that there is a perpetual communication between the different lacteal currents, and between the different lacteals. In the cellular coat of the intestines, at the common spring and fountain-head, the communion is absolute. Hence the veins as well as the lacteals are enabled to satisfy their thirst, according to every requirement of the blood, and intimation of the sovereign animus. The same is true of the vessels throughout their course; for the lymphatics continually incline to each other, and intertwine; and each gland, therefore, is enabled to drink as much lymph or chyle, and precisely such a quality thereof, as it may happen to desire.

(*n*) The truth is, that the arteries and veins enter the glands in a body, as may be seen from the Tabulæ of Eustachius and others. "The principal trunks of these arteries and veins," says Winslow, "accompany the cellular tissue between the laminæ of the mesentery, and there ramify and form belts and arches. It is also worthy of note, that as they enter, and just after they leave the glands, they pass through the cellular coats of the mesentery, and either bathe in their humor, or else purify themselves there of some fatty humor which disturbs them (*o*). The sum, then, of the whole matter is this, that by virtue of such an organism, the lowest, the last and the outermost things necessarily provide and supply, whatever the highest, the first and the innermost, in their mysterious longings, demand, or desire.

the arteries and veins pass through the mesentery, they supply branches to its glands, laminæ, and cellular tissue" (n. 118). It is an observation of Morgagni, that the lacteals follow the course of the larger, but not of the small blood-vessels. *Advers. Anat.* ii., Anim. 1. Since therefore, the arteries, veins, and nervous fibres enter these glands, it seems sufficiently evident that the glands are not only constructed and excited to action by them, but also that the arteries pour in something, and that the veins absorb something, exactly according to nature's wants: and that by these means there can never fail to be a proper tempering of the chyle, according to the demand of the interior animus. See the next chapter, n. 167.

(*o*) Hence the fat in the cells; which is a secretion of an oleaginous matter from the vessels; consequently a purification from those things which obstruct them.

CHAPTER VII.

THE THORACIC DUCT AND THE LYMPHATICS.

158. HEISTER. "The thoracic or chyliferous duct is a very delicate canal, which receives the chyle from the lacteals, and the lymph from the lymphatics, and conveys them through the thorax, usually to the left subclavian vein. This duct was discovered by Pecquet, in 1651, but Eustachius had noticed it long before; though he described it too obscurely. The beginning of this duct is in the reservoir or receptaculum chyli, which is placed on the left side of the superior lumbar vertebræ, under the aorta and the vessels of the left kidney. It is somewhat more capacious than the rest of the duct, resembling a kind of sac or bag, and being frequently double, and irregular in figure. *(Comp. Anat.*, n. 214.) Many anatomists have denied the existence of the receptaculum chyli in the human subject; and some have asserted that the lumbar glands supply its place: but I have sometimes beautifully filled it and the thoracic duct with quicksilver, and demonstrated it as a fine membranous sac, sometimes single, sometimes double; and numerous writers have given figures of both, from the human subject; as Cowper, in his *Anatomy;* Bidloo, in his dissertation, *De Organo Chylifero;* Drake, in his *Anthropologia*, tab. ii., fig. 3; Salzmann and Wium, in particular dissertations on this duct; and Henninger, in *Ephem. Nat. Cur.*, cent. iii., App., p. 120; as well as Cant, tab. vi. Morgagni shews that Ol. Rudbeck was the first who gave a delineation of this receptacle from the human subject. *Epist. Anat.* i., n. 83. *(Comp. Anat.*, not. 15.) The termination of the thoracic duct is generally in the left subclavian vein, or else in the jugular. In dogs, it passes under the aorta, but in the human subject, it ascends on the right side of the dorsal vertebræ, between the aorta and the vena azygos; sometimes with a single, sometimes with a divided trunk.

When undivided, it is about as broad as a thick straw. It is composed of a fine, thin, pellucid membrane. Throughout its whole extent, there are valves, as in the lacteals and lymphatics: these valves prevent the reflux of the chyle. They are more numerous in man than in beasts. There is a semilunar valve, covering its termination in the subclavian vein. (*Comp. Anat.*, n. 214.) This valve is generally considered to be for the purpose of closing the thoracic duct, to prevent the blood of the subclavian vein from entering it. But Jo. Adolph. Wedelius dissents from this; and demonstrates, that no blood could enter it, even if the valve were absent. He shews that the position of the valve is such, that it is by no means calculated for preventing the ingress of blood, if the nature and constitution of the duct would otherwise permit it: but on the contrary, that this valve, with the assistance of a similar one on the opposite side, has the same office as the valves of the veins, namely, of preventing the blood from easily going this way back again from the vena cava, during the contraction of the heart. *Comp. Anat.*, not. 16.)

159. "The lymphatic vessels, from their similarity in structure to the lacteals, are the next we shall speak of. They are thin, fine, pellucid vessels, which convey an aqueous fluid, called lymph; but in the intestines, and particularly in the small intestines, during digestion, they also convey chyle; and at such time, they are called lacteals; the lymphatics and the lacteals being indeed the same vessels in the intestines.

"The discovery of the lymphatics which are outside the intestines, was made by Bartholin and Rudbeck, in 1651. English writers, however, as Glisson, Charleton, and others, have claimed the honor of the discovery for their countryman, Joliff. The situation of the lymphatics is on the surface of most of the internal parts, particularly on the concave side of the liver, the neck of the gall-bladder, and the trunks of the vena portæ and vena cava, &c. They are composed, like the lacteals, of a very fine and delicate membrane. They have great numbers of double semilunar valves, which have been excellently described by Ruysch, in his *Dilucid. Valvul. in Vasis Lymphaticis et Lacteis.* They begin in most parts of the body; but they have not yet been demonstrated in all, although they are probably present everywhere. Their terminations or insertions are in several of the larger veins, particularly in the cava and porta, in the cistern or receptacle of the chyle, and in the thoracic duct." (*Comp. Anat.*, n. 215.)

160. Winslow. "The thoracic duct is a thin, transparent canal, which runs up from the receptaculum chyli, along the spina dorsi, between the vena azygos and the aorta, as high as the fifth dorsal ver-

tebra, or higher. Thence it passes at the back of the aorta towards the left, and rises behind the left subclavian vein, where it terminates; in some subjects, in a kind of ampulla; in others, by several branches united together; and opens into the back part of the subclavian vein, near the left side of the internal jugular. This canal is plentifully furnished with semilunar valves. Its insertion and opening into the subclavian vein in the human body, instead of having a semilunar valve, is covered by several pellicles, so disposed as to permit the onward passage of the chyle towards the vena cava, and to hinder the blood from flowing into the duct. The canal is sometimes double, one lying on each side, and sometimes it is accompanied by pampiniform appendages." *(Exp. Anat., Traité de la Poitrine,* n. 163, 164.)

161. Bartholomæus Eustachius. "In the horse, from the great left trunk, near its junction behind with the internal jugular vein, a kind of large offset proceeds, which besides having at its origin (or insertion) a little semicircular orifice, is also white, and full of an aqueous humor, and a little way from its origin divides into two parts, which again shortly afterwards reunite into one, that puts forth no branches, but penetrating the diaphragm, passes downwards at the left side of the vertebræ, as far as the middle of the loins, where becoming broader, and embracing the aorta, it ends in the greatest obscurity, so that I have not hitherto been enabled to make out its precise termination." *(Lib. de Vena sine Pari,* p. 301. Venet. 1564.)

Further particulars respecting the thoracic duct and the lymphatics, from Nuck and Malpighi, may be seen below, n. 172, 173.

162. The chyle and its substitute the lymph is as eagerly desired and imbibed by the blood of the jugular veins, as water by the parched and craving throat, as refreshing showers by roots in a thirsty soil, and as the favors of a husband by his bride. For the blood bereft of its life and spirit in the cortical and fibrous substances of the brain (*a*), and recruited and reanimated in the receptacles of the sinuses, descends by the internal jugular vein to the threshold of the trunk, and there joyfully meets the chyle carried up by the thoracic duct, and bears it without delay, as in a manner plighted and betrothed to it, into the right chamber of the heart, where the two celebrate their nuptials, and enter the marriage-bed, and conjointly generate a new sanguineous offspring (*b*). The canal that takes the

(*a*) That the cerebrum attracts the purer, lighter, and more easily-divisible blood, for the purpose of extracting and as it were recovering the spirit which inhabits it, and of circulating this spirit through the fibres, is a point we shall prove in detail in the Chapter on the Cerebrum. Such being the case, therefore the residue of the blood, (after its division or resolution has taken place,) and the harder and more refractory globules, which do not admit of easy resolution, are rejected as recrementitious, into the sinuses of the dura mater; and hence the mass of blood contained in these sinuses would be exceedingly dry, sluggish, obsolete, and apt to coagulate, unless the new spirit put forth from the pituitary gland in their ultimate receptacles, that is to say, in the terminal retorts of the lateral sinuses, roused it again from its lifeless state, and revived it for a time.

(*b*) That sanguification itself is primarily performed in the heart, where, as in a whirlpool or cauldron, are mingled all kinds of elements which the soul selects by means of the cerebrum, and applies by means virgin chyle to its future husband, the spirit, either joins the jugular vein immediately, or else the left subclavian, in which latter case it communicates with the jugular by means of a sponsor or bridesmaid (*c*).

The soul gives the blood its life, the spirit its bond of union, and the chyle its embodiment (*d*); thus every globule of the blood has both a soul and a body; and what is true of all the parts is also true of the whole that lives from the blood (*e*).

of the spirit, as suitable for the generation of the blood, see the Treatise on the Heart, in my *Economy of the Animal Kingdom*. That the heart is a chemical vessel for preparing liquids to enter into the composition of the blood, *see ibid.*, n. 453, 454, 455, 457.

(*c*) A clear idea may be obtained from anatomical plates, of the manner in which the jugular vein is implanted in this situation into the subclavian vein. It appears that the blood which comes down from the cerebrum, flows instantly, with scarcely an interval, into the chyle thrown out by the thoracic duct; and then that it and the chyle go headlong to the right auricle of the heart, or into the superior vena cava. Whence it is absolutely certain, that the insertion of the thoracic duct close to the jugular vein, is in some way salutary and necessary to the body; but in what particular manner, it is quite impossible to know, without first exploring the nature of the contents of the jugular vein. That this vein draws off the blood thrown out by the cerebrum, and likewise conveys the fresh spirit, will be seen in our analysis of the cerebrum. This is the reason why the thoracic duct is sometimes observed to pass into the jugular vein itself. "The termination of the thoracic duct," says Heister, "is generally in the left subclavian vein, or else in the jugular" (n. 158). And Winslow says, "It opens into the back part of the subclavian vein, near the left side of the internal jugular" (n. 160). Eustachius places its termination at the back of the internal jugular. See n. 161.

(*d*) We shall treat of these subjects in the Chapters on the Blood, on the Animal Spirit, and on the Soul.

(*e*) By the body we mean particularly that which lives from the blood, or which consists of elements borrowed from earth, air, and water; and which are carried by the food and the chyle, and supplied to the blood; consequently, the body is all that which consists of ultimate, or angular and circular forms, which are the means summoned by the soul to enable it to live in the body, and, in this ultimate region, upon the earth.

Consequently when the spirit descends from the brain, in company with the exhausted blood with which it can no longer live united, it burns to effect an espousal (*f*), and to give the one to the other, (that is, the soul to the body,) in marriage. Hence its chief aim on meeting the chyle, is, to exert on it a secret attraction, and imperceptibly and with mystical rites to unite and to marry it (*g*).

163. The chyle is carried upwards from the receptaculum to the subclavian vein, by a number of appliances and powers; that is to say, it is allured, it is attracted, it is promoted, and it is impelled. It is *allured* by the spirit, which desires conjunction and alliance. It is *attracted* by the heart, which like a vortex or whirlpool sucks in the waves of blood, and attracts them in a fresh and constant stream; particularly the blood of the jugular vein (*h*), which renovates its vital motions, vivifies

(*f*) That is to say, between the soul and the body; for what is called the animal spirit is their uniting medium, and owes its principal essence to the soul, of which it is the habitation.

(*g*) It cannot I think be doubted, that the fresh spirit descends from the cerebrum towards the chyle, with a kind of mystical effort and affection; and that the chyle rises to meet it with the same: nor can the means be wanting, inasmuch as the soul possesses life, and is omniscient with respect to all things done in the body; and consequently is enabled to dispose all things to this end. Furthermore, a kind of auxiliary or impulsive force, arising from causes in nature, must also be present, both externally and internally; for attraction never can exist by itself. What other properties exist in the chyle and spirit, besides their homogeneousness, harmony, and desire of conjunction, it is not easy to explain, until after we have thoroughly examined the blood and the spirit. A similar attractive force, but of a physical kind, is observed in the vegetable kingdom; where all the fibres of roots attract their own proper species of humor, and impregnate themselves therewith; so that a kind of magnetism pervades the world. This force then must be called mystical, before it is explained. The way in which the magnet attracts iron is also mystical; yet still manifest, so far as the fact is concerned.

(*h*) That the cardiac motions are principally kept up by the afflux of blood from the brains, and from the medulla oblongata and spinalis; but not so by the blood of the body, which reaches the heart through the inferior cava, see my *Economy of the Animal Kingdom,* treatise i., the sanguineous system with spirit, and sustains it with chyle. It is *promoted* by its own active fibrillæ, in every cell and in every point; for the tube diminishes to a mere thread of the smallest possible dimensions, and continually urges it onwards (*i*). It is *impelled* by the receptaculum, which contracting alternately, is always extruding the lacteal stream, and urging it into the thoracic duct (*k*).

164. The duct passes through as it were the general axis of the body, and runs indeed through mere centres. For the vertebral column, on which it rests, consists of mere centres of motion: each rib respects and keeps to its own vertebra as its proper centre, whence it runs forth into its circumferences: all the ribs together, and the muscles and the viscera, respect two most general centres (*l*); one, where the thoracic duct begins,

n. 553—558, and n. 570—577. The brain, in a general sense, may be considered as the living fountain of all the motions of the body, and the fibres of the brain as their intimate causes. But the cardiac motions themselves are produced by the blood of the brain, which flows into the heart by its own proper living force; thus they are produced by primary causes, and not, as in the body, by secondary causes.

(*i*) When the lymphatics are emptied, they become invisible; and almost the same remark applies to the thoracic duct; a sign that these vessels, like the arteries, contract, and tend to obliterate their canals; also, that they are furnished with the finest motive fibres; for nervous fibre acts in the same manner as muscular fibre, only more perfectly.

(*k*) Respecting the propulsive force of the mesenteric glands and of the receptaculum, see above, n. 153. The fact stated may be demonstrated by injections, for, according to Verheyen, "When air is forced into the receptaculum, it readily passes through the thoracic duct and the vena cava, all the way to the right cavity of the heart, which it distends, presenting a beautiful appearance. When the duct is compressed, it swells up towards the receptaculum." (*Corp. Hum. Anat.*, tract. ii., cap. xiii.) The extreme gentleness of the impulse communicated by the receptaculum, will be seen presently. And inasmuch as the chyle is promoted also by the peculiar compression of the duct, hence the latter is necessarily furnished and divided with a great number of noduli or contractions, and valves.

(*l*) We shall speak of these subjects in the Part on the Spinal Marrow, where we shall treat also of the vertebral column, and of its levers the ribs, as well as of their determinations of motion. In the other, where it terminates. Now, as the duct passes through continual centres, and consequently through continual equilibria, it is, therefore, necessarily obedient to every power, whether attractive or propulsive.

165. The thoracic duct is subject to the same systole and diastole as the contiguous and continuous viscera, from which the lymphatics proceed; that is to say, to the rhythmical movements of the respiration. It rises along the vertebræ, which constantly receive the action communicated to the ribs by the lungs, and vibrate, twirl, and rub the duct, during each alternation. It is enclosed in the pleura, which unfurls and contracts with synchronous reciprocations. The receptaculum likewise is enclosed in the peritonæum (*m*), which, by the coöperation of the muscles, and of all the abdominal members, keeps time with the general movements. The diaphragm also surrounds it at its commencement, and governs these alternate changes. The vena azygos, which is in contact with the duct; the intercostal vessels, which lie close to it (*n*); the jugular vein, which termi-

all parts of the body, there are particular, special, and general motions; each referrible to its own centre. The centres of the particular motions respect the centres of the special motions, as general centres. The centres of the special motions similarly respect those which are the centres of the general motions. The axis is the line through which these centres run, and which they constitute. Thus there are various axes, more or less universal. Hence we see what is the nature of the form of motion.

(*m*) "The thoracic duct," says Verheyen, "is defended by being placed under a second lamella of the pleura: the receptaculum consists of a thin membrane, which is protected against rupture by its peritonæal covering." (*Corp. Hum. Anat.*, tract. ii., cap. xiii.) That the pleura and the peritonæum cannot expand and contract otherwise than synchronously with the lungs, will be shewn in the proper places.

(*n*) Respecting the manner in which the intercostal vessels are in contact with the thoracic duct, see Cowper's Tabulæ and descriptions. The blood of the aorta is not poured into these vessels synchronously with the pulse or action of the heart, but synchronously with the respiratory movements of the pleura, the sternum, and the ribs; at the time when the costal muscles, and the others which take part in respiration, are filled with blood. On this account, they come off perpendicularly nates it (*o*); and the aorta itself, which accompanies it, and sometimes perforates the receptaculum (*p*), all contribute to dispose it to similar reciprocations.

166. The orifice or osculum by which the thoracic duct is inserted into the subclavian vein, and by which the chyle is committed to the blood, is furnished with membranous pellicles, not, however, of a valvular nature, but rather resembling inclined planes, or little bridges, which direct the chyle along the stream of blood, towards the right auricle of the heart (*q*).

from the trunk of the aorta. That the vena azygos imbibes the blood, and pours it into the superior cava, synchronously with the respirations, see my *Economy of the Animal Kingdom,* treatise i., n. 283, 566.

(*o*) Inasmuch as the brains and the lungs coincide in the times of their animations, therefore the internal jugular vein, which carries away the blood from the brains and their sinuses, must inevitably be actuated by the same motions.

(*p*) It may seem as if the aorta did not obey the pulmonic motion, inasmuch as the pulse in all its superficial branches is manifestly synchronous with the systole and diastole of the heart. But if the pulmonary and the cardiac motions be more deeply considered, it will be evident, that they both coëxist in the same tube and plane, and both perform their respective reciprocations harmonically; as in the diaphragm, in the pericardium, and even in the heart itself, which lies imbedded in the lobes of the lungs, and connected and inserted into them by the pulmonary artery; and consequently, without doubt, in its volume or body, obeys the traction and respiration of the lungs, and nevertheless at the same time has its own peculiar systole and diastole. The same may be said of the dura mater. Nothing is more common or better known, than the coëxistence of two, three, or four motions, simultaneously in one compound; this being continually the case in the organic body, as well as in practical mechanics. Thus the trunk of the great artery obeys the pulmonary movements, wherever in its descent it comes in contact with either the vertebræ, the pleura, the diaphragm, or the peritonæum; which yet does not by any means hinder it from having its own proper reciprocations, synchronous with those of the heart.

(*q*) Wedelius has treated this subject ingeniously in a particular tract on the valve; where he demonstrates, that no blood could possibly enter the thoracic duct, from the subclavian vein, even if the valve were totally wanting. This is sufficiently confirmed, by the perpetual

These membranes also withhold and separate the chyle from the spirit, and prevent it from being ravished before the nuptial hour, and before it reaches the bed-chamber of the heart, and diffused promiscuously through the blood; for were this to happen, the laws of the legitimate bed would be violated inevitably, and no binding marriage could take place (*r*).

167. The chyliferous vessels, the lymphatics, the receptaculum, and the duct, with their transparent and delicate membranes, cells and valves, clearly indicate the nature of the chyle which permeates them, and of the lymph which supplies the place of the chyle when the latter is deficient (*s*): they indicate that this lymph is of an illustrious stock; the most refined lymph in the body; the noblest progeny of the blood and the serum; like the chyle, but still more purely, the nurse, the vehicle, the bed and throne of the spirits, with which it is richly endowed and vivified; the true purer blood (*t*).

direction of the stream of blood towards the heart; and by the attractive and propulsive powers which have been before shewn to act upon the chyle. The so-called valves then, (which, according to Winslow, are "pellicles so disposed as to permit the onward passage of the chyle towards the vena cava," n. 160,) are merely directing planes, to guide the chyle along the stream, and to prevent its nuptials with the spirit from taking place, before it has entered the chambers of the heart.

(*r*) For, as shewn above, the spirit as eagerly desires, and as greedily imbibes the chyle, as arid ground or thirsty roots imbibe refreshing showers; as a bride receives the favors of her husband; or as sulphur catches flame. Consequently, if the chyle were allowed to enter promiscuously, coition would evidently precede legitimate marriage, (which ought properly to take place in the heart,) and the fruit would be an adulterous offspring, that is to say, spurious blood.

(*s*) According to a general law, the coats of the vessels, and their fluids, are mutually determinant of each other. The one requires to be absolutely accommodated to the other, in the same manner as active and passive, instrumental and principal; wherefore the one may be judged of from the other.

(*t*) It would be tedious were we to comment at full length upon all the particulars of this description of the lymph which permeates the lymphatics: for a knowledge of the blood, of the spirits, and of the soul itself, is presupposed, to serve as the premiss to such comment as

168. The subtle lymph which glances through these vessels, tempers, edulcorates, adapts, and informs the chyle, and renders it suitable and marriageable; and serves as the link between the chyle and the spirit, as the spirit is the link between the body and the soul (*u*). This is the reason why such numerous stream-

a conclusion. But thus much experience assures us, that this lymph is of some important use to the blood, and of some important use to the chyle. Its extreme tenuity and volatility, is clearly evidenced by the membranes which take it up, and impel it into circulation: the lymph and the membrane being proportionals and correspondents. Verheyen asserts, "That he has stretched the receptaculum of a dog to the full size of a pullet's or bantam's egg." (*Corp. Hum. Anat.*, tract. ii., cap. xiii.) From which circumstance we may conclude, how slight a propulsive force the chyle, or the lymph requires, to drive it upwards through the duct, divided, as it is, by such numerous nodes and locks; for a membrane so extremely extensile and elastic, must of course impel its liquid, necessarily of corresponding levity and nature, with the softest touch, and the gentlest force. "The receptaculum chyli," says Winslow, "is composed of very thin coats" (n. 145). "The thoracic duct," says Heister, "is a very delicate canal" (n. 158). Nuck delineates the valves and coats of the thoracic duct, with certain orbicular granules scattered over it, in his *Adenographia*, fig. xx. xxi. Respecting this lymph, Boerhaave says, "This liquor, abounding in water, spirits, and the most subtle salt, is the finest and purest part of the blood; as its place of preparation, excretory vessels, and sensible properties demonstate." (*Inst. Med.*, n. 123.) He also states that it may be evaporated to dryness, without leaving a residuum[x]: others who have exposed it to heat, declare that it was converted into a substance resembling the boiled white of egg. Thus much at least is certain, that this lymph is more pure than the blood itself; inasmuch as it flows back into the blood, and is implanted within it.

(*u*) The chyle itself is not always defecated, but sometimes contains crude vegetable emulsions, effluvia, juices, and vapid portions of new wine, of sour fruits, or of unfermented or unclarified beer; and is mixed with those matters which are ultimately thrown out by the bladder: wherefore the lymph is required, to purify, correct and qualify it. Were it not for the lymph, it might perhaps stick in the constrictions, strainers, and noduli of the duct. Hence it may be concluded, that this lymph is a kind of ultimate saliva, whose office it is to digest the chyle itself, in the same way as the common saliva digests the food lets of it run to the receptaculum at the beginning of the duct, and to the court at the end, where the duct enters the vein (*x*). It seems reasonable to suppose, that the mesenteric glands qualify the chyle with some secretion, and confer upon it some peculiar property, to facilitate its introduction into the blood (*y*).

169. Every time the blood is resolved into its principles, and throws off its useless parts, the finer and richer humor, and the spirit of the blood, betake themselves into the cellular coats of the viscera, and thence into the lymphatics, with a view of returning into the blood (*z*). Thus no portion of the nobler essence of the blood is suffered to waste, before it has performed its destined use; and this is the reason why the circulation of the lymph continues for a time, after that of the blood itself has ceased (*a*).

in the stomach. Moreover, it issues in many places from the glands, as the sacral, the subclavian and the thymus glands. Thus it inaugurates the chyle into the blood, and is a kind of bridesmaid.

(*x*) Lymphatics flow into the receptaculum, from the stomach, the liver, and many of the members and septa of this region; and enter it, comparatively as subterranean veins enter a fountain, at its lowest part; in this situation evidently sprinkling in somewhat of a diluting and tempering nature. The receptaculum is also bilocular or trilocular, in order to the thorough mixture of its contents. Lymphatics also proceed to the upper end of the thoracic duct—the threshold and court, as I term it, of the subclavian vein; and which sometimes has a number of divisions, and sometimes dilates for a short space into a kind of ampulla. The lymphatics which enter it here, come from the subclavian and thymus glands, from the lungs, &c., (see Cowper's Tabulæ, and the observations of our authors,) in order that here also the chyle may be prepared, and enriched with some peculiar property, before it comes up into the blood.

(*y*) Innumerable arterial vessels penetrate into these very glands, and by a lymph of their own appear to correct the lymph which flows into the glands; particularly in man, for the reasons mentioned in the last chapter. Heister also is of this opinion; for he classes it among the uses of these glands, that "they secrete a fluid, wherewith to dilute the chyle" (n. 141).

(*z*) Just as in the intestines; for the lymphatics of the viscera are always seen to arise from the coat next to the external one.

(*a*) The chyle and the lymph are sometimes found in their vessels,

170. From the lymphatics we may infer respecting the thoracic duct; from the duct, respecting the mesenteric glands, and the receptaculum chyli; and from both together, respecting the cellular tissues; for all these parts are continuous, and identical in use, structure, and nature. The lymphatics may be regarded as lines or threads; the cellular tissues, as areas, planes, or webs of these lines; and the glands, as masses or volumes thereof, or bodies of trine dimension (*b*). Thus each shews the nature of the others. They all admit and absorb such lymph (*c*),

for several hours after death, or even for a whole day: thus they continue to circulate long after the circulation of the blood has ceased. "The passage of the lymph," says Verheyen, "does not cease with life, like that of the blood; but it continues for some time after death; as may be proved by putting a ligature upon the thoracic duct." (*Corp. Hum. Anat.*, tract. i., cap. iv.)

(*b*) The same texture prevails in the cellular coats as in the lymphatics themselves; there being an open passage from cell to cell, or from loculus to loculus: so that it is evident at a glance that the cellular coat is as it were a lymphatic projected into a plane, or extended into length and breadth: which also is a necessary consequence, if the lymphatics derive their origin from the cellular coat. That the mesenteric glands are also of a similar nature, is shewn by Winslow, where he says, "They appear to be cellular, surrounded by a thin coat, that under the microscope seems to consist of an interlacement of filaments" (n. 143): and by Nuck, in many of his figures, where he represents these and similar glands as made up of very minute follicles or cells.

(*c*) Respecting these particulars, see the preceding chapter, n. 156, 157. Inasmuch as the state of each fibre corresponds to the state of its principle in the brain, it follows, that desire and aversion may be predicated of the fibre, in the same manner as, in certain cases, of the brain: consequently that all the fibres, just like the tongue and stomach, either receive with eagerness, or reject with aversion, what is offered to them; for whatever is a property of the tongue and stomach is a property of the very fibres. By remembering this, we may explain the causes of innumerable effects; and shew in general how it is, that the minute venous orifices take up nothing but what is suitable for the blood, and reject everything heterogeneous. This is not to be accounted for by fitness between the pores and the particles; for those things are sometimes received, which were formerly rejected; and those are some-

or such chyle, and such only, as the animus of the fibres desires and demands.

times rejected, which were formerly desired: just according to the changes in the state of the animus. The subject of the lymphatics will be resumed in the next chapter.

CHAPTER VIII.

THE GLANDS GENERALLY.

171. THE viscera that now present themselves for analysis, are the liver, the pancreas, the spleen, the kidneys, the succenturiate kidneys &c., which appear to be made up of glandular forms and corpuscules. In order to further our insight into these viscera, I have found it necessary to premise an adenology or analysis of the glands, by way of introduction. The thoracic duct, the receptaculum chyli, and the cellular and particularly the glandular coats of the viscera already described, afford some preliminary light. But the vastness as well as the extreme difficulty of the subject will be seen from the following attempt.

172. NUCK. "The glands usually reckoned conglomerate, are such as are composed of a congeries of lesser glands, and surrounded by a common coat; and each of the lesser glands puts forth an excretory duct; and these ducts again form a canal, through which a certain liquid flows, either into some considerable cavity, or to the outside of the body. Of this kind are the glandulæ innominatæ, the parotid glands, the pancreas, the glands of the mammæ, &c.; which, with all the others of this description, are nothing more than a texture of very minute vessels, composed of the smallest arteries, which carry blood laden with various particles, and terminate in two orders of vessels, of which some are venous, and admit venous blood, and some excretory, as we observed before. To get a clear view of the latter vessels, they must be made the subjects of certain experiments. When the membrane investing a conglomerate gland is removed, and any liquid injected through either its excretory duct, or its arteries, we then observe the lesser glands receding from each other; their little vessels turgid with injection; and the glands themselves all rising up. *(Adenog.*,

de Gland. in Gen., cap. ii., p. 8, 9.) Having obtained a mamma, the nipple of which was perforated by many excretory ducts, I pressed out of it all the milk, and selecting one of the largest orifices, I injected my mercury into it with as much dexterity as I could command; and immediately I had the pleasure of observing, not only that the lactiferous ducts were beautifully filled, and presented an arborescent appearance, but that several of the lesser ones had so far admitted the mercury, that it had passed into certain arterial vessels, continuous with the lactiferous ones. (*Adenog.*, *de Gland. in Gen.*, cap. ii., p. 12, 13.) The [conglobate] glands have two membranes, an exterior and an interior. The exterior is the thinner membrane, and encloses the whole gland, and adheres in general so firmly to the membrane underneath it, that it cannot be removed without laceration; although in some cases its connexion is looser, and it may be separated more easily. It consists for the most part of circular fibres, without either beginning or end, except about the entrance of the vessels, where the circular arrangement is usually disturbed. The glands have four vessels, two adducent or afferent, and two abducent or efferent. . . . When the exterior membrane is removed, we come to another and a thicker and more compact membrane, immediately covering the substance of the gland, and which is perforated throughout by both the afferent and efferent vessels. Its pores, however, are so minute, that air does not ordinarily escape through them, notwithstanding it is forcibly driven through the vessels; a result to which the exterior membrane contributes in no slight degree. It is furnished with various fibres, longitudinal, circular and oblique, which touch each other in innumerable places, but are arranged quite irregularly. It is also supplied with both afferent and efferent vessels, arising from the same ramifications as the vessels of the exterior membrane. (*Adenog.*, *de Gland. Conglob.*, cap. i., p. 27, 28, 29.) Besides the investing membrane, I found that the glands have also a fibrous tissue, consisting of a number of fibrillæ of different species and figures, and all united together. The exterior ones, placed immediately under the internal membrane, (to which also the fibres were firmly connected,) by their extremities, which represented so many little heads, were connected to the membrane in a certain regular order; so that the intermediate spaces, being ordinarily filled with lymphatic juice, rendered the coat uneven, and dotted with great numbers of little hemispherical tuberosities. The fibres immediately under these, and which were turned inwards towards the gland, formed various angles, some being hexagons, some pentagons, and some of different other figures, according as best suited the other part of the fibrous texture. The rest of the fibres, as far as the centre of the gland, still followed no other arrangement, but kept the same irregular order; some being larger, some smaller, some longer, some shorter, than others, but always firmly connected together, and nearly resembling the moss on a tree: wherefore the conglomerate glands might not inaptly be termed, glandulæ vasculosæ, and the conglobate, glandulæ muscosæ. See fig. xiii. After a careful examination, I have never been able to find canals in these fibres; the whole of them, excepting the blood-vessels, are destitute of cavities. And here it may be observed, that the compages of the glands was not the same in all subjects; but more lax in some, more tense in others, according as the glands were distended by a greater or lesser quantity of lymph. The glands also vary greatly in color; some being grey, (which is the usual color,) some yellowish, some blackish, and others again, variegated. These glands, indeed, are naturally full of a certain transparent and limpid fluid; but this is often changed into a viscid humor; and I have seen the interstices of the glands filled and obstructed with tartareous and sabulous matter: and in some instances I have extracted as many grains of sand as there were cavities between the fibres, and occasionally have found the whole gland in a calculous state. (*Adenog., de Gland. Conglob.*, cap. ii., p. 35—38.) Besides the fibres already mentioned, I have also seen various others, extending from one side of a gland to the opposite side; some being nearly rectilinear; others departing from the right line, and forming obtuse angles. I half suspect that these fibres, so called, are vessels, either sanguineous or nervous. But having often previously observed, that each gland receives at least one, frequently two, and not seldom three or four branches from the neighboring arteries, hence I selected one of these branches for injection with mercury; which finding a ready passage, made the little artery with its various ramifications conspicuous through the tissue of the gland; nor were the corresponding veins wanting, divided into a number of lesser twigs: but the blood-vessels were greatly exceeded in number by the nerves, which kept along their sides. (*Adenog., de Gland. Conglob.*, cap. iii., p. 39, 40.) I admit that these glands cannot be the origins of the lymphatics; for more than once I have observed so close a connexion between the lymphatics and the arteries, that when a little artery was inflated, the lymphatic of some gland was inflated at the same time, and its lymph seemed often bloody, like the washings of flesh. A lymphatic sometimes puts forth lateral twigs, its principal branch only lying on the gland, and taking up again the returning twigs. See fig. xxvii. No lymphatic can possibly finish its course, without being obliged to salute the glands in some part thereof; as we see particularly in those parts where there appear to be no glands, and yet where there are lymphatics;

which latter, in such case, by ways hitherto unnoticed by anatomists, and as it were, by secret paths, seek out some gland, and discharge into it their lymph; there to be mingled with some other lymph proceeding from a different source; and thence propelled by the common efferent vessel, towards the receptaculum chyli, or some other destination. See fig. xxix. *(Adenog., de Gland. Conglob.*, cap. iii., p. 47, 48, 49.) While inspecting the abdomen, I have seen the lacteals still turgid with chyle in perhaps a hundred places; and the glands of the mesentery full of a certain juice: and being desirous of knowing what intercourse there was between the lacteals and these glands, I injected mercury into one of the vessels proceeding from the intestines, and which was conspicuous on the outside for the number of its valves; and I had the unexpected satisfaction of observing, that not only was the lacteal filled by the injection, but the gland, to which it went straight, assumed a new figure and appearance; changing from flat, even and smooth, to globular, rough and tuberous." *(Adenographia Curiosa, de Gland. Conglob.*, p. 31, 32. Ludg. Bat. 1691.)

173. Malpighi. "There are many species of conglobate glands, differing from each other in their exterior configuration and internal structure, as well as in the quality of the juices they excrete, and in their excretory vessels. The gland with which the palate, the œsophagus, the intestines, and other similar parts are plentifully furnished, is the simplest of all, and the model of the rest of the glands. It consists of a membranous follicle or loculus, which is sometimes oval, sometimes round, and sometimes lenticular, or oblong. Around this loculus or follicle, blood-vessels and nerves ramify, and we may conjecture that it is also surrounded by fleshy fibres, or at any rate covered by a muscular extense, something like what we see in the stomach and œsophagus. Next in order to this simple gland, are glands consisting of a number of loculi; as those in the face, in the lips, in certain parts of the skin, about the pudendum and the palate; in which the excretory vessel, (which is sometimes oblong,) has numbers of membranous loculi appended to it; and which open into it, and are surrounded by ramifications of nerves and blood-vessels. The conglobate glands are situated in the axillæ, in the groin, in the fat, in the mesentery, and in almost every other part of the body. They are covered externally with a very dense membrane, which is supplied by lateral branches from the blood-vessels. Under this are placed circular fleshy fibres, which penetrate the body of the gland horizontally. This membrane is also frequently studded with multitudes of minute, round tubercles, caused by the turgid corpuscules contained within the gland. When the membranes are removed, or what is better, when a longi-

tudinal section of the gland is made with a scalpel, and the part is well macerated in water, we meet, on a careful examination, with the following phenomena. In the first place, transverse fleshy fibres proceed from the investing membrane; but they do not lie parallel, but slant and cross each other; and by their inosculation, or at any rate, their interweaving, they form numerous and indeed almost innumerable reticular areolæ, generally roundish, though sometimes angular, and of different sizes. In the middle of each areola there is a glandular loculus or follicle, which is either round or oval; and larger or smaller according to the larger or smaller quantity of substance which it contains. It is composed of a soft, white membrane, which collapses when the humor is evacuated; and if it be cut, a cavity is seen. These loculi are very similar to the glands of the spleen. But when the glands are handled, the loculi are emptied, and become contracted and indistinct; and hence it is not always easy to see them, and for a long time, both my mind and eyes were puzzled by this circumstance. At length, however, I found what I wanted, in diseased oxen, and other animals of the same kind; where the glands were sometimes increased to a hand-breadth, and their loculi so full of tartar, that a slight examination was sufficient to shew them. In these obstructed glands, the membrane of the follicles was more solid and thick than in the normal state, and near the glands were unusually large varicose and reticular productions of blood-vessels, such as I have sometimes observed in the larger and redder loculi. But in order to see the loculi in all the glands, the exterior portion of the latter must be examined, for corresponding swellings are caused on the outside by the turgid loculi within, and when the gland is laid open, we find rows of loculi, surrounded by areolæ of fibres, under the membrane itself. The follicles of the glands are full of a certain cinereous humor, which is but little transparent in the natural state; but when the animal is diseased, they contain other matter; generally tartareous and mucous humors, or concreted juices. The loculi are appended to the blood-vessels which pass over the fasciculi of the fleshy fibres, and form the areolæ; wherefore the body of the gland is composed of various layers of areolæ and spaces of the kind, placed one upon another. The blood-vessels, that is to say, the arteries and veins, enter these glands by many branches, mounting over their sides; and in some of the larger glands which are connected to the trachea, the vessels enter the belly of the gland, or the concavity formed by the doubling in and approximation of its two ends. After entering, their larger branches form a network, and their last twigs appear to terminate on the loculi and parietes of the areolæ. The glands have many nerves, and sometimes a single nerve enters their substance. And although in the interior of some of them there is a reticular tissue resembling a thin membrane, yet the filaments of which it is composed are not all nervous, but are frequently portions of fine fleshy fibres, and of the fimbriated membrane which covers the gland exteriorly. The conglobate glands certainly have lymphatics appended to them, and which communicate with their innermost substance; so that a single lymphatic penetrates the smallest glands, and very frequently many enter the larger ones: thus we may declare that every conglobate gland is supplied with lymphatics. *(Op. Post., De Struct. Gland. Conglob.*, p. 139, 140, 141.) So great is the abundance, and such is the minute division of the lymphatics in the spleen, as to admit of neither description nor representation. Their trunks pass out about the thicker portion of the spleen, where the blood-vessels have their ingress and egress. *(ibid.*, p. 142.) In the fat and the interstices of the muscles, and within the conglobate glands themselves, we frequently find little glands, not bigger than vetch-seeds, and of a red color; these when laid open are seen to contain coagulated blood, extravasated from the lòculi. . . . With respect to the glands called renes succenturiati, I have clearly made out that they are supplied by a beautiful network of vessels, and particularly with vast numbers of white nerves, reticularly interwoven; whence it seems probable, that a further separation is carried on by the nerves in the cinereous substance of these glands, or rather that this substance is an appendage to the excretory vessels, and their extremities; inasmuch as it is immediately connected to a broad and capacious duct, which passing lengthwise, opens externally, and discharges itself into the emulgent veins; while at its other extremity it gives off copious branches, whereby it receives the humor separated by the glands and tubuli. This duct or cavity is lined by a fine membrane, which is studded with innumerable foramina of irregular figures; whence it is probable, that great numbers of excretories open into it, and that there is a passage from it into the foramina; as we see exemplified in the kidneys. I have often noticed an analogous structure in certain glands situated in the intestine that adheres to the fleshy stomach in hens and chickens. *(ibid.*, p. 144, 145.) Elsewhere we have pointed out, that nature makes use of the same method in the composition of the viscera, as of the glands, and that the liver, the cerebrum and the kidneys, are glands: the compages of follicles in them, appended to the excretory vessels, clearly proves this to be the case. . . . The glandular nature of the cerebrum, made up, as it is, of membranous follicles, is forcibly shewn by the remarkable case which Wepfer relates, where the cranium was entirely filled with a congeries of vesicles, from which arose fibres that passed towards the base of the skull. In more than one instance a large vesicle has supplied the place of the brain." *(Opera Posthuma—De Structurâ Glandularum Conglobatarum,* &c., p. 145, 146. Amstelod. 1598.)

174. Boerhaave. "Some of the glands are simple, others are compound; the latter commonly consisting of the former, aggregated and enclosed in a common membrane. The simple glands discharge a peculiar humor through their lymphatic ducts, either into the chyle or the venous blood; or exhale it from the exterior of the skin, or from the surface of the free membranes that are found in all parts of the body. The compound glands, on the other hand, discharge their humor, (which is prepared in every part of the gland,) through little canals belonging to each part, into a larger canal, and through this common excretory duct, into the great cavities of the mouth and intestines, or else out of the body; for particular uses. The first kind are called conglobate, and the latter, conglomerate glands. The simple glands are composed of a thin external membrane, and of an internal one which closely adheres to it. The first, composed of elastic circular fibres, entirely surrounds the glands, and contracts and compresses them; it chiefly consists of a contexture formed by the small vessels which enter into and pass out of the glands. The internal membrane is thick and more dense, being formed of fibres passing in all directions, and of an interlacement of vessels; and serving almost for the same uses as the external membrane. These glands are furnished with arteries, the branches of which are supported by and distributed in the membranes, in a firm and regular order, and so accurately conveyed to every minute particle of the gland, that if wax or quicksilver be injected into the small arteries, it increases them and compresses the other vessels to such an extent, that we might thereby be led to the false conclusion, that the whole fabric is arterial. They are also furnished with veins, which are distributed much in the same manner as the arteries. They have more and larger nerves than any other part of the body of the same magnitude; and these are so divided in the gland, that they seem to be present in all parts thereof. They have also lymphatics, which go to, and return from them. Their arteries are tubes—conical, curved, branching, elastic, and convoluted; at the extremities, cylindrical, no longer branching, but now changed into veins; but before they suffer this change, the small arteries communicate with one another by numberless anastomoses, in various positions, and at innumerable angles; so that their extremities are distributed in various ways in the different glands. The arterial blood, therefore, when driven into the glands, experiences rapid motion; strong resistance and compres-

sion; pressure of its parts against each other; oblique pressure; perpetual change of the points of contact; application many times, and in every direction, against all the minutest points of the canals; momently varying rotation in every particle; squeezing together of the particles; onward and backward movement in the tubes; attenuation; attrition; preservation of fluidity; solidity; polish; secretion; and thorough mixture. Now as the branches which arise from an artery are generally narrower than the trunk from which they rise, so in these minute vessels the last branches are less than the last trunk. The last trunks convey the red, thick part of the blood, and pour it into the beginnings of the veins; the narrower branches receive the finer, more fluid, and pellucid parts, which are less in diameter than themselves, and which have been pressed by a strong, oblique, opposite force. But this thin humor, thus separated from the grosser part, is no longer blood, but another fluid, and that, too, various,—as sweat, perspiration, matter in the pores, tears, fatty wax, cerumen, mucus, saliva, sputum, linimentum, lymph, serum, bile, semen, oil, milk, fat, &c. Therefore the last branches, ceasing to be named arteries, are differently called, according to the nature of the humors they carry; and as they often again put on all the properties of arteries, they must have also their smaller branches and their veins: hence the arteries and veins are as much vehicles of serum, lymph, water, and spirit, as of blood; nor do we know where these vessels terminate; but hence at least we see the origin, progress, termination and office of the lymphatics; which not only consist of veins furnished with valves, and large enough to be visible, but also of arteries without valves, and which, by reason of their thinness and transparency, are invisible. This we learn by the artificial means invented by Ruysch. Yet the branches, perhaps, of every such artery, now no longer branching, but straight, and distributed in the delicate little membrane of the finest glandular follicle, discharge their humor by open mouths into the common cavity formed by that membrane; where, being collected from all quarters, it in a manner stops; and constitutes the glandular lymph, there prepared and reposited. It is by no means improbable, that the nerves of the glands also discharge the spirits into them by a similar apparatus, mixing the spirits with this lymph, and thus supplying it with the requisite qualities. And the lymphatic arteries frequently bring their lymph, discharged into their valvular veins,—a lymph which we call vascular,—to those glands; and after a different manner pour it into the same follicle, and mixing it with the spirits and the glandular lymph, they subtilize it afresh. When the abdomen of a healthy, living animal is opened, the lymph tends rapidly from all parts thereof, towards the receptaculum chyli; and even after death, when the lymphatics are wounded, it tends to pass out of them, a contraction of the body taking place from the cold of death. Then this compound liquid passing out through the lymphatic veins, is driven by the contractile power of the fibrous membrane, by the motion of the artery and the pressure of the muscles, into other glands, there to undergo the same alterations; and thence into the receptaculum chyli, the thoracic duct, or the sanguiferous veins. And these seem indeed to be the universal conglobate glands of the body.

175. "But in other glands the case is different; for the follicle directly expels the liquid it receives, by its emissary vessel into a common cavity, as into the frontal sinuses, the osseous cavities of the superior maxilla, the cells of the sphenoid bone under the sella turcica, the recesses of the spongy bones, the cavities of the nares, the lacunæ of the tonsils; where the mucus secreted, is deposited, collected, and altered. This is the case with the mucous glands of the mouth, of the posterior part of the tongue, of the exterior and interior of the epiglottis, of the internal nares, of the meatus auris, of the fauces, the larynx, the trachea, the bronchia, the œsophagus, the stomach, and the intestines; which glands may be called simple excretory glands. Others again, in the same way, discharge the humors they have prepared, to the external surface of the body, through peculiar emissary ducts arising from a cavity, as in the meatus auditorius, the pinnæ nasi, the exterior of the nose, the beginning of the internal nares, in the face, the neck, the axillæ, the scapulæ, in the areolæ of the mammæ, the areola of the umbilicus, in the nates, the margin of the anus, the perinæum, the pubes, the mons pubis in both sexes, in the scrotum, the integuments of the penis, in the labia vaginæ, and the knees; the glands of which parts are now styled, sebaceous glands. (*Inst. Med.*, n. 241—252.) But the simple glands already described, or others similar to them, when united together by common vessels, and connected and enveloped by a common membrane, generate certain compound, or, as they are called, conglomerate glands. These have usually one common excretory duct, which receives the humor sent into it by emissary ducts from all parts of the gland, and collecting it, pours it into some larger cavity. The glandula innominata of the eye, the parotid gland, the pancreas, &c., are glands of this description. Moreover, the excretory duct of the common receptacle just mentioned, 1. often changes into a kind of arterial, sinuous vessel, which alters the humors; and then, by its arterial apparatus, discharges them into some open channel; as in the testis, the ductus Highmorianus, the epididymis, the vas deferens, and the vesiculæ seminales: or, 2. discharges the humors direct into

some common emunctory. Hence, we know with certainty that water, lymph, thin serum, and salts, spirits, and the most subtile parts of oils mingled therewith, are separated by means of the glands from the arterial blood; and that all these are either collected, altered, and accumulated in certain places; or else driven through the small vessels into the minutest parts of the body, there to serve for motion or nourishment; whence they either return through the proper vessels to the heart, or else are exhaled; and lastly, that the part of the blood which remains in the arteries after this process is concluded, passes by degrees into the larger veins, to be mixed with similar blood, diluted with lymph, and returned to the heart. *(Inst. Med.*, n. 257, 259.) Other glands again seem to be of a different structure. In these, the artery which conveys the humors, communicates the grosser blood to its accompanying vein, through anastomoses between it and the vein; and then proceeding alone, and wreathing and gyrating, discharges from its extremity, into some common receptacle, a peculiar humor, prepared from the blood, though differing from it in nature. If we consider the stomach as a glandular cavity, the small intestines as a continual excretory duct, further altering, refining, secerning, and mingling their contents, and the large intestines as also an excretory duct; and if we apply the same to the testis, the epididymis, the vas deferens, the vesiculæ seminales, the urethra, and the prostate gland, we shall not perhaps be disposed to question that similar operations may go on in even the minutest parts of the glands. Who shall say what is accomplished by the elaborate structure of the cortex cerebri, cerebelli, and medullæ spinalis, in their invisible initiaments?" *(Inst. Med.*, n. 262.) See also Sylvius, Steno, Wharton, Graaf, Bellini, Borelli, Peyer, Leal, Cowper, Ruysch, Heister, &c.

176. "A list of the glands. 1. The pineal gland. 2. The glands of the choroid plexus. 3. The glands which constitute the cortical substance of the brain. 4. The pituitary gland. 5. The glandular membrane of the inner canthus, and of the exterior of the eye. 6. The glandula innominata. 7. The glands of the ears, or of the meatus auditorii. 8. The glandula subpalpebralis. 9. The parotid gland. 10. The superior and inferior labial glands. 11. The internal and external maxillary glands. 12. The amygdalæ or tonsils. 13. The glands of the epiglottis. 14. The glands of the uvula. 15. The lingual glands. 16. The palatine glands. 17. The occipital glands. 18. The cervical glands. 19. The glands of the larynx or thyroid cartilage. 20. The œsophageal glands. 21. The supra-scapular glands. 22. The glands of the sternum. 23. The thymus gland. 24. The glands placed at the bifurcation of the subclavian veins. 25. The axillary glands. 26.

The mammary glands. 27. The glandulæ submammillares. 28. The glands of the lungs. 29. Of the diaphragm. 30. Of the liver. 31. The cystic gland. 32. The glands of the vena portæ placed at the concavity of the liver. 33. The glands of the spleen. 34. The pancreas. 35. The upper and lower glands of the stomach. 36. The renes succenturiati, or suprarenal capsules. 37. The glands of the intestines. 38. The mesenteric glands. 39. The glands of the mesocolon. 40. The lumbar glands. 41. The omental glands. 42. The glands of the kidneys. 43. The iliac glands. 44. The sacral. 45. The inguinal or crural glands. 46. The glands of the testes. 47. The prostate glands. 48. The uterine. 49. The glands of the sartorius muscle. 50. The glands placed about the elbow and knee-joints. 51. The glands of the hands. 52. Of the feet. 53. The subcutaneous glands." (Nuck, *Adenog. Cur., de Gland. in Gen.*, cap. i., p. 5—7.) Also the glands of the fat; the glands of the muscles; and the little glands situated inside the larger conglobate glands.

177. Heister. "The fluids of the body are the following: chyle, milk, blood, serum, lymph, animal spirit, saliva, the mucus of the mouth, bile,—the liquor of the pancreas, stomach, intestines, œsophagus, brain, eyes, thorax, pericardium, abdomen, tunica vaginalis testis, —semen, the liquid of the prostate glands, the mucus of the nose, tonsils, joints, urethra, uterus, vagina, and Fallopian tubes,—the humor of the ova, and that in which the fœtus swims, cerumen of the ears, urine, and sweat." (*Comp. Anat.*, n. 34.)

ANALYSIS.

178. To attempt to examine specifically the stupendous modes and processes whereby nature operates in the animal world, particularly the chemical processes of the different glands, is like attempting to sound the abyss of the ocean, or to comprehend by sight an immense and unlimited space, where the visual rays have no termination or resting-place, but fall and vanish at the outset. For wherever the mind penetrates, it comes against closed doors: and if a chink be found, it reveals the most ample halls within, but without discovering those secret recesses where nature is sitting at the fires in the depths of her laboratory. It is enough, then, if we elicit from effects, and these, ultimate and penultimate effects, some little that has a common or general agreement with the laws of any of the recognized sciences (*a*).

(*a*) Animal nature is almost universally occupied in her peculiar chemistry or alchemy; that is to say, in preparing series of menstrua, more and more universal, whereby to prolong the life of the body; and indeed, to perpetuate it; since she also prepares a prolific humor, by which she transplants life from one being into another. All the glands are so many workshops. No two of them produce a lymph or humor of the same class, nature or use; consequently, by the examination of any one specific gland, we can never discover the specific product of any other: each therefore demands a separate consideration. All the viscera, of both the abdomen and the thorax, as well as the infra-abdominal viscera—I mean the genital members—aye, and the very brain itself, are chemical organs: the motive fibre, and the gland, are almost their only ruling principles. Each viscus, as the liver, the pancreas, &c., is covered with both muscular and glandular membranes. In order, therefore, that we may elicit some secret from each, we must prepare the way by certain general principles.

179. The blood is the complex and seminary of all the fluids and continents of fluids in the bodily microcosm. The blood-globule contains, besides the spirit which inhabits it, the first elements, simples, and unities of all kinds of compounds: these are so arranged in it, that it admits of being readily resolved into each. And such being the hidden contents of the globules, there is, therefore, no possible fluid, and no possible formation from fluids, either hard or soft, whereof the blood cannot furnish the principles, and which it cannot compound therefrom, by means of the glands: hence these infinite varieties from a single source. Thus, in this ultimate nature, there is nothing more fertile, more perfect, or indeed more simple, than the blood; because it comprehends mere simples (*b*).

180. During almost every round of the circulation, the genuine blood, in the extreme capillaries, is resolved into its constituent elements, that is, into its origins; and when the gyre is completed, it is recompounded; to flow with everlasting newness and freshness into the minute laboratories of the body: consequently the blood, in its parts, is subject to the same fortunes as the body in its compound; it is born, it dies, and it is born again: such is its circle of life (*c*). Every time, then, that it is decompounded, the veins claim a part of it; the glands, engaged in preparing the most divers liquids, claim another part; the corporeal fibres appropriate another part; the cerebral fibres another; and the cerebrum demands its spirit. The residue, which has done its office, and become inept and intrinsically useless, or involved and entangled in matter of the kind, is separated, and thrown out by myriads of pores and foramina: whence perpetual hunger and thirst, perpetual want and restitution (*d*).

(*b*) On these subjects, see the Analysis of the Blood. By the simples of the blood we do not mean the simplest elements of nature, but the simples and unities of particular compounds; for there are degrees of simples as well as of compounds. The blood contains simples, not only of one, but of several degrees; whence it is adapted for producing all things which can possibly be produced from elements. The form is the only thing that varies the essence of the compound.

(*c*) See the Analysis of the Blood, where this circle of dissolution and recomposition will be explained in full.

(*d*) We shall treat of all these particulars in due course.

181. While the blood runs through these its changes, or lives, and every time the bond is dissolved, its genuine, noble, and spirituous portion—instead of being spilt, evaporated, and lost—is carried away, by determinate channels, into certain chambers, receptacles, and cells, and taken up by the lymphatics (*e*); which swollen with their multitudinous contents, hurry to the glands, and from one gland to another; or else directly from the first station of the kind to the receptaculum, at either the beginning or the end of the thoracic duct (*f*). Thus this noble progeny is rescued from the greedy viscera, and rendered back, a rich present, to the blood, by the blood to the brain, by the brain to the fibres, by the fibres to the arteries, by the arteries to the veins; in these the blood is born again.

182. The CONGLOBATE GLANDS—uniform and cellular in structure (*g*), and more simple than the conglomerate glands—

(*e*) It is carried away towards the cellular coats, and there absorbed by the lymphatics; see above, n. 149, &c.

(*f*) The lymphatics issuing from the surfaces of the viscera, proceed to some neighboring conglobate gland, and after diffusing themselves or ramifying through it, their branches unite again, and emerge, and then go either to a second gland, or else directly to the receptaculum chyli, and thence to the dilatation at the top of the thoracic duct. See those authors who have so beautifully figured the course of the lymphatics. "No lymphatic," says Nuck, "can possibly finish its course, without being obliged to salute the glands in some part thereof; as we see particularly in those parts where there appear to be no glands, and yet where there are lymphatics; which latter, in such case, by ways hitherto unnoticed by anatomists, and as it were, by secret paths, seek out some gland, and discharge into it their lymph; there to be mingled with some other lymph, proceeding from a different source; and thence propelled by the common efferent vessel, towards the receptaculum chyli, or some other destination." (n. 172.) Very much in the same manner as the lacteals pass through the mesenteric glands, which in structure, nature and use, are similar to these conglobate glands.

(*g*) See the descriptions prefixed to the chapter. The conglobates, although their constituent glands are not conglomerated, are nevertheless full of follicles and cells. These follicles and loculi themselves are deciduous, and collapse after death, as Malpighi pointed out. "The take up this choice lymph, and diffuse it through their filters, cells, and follicles, and then collect it again, and transmit it to its channels, through vessels of a similar structure to those by which it entered; and this, from all the viscera; just as the receptaculum chyli and its auxiliary glands transmit the lacteal fluid from the intestines (*h*). For these glands likewise, by their alternate expansion and contraction, invite, allure, and attract a lymph, only purer than that of the lacteals (*i*); this they send forth in a new and narrower circle; and thus cohobate, rectify, and refine it (*k*); and by a sprinkling of fresh lymph from the numerous arteries, and an absorption of a part of the old lymph

body of the gland," says he, "is composed of various layers of areolæ and spaces, placed one upon another. When the glands are handled, the loculi are emptied, and become contracted and indistinct; and hence it is not always easy to see them, and for a long time both my mind and senses were puzzled by this circumstance. At length, however, I found what I wanted, in diseased oxen, and other animals of the same kind; where the loculi of the glands were so full of tartar, that a slight examination was sufficient to shew them. . . . When the gland is laid open, we find rows of loculi, surrounded by areolæ of fibres, under the membrane itself. The follicles of the glands are full of a certain cinereous humor, which is but little transparent in the natural state." (n. 173.) Nuck, who styles the conglobates, glandulæ muscosæ, and has often observed them distended with tartar, found that the mesenteric glands, by injection with mercury, "assumed a new figure and appearance; changing from flat, even and smooth, to globular, rough and tuberous." (n. 172, ad. fin.) These glands, therefore, are evidently cellular, with various differences; see also, n. 170 (*b*).

(*h*) See the preceding chapter, n. 170, &c.

(*i*) Just as in the mesenteric glands and the lacteals, into which the chyle is attracted by general, specific, particular, and individual forces. (n. 151.)

(*k*) For the septa and areolæ in the glands (see Malpighi, n. 173) clearly act as so many filters or strainers, (and may be compared to bibulous paper, or to glass vessels [vitra hypocleptica] contrived for separating between different liquids,) and instantly diffuse the lymph through suitable pores. Thus they evidently rectify it, and allow nothing to pass them excepting what is akin to, and homogeneous with, the blood. The little veins, which with the corresponding arteries form these septa and areolæ, absorb the heterogeneous matters.

by the veins, they temper, proportion, and adapt it to suit the chyle, and the chyle to suit the blood (*l*); and ultimately qualify it for being moved forwards, and driven home into its receptacles (*m*). Thus they regulate and proportion the quantity, quality, fluidity, and essence of the supply, to what the state of the chyle, the blood, the body, the cerebrum, and the spirits requires and demands (*n*). This is the reason why such multitudes of fibres, arteries, and veins enter these glands; why they are divided by so many little septa, areæ, and loculi; and excited by so many motive fibres: by all which means it is provided, that everything shall pass into act, agreeably to the guidance and nature of the soul.

183. The CONGLOMERATE GLANDS are so many perfect antetypes of chemical works, and models of laboratories; preparing infinite species of humors, adapted to every conceivable use. They stand in a kind of continual series (*o*), as the subsidiary

(*l*) It is the common opinion, that the conglobate glands alter and refine the lymph: and we shall shew presently, from the observations of anatomical authors, that the quantity of vessels which enter the glands is so great, as not only to fill them, but also to constitute their very structure.

(*m*) Just like the mesenteric glands (n. 153), and the receptaculum chyli (n. 163).

(*n*) The principal office of these glands appears to be, to temper and qualify the lymph to suit the chyle, and the chyle to suit the blood; also to let out no more lymph than the requisite quantity. No intrusion is permitted: leave to enter must always be given; and it is given according to requirement; and this, in many ways, some of which have been enumerated already, and some have not. The economy of the body is ruined and destroyed, if anything is obtruded unawares, without the previous permission of the brain. Wherefore these glands retain so much of this lymph as is not demanded, and absorb it by their little veins; and pour in by their little arteries so much as is required. They increase or relax their forces to suit every alteration of condition. Thus they are regulative with respect to quantity and quality.

(*o*) I use the expression, continual series, because the use of one humor is a means to the use of another; and the former serves to perfect or complete the latter. For instance, the salivæ in the mouth are watery, limpid, and mild; in the palate and œsophagus they are denser, and instrumental causes of effects, consequently of uses; which are universal, superior and inferior, or general, specific and particular. The *general* respect the generation, circulation, and life of the blood and the spirits (*p*). The *specific* respect the formation (*q*), renovation, and preservation of the whole and the parts, and the prolongation, and even the everlasting perpetuation of life (*r*). The *particular* respect any incidental uses, re-

stronger, and have additional aperitive and solvent powers; in the intestines, they are still more potent and acrid, (as we see in the biles); thus they are in a continual series. The saliva moreover serves as a means to the chyle; the chyle as a means to the blood; and the blood as a universal means for preparing the humors; so in all other cases: wherefore there is a kind of tenor, chain, and circle of means. The humors, and the glands which produce them, are correlatives with respect to use; the same thing is predicable of both producer and product.

(*p*) Inasmuch as there is such a series, tenor, and circle of mediating causes, therefore some superior universal use is always ultimately respected. The salivæ, although they are derived from the bosom of the blood, nevertheless tend to its restoration; for they imbibe and extract from the food the essential juices; and afterwards enfold and alter them,—at length into chyle; which they insinuate into the little pores, and return through the lymphatics into the veins, that is, back into the blood. So likewise the pancreatic juice and the hepatic bile. That the simpler glands also—the papillary villi in the stomach, the skin, and other parts—have a similar office to perform, will be shewn in the proper places, when the requisite experience has been premised.

(*q*) As the fluid in the uterus and Fallopian tubes, the liquor amnii, the liquor ovuli; likewise the nutrient juice prepared by means of the glands of the stomach, intestines, mesentery, liver, &c., after formation is completed.

(*r*) The animal juices, particularly the purer, aspire not only to effect the prolongation of life from infancy to old age, but also to effect its perpetuity after the body dies, and the existing machine perishes. This is the case with the semen or generative fluid, which is elaborated by the testicles, refined by the vesiculæ seminales, and sheathed by the humor of the prostate—all so many glands, which transplant life successively into new beings, as it were other selves. The uterine and other humors of the female likewise perform their part, and lend their aid. So also the liquor ovi in oviparous creatures, the milk in mamquired by alterations in the state of any viscus, organ or member, or integral or individual part thereof (*s*).

184. The conglomerate glands elaborate and produce the animal and vital juices by innumerable methods; that is to say, by extraction, filtration, decomposition, commixtion, decantation, secretion and excretion; sometimes by repeated courses of similar processes; not to mention other most occult methods, hitherto altogether unrecognized in human art and science. And such an organism is bestowed on the glands; such a power implanted in the organism; and such a direction in the power; such a place is appointed, and such a connexion with other organs, near and remote, is assigned them; such a kind of blood is invited; and such an essence thereof elected and allured, or refused and rejected, as the foregoing processes require (*t*). The blood supplies from the common store and mass exactly what is wanted by the gland (*u*), to which it opens and discovers its very

malia, and all the rest of the fluids which continue the thread of life from being to being, in succession.

(*s*) The glands change their state, consequently the nature of their humors, for the most part to suit the states of the body. New glands also are produced, particularly beside old, or dried-up collections of humors; as in the membranes, meninges and sinuses of the cerebrum, in the fat, at the groins; in short, all over the body. These new glands constitute so many tubercles, steatomata or atheromata, pseudoschirri, erysipelatous swellings, &c.; which increasing in size, number and power, endeavor to repair the mischief, by diverting malignant, and absorbing imprisoned matters, clearing away superfluities, restoring the tone to flaccid parts, and thus preventing disease and destruction from extending. Such glands, however, are so many proofs of a changed constitution and a morbid state.

(*t*) I have brought together these particulars, in order to afford a general idea of the operations of the glands. The specific description of each gland,—as of the liver, the pancreas, the spleen, the kidneys, the thymus gland, the testicles, the epididymes, the vesiculæ seminales, the prostate glands, the cerebrum, the pituitary gland, &c.,—will be found in agreement with this idea. But the structure of no one conglomerate gland has been exhibited specifically in the foregoing pages; wherefore I cannot venture beyond this outline at present.

(*u*) Every gland throughout the animal kingdom enjoys a plenary marrows and recesses every time it runs through its circle of life (*x*); thus it spends itself and its life in the service of the commonwealth, and dies that it may rise again, and thereby renew its services, and provide for the kingdom in perpetuity.

185. The soul, which embraces series and intuitions of ends, disposes all these things by the simplest fibres, which are so many determinations of her final ideas. The simple fibre exclusively, is what forms and acts in the kingdom of the soul. This fibre rushes into modes corresponding to the nature and decrees of the soul, infinitely faster than the muscular fibre into the different forms of action, under the determination of the conscious will; readier than words rush into speech, according to the meaning of the speaker; swifter than a bird's wings into flight, or than its little throat into song. There is nothing but ministers most humbly, and submits most willingly to the soul: thus in the glands, every conceivable modification of fluid is supplied to it, according to its intentions. Hence we may have some idea of the nature of the soul, and of the entire obedience of the body (*y*).

communion of its goods and fluids, as well as of its blood. Not the minutest drop floats in any artery, wherever situated, be it even in the heel or in the sole of the foot, that is not communicated, if required, to any other artery, whether in the crown of the head, or in any of the viscera, or in any of the glands. That there is such a communion of blood in the brain, and in the chest, throughout, and that there is a similar communion of chyle in the mesentery, is perfectly evident from the everlasting anastomoses between the different vessels. And we shall shew further, in the Part on the Blood, that that fluid is various in every artery and vein, in every minute branch, and in every twig. It is by virtue of this communion that the glands supply their wants and necessities from every part.

(*x*) That the blood is perpetually undergoing dissolution and new birth, consequently opening and discovering itself to the glands, see above, n. 180. And that there is no possible humor to which the blood cannot give genesis, origin and existence, see n. 179.

(*y*) We shall treat of the government of the soul in the body, and of the obedience of the body, in the Part on the Medullary Fibre of the Cerebrum and the Nervous Fibre of the Body: and of the government of the will, that is, of the human rational mind, in the Part on the

186. The fibres which compose the conglomerate glands are fourfold in origin; consequently also in nature, use, and determination. The *first,* which is the principle, the form, and the power, of all the others, is the fibre of the brain, produced from the cortical or cineritious substances, as matrices: this, in the brain, is called the medullary fibre; in the body, the nervous fibre. The *second,* is the fibre of the body or inferior system—the corporeal fibre—governed and fashioned by the former, or cerebral fibre: it constitutes the blood-vessel, or the artery and vein (*z*). The *third* is the union of the cerebral and corporeal fibres—the offspring of nerve and vessel—and is called the muscular or motive fibre. The *fourth* is the fibre of the conglomerate gland, generated by the apposition and unition of the granules and acini thereof (*a*). This latter, on account of its tenuity, form, color, and general appearance, sometimes resem-

Cerebrum, the Motive Fibre, and the Rational Mind. In those treatises, I hope shortly to be enabled to explain the miracles which occur in forms, powers of action, and actual operations, particularly in such operations as are of a chemical nature.

(*z*) That the nervous fibres generate the vessels, constructing the latter by their circumvolution, and that thus they compose a kind of new fibre, which carries blood in the same manner as the fibre of the cerebrum carries spirit, will be shewn in the due course of our analyses. This fibre, by means of the third class of fibres—the motive fibres—realizes the powers of the primary fibre, and brings them forth into the world of visible effects.

(*a*) Our authors speak, from much experience, of the immense number of nervous and muscular fibrillæ which enter the glands. All the conglomerate glands "are nothing more," says Nuck, "than a texture of very minute vessels, composed of the smallest arteries" (n. 172). The manner in which the vessels and fibres enter, construct, and connect or glomerate the glands, is described by Malpighi, n. 173. "These glands," says Boerhaave, "are furnished with arteries, the branches of which are supported by and distributed in the membranes, and accurately conveyed to every minute particle of the gland, &c. They have more and larger nerves than any other part of the body of the same magnitude" (n. 174). Ruysch discovered, by means of his wax injection, that this class of glands is entirely vascular: and many other anatomists have come to the same conclusion by using mercurial bles either the fibre of the brain, or the vascular capillary of the body (*b*). Besides which, there are fibres produced and compounded from, and growing out of these, united and altered in various ways (*c*). So numerous then are the fibres which build

injections. For, as we before said, the fibres flow into the vessels, and reciprocally do and suffer whatever is necessary, by means of the vessels (*z*).

(*b*) As in the stomach, the intestines, the skin, and all over the body. For the first threads and initiaments of the tubuli arising from the minutest glandular granules, resemble the fibres, or the vessels of the purer or colorless blood, or exsangueous villi composed of such vessels. The microscope detects no difference between the two. Leeuwenhoek mentions these glands in their principles, and the threads proceeding from them.

(*c*) As the first and simple threads emerge more and more into the visible world, or increase by apposition and composition, they are no longer regarded as fibres, because they begin to assume a new garb and appearance, but they are named according to their differences in point of size, figure, and use. The simple fibres of the cerebrum, when fasciculated, or formed into cords, are called nervuli, and the compounds of these, nerves. The vascular capillaries are at length termed arteries and veins, and further distinguished by particular names. The motive fibres, when disposed into a form suitable for action, are termed muscles; wherefore a muscle is a compound motive fibre. The glandular fibres, when a number of them are put together to form fibres of larger dimension, afterwards conglomerate a vessel,—either a tube, duct, canal, or emissary; and ultimately an infundibulum or bladder. The cavities of the glands derive their names from the same circumstances; in the first instance they are termed follicles, loculi, vesicles, excipula, or cells; next, receptacula, cisternæ, antra, ventres, &c. There are also certain fibres identical with others in origin, but the nature of which is changed; for instance, tendinous fibres, which are motive fibres deprived of their living force, and become passive; and which successively form cartilaginous and osseous fibres. There is also a kind of tendinous, sinewy, and dried-up filaments, which were formerly blood-vessels, as the umbilical ligaments, the ductus arteriosus, and an infinity of others, in the skin, the membranes, and the meninges of the brain; but these are degenerate fibres and spurious vessels, of a passive character, yet springing from the same stock as the true vessels themselves. We shall recur to these subjects continually in the progress of our analyses.

and compose the glands. The cerebral fibre gives them essence, power, and life: the corporeal fibre, or the vessel, strengthens them, and superadds and provides the fluids: the motive fibre supplies the ultimate forces, and excites a motion corresponding to the animus of the cerebral fibre. The glandular fibre, arising and existing from these three principles, thus subordinated and coördinated, takes up, emulges, discharges, and determines the fluids, which consist of different species of humors essential to life. These glandular glomes are named from their office, as their primary feature; inasmuch as they are glands in a more especial and eminent sense than the conglobates (*d*).

187. The conglomerate glands sunder, slay, and destroy the blood: the conglobate glands gather its members together again, and revive and reinstate it (*e*). Thus the conglomerate glands

(*d*) It has been disputed whether any vascular structure, or any viscus or part thereof of a vesicular composition, could properly be termed glandular; but the dispute is verbal only, and has no connexion with true science. If those structures (whether composed of fibres or vessels, it matters not), which by peculiar and eminently chemical organs elaborate and produce all the amazing varieties of animal juices, by means of extraction, filtration, decantation, commixtion, secretion, and excretion, be glands, (according to the description given above of the offices of the glands, n. 184,) then of course the conglomerates are glands in a high sense and an especial degree. But if they are not to be named from their office, but from their form and appearance, then some other name with a different sound must be given them; only let it signify the same as the word gland. But in grave matters, critical trifling should be avoided.

(*e*) Respecting the conglobate glands, see n. 182; respecting the conglomerate glands, n. 183. seq. The blood is the single source whence the latter derive the principles and elements of the humors which they discharge into the viscera and cavities of the body. From this source, the parotid and maxillary glands, the tonsils, the thyroid gland, the dorsal and œsophageal glands, &c., derive their different kinds of saliva and mucus; the pancreas, the pancreatic juice; the liver, the different kinds of bile; the testicles and epididymes, (through the vas deferens, to the vesiculæ seminales,) the semen; the uterus, by means of the placenta, the liquid which surrounds the fœtus. Thus these glands transform the blood into a fluid of a different kind; conseplace the blood, like a phœnix, upon the pyre, and burn and bury it: but the conglobate glands collect and irrigate its ashes, and resuscitate the phœnix. The conglomerate glands live on the borrowed and precarious means lent them by the conglobate, and which the conglobate glands require again after use: thus the two are confederate, and yet antagonist: the conglomerate would be useless and void of office without the conglobate; and the conglobate without the conglomerate (*f*).

188. No two glands are exactly and absolutely similar; there is always more or less of difference between them, and indeed between their very acini and miliary elements; but from the equation of disparates which yet are proportionals, a suitable compound results in harmony with the intentions of nature. Each gland, however, has its generic or universal similarities (*g*); but in order to discover these, the glands must be examined severally and specifically (*h*).

189. Now in order to shew the general affinity and analogy which pervades the whole family of glands, and thus to further

quently, they may be said to slay it, and to carry it to the viscera as to the tomb. But the conglobate glands, on the other hand, from these very humors, after they have served their purpose, recover a lymph which is the purest lymph in the body, the nobler extract of the blood and the serum, the vehicle, bed, and throne of the spirits, with which it is richly endowed and vivified; the true purer blood (n. 167); and by this lymph they recruit the blood, and perpetuate its circulation and life.

(*f*) From these particulars we may now see what is meant by the blood's circle of life (mentioned above, n. 180), and by what means it is kept up; namely, that the conglomerate glands destroy the blood, and the conglobate glands restore it. Their mutual correspondence is requisite to perpetuate the circle.

(*g*) Although there are what are called generic differences between the universal offices of the glands; as there are specific differences between their specific offices; yet similarity is not sameness or equality; it only involves harmony of relations. But we do not dwell on generic differences, because they are hardly discernible between parts that are harmonically correspondent.

(*h*) As the liver, the pancreas, the spleen, the succenturiate kidneys, &c.; and those glands also which are not viscera, as the parotids, the maxillaries, and the others enumerated above, n. 176.

our analysis, it is important to institute a comparison between the glands of the body, and the cerebrum, which not only is a gland, but the prince of glands (*i*), and the pattern and head of the family; for it derives its principles, educes its fibres, and conceives and produces its spirit, from itself; and not like the glands of the body, from preëxistent organs (*k*).

190. The CEREBRUM is proximately covered and surrounded by two membranes—the arachnoid, a thin membrane, and the pia mater, which is somewhat thicker: the *glands of the body* likewise are covered by a thin common membrane, and by a thicker one underneath it (*l*). The CEREBRUM has its peculiar artery—the internal carotid—which enters it by two trunks: the *glands of the body* likewise have for the most part two arterial

(*i*) All the ancients, with Hippocrates, regarded the cerebrum as a great gland. The moderns, with Malpighi, have done the same, and in fact class it among the conglomerate glands, regarding the cortical substances as so many conglomerant glands. "The glandular nature of the cerebrum," says Malpighi, "made up, as it is, of membranous follicles, is forcibly shewn by the remarkable case which Wepfer relates," &c. (n. 173.) That the cerebrum takes the lead of all the glands of the body, is conspicuous from its exquisite organism, from the universal lymphs of the body elaborated therein, and from its elevated office.

(*k*) In the subsequent analyses, a further comparison will be instituted between the different glands of the body; in this place, however, between them and the cerebrum, in order that we may explore generals, in regard to which the cerebrum is peculiarly illustrative.

(*l*) The most general membrane of the cerebrum is the dura mater; which, however, is not proper to the cerebrum, but common to it and the cerebellum, and even to it and the medulla oblongata and spinalis; as the pleura is common to the viscera of the thorax, and the peritonæum, to the viscera of the abdomen. Moreover, the dura mater is held off from the cerebrum, and adheres throughout to the bones of the cranium, forming their internal periosteum; consequently, it ought not to enter the comparison in the same way as the proper coats of the glands. The arachnoid is the thinnest of all the membranes of the cerebrum, and invests the pia mater, which is thicker. These two membranes at first appear to be separate around the cerebellum and medulla spinalis, and to be united more or less closely in different places. "The glands," says Nuck, "have two membranes, an exbranches and two venous ones (*m*). The ARTERY OF THE CEREBRUM, dividing into twigs, winds and branches between the two membranes, and penetrating their folds, is dispersed about, and ramifies through the recesses of the viscus: *the arterial branches of the glands,* and their twigs likewise, are entirely expended in forming the body of the glands (*n*). Indeed the ARTERY OF THE CEREBRUM, dividing at last into the minutest threads, enters

terior and an interior. The exterior is the thinner membrane, and encloses the whole gland, and adheres in general firmly to the membrane underneath it; in some cases its connexion is looser" (n. 172). "The simple glands," says Boerhaave, "are composed of a thin external membrane, and of an internal one, which closely adheres to it" (n. 174).

(*m*) The internal carotid is the only artery which ascends to and enters the cerebrum: the vertebral artery goes to the cerebellum and medulla oblongata, and communicates with the carotid. "The glands," says Nuck, "have four vessels, two adducent or afferent, and two abducent or efferent" (n. 172). But this is a point too well known to require further proof.

(*n*) The carotid artery ramifies so luxuriantly round the membranes of the cerebrum, between the arachnoid and the pia mater, that this surface of the cerebrum may be said to be purely vascular. The ramifications penetrate the duplicatures and processes of the meninx, and insinuate themselves into the windings, convolutions and masses of the cerebrum, mingling with the genuine fibres in such a manner, that the body of the cerebrum is nothing but a mixed compages of arterial twigs and fibres. That something similar is the case in the glands of the body, is shewn by Boerhaave. "The internal membrane [of the gland]," says he, "is thick and more dense, being formed of fibres passing in all directions, and of an interlacement of vessels. These glands are furnished with arteries, the branches of which are supported by and distributed in the membranes, and so accurately conveyed to every minute particle of the gland, that if wax or quicksilver be injected, we might be led to the false conclusion, that the whole fabric is arterial" (n. 174). Malpighi teaches that the networks and areolæ in the glands are entirely formed of branches of vessels (n. 173); and Nuck, that "the interior membrane is perforated throughout by both afferent and efferent vessels; yet that its pores are so minute, that air does not ordinarily escape through them, notwithstanding it is forcibly driven through the vessels" (n. 172).

the very cortical glands or spherules, grows to them, weaves their substances, and seems to vanish or terminate there in extreme tenuity. The *arteries of* all *the glands of the body* likewise, are entirely expended, and in a manner consumed, in their minute conglomerate or miliary seeds or corpuscules (*o*). The CORTICAL GLANDS OF THE CEREBRUM put forth fibres, which when collected into fasciculi, make up the medullary substance, or corpus callosum cerebri; and a part entwined with the fibres of the medulla oblongata, is finally continued to form the nerves: the *conglomerate glands* likewise, from their several corpuscules or primitive granules, put forth a kind of fibrous or medullary productions, which compose the body of the gland; and these they ultimately fold into a vessel or emissary duct, which they determine to its proper field of use (*p*). The CORTICAL GLANDS OF THE CEREBRUM elaborate the universal essences, that is to say, the vital spirits peculiar to different animals, and send them forth in all directions into the kingdom,

(*o*) Ruysch discovered by means of his wax injections, and by frequently immersing the capillaries in water, that the cortical glands of the cerebrum are only organic expansions and forms composed of vessels. Of this indeed there is ocular evidence; for these glands hang from the arterial twigs, like bunches of grapes from the branches of the vine: shewing that the arterial branches are in a manner grafted upon the cortical glands. Nuck asserts the same of the glands of the body: "The conglomerate glands," says he, "are nothing more than a texture of very minute vessels, composed of the smallest arteries" (n. 172); as he also shews experimentally by means of injections, which cause "all the glands to rise up." (*ibid.*) Ruysch observed the same thing in several other glands, as those of the stomach, intestines, &c.; wherefore he regards these corpuscules also as purely vascular.

(*p*) It is evident from the delineations of the cortical spherules of the cerebrum, that a particular fibre proceeds from each, and that this fibre immediately unites with its fellows to form a fasciculus; and that the fasciculi, and the extremities of the arteries together, make up the corpus callosum. That the remaining part of the fibres of the cerebrum sinks through the corpora striata into the medulla oblongata, will be shewn in our Treatise on the Cerebrum. "The fibrous tissue of the glands," says Nuck, "consists of a number of fibrillæ of different for the maintenance and support of the parts and the whole: so the *glands of the body* elaborate their essences, humors which are less universal, indeed, but nevertheless essential to life (*q*). The CEREBRUM internally has two large cavities or ventricles, and a general intermediate one full of clefts or chinks: the *glands of the body* have similar cavities—their follicles, loculi, cells, vesicles, cysts, cisternæ, &c., more or less numerous. The CAVITIES or ventricles OF THE CEREBRUM communicate with each other by little foramina: so do the *loculi* or follicles *of the glands of the body* (*r*). The CEREBRUM distributes the branches

species, all united together" (n. 172). In the kidneys, and in many other tuberous and granulated bodies, each glandular granule is evidently the beginning or little head of its fibre; just as the cortical substance is the beginning and head of the medullary fibres of the cerebrum. By the arrangement of such fibres, one or more canals are formed; this is evident in all parts of the body,—in the mammæ, the pancreas, the kidneys, &c. These fibres belong to the fourth class specified above; see n. 186.

(*q*) That the cortical glands generate and produce what is called the animal spirit, is a point upon which the highest authorities are agreed: also, that this spirit is sent away into the provinces of the body by the fibres, as being the purest channels. This spirit is the most universal essence of corporeal life. So the humors are elaborated by the conglomerate glands, and are also essences, only not so universal; notwithstanding which they perform a kind of analogous function: and hence the glands are excited to action by the very fibres of the cerebral glands; their own fibres being unable to effect this, which is the office of the cerebrum alone. Nor are the glands of the body excited to their reciprocations of expansion and contraction, by any motive principles of their own, as the cerebrum is; but by moving fibres; in short, by the fibres of the cerebrum united to the fibres of the body; that is, by vessels; consequently, by moving fibres, exactly according to their analogical relation with respect to power and office.

(*r*) The communicating foramina in the cerebrum, are the anus and vulva, in addition to which there is a passage or emissary for the lymph, by which the lateral ventricles communicate with each other, and with the third ventricle. Something similar occurs in the liver, the pancreas, the kidneys, and in other parts; as may be seen and known, not by injections only, but also by effects themselves, and the chain and tenor of causes.

of its artery through the ventricles in the form of a web, (the choroid plexus:) so also do the *glands of the body* (*s*). The CEREBRUM from its ventricles puts forth a hollow tubulus, the infundibulum: the *glands of the body* likewise put forth their canals, emissaries, and vessels; the kidneys put forth the ureters; the testicles, the vasa deferentia; and the mammæ, the tubuli lactiferi. The CEREBRUM inserts the tip of the infundibulum into a certain gland, called the pituitary gland—one of the conglobate kind; into which it pours the lymph, there to undergo further separation and defecation. Just so do the *conglomerate glands of the body*, only not immediately into a conglobate gland, but after a circuit; for the juice elaborated, after serving its purpose in the viscera, is at length presented to the lymphatics, and at last to some conglobate gland (*t*). The PITUITARY GLAND—the conglobate gland of the CEREBRUM—sends out and expels these essences and spirits by distinct channels, and indeed, into the ultimate receptacles of the venous sinuses, and into the jugular veins. The *conglobate glands* likewise send their lymph into the veins, and particularly into the subclavian vein. In this place, that is, at the junction between the jugular and

(*s*) The choroid plexus passes over the thalami nervorum opticorum, from one end of the great lateral ventricles to the other; and is continued to the pineal gland and the orbicular protuberances, or into the isthmus cerebri. Similar reticular plexuses are seen in the glands of the body, passing through their follicles and cavities; as constantly described by Malpighi from his own observations. "After the vessels enter," says he, "their larger branches form a network, and their last twigs appear to terminate on the loculi and parietes of the areolæ" (n. 173): which he again remarks in speaking of the succenturiate kidneys. *(ibid.)*

(*t*) It is extremely worthy of notice, as a principal point in the comparison, that the cerebrum pours the spirit of its fibres, in union with the arterial juice of the glandular choroid plexus, through a particular efferent vessel, the infundibulum, into a certain conglobate gland. For the pituitary gland exactly resembles the conglobates, in both form, texture, membranes, influx of fibres and arteries, office and modus operandi; as will be clearly proved in our treatise on that gland. Therefore, the infundibulum represents a kind of preëminent and complicated lymphatic.

subclavian veins, THE LYMPH OF THE CEREBRUM, full of spirit, and the corresponding *lymph of the conglobate glands of the body*, meet, mingle, and salute; and openly proclaim their glandular origin, and their common appointment and determination to the performance of accordant uses (*u*).

(*u*) This last or crowning fact, that the internal jugular vein is fitted to the left subclavian, just where the thoracic duct is inserted, (which latter is sometimes planted on the end of the jugular vein itself,) demonstrates that a similar spirituated lymph comes down from the cerebrum, to what comes up from the conglobate glands, by means of the thoracic duct; and that thus the new spirit meets the old, for the maintenance and restoration of the blood, and of the life of the body. *Q. e. d.*

CHAPTER IX.

THE LIVER AND THE GALL-BLADDER.

191. Heister. "The liver is a very large viscus, of a red color, situated in the right hypochondrium, and designed for the secretion of bile. The points we are to notice in it are, its size, which is very great, in order that the bile may be secreted in abundance, for the purposes of digestion: its figure, which is irregular; the upper surface convex and even, the lower, hollowed and uneven, and containing the gall-bladder and certain glands. A remarkable prominence is seen in the liver, where the vena portæ enters it; this is called the porta; and there is also another, named lobulus Spigelii by some anatomists. A great fissure divides the liver into a right and thicker, and into a left portion; but not into such lobes as we find in the liver of the dog. The liver is connected, 1. By the right and left suspensory ligaments, with the diaphragm; by the round ligament, with the navel; this latter ligament in the fœtus, was the umbilical vein. 2. By its proper membrane with the diaphragm posteriorly. 3. And it also has connexions by the vessels, particularly the vena cava and vena portæ. The membrane which surrounds the liver is thin, and continuous with the peritonæum. The capsule of Glisson is a membrane continuous with the peritonæum, enclosing the branches of the vena portæ, the arteries, and the biliary ducts, as they approach to, and after they enter the liver. The arteries for the nutrition of the liver, are, 1. from the cœliac; 2. from the cystic arteries; 3. from the diaphragmatic; 4. and sometimes from the superior mesenteric artery. The veins are, 1. from the vena portæ, entering the concave surface of the liver: this vena portæ performs the office both of vein and artery, bringing in the blood for the secretion of bile; but of this we shall speak in another place: 2. from the vena cava: and 3. from the diaphragmatic vein; for the return of the superabundant blood to the heart. The nerves of

the liver arise from the hepatic plexus of the intercostal nerve. The biliary vessels, which often have glands near them, are the ductus cholidochus communis, which opens obliquely into the duodenum. The ductus cysticus, which goes from the gall-bladder to the ductus cholidochus, and unites with the hepatic duct at an acute angle: in man the cystic duct is tortuous, and often furnished with valves of various kinds. *(Comp. Anat.*, n. 217.) In the *Ephemerides,* cent. v. and vi., I described two human gall-bladders, in the cystic duct of which there were certain beautiful spiral valves. In another subject, I also saw a number of transverse and oblique valves or membranes, which divided the duct as it were into cells: these, however, did not entirely close the duct in any part, but were disposed much like the valvulæ conniventes in the jejunum or colon. Glisson and Bianchi deny the existence of these valves, but Bauhin, Bidloo, and Vestus assert it. And Schelhammer has remarked, that the cystic duct will not admit a style, either from the direction of the duodenum or of the gall-bladder; but that air may be very readily blown through it. Plancus has seen these valves, and describes them as perfectly resembling the turns of a snail-shell. *(Comp. Anat.*, not. 18.) The hepatic duct runs from the liver to the ductus cholidochus; its branches are distributed throughout the liver, and are called pori bilarii. The hepatico-cysticus, and the radices felleæ are observed in cattle. *(Comp. Anat.*, n. 217.) It has been much disputed, whether there are, or are not, ducts which convey the bile immediately from the liver into the gall-bladder. Bohn relates some very remarkable experiments, by which he proves that there are such ducts in the ox; and adds, that nearly all the most celebrated anatomists have allowed their existence in various animals; though not in man. Cheselden thought he saw these ducts, and described them as most minute; but he afterwards altered his opinion. Bianchi contends for them, and divides them into hepatico-cystic and cystico-hepatic. Morgagni denies them altogether. For my own part, I happened to find a duct, near the back of the neck of the gall-bladder, and on making a small aperture into it, and inflating it, I found it distributed in an arborescent manner over the gall-bladder, and that its trunk was one of the cystic blood-vessels; and therefore I conclude, that if these ducts really exist in the human subject, they are not universal, but very rare indeed. *(Comp. Anat.*, not. 18**.) The lymphatics of the liver are shewn, either by putting a ligature on the vena portæ in living animals; or by the inflation of the artery, or of the hepatic duct. The substance of the liver was supposed by the ancients to be formed merely of blood, coagulated about the blood-vessels. Malpighi and many of the later writers have considered it to be glandular: and Ruysch makes it vascular, formed of a congeries of very minute vessels. (*Comp. Anat.*, n. 217.) Those who have regarded the substance of the liver as glandular, have done so principally on the ground, that in diseased subjects, particularly in such as have died in dropsy and atrophy, a multitude of spheroidal bodies have been observed in it; but this is not the case in healthy subjects. Others again have found morbid vesicles in the liver, which some will have to be glands. (*Comp. Anat.*, not. 19.)

192. "The gall-bladder, called also vesicula and cystis fellea, in some measure resembles a pear in shape, and is situated in the concave part of the liver. The points it presents for notice, are, its size, which is about that of a small hen's-egg. Its neck or cervix, which is furnished with a kind of sphincter. Its fundus; which is the lowest part when we stand upright: its neck being its upper part. It lies on the colon, and tinges it with its own color. We must also observe its connection, by means of the common membrane, and of the vessels: the ductus cysticus, going to the ductus cholidochus; and the frænulum, by which the ductus cysticus is joined to the gall-bladder. It is composed of four membranes: 1. A common coat. 2. A cellular and vascular coat. 3. A muscular coat, consisting of minute, straight, oblique, and transverse fibres. 4. A nervous coat, which has rugæ or reticulations on its inner surface, is covered with an unctuous moisture, and sometimes has minute glands upon it. The foregoing structure seems also to extend to the biliary ducts. The gall-bladder has its vessels in common with the liver, but in the former they are called cystic vessels. The radices felleæ I have found in the ox, but never in the human subject. The mode in which the bile is carried into the gall-bladder: in the human subject, the greater part of it ascends from the ductus cholidochus through the ductus cysticus; but its bitterest portion seems to be secreted by the gall-bladder itself. In the ox, a part also enters by the hepatico-cystic ducts and the radices felleæ. (*Comp. Anat.*, n. 217.) Verheyen and many of the greatest anatomists think it impossible that the bile should pass both ways along the cystic duct; but Bohn and Ortlob have proved this to be the case, by the weightiest reasons and experiments; which experiments I have myself repeated. Cole, also, J. Maur. Hoffman and a multitude of others, are of the same opinion: and Bianchi has abundantly proved its truth. (*Comp. Anat.*, not. 19**.) The use of the liver is, to secrete bile or gall from the blood of the vena portæ. The ancients supposed its use to be sanguification. The use of the gall-bladder is, to collect the bile, to refine it, to retain it for a certain time, and then to expel it. The use of the bile is, to attenuate the chyle; to mix oleaginous with aqueous

parts; to stimulate the intestines; and in part to alter the acid of the chyle. The bile, however, is of two kinds; the hepatic, which is thin, almost insipid, and scarcely colored; and the cystic, which is thicker, more colored, and very bitter. *(Comp. Anat.*, n. 217.)

193. "The vena portæ resembles a tree. Its roots (or inferior branches) are divided into right and left. From the right arise all the mesaraic veins of the intestines, the internal hæmorrhoidal vein, and the venæ epiploicæ dextræ. The left is called the splenic vein; and from it arise the various gastric veins, which form the coronary and the other veins of the stomach; the vasa brevia, the epiploic and gastro-epiploic veins, many of the pancreatic veins, and sometimes also the internal hæmorrhoidal vein. The trunk of the vena portæ, which goes to the liver, gives off the cystic veins, the right gastric vein, the duodenal vein, and this latter often the pancreatic vein. The branches, where the trunk begins to unfold, constitute the sinus portæ of the liver, and from this the porta is distributed by innumerable branches through the whole of the liver, and through the liver alone." *(Comp. Anat.*, n. 297.)

194. Winslow. "The liver is a large and pretty solid mass, of a dark red color inclining to yellow, situated immediately under the arch of the diaphragm, partly in the right hypochondrium, which it fills almost entirely, and partly in the epigastrium, between the xiphoid cartilage and the spine, and terminating commonly towards the left hypochondrium. The figure of the liver is irregular, being convex above, unequally concave below, and very thick posteriorly and on the right side, but gradually decreasing in thickness anteriorly and towards the left side. The liver may be divided into two extremities, one large, the other small: into two edges, one anterior, the other posterior; into two surfaces, one superior and convex, which is even, smooth, and proportioned to the arch of the diaphragm; the other, inferior and concave, which is uneven, and has several eminences and depressions. It may likewise be divided into two lateral portions, called lobes,—a great or right lobe, and a small or left lobe. These two lobes are distinguished above by a membranous ligament, and below, by a considerable fissure, which follows the direction of the ligament above. The eminences on the concave side of the liver belong to the great lobe. The principal eminence is a sort of triangular or pyramidal apophysis of the great lobe, placed behind the great fissure: this eminence is called the lobulus Spigelii. The ancients named these eminences, portæ. The depressions on the concave or lower side of the liver, are four in number. The first is the fissure that separates the two lobes: this is termed the great fissure of the liver; in some subjects a part of it is an

entire tube. The second depression is situated transversely between the two eminences of the great lobe, and filled by the sinus of the vena portæ. The third is behind, between the great lobe and the lobulus Spigelii, and the vena cava passes through it. The fourth is a kind of sulcus between the greater and lesser lobes of the liver, and in the fœtus affords a passage to the ductus venosus, which afterwards disappears, nothing but a kind of ligament being left. This sulcus is in some measure a continuation of the great fissure, and joins the vena cava at an acute angle. Besides these four depressions, there is one in the great lobe, which receives the gall-bladder, and which sometimes extends to the edge of the liver, where it forms a little notch. There is still another, in which the right kidney is placed; and several small depressions, corresponding to the spine, and the extremity of the œsophagus, not to mention others, which are not ordinary. The convexity of the liver is connected to the diaphragm by three ligaments, which are only continuations of the membranous lamina of the peritonæum. One lies near the extreme edge of both lobes, and one in the middle: these are termed the right, left and middle ligaments. Their duplicatures contain a cellular substance, in which the blood-vessels and lymphatics run, and which sends off a kind of lamina into the substance of the liver. The right ligament sometimes connects the great lobe to the cartilages of the false ribs; and the left ligament, or that of the small lobe, is often double, and advances towards the middle ligament. The middle ligament begins below, in the great fissure of the liver, near the eminences called portæ, and passes through the anterior notch, over the union of the two lobes, to the convex part of the liver, and is obliquely attached to the diaphragm. It is likewise fixed all along the upper and inner part of the sheath of the right rectus abdominis muscle, in such an oblique manner as to be nearer the linea alba below than above. The liver is likewise connected to the diaphragm, not by ligament, but by a broad and immediate adhesion, without the intervention of the peritonæum, which is only folded round this adhesion, to form the external membrane of the body of the liver. This adhesion is commonly but improperly called the coronary ligament. The middle ligament, called improperly the suspensory ligament, contains in its duplicature a white cord, which was the umbilical vein in the fœtus. All these ligaments serve as supports to the liver, but none of them suspend it; it is principally supported by the stomach and intestines. The passage of the vena cava, between the body of the great lobe and the lobulus Spigelii, may serve in some measure as a guide in placing a detached liver in its true situation. The liver is composed of several kinds of vessels, the ramifications of which are multiplied in a stupendous manner, and by the intertwinement of their capillary extremities, form an innumerable quantity of friable and pulpy granules, which are looked on as so many peculiar organs, whereby a particular fluid called bile is secreted. The chief part of these vessels, from one end to the other, is enclosed in a membranous sheath, called the capsule of the vena portæ, or Glisson's capsule. The vessel which carries the blood to the liver is called vena portæ. This vein may be considered as two large veins, the trunks of which are joined endwise, and send out their ramifications in opposite directions. One of these trunks is connected to the liver, and ramifies in it; the other lies without the liver, and sends its branches to the viscera of the abdomen; so that the first of these trunks may be called the vena portæ hepatica, the other, the vena portæ ventralis, &c. The particular trunk of the vena portæ hepatica is situated transversely between the broad anterior eminence of the great lobe of the liver, and the root of the small lobe, in a particular fissure, and forms what is called the sinus of the vena portæ. From this sinus, five large principal branches go out, and are distributed by myriads of ramifications through the whole mass of the liver. At this place, the vena portæ lays aside the office of a vein, and becomes a kind of artery, as it enters and circumramifies in the liver. The extremities of all its ramifications end in pulpy, friable granules, which when examined through a microscope in clear water, seem to be thick villous follicles. In these follicles, the bile is secreted, and it is immediately collected in the same number of extremities of another kind of vessels, which unite by numerous ramifications into one common trunk. These ramifications are termed pori bilarii, and the trunk is called the hepatic duct. The ramifications of these two kinds of vessels are enclosed together in the capsule of the vena portæ. The blood deprived of this bilious fluid, is conveyed away by a great number of minute venous ramifications, which afterwards unite into three principal branches, and several others less considerable, that terminate in the vena cava. These in general are simply called, hepatic veins. The capillary extremities of the ramifications of the vena cava join those of the vena portæ, and accompany them through the liver. And yet the great branches of both veins intersect each other in several places. It is remarkable, that the branches of the vena cava are thinner and more capacious, and more closely connected with the substance of the liver, than those of the vena portæ; which, on the other hand, are invested by a cellular capsule, and appear to be a little ruffled and corrugated when empty: because the cellular capsule subsides and retracts, when it is cut; but the other veins remain open. The liver receives from the cœliac artery a peculiar but very small branch called the hepatic artery. The plexus hepaticus, formed by the sympathetic nerves, furnishes a great number of nerves to the substance of the liver. The ramifications of the hepatic artery and nervous plexus, are enclosed in the cellular capsule, together with the branches of the vena portæ, and the pori bilarii. The common coat of the liver is a continuation of the peritonæum. The substance of the liver appears to contain also a membranous or filamentary tissue, which connects the ramifications and extremities of all the vessels to each other. This tissue seems to be a most multiple prolongation of the capsule, and of the external membrane of the liver. The outer surface of this coat is very smooth; but its inner surface is uneven, being made up of a great number of fine membranous laminæ or leaves, between which we observe very distinctly numerous lymphatic vessels, on both the convex and concave sides of the liver: but it is more difficult to trace those which accompany the filamentary substance through that viscus. Each glandular granule is bounded and in a manner invested by a particular expansion of the capsule of Glisson: and all these expansions are held together by common septa, almost like the cells of a bee-hive. In the inside of the liver, these grains are angular and polygonal, but near the surface they are more raised, and resemble little tubercles. They are of a pulpy texture, like radiated villi, with a small hollow in the centre of each. If we blow through a tube into the vena portæ, vena cava, hepatic artery, or trunk of the pori bilarii, but especially into the veins, the mass of the liver immediately begins to swell, and the granules next the surface are raised, and become more perceptible. If we blow with much force, they burst, and the air escaping between them and the external membrane, detaches and raises the latter in the form of blebs.

195. "The ductus cholidochus.—The hepatic duct, or trunk of the pori bilarii, having run a little way, joins the cystic or vesicular duct which comes from the gall-bladder. These two ducts unite to form a common trunk, named ductus cholidochus. This duct goes to the curvature of the duodenum, insinuates itself between the coats of that intestine, and opens into its cavity, not by a round papilla, but by a longish orifice, rounded above, contracted below, like the spout of a ewer, or a common toothpick. The edges of the orifice are raised, broad, and corrugated; as we may see by immersing this portion of the duodenum in clear water. At the entrance to this orifice, we see another smaller opening, distinct from it, which is the orifice of the pancreatic duct. The gall-bladder is a kind of pouch, shaped like a pear, narrrow at one end, broad at the other. Its wide extremity is called the fundus; the narrow extremity, the neck; and the middle portion,

the body. About one third of the body of the gall-bladder lies in a depression on the concave side of the liver. When we lie upon the back, the gall-bladder is almost inverted. Its external coat is a continuation of that of the liver, and consequently of the peritonæum. The second coat is muscular, and made up of two layers of fibres, one longitudinal, the other transverse; but irregular in their direction, almost like the muscular fibres of the stomach. These two coats are connected by cellular tissue, continued between the body of the gall-bladder and the liver, all the way to a whitish layer, which is looked upon as a third coat, corresponding to the nervous coat of the intestines. The internal or fourth coat has on the inside a great number of reticular folds, covered with small lacunæ, like perforated papillæ, particularly near the neck of the gall-bladder, where the folds are longitudinal, and afterwards form a kind of corrugated pylorus. These lacunæ are regarded as glands. The ductus hepatico-cystici.—That side of the body of the gall-bladder which lies next to the liver, is connected to that viscus by a vast number of filaments, which penetrate into the substance of the liver; and among these filaments there are some ducts which form a communication between the pori bilarii and the gall-bladder. For a long time these ducts could be found in brutes only, but lately they have been seen in the human body. They are most numerous near the neck of the gall-bladder, and they are named, ductus cysto-hepatici, and hepatico-cystici. The neck of the gall-bladder bends in a peculiar manner, and produces a narrow canal, called ductus cysticus. The curvature it makes resembles the head of a bird, the cystic duct forming the beak. The neck of the gall-bladder is furnished on the inside with several reticular rugæ and some folds, which appear like fragments of valvulæ conniventes. These are situated very near each other, from the neck all the way to the cystic duct. The first of these folds is raised, large, and almost circular; the next is more oblique and smaller, and the rest diminish one by one in the same manner. Taken all together, they form a kind of spiral flight. For this observation we are indebted to M. Heister. The internal surface of all the biliary ducts, that is, of the ductus hepaticus, cysticus, and cholidochus, when examined through a microscope in clear water, appears to be nearly of the same structure, through their whole extent. The cystic and hepatic ducts, where by their union they form the ductus cholidochus, do not resemble the letter Y. After the curvature of the neck of the gall-bladder, they keep close to each other. I have observed that the hepatic duct runs for some space, united to the body of the cystic duct, before it opens into it; and in the opening I have seen a floating valvular membrane. The hepatic bile flows continually through the ductus cholidochus into the duodenum, but the cystic bile only when the gall-bladder is full.

196. "The trunk of the great vena portæ, the hepatic artery, the hepatic duct, and the nerves of the hepatic plexus, form all together a large bundle, before they enter the substance of the liver. The trunk of the vena portæ hepatica is in the middle of the bundle; the hepatic arteries lie on the right and left sides of this trunk; the nerves surround it on all sides, and communicate with the superior mesenteric plexus. Afterwards, the first branches of these arteries, nerves and biliary ducts, leave the trunk of the great vein, and join in the same manner respectively, with the trunk of the small or hepatic vena portæ, and its ramifications in the capsule of Glisson. All these branches of the vena portæ, and of the arteries, nerves, and pori bilarii, accompany each other by ramifications through the whole substance of the liver, forming everywhere small fasciculi, in the same manner as the large bundle is formed by their trunks. Every branch of the vena portæ, artery, nerve, and biliary duct, has its own capsule, and all the four have a common capsule, distinguished from the former by cellular septa, which are nothing more than continuations of both the common and particular capsules. The convex side of the common cellular capsule, round the substance of the liver, gives off a number of filaments, which form a cellular tissue that insinuates itself between the glandular granules. The concave side produces the cellular septa above mentioned. In this common sheath, the vessels, ducts, and nerves are disposed in such a manner, that a branch of the vena portæ ordinarily occupies the middle of it, and is there placed laterally; the arterial branch and biliary duct lie at the side of the vein, and the nerve is divided into several filaments, which run and twine between the vessels and ducts, and chiefly accompanies the artery and biliary duct, but not to so great an extent the vena portæ." *(Exp. Anat., Traité du Bas-Vent.*, n. 250—317.)

197. MALPIGHI. "The graphic description given by Harvey of the first development of the liver, seems clearly to indicate that it is a conglomerate gland. He states that on the seventh or tenth day, the parenchyma of the liver grows from the ramifications of the umbilical vessels, much in the same manner as a bunch of grapes grows from the vine, as the buds of trees from the twigs, or as the young ears of corn grow from the blade. The comparison with the vine is particularly apposite. For as all the fibres of the bunch, and all the ends of the vessels, have a kind of round or conglobate mass appended to them; and this mass, by vessels which pass through the branch, and are continuous with the mass itself, receives and filters a certain humor, and by other different and new textures of vessels communicates it to the

seed or embryo plant,—so it is probable, that all the last twigs of the vena portæ have glandular acini appended to them, and that these acini receive and filter the humor driven thither. And as each entire bunch of grapes consists in a manner of lesser clusters, communicating and connected with each other by vessels, and these lesser clusters are composed of single grapes,—so the whole liver consists of manifold globes, which are themselves made up of glandular globules. . . Assuming these statements as probable, we may then with reason decide that the liver is a conglomerate gland; inasmuch as it may be divided into different pieces and acini; has a single excretory duct; and separates a determinate juice, to be discharged into the cavity of the intestines. And if we might judge from the obvious sensible appearances presented by the livers of fish, the question would be settled at a glance; for these livers externally bear an exact resemblance throughout to the pancreas and the thymus gland; particularly if the blood which they contain be suffered to escape, and then washed away." (*De Hepate*, cap. iii.)

198. SWAMMERDAM. "Nearly in the middle of the stomach of the louse there is a certain corpuscule, which Hooke apprehends may be the liver; but I should rather take it to be the pancreas. Its color is somewhat inclining to citrine yellow. It is so strongly connected to the stomach, that it cannot easily be separated therefrom. If it be placed under the microscope, it may easily be divided into many little grains, like glands, but these are not very transparent. Pulmonary pipes also appear in it. The substance of this part is more firm than that of all the rest; for when it is extracted from the body and dried, it is but little diminished. It is of a very irregular figure." The author has given five different views of it. (*Bib. Nat.*, p. 76, tab. ii., fig. 3, 5.)

"The liver of the covered snail, with the intestines, completes the spiral of the body. It is divided into divers lobes, according to the different course of the intestines, which make as many depressions on it as they have turnings and windings. The liver is very full of vessels; and it has its particular ductus cholidochi, which discharge themselves into the intestines without any intermediate gall-bladder; in the same manner as in horses, pigeons, &c., which are erroneously supposed to have no bile. The bile of the snail is not remarkably bitter. The liver seems to consist of small equidistant granules. It is of a dark brown color, a little approaching to green; and it abounds with a juice of the same color. Its outer coat is granulated with minute, whitish points. It has very few lobes in the inside. (*Bib. Nat.*, p. 123, tab. v., fig. 6, 8, *a. g.*)

"In the cuttle-fish there is a loose and fungous body, called by naturalists, the mutis, and thought to have a great resemblance to the liver. I have always found this mutis regularly divided into two distinct parts. The upper part is very thick, and with gentle treatment and a little pains it may on each side be parted into two lobes. Otherwise, on wounding the coat that invests it, its substance readily flows away, being exceedingly soft, and like a liver bruised in order to extract its parenchyma. In the middle the mutis is very thick and spongy; and at its lower extremity constantly ends on each side in an obtuse appendage, which extends to the belly. The gula, resting on the salivary glands, always runs under the mutis to the stomach. *(Bib. Nat.*, p. 888, tab. li., fig. 3.) The heart of the cuttle-fish lies in the abdomen, and the great artery sends two considerable branches to the mutis. The substance of this part is entirely composed of an infinite number of granules, very loosely connected by means of the blood-vessels. Its color is intermediate between red and yellow, somewhat approaching to a brown. It lies by itself within a peculiar membrane, and consists of two distinct bodies, without having the least resemblance to the liver. Nevertheless, I cannot take upon me to deny its performing the same office as the liver. *(Bib. Nat.*, p. 888, 889.) The white bag which contains the ink of the cuttle-fish, lies principally on the left side of the body, and communicates by a slender canal with the upper extremity of the rectum; so that both the fæces and the ink are discharged through the straight gut by one and the same orifice. This bag is in part membranous, and in part muscular; and is furnished with blood-vessels, which run over its surface. Within it I discovered a small glandular body. When the bag is cut, such is the efflux and intense blackness of the ink, that it obscures all the other parts of the fish. The ink is insipid to the taste, and without the least bitterness. Near the stomach, and between the convolutions of the pancreas, there is a glandular body, which runs as far as the gula: I am in the dark as to its nature." *(Bib. Nat.*, p. 890, 891, tab. li., fig. 5.)

199. For the rest, the reader may consult Boerhaave, on the action of the liver, *Inst. Med.*, n. 338—350. Malpighi, *Epist. Posth.*, i. ii. Ruysch, *Thes. An. Max.*, n. 86; *Epist.*, v., ix.; *Observ. An. Chir.*; and *passim*. Lancisi, *Epist. de Bilis Secretione ad Joh. Bapt. Bianchi*. Glisson, *De Hepate*. Bohn, in *Act. Erud.*, 1682 and 1683. Reverhorst, ap. Nuck, *Sialog. et Adenog. et de Inventis Novis*. (Lugd. Bat., 1723.) Verheyen, *Corp. Hum. Anat.*, tr. ii., cap. xvii. Littre, in *Hist. de l'Acad. Roy. des Sciences*, an. 1701, p. 51. (Paris, 1719); where he relates, that in a healthy liver, he had demonstrated certain glands which are not ordinarily visible, but which were of nearly a line in diameter, and the extremities of the arteries, and of the vena portæ,

as well as the roots of the vena cava, and biliary ducts, proportioned in size to those glands, terminated in them, and were visible to the naked eye. Especially let the reader consult the plates of different authors, as Eustachius, *Tabul. Anat.*, tab. ii., fig. 3; where he exhibits the concave part of the liver, with the gall-bladder, ductus cholidochus, hepatic arteries and umbilical vein. Tab. xi., fig. 2, shewing the cœliac artery and the vena portæ: fig. 1, the trunk of the vena cava inferior, with the branches of the vena portæ distributed through the liver. Heister, *Comp. Anat.*, tab. iii., fig. xii.; representing the liver of a new-born infant, and the gall-bladder; the vena portæ, with its branches distributed through the liver, the trunk being cut off; also the vena cava and the umbilical vein, and the insertion of the latter into the vena portæ: likewise the ductus or canalis venosus, going to the cava, and which is obliterated after birth. See also Ruysch's plates: as 1. *Epist.* v., tab. v., fig. 3; where he exhibits the inner surface of half the human gall-bladder, furnished with glands, which are adherent not only to the lower portion connected to the liver, but also to the opposite portion, and to the neck of the gall-bladder, the ductus cysticus and the ductus hepaticus: also, the valvular and cellular structure of the ductus cysticus. He compares the inner surface of the gall-bladder with the surface of a reticulated melon. 2. *ibid.*, tab. v., fig. 4; shewing the human gall-bladder, its arteries and lymphatics, the cystic artery, the tortuous course of the cystic duct, &c. 3. *ibid.*, tab. v., fig. 5; a portion of the liver with the suspensory and umbilical ligaments connected to it, and certain branches arising from the diaphragmatic arteries. 4. *ibid.*, tab. v., fig. 6; the neck of the gall-bladder, and the serpentine course of the cystic duct. 5. *ibid.*, tab. vi.; the membrane investing the dorsum of the liver, with myriads of lymphatics passing through its folds, together with small branches from the hepatic artery. 6. *Thes. Anat.* ii., tab. vi., fig. 1; a portion of the vena portæ, from the liver of a calf. 7. *ibid.*, tab. vi., fig. 3; one of the twigs of the vena cava, distributed through the liver. 8. *ibid.*, tab. vi., fig. 4; a branch of the vena portæ. 9. *Thes. Anat.* ii., tab. ii., fig. 3; a portion of the liver of a dropsical subject, exhibiting certain cells and vesicles. The reader may next consult several very beautiful plates by Bidloo: 1. *Anat. Hum. Corp.*, tab. xxxvii., fig. 1, 2. (Amstel. 1685); the concave part of the liver, and the hepatic artery in that situation, terminating, as he says, on the common capsule of the hepatic vessels, also on the coats of the pori bilarii, and on other parts, but principally on the coat of the vena portæ; and surrounded with a plexiform interlacement of nerves, as the vena portæ is surrounded with lymphatics: he also shews the tunica vaginalis, the trunk and branches of the vena cava, the gall-bladder, the ductus cysticus, the porus bilarius, and the ductus cholidochus going towards the intestine. 2. *ibid.*, tab. xxxviii., fig. 5; the liver, deprived of its parenchyma on its concave part, shewing the hepatic artery, and its ramifications, the trunk of the vena cava and its branches, and the trunk of the vena portæ and its ramifications, &c. 3. *ibid.*, fig. 1; a portion of the liver boiled. 4. *ibid.*, fig. 2; the ramifications of the vena cava and of the vena portæ. 5. *ibid.*, fig. 3; the ductus cysticus, with the valve, the biliary passages, and the ductus cholidochus. 6. *ibid.*, fig. 4; the gall-bladder, with the radices felleæ, their trunk of some considerable size, and the valve in the cavity of the gall-bladder. Reverhorst, *Diss. de Motu Bilis Circulari*, tab. i., fig. 1; shewing the liver, and its attachment to the sternum by the suspensory ligament, the gall-bladder, the umbilical vein, and numberless lymphatics. *Ibid.*, fig. 2; the internal surface of the liver, the small lobe, the fissure, the hepatic artery, the vena cava, the ductus cysticus, the ductus hepaticus, and the lymphatics. *Ibid.*, tab. ii., fig. 4; the cystic gland, placed at the porta or near the ductus cysticus. Verheyen, *Corp. Hum. Anat.*, tab. xi., fig. 3; the convex surface of the liver, with the suspensory ligament, the vena cava, and the fundus of the gall-bladder. *Ibid.*, fig. 4; the concave surface, with the lesser lobe, the gall-bladder, the ductus cysticus, hepaticus and cholidochus, also the ductus hepatico-cystici, the hepatic artery and the vena portæ. *Ibid.*, fig. 5; the coats of the gall-bladder, vascular, muscular and nervous. *Ibid.*, fig. 6; the hepatico-cystic ducts. *Ibid.*, fig. 7; the capsule of Glisson, with the vena portæ, the gall-bladder, &c.

200. The liver is the general purificatory and defecatory organ, both of the chyle, of the blood itself, and of the serum. Its peculiar vessels, the biliary passages and ducts, are as it were so many sieves or instruments for sifting and separating different fluids. The stomach and intestines continually supply it with the chyle clogged with cruor; the spleen, the omentum, and the pancreas pour into it all the blood of this region; with a view to its fining both these liquids to the utmost, and reducing them to extremities, and so purifying them (*a*). Since so many

(*a*) That the functions of the liver are most comprehensive, and form a kind of complement or conclusion to those of the other viscera, may be inferred with certainty from the immense number of vessels which come to this organ; namely, all which arise from the cœliac and its trunks, and which pass through the spleen, the pancreas, the omentum, the stomach, the intestines, and the mesentery; forming the venous channel of the porta; not to mention the particular branch of the same artery, which gives off the hepatic and the cystic arteries. Hence it appears that the liver, or the vena portæ, is a kind of port to the blood of this region, that is to say, of the abdominal viscera, and the common gate to the vena cava, and to the right side of the heart: and still more evidently so in the fœtus than in the adult; for in the fœtus the liver is extremely large; but after birth it does not increase in size in the same proportion as the other members, although even in adults it occupies a large portion of the abdomen. So likewise in brutes. "In the mouse," says Lancisi, "the liver is very large, wrapping over the stomach, and concealing it in a manner in its centre. It is also very large in the viper; reaching not merely over and about the stomach, but as far as the duodenum[x]," &c. The same remark is applicable to insects, which are destitute, as it seems, viscera assist the liver, therefore we can never have any specific knowledge of its offices or their nature, without taking account of the organs which act in consort with it, or which lend their aid and contribute their share to the common object. To examine and consider the liver alone, separately from its compeers and associates, would be like examining one cylinder in an automaton, one little wheel in a machine, one link in a chain, or one municipality in a kingdom; from which, taken singly, we should be enabled merely to conjecture and suspect the use of each to the whole. The nature of every member, and the part it plays, can be learnt only from the whole, and from the connexion of the member with the whole (*b*). We must therefore explain, but briefly and slightly, the intention and office of the several viscera which assist the liver; we can then ascertain what is the office of the liver, as their result and complement.

201. The members enclosed in the peritonæum are so many most elaborate chemical apparatuses, or organs, designed and fashioned entirely with reference to the blood (*c*). Each coöpe-

of spleen, pancreas and mesentery, and which have only a liver, that thus includes and embraces the offices of all those viscera. In the snail, the liver is placed in the very apex, or at the top of the spiral; so that it may regulate, and provide for, all the members underneath it. See Swammerdam, n. 198. Such then being the importance and dignity of the liver, it is therefore essential to explore its offices with close attention.

(*b*) A common source of the errors of the human mind appears to be, that we generalize hastily from some one particular, and rush headlong into conclusions, after a very slight experience of facts; and those principally isolated facts, of our own finding out, or the fruits of our own endeavors. Then for the most part some inherent self-love leads us into high reasonings, and spurs us onwards, so that the mountain labors in bringing forth a mouse. To conclude from a part to the whole, is like concluding from a branch to the state of the whole arterial system; from a single province respecting a whole empire, or the globe itself; or inferring the propriety of a speech from a single felicitous expression. In this manner we form prejudices and preconceived opinions, which inhere with such tenacity, that they are not to be eradicated by even the plainest facts. But enough of this for the present.

(*c*) That not only the abdominal but also the thoracic viscera, with-rates with each in a harmonious division of labor. Some of them prepare the chyle; some export it; some defecate and refine it; some marry it to the blood; some cleanse the serum; some purify the blood itself; some lay it open; distribute, and bestow it for use; some restore it and bring it to itself again. Hence it may be seen how difficult it is to unroll this chain of operations distinctly, so as to submit it to the understanding; for the offices are divided and apportioned among the organs in such a manner, that they go forth and return as it were in a circle; that is to say, they commence where they cease, and where they are in the end, there also they are universally in the beginning. Nevertheless it is necessary both to unfold them, and to understand the uses of the viscera; but then this metachymic process, inasmuch as it transcends all commonly recognized processes, must not only be clearly divided and subdivided

out exception, are entirely for the sake of the blood; in order, that is, that the blood may exist, and constantly subsist in its primitive integrity, will be shewn in the analysis of each viscus, and in the Epilogue at the end of the Part. In the case of the stomach, the intestines, and the mesentery, the fact is undoubted; for they prepare the chyle, and convey it into the blood, to constitute the serum. Nor will it be disputed, that the kidneys, the ureters, and the bladder purify this serum. That the liver, the pancreas, and the spleen are intended to defecate the chyle, serum, and blood, remains yet to be shewn. That the lungs inspire motion into all the viscera or organs of the blood-laboratory, to enable them to perform their offices aright, can hardly be a matter of uncertainty; nor yet that they refine and exalt the blood, inasmuch as all the blood is driven through their pipes, and comes out brighter than it went in. For the blood is the essential life, and, as it were, the soul of the body; the repository of all that belongs, or that ever will belong, to the body; the source of all the vital humors; of the motion of the whole and the parts, (for the muscles lie idle without the blood): it is the source also of the life of the senses, and even of the life of the interior senses of the cerebrum itself; for when the blood fails, the very mind labors, gropes, and faints. The members dedicated to generation derive their prolific juices from the blood. Consequently, the life of the body is the life of the blood. Hence then it follows, that the functions of both the abdominal and thoracic viscera, are performed on account of the blood. How these functions are apportioned among the viscera, will be pointed out presently.

into its proper stages, but these must also be expressed by new significative terms (*d*).

202. I. The first of the chemical operations of the body, is CHYLIFICATION; which is carried on by trituration, percolation, maceration, elixation, extraction, secretion, and excretion; in the intestines, the stomach, the œsophagus, the fauces, and the tongue; that is, in the great alimentary canal (*e*). The next operation is the PURIFICATION OF THE CHYLE; which is effected in the liver, the mesenteric glands, the receptaculum chyli, and the thoracic duct, by an afflux of lymph from the conglobate glands; and even in these glands themselves. This purification is a kind of perpetual chylification, performed with increased expedition and perfection (*f*); namely, by the *trituration, filtra-*

(*d*) Absolute necessity demands new terms, or new formulas, for expressing the chemical operations of the animal body. These are ordinarily expressed by terms borrowed from retorts, cupping-glasses, phials, bladders, baths and fire; the phrases connected with which, however,—as distillation, rectification, cohobation, decantation, clarification, inspissation, &c.,—are not appropriate, excepting in an indirect and metaphorical sense. Some, indeed, are more suitable, as extraction, secretion and excretion. But when these operations are subdivided as it were into parts, then the terms can no longer be borrowed from their field, but we are obliged to resort to our social field. For as soon as generals are dismembered, and distributed into parts, they become so much attenuated as in a manner to disappear; wherefore the terms by which they are expressed undergo the same attenuation, and seem to vanish and fail. In short, all new arts and sciences, at their first appearance on the world's theatre, and all parts of such, require a new wardrobe of terms; only let them not be barbarous, but as far as possible akin to those already in use.

(*e*) Respecting all these parts, see the foregoing chapters, particularly that on the Stomach, n. 94.

(*f*) The chylification of the intestines and stomach is only rude and preliminary: their chyle does not enter the blood until it has passed through certain mediating viscera, and been constantly purified in its passage; as first prepared, it is nearly crude and undigested, and therefore it requires to be driven through certain filters. That the liver is one of the requisite filters and sieves, is now to be shewn; as such, its operation deserves to be termed, perpetual chylification.

tion, and *sequestration*, or expulsion, of coarse, incongruous, and worthless particles, in the pori bilarii and hepatic duct; also in the mesenteric glands, and in others of the same kind, external to the mesentery (*g*). II. Next it is requisite that the chyle should be suited to the blood, and be made in its likeness, and become homogeneous with it; and with a view to its being properly insinuated or inaugurated, it requires to be impregnated by the blood, or with certain products of the blood and serum, such for instance as the salivæ or lymphs, all of which, to fit them for the purpose, must be previously tinctured and enriched with the spirit of the blood. This operation may, I think, be called, the INAUGURATION of the chyle; and it is effected by IMPREGNATION, which is followed by MARRIAGE; and then, by CONJUNCTION or copulation. These two operations, of chylification and inauguration, together constitute SANGUIFICATION. They are performed, first, in the whole of the alimentary canal, in the mesentery, the thoracic duct, and the conglobate glands; principally in the liver (*h*): next in the large veins, as in the subclavian, at the foot of the jugular (*i*); in the vena cava inferior, and its hepatic branches: and in all the rest of the veins, even to the very least: lastly, in the heart, by which the circle of this operation is generated. III. When the chyle is initiated into the blood, and in the arteries and veins, it is no longer called chyle but serum: now therefore instantaneously succeeds the PURIFICATION OF THE SERUM. This is effected by a continual process of separation carried on in the capillary ves-

(*g*) That is to say, in the conglobate glands, as the iliac, the sacral, the subclavian glands, &c.; respecting which, see n. 176.

(*h*) The ancients, therefore, ascribed sanguification to the liver; although sanguification is the work of no one viscus, but of many combined. For the blood is not a simple substance, but is composed of simples of many degrees; consequently, many different appliances are required to wring these elements from the ingesta, and afterwards to insinuate them into the blood: particularly, to enable it to subsist perpetually in its primitive state of integrity, and perennially to maintain its circle of life.

(*i*) The lymph of the cerebrum or superior region, impregnated with fresh spirit, pours in at the very same place as the fresh chyle of the thoracic duct.

sels and miliary glands (*k*); consequently, by SECRETION. Of the various sera thus separated and secreted, the finer are either sent back into the blood (*l*), or adopted by the conglomerate glands for combination with the humors; the worthless kinds, which pollute their receptacles, are thrown away. The first-mentioned process may, I think, be called RESTITUTION; the second, ADOPTION; and the third EXCRETION or extermination; which last is carried on by divers emunctories, as the skin, the nares, the fauces, the kidneys, and the bladder. The secretions and excretions are either *profitable*, as the more impure kinds of saliva, the pancreatic juice, and the hepatic and cystic bile (*m*); or they are *unprofitable*; in which latter case they are rejected as fæces. The secretions and excretions consist either of *obsolete* or of *superfluous* materials (*n*). IV. And not only the serum,

(*k*) The red blood, entirely cut off from the serum in the extremes of the arteries, and divided into its very principles in the minute capillary vessels, circulates under the form of white blood or lymph; thus every time it runs through its circle, it is sundered into its constituent parts. Arterial extremities of the kind exist in all the glands, particularly in their primitive acini, and in the papillæ: these extremes expire the divided blood, each according to its organization and the state of its parent fibre. Not to mention that the same thing obtains in all the other extremes, as the skin, the interiors of the viscera, and their coats; in the brains, &c.

(*l*) The little veins everywhere open their mouths for the fluid which comes to them, and what goes past the veins comes to the orifices of the lymphatics. Thus the choice parts of the blood and serum, are conveyed home again into the blood, both by way of the veins and of the lymphatics.

(*m*) The biles are, indeed, excretions, as we shall shew in the present chapter; but this does not prevent them from being of use before they are excreted; nor from being obliged to give up any rich and nutrient matter which they may happen still to contain.

(*n*) As the milk, the semen, &c., which are thrown off, or determined outwards, not as obsolete, but as superfluous; and when they are not thrown out, the body labors with superfluity, and maladies, of both the body and the animus, may be the result; as also may be the case from the presence of too much blood, even of the better kind; for instance, in plethora, which is the frequent cause of fevers, as well as of uneasiness and ferment in the animus.

but even the blood itself is purified, during every round of the circle. For its globules are either spurious, and stuffed with heterogeneous particles, or they are covered with foreign dust, or clotted into strange shapes and irregular pieces, or too soft, or too hard; in all these cases, worthless (*o*). The process of dividing and sifting them, may, I think, be called the LUSTRATION OF THE BLOOD. It consists in constant acts of division and separation between the globules and the serum, carried on in the finest filters, that is, in all the extremities of the arteries, and in the minutest acini or seeds of the conglomerate glands (*p*). But in the adult, the general laboratory for the lustration of the blood, is the spleen, assisted by the pancreas, the omentum, and particularly by the liver: in the fœtus, the succenturiate kidneys, also assisted by the liver. The sundering and removal of the blood from the serum in the course of the circulation, may perhaps, be named, SEPARATION, as being a species of divorce (*q*). The elimination of matters from the mass of the blood, PROSCRIPTION; which consigns these matters to the gall-bladder, and thence to the duodenum. The laying open of the blood which either is becoming, or has already become, obsolete, in which its LUSTRATION properly consists, is carried on in the pori bilarii and hepatic duct. V. The blood, which is the storehouse and repository of all things which either do or will belong to the body, lays open its bosom, and thereby expends itself in supplying the necessities of the kingdom, and as it were immolates itself for the prosperity thereof. It is not easy to express this opening of the blood, its self-devotion for the common good, or how it expends itself on uses, by an appropriate term, and no periphrasis would answer the purpose; the first, therefore, we may call LAYING OPEN; the second, IMMOLATION; and the third, DISTRIBUTION (*r*). Each is performed

(*o*) On these subjects, see the Part on the Blood, where we shall treat of its various conditions arising from external and internal causes.

(*p*) See (*k*) above.

(*q*) See n. 180, above; and our Analysis of the Blood.

(*r*) Inasmuch as the blood lays itself open, in order to give birth to the vital fluids, (which renovate the organs of the body, and enable them to perform their labors,) thus offering itself in a manner as a kind in the conglomerate glands of the body, but preëminently in the cortical spherules of the cerebrum, and in the cineritious spherules of the cerebellum, medulla oblongata and medulla spinalis (*s*). Of all the operations of the blood, this is at once the largest and the most mysterious. VI. The last stage of these processes is the RETURN of the blood which has been laid open and distributed. For whatever of genuine blood and spirit is dispersed in the salivary and other vital juices, is again collected by the beginnings of the veins, by the lymphatics, and by the conglobate glands of the body and of the cerebrum (*t*); and re-introduced, that is, returned into the blood, under the name of lymph. This is attended with a perpetual acquisition both of new chyle and new spirit; and thus with gain and profit (*u*); which perhaps we may be allowed to name, LENDING ON INTEREST. This is again succceeded by *chylification;* or the *inauguration,* impregnation, and marriage of the chyle (*x*); and

of victim, and devoting itself for the welfare of the kingdom,—inasmuch, I say, as it thus suffers itself and its very marrows and entrails to be opened, therefore the word, *immolation,* seems not unsuitable to the case. But the manner in which it bestows itself, constitutes an office of vast extent; and if it be subdivided, specifically and particularly, then each subdivision of the office must be named from its analogy of function.

(*s*) The cortical parts of the cerebrum demand back the spirit from the blood, and transmit it into circulation through the fibres; wherefore the cerebrum invites the softer, purer and more spirituous blood; and what it does not resolve and expend, it sends off into the veins and sinuses of the dura mater; in the extremities of which sinuses it is again recruited and vivified with new spirit.

(*t*) The pituitary gland is the conglobate gland of the cerebrum. See n. 190 (*t*).

(*u*) The chyle itself expressed from the food, and the spirit generated in the cortex cerebri, meet at the place where the subclavian vein is inserted into the jugular: both are new comers, and, as it were, fresh guests, waiting to be inaugurated into the blood; consequently, the old inhabitants return with profit and interest, and introduce the strangers to their chambers.

(*x*) Chylification begins by the extraction of the juice from the food, and by the cohobation thereof with the saliva: this produces the chyle, lastly by its REGENERATION, or its perpetual generation. The offices of the viscera, and actual phenomena, prove that such is the *gyre of operations,* and such the *circle of life* of the blood.

203. All these things proceed in the most constant course, to nature's, or the soul's, appointed end, by means of powers and forces, imparted to and so implanted in all parts of the organs. Each is excited to its labors and services, by a certain propulsive, or active and living force, which we may term INCITATION (*y*); and also by an attractive physical force corresponding thereto, which we may term INVITATION. The general cause of invitation, is the unanimous tendency or effort of the fluids of the body, and particularly of the blood, to their state of equilibrium; which, if designated by a single term, may perhaps properly be called EQUATION; referring both to the quantity and to the quality of the blood (*z*). The general cause

which is thus inaugurated and introduced into the blood by means of the saliva. Thus the end of these processes is coincident with the beginning.

(*y*) We before treated of the attractive power of the tongue, the mouth, the stomach, the mesentery and the lacteals, also of that of the thoracic duct and the conglobate glands, in our analyses of those parts. This, however, is a physical or mechanical attraction, very similar to that of a syringe or syphon. Thus the chyle is allured or invited from the peripheries of the intestines into the mesentery, and from the mesentery into the subclavian vein. But that there is also an excitative or propulsive force corresponding throughout to this invitative force, see n. 153 and 163.

(*z*) While I have been dwelling on these stupendous mysteries of the animal kingdom, and endeavoring to reduce its particular and specific modes of operation to a few general heads, and these to one universal principle, the idea has offered itself, of *a certain equation of quantity and quality* of the fluids, pervading the system, and to which nature, as if for the sake of equilibrium, tends and aspires with all her might. But since in the body there is a perpetual loss and restitution of equilibrium and rest, and consequently, a change of equation; therefore, from this source results the diversity of the blood and the serum in the different viscera. For when more of one species of liquid is demanded, consumed, or eliminated, in one extreme than in another, a new liquid of the kind must immediately run thither to supply the want, from all parts, corners and provinces of the kingdom, near and of incitation or propulsion, is a similar effort on the part of the spirits to their state of equilibrium (*a*); this is brought about by

remote: and this, by reason of the equation of quantity and quality, which nature has ordained throughout, for the sake, as we before said, of maintaining her equilibrium. Wherever, therefore, from any cause, an excess of one species is consumed, thither of course a supply of the same species must flow, from other parts, near, intermediate and remote; and with greater rapidity than to a part where little or none of this species is required. Thus where such fluid is rapidly supplied, the quantity appears great; where slowly, it appears small; when notwithstanding, the equation is maintained, modified by mere differences of celerity: on which account, there are many paths which lead thither, and which are so adapted, as to provide a more copious afflux, corresponding exactly to the efflux. This appears to be the reason why the blood full of the better substances of the kingdom, rises to the brains; why the blood full of its more worthless and outcast substances, goes to the kidneys and the gall-bladder; why the blood, dilute in the first instance, is loaded with substances in a graduated series, from the lips, the gums, the fauces and the œsophagus, all the way to the stomach and the intestines; why the spirit of the mother is emulged by the embryo; why the power of venery is encreased by its exercise; why the breasts and teats overflow while the infant sucks; and why habit becomes second nature; with many other things, which proceed from this single, simple and universal source and law. But this is a subject requiring deep investigation; consequently, if it be laid down briefly, and not illustrated constantly by examples, and proved by a continual reference to facts, its details must appear for the most part entirely hypothetical.

(*a*) As long as animal life continues, there is a perpetual destruction of equilibrium: so the higher powers ordain, for by this means effort is perpetuated. The appetites and pleasures of the body, the passions of the animal mind, and the desires of the rational mind, arising from infinite causes and incentives within and without the microcosm, disturb the animal world, as tempests disturb the quiet of the atmosphere, and constantly call forth new and contending inclinations; whence I have often termed the passions, the incentives of corporeal life. A similar effort to equilibrium on the part of the spirits, cannot fail to produce this active and propulsive force of the blood-vessels; for the spirits actuate the very fibres; the fibres actuate the vessels, consequently the organic forms constructed of vessels and motive fibres. But turmoil and danger of absolute destruction arise, whenever contentions and means of the organism of all parts of the body; also by means of the animation of the brains, the respiration of the lungs, and the corresponding movement of the fibres, in particular and in general. Now in order that all these causes may constantly pass into their effects along the appointed channels, similar powers and forces are almost everywhere indefinitely multiplied; and this may be called, the UNANIMOUS CONSPIRING OF EFFICIENTS (*b*). The powers and forces so conspiring, proceed in a constant series from greatest to least: whence arises a CONTINUOUSLY SUCCESSIVE PROGRESSION of efficients and effects (*c*). Such

struggles spring up between the spirits and the blood, with respect to their different equations of motion.

(*b*) The vessels are so multiplied, and so united by infinite anastomoses; likewise the lymphatics and the lacteals; also the motive fibres, the glands and papillæ; the cells in the tissues; the pori bilarii in the liver; the cortical substances in the cerebrum, &c.; that even were a large part of the myriads to be wanting, the effect would still follow, although not so fully.

(*c*) Every series has its greatest and its least, or its *maximum* and its *minimum;* its congregate of many parts, its congregate of few, and its positive unity. The greatest or *maximum* of the vessels is the aorta and the vena cava; the least or *minimum,* is the finest arterial and venous capillary. The hepatic duct is the largest porus bilarius: the muscle is the largest fibre; the several motive fibres are its unities. The nerves are congregates of fibres; the fibres are the simples of the nerves. The passage by a continuous series from compounds to the simples thereof, is called successive progression; and as in substances, so in all their predicates and accidents. But there is no such progression from posterior to prior things—*a posterioribus ad priora:* for continuously successive progression itself, terminates in the unity of the compound; from this unity as a new compound, another progression commences and continues: thus we leap from posterior to prior things, but do not progress. These explanations, however, are only rudimentary and superficial; that is, they are the most general doctrines of the chemistry of animal nature: but the real principles and more internal doctrines, can scarcely come forth at all from the deep recesses of nature's organic laboratory, or be revealed to the human mind, excepting by a kind of indirect and reflex view, caught as it were from the mirror of generals: which is the reason of our declaration (n. 178), is a general view of the metachymic processes of animal nature, which we shall unfold and elaborate more distinctly, and in successive order, in the analysis of this and the other viscera. The state of the human mind requires, for the sake of distinct perception, that we educe every thing from a common source and a universal idea, and also refer it thither: which is the reason of this prologue. We now proceed to the liver.

204. The LIVER is the general laboratory for the defecation of the chyle, and for the lustration, and hence also for the regeneration of the blood: it performs a kind of ultimate and conclusion of the functions of all the abdominal viscera (*d*). Consequently, it is the place of preparation for the hepatic bile (*e*), which, with the cystic bile, is the ultimate salivary menstruum.

that the rays of rational sight here fall, disperse, and vanish in midway, in the same manner as the rays of ocular sight in the universe.

(*d*) That the liver gives the last touch and finish, that is, perfects and completes the works of all the abdominal viscera, is very conspicuous from the influx of blood into the great sinus of the porta; that is, from the stomach, the intestines, the mesentery, the pancreas, the spleen, the omentum, and even from the peritonæum and the diaphragm. For the veins which form the porta, are the veins of the intestines and mesentery, (commonly called the mesaraic veins, because their branches unite in the mesaræum,) and the internal hæmorrhoidal vein; also the veins constituting the coronaria ventriculi and pylori, and the vasa brevia of the stomach; besides which there are the epiploic or omental veins; the pancreatic and the splenic veins. Inasmuch as all these viscera, particularly the pancreas and the spleen, are entirely made up of arterial and venous ramifications; and inasmuch as the arteries, according to their nature, constantly secrete and excrete something; and the veins as constantly absorb something; and inasmuch as the venous reticulations of the stomach and intestines are continually opening their mouths for the fresh chyle; and inasmuch as all this crowd of vessels, and in the fœtus, the whole mass of the blood, goes to the liver alone, and is wonderfully distributed therein; and inasmuch as there is no other door or gate through which this blood can enter the vena cava, and go towards the heart, but through the liver and the mesentery alone,—therefore, of necessity, the liver must perform some ultimate office, which is also the complement and the conclusion of the offices of the abdominal viscera.

(*e*) It follows from this, that the purer and clearer chyle goes into

205. To enable it to discharge these crowning functions, the liver receives the semi-digested and crude chyle from the stomach and intestines (*f*), but by way of the veins, as the mesentery by way of the lymphatics (*g*): also the impure blood from

the blood, and the unclean or feculent portion, into the hepatic bile; hence we call the liver, the *defecatory of the chyle:* likewise, that the harder and more intractable blood is here sundered into its parts, the finer of which are sent into the blood, and the worthless, into the same bile; hence we call the liver, *the lustratory of the blood.* But the blood-globules which are so indurated and resisting as to refuse to be resolved or laid open, are thrown out into the gall-bladder, and constitute the cystic bile. From these two functions results this third, that the liver is *the regeneratory of the blood.*

(*f*) The stomach and the intestines are only the first vessels, and as it were receivers and retorts, into which the aliment, dry and moist, is poured; consequently, they merely begin the true work of digestion and chylification, and perform the rudiments thereof. Wherefore the ventricular and intestinal chyle must be considered as only crude and scarcely half digested; and therefore as requiring to be rectified and refined in the liver. See n. 168 (*u*).

(*g*) The chyle of the stomach and intestines is partly imbibed by the veins, partly by the lacteals: that absorbed by the venous orifices goes by a venous path, which conducts it towards the liver: but the chyle imbibed by the lacteals goes by a lymphatic path, which conducts it to the mesentery. The quantity of chyle which goes by way of the veins from the stomach and intestines to the liver, may be inferred from the vast number of little veins and reticulations of veins which exist in both; retiform plexuses having been discovered and fully described by many anatomists, in the internal coat, and particularly in the nervous as well as in the cellular coats, in both the stomach and intestines. Moreover, the very glands and papillæ of the intestines, and their ducts even, are mere tissues of arterial and venous capillaries; whence we may conclude what a quantity of chyle they everywhere absorb, and transmit towards the liver: as may be concluded also from the mesenteric glands themselves, which the vessels enter and leave in abundance, pouring their blood, impregnated with chyle, into the veins which are flowing to the gate of the liver. The peculiar office of the veins consists in imbibing the fluid applied to them: that in the present instance they imbibe the chyle, is proved by the case of all those animals that are destitute of mesentery and thoracic duct, (as the whole the spleen, the pancreas, and the omentum, and from the mesentery itself (*h*). This chyle and blood it brings together into the sinus of the porta; and therein mixes and confounds clean things with unclean, soft with hard, shapely with shapeless, and living with dead (*i*). Such is nature's usual mode of working in all her universes (*k*).

tribe of birds and insects,) in which the chyle manifestly passes into the blood by way of the veins alone.

(*h*) In the chapters on the Spleen, Pancreas, and Omentum, we shall shew, from experimental evidence, that the spleen purifies the blood itself, or purges away its heterogeneous parts, and remits what is still not properly purified to the pancreas and the omentum: and that the pancreas and the omentum send the residue that is still not defecated, but which requires further purification, to the liver: thus that the liver completes and concludes the work begun by the spleen, and continued by the pancreas and omentum.

(*i*) I do not think there is in the whole body another such sewer and turbid gulf of humors, as the vena portæ: for the chyle rushes in thither with all its impurities, and the blood with all its sediments. In order for the liver to separate all these materials, in short, to defecate and purify both the chyle and the blood, it absolutely requires to be of considerable size.

(*k*) It is well worthy of observation, that nature, in the ultimates of her world, first confounds and crowds the elements into one, before she separates them, or calms her world. In the chambers of the heart, particularly in its right chambers, she brings together the blood, the serum and the spirit; in short, whatever the veins can possibly scrape together from every source. In the stomach, she heaps together all kinds of food and aliment, dry and moist; fruits, fermented liquids, mucus, sputa and salivæ. In the ventricles of the cerebrum she brings together various things in a similar manner. And not only is this the case in her animated, but also in her inanimate worlds. Thus, by tempests, winds, clouds, rain and lightning, and by myriads of effluvia streaming like a torrent from her three kingdoms, she confounds the atmosphere; and yet these very tempests, showers and thunders, are the means of its purification. In this manner she is continually reducing her universe to a kind of chaos, in order that she may select all things therefrom, and distribute them into their proper places. Similar is the condition of human minds: generals enter them at first, without distinction or digestion, and induce a kind of dark, obscure and mid-

206. This medley, turbid, and impure stream the sinus sends out by five doors or gates, and afterwards by a variety of passages, and diffuses throughout the mass of the liver (*l*); and drives it successively through finer and finer strainers, into myriads of glands, which are so many little sinuses, least organs for the defecation of the chyle, and for the lustration of the blood (*m*). In these it sifts and divides the muddy current into

night state; but from this chaos, more and more distinct ideas are gradually evolved, and produce understanding and clarity. Just as the liver evolves order and distinctness from the chaos of the vena portæ.

(*l*) In the whole animal body there is no worthier subject of enquiry than that presented by the liver; which is the reason why we devote so much space to it; for we are here instructed respecting the order and the art whereby nature distinguishes particulars, and distinctly distributes them for their various uses. First of all she throws from the sinus the obscure and muddy stream, full of all kinds of impurities, sanguineous and chyliferous; and sends it by several branches into the substance of the liver: from the trunks thus put forth she educes other branches; from the branches, twigs; and from the twigs, villi; and finally threads of inconceivable minuteness. (Ruysch shews, in one of his plates, the manner in which the first venous, or, if you please, arterial stems, (they are, in fact, neither, or both,) are continually splitting and subdividing into offsets or lateral ramifications; never stopping until they arrive at their unities, in the minutest threads.) Then first, at the very extremes, she conglomerates them, and fashions new sinuli or glands, as it were new portæ; in such vast numbers, that myriads of them are not equal to the grand sinus of the vena portæ. Thus these glands are placed at one extreme, the great vein at the other, but the two are closely connected by continuous branches; and they are mutually respective and correspondent, as the two extremes of one thing, or as the first and the last, the *maximum* and the *minima*, the composite and the simples, the sum and the unities thereof.

(*m*) That similar types of the large occur universally in the leasts, see n. 100. These least types must be considered as the unities of the large, and as the ultimate divisions of their compound or concrete; consequently as similar to it in essence and nature; the sole difference consisting in magnitude and multitude. "The liver," says Winslow, "is composed of several kinds of vessels, the ramifications of which are multiplied in a stupendous manner, and by the intertwinement of their capillary extremities, form an innumerable quantity of friable and pulpy three species: one it assigns to the veins; another to the lacteals; and a third to the pori bilarii: that assigned to the veins it determines to the cava; that to the lacteals, to the receptaculum chyli; and that to the pori bilarii, to the duodenum (*n*).

207. The convoluted twigs of the vena portæ are the principal constituents of these little glandular sinuses or noduli, and prepare a way to them and open into them (*o*). The ultimate

granules, which are looked on as so many peculiar organs, whereby a particular fluid is secreted" (n. 194). Malpighi dissertates with much elegance on these glandular bunches and clusters. See n. 197. Littre has demonstrated these glands, of nearly a line in diameter, and the extremities of the arteries and of the vena portæ terminating in them (n. 199). "Wherever," says Boerhaave, "the last ramifications of the veins meet, they become so small, that they appear to terminate in a singularly fine substance, made up as it were of little brushes arranged in bundles in which distinct globules seem to be formed, composed of almost invisible vessels, and separate from each other. These corpuscules are very similar to what are called simple glands." (*Inst. Med.*, n. 342.) The liver of the louse, according to Swammerdam, "may easily be divided into many little grains, like glands. The liver [of the covered snail] seems to consist of small equidistant granules" (n. 198). So also the mutis of the cuttle-fish (*ibid*). Nor do Ruysch and Heister deny the existence of these parts, only they will not call them glands; nor do they deny the existence of the cortical substances, which are, indeed, perfectly visible to the microscope.

(*n*) Each of these particulars will be considered separately in the sequel.

(*o*) As appears by the divarication of its branches from their trunks; see Ruysch's Tabulæ. But the question occurs, In what way does the vena portæ contribute to the formation of the gland? and in what way do the hepatic artery and vein contribute to it? With respect to the vena portæ, its very continuity shews, that its coat, ramifying and becoming thinner and thinner to this point, forms the gland by a process of conglomeration or convolution: very much in the same manner as everywhere else in the animal kingdom; and even as in the vegetable kingdom; in which latter the inner bark expands ultimately into a tuberous nodule, where it terminates in circles and gyrations. "The extremities of all the ramifications [of the vena portæ]," says Winslow, "end in pulpy, friable granules" (n. 194). The coat of the vena portæ is a continuation of the common coat of the liver, and this latter is a ramifications of the hepatic vein and artery, also enter and connect them, and insert into them their little lips; hence their glandulo-organic form, woven of the ends of these three ramifications (*p*). All these interlacing vessels, namely, the portal

continuation of the peritonæum, which, where it accompanies or encloses the vessels, is called the capsule of Glisson. "The capsule of Glisson," says Heister, "is continuous with the peritonæum, enclosing the branches of the vena portæ, the arteries, and the biliary ducts, as they approach to, and after they enter the liver" (n. 191). Thus not only does the vena portæ concentrate itself in these places, and produce a representation of itself in the least forms, but this is even the case here with the liver itself, the peculiar nature of which commences in these glands. The peritonæum also here at length forms a knot, and buds forth in like manner: for whatever is the cause of the cause, is also the cause of the thing caused. Consequently, these spots [the glands] are the centres and meeting-places of the abdominal viscera.

(*p*) The hepatic vein, arising in these cradles of the liver, conveys the blood into the branches of the vena cava: nay sometimes this vein is reckoned among, and confounded with, the branches of the vena cava. "The blood," says Winslow, "is conveyed away by a great number of minute venous ramifications, which afterwards unite into three principal branches, that terminate in the vena cava. These in general are simply called, hepatic veins." (n. 194.) The hepatic artery arises from the right branch of the cœliac, and entering the liver not far from the vena portæ, ramifies through the viscus in all directions: see Bidloo, *Anat. Hum. Corp.*, tab. xxxviii., fig. 5; where the trunk or channel, and its insertion into the liver, are beautifully delineated. That both the hepatic vein and artery penetrate all the way to these glandular and central forms, see above, schol. (*m*) and (*o*): that they also insert their extremities therein, is very evident from the down and villosity wherewith the interior cavities of these glands are covered; as shewn by the microscope. "The extremities of the ramifications," says Winslow, "end in pulpy, friable granules, which when examined through a microscope in clear water, seem to be thick villous follicles:" and again, "They are of a pulpy texture, like radiated villi." (n. 194.) Hence it is, that if we inflate either the biliary duct, the hepatic vein or artery, or any of the branches of the vena portæ, the glandular follicles are distended; which would not be the case unless the extremities of these vessels either terminated or commenced therein. Winslow says, "If we blow through a tube into the vena portæ, vena veins and the hepatic veins and arteries, are here in their leasts; or in coming hither, proceed from greatest to least; and when they are in their leasts and as it were their unities, they then commence their play, that is, exert their powers, and exercise their forces and peculiarities, with the most perfect distinctness (*q*). But where they end there they also begin, so that their last term is their first.

cava, hepatic artery, or trunk of the pori bilarii, but especially into the veins, the mass of the liver immediately begins to swell, and the granules next the surface are raised, and become more perceptible." (n. 194.) What further than this may be the implantation and distribution of the extremities of the vein and artery in these glands, must at present be a matter of mere conjecture. The fine villosity plainly shews, that they are inserted therein, just like the arterial extremities in other parts, which eructate some serous humor into a particular cavity; and like the venous extremities, which absorb some humor from the cavities; it shews, in short, that the extremities do certainly open therein. Furthermore, a surface composed or conglomerated of a branch of the vena portæ, seems to serve the glands as a basis and wall, or parietal covering; as I shall endeavor to prove presently by many considerations. Hence these forms are not made up entirely of vessels, but also of a common production of the vena portæ and of the liver, as well as of the nervous fibrillæ, which closely accompany the hepatic veins and arteries. Thus, inasmuch as they are not mere convolutions of vessels, they seem to merit the name of simple glands.

(*q*) This we before named, the *successive progression of efficients*, which progression is continuous, and takes place, when a trunk decreases constantly into branches, and ends in the most minute and imperceptible twigs. Every artery in the body progresses in this manner; from the heart, namely, and the aorta, to the minutest twigs, which may be considered as the unities of the arteries. The veins likewise progress from their least to their greatest; from the least venous threads to the vena cava, which terminates in the grand venous reservoir, that is to say, in the right auricle of the heart. But the veins which go to the liver, ramify thus, not once only, but twice: for the cœliac artery divides into the smallest capillaries in its way to the liver; first, namely, in the pancreas, the intestines and the stomach: thence as from new beginnings, the vessels again enlarge towards the vena portæ; from the vena portæ they gradually again diminish, until they reach the hepatic glands: and from the glands, they once more successively

208. Into these little sinuses the vena portæ spews its muddy wave; the hepatic artery sprinkles the new essences of the serum and the blood (*r*): the hepatic vein elects and absorbs from the

increase in calibre: and this, in fact, thrice or four times; if we consider them as first contracting in the spleen, and dilating again towards the stomach and the vena coronaria ventriculi; and as then again subdividing into small branches, and emptying themselves into the vena portæ: as they necessarily must do, in order that the blood and the chyle which they carry, may be purified. But in these glands they are not the same veins as entered through the branches of the porta; nor are they absolutely continuous with the latter; but they are new venous germs or roots, growing up in these places. The reason why the vessels, when they reach their leasts and unities, come into possession of their true powers, appears to be, that they are then nearer to their parent fibres, in which a higher power resides: for the perfection of everything is in proportion to its simplicity: they are then under the government of intrinsic accidents, and not at the same time of extrinsic accidents. But we shall speak further on these subjects elsewhere.

(*r*) Were it not for the hepatic artery, the defecation of the chyle and the lustration of the blood could never be accomplished by the liver. The hepatic artery sprinkles in first the blood, and the essentials of the blood when in a state of division: next, particularly in the hepatic duct, it pours in the obstinate, resisting, and antiquated blood, in order that it may be sundered into its parts by the grinding action of the duct. With respect to the first point, the chyle in the vena portæ appears to be so clogged with impure blood, that it does not readily allow itself to be separated in these most minute and delicate glands; wherefore it is necessary to have fresh blood, to dilute it, and like a kind of sponge or pumice, or like a kindred and kindly menstruum, to absorb the purer parts of the viscid chyle; and thus render it suitable to be applied to the outstretched teats or lips of the little veins. This I have before termed, the INAUGURATION of the chyle into, and its MARRIAGE with, the blood, or the essential serum and spirit thereof. See n. 202. The same thing is accomplished in these glands as in other parts, and in the most general manner in the left subclavian vein, where the new lymph and the fresh spirit of the cerebrum flow to meet the chyle carried up by the thoracic duct. See n. 190 (*t*). It would be tedious in this place to bring forward all the proofs of the fact, that there is such a nature as this in the blood and its spirit. With respect current the chyle married to the blood: the fine cellular tissue, stretched round in all directions, imbibes the lymph driven out through the pores (*s*) : the fluid refuse is thrown out into the pori bilarii, or roots of the hepatic duct (*t*). All these results

to the second office of the hepatic artery, of conveying the hard and unyielding blood to the hepatic duct, or even to the gall-bladder, it will be explained presently. The office of nourishing the liver is very generally attributed to this artery ; but let us understand first what nutrition is. If it consist in the impletion of the veins and arteries with blood, I grant the point. But inasmuch as the veins and arteries, in their intimate structure, consist of nothing but fibres, and inasmuch as the tissue of the liver consists of arteries and veins, therefore it follows that nutrition is primarily owing to the fibres, consequently not to the body but to the brain.

(*s*) That the little membrane which covers each gland is permeable and porous, may be conjectured not only from the mode in which the minute vessels are connected together, as leaving interstices between them ; but also from analogy with certain other glands, as the tonsils, &c., and particularly with the whole spleen and its tissue of vessels ; but principally from the cellular substance with which the gland is here surrounded. The truth is, that a cellular tissue of the finest kind surrounds not only the hepatic duct, and the fasciculus, but also each of the vessels separately. "The convex side of the capsule," says Winslow, "gives off a number of filaments, which form a cellular tissue that insinuates itself between the glandular granules" (n. 196). "Each glandular granule is invested by a particular expansion of the capsule of Glisson" (n. 194) ; and to this capsule a fine cellular tissue is constantly adherent. This glandular texture, and the lymphatic vessels which absorb such a quantity of fluid from this very tunic, seem to warrant the conclusion, either that the extremities of the cœliac artery throw out thither some serosity, or that the capsule transmits some exudation through its pores: no third supposition is possible. It is very probable that there is some transudation, both for the reasons alleged, and for this further reason, that a cellular down of the same kind as that mentioned above, universally covers even the very veins and the pori bilarii. But let us merely assume the point for the present ; in the sequel we shall find it supported by an abundance of experimental evidence.

(*t*) As there is an ingress of the blood of the vena portæ into the glands, so there is also an egress of the blood from them ; just as in all are accomplished by the alternate expansion and contraction of the sinuses, ducts and vessels, large and small (*u*).

209. These little vascular balls, or glandular sinuses, are the very principles of the liver and its functions (*y*). From these

the other glands, both conglomerate and conglobate. This may be demonstrated by the inflation of the biliary duct, which distends the gland, in the same manner as inflation of the branches of the vena portæ, or of the hepatic vein. See above (*p*). The feculent matter which is intruded by the branch of the vena portæ, and which has no opportunity of escaping through any lesser porta, into the pori bilarii, produces varices and tumors, such as have been observed by Ruysch and others in diseased livers.

(*u*) Nothing of work can be performed in any part, or in any compound, without an active and living force, or a motion of expansion and contraction. This is the vital principle of all operations. We conclude from the chain and tenor of causes, that while the vena portæ is disgorging its turbid stream, the hepatic artery is also disgorging its liquid, and that these circumstances produce expansion or diastole of the glands, during which nothing is expressed from them; but that when systole or contraction supervenes, the veins imbibe their portion, and the limpid residue is expressed through the pores of the little tunic, and the sediment that remains, into the porus bilarius: just after the same manner as in the heart itself, concerning which, see my *Economy of the Animal Kingdom*, treatise i. Respecting the times of the expansion and contraction of the glands, see below, n. 214 (*q*).

(*y*) The least glands of the body, wherever situated, are the principles of all operations. The little vessels and ducts of which the glands are the matrices or uteri, and which proceed from the glands, are the continuations of those principles, consequently, of the operations: so that these principles, like signs, lead consecutively to the things signified, or like efficient causes, to results and effects. Thus it is with the cortical glands of the cerebrum and cerebellum, which are the sources of the fibres; consequently the principles of their operations; and which, therefore, produce the living forces and actions in the body. But as the least glands are beginnings or principles, so also are they ends; consequently they resemble centres. As for instance the cortical glands, which are the ultimate termini of the blood-vessels; and of the sensations also, which latter rise along the same fibres to them, as at once their ends and their beginnings. Thus the relations of influx and efflux are concentrated therein. "It is probable," says Malpighi, speaking of the glands of the liver, "that all the last twigs of the vena principles spring the hepatic veins (*z*); also the roots of the biliary duct (*a*): and the capsule of Glisson here becomes a mere shadow, and almost vanishes away; but only to reappear, and return to the general capsule (*b*). In these principles also the defecation of the chyle and the lustration of the blood com-

portæ have glandular acini appended to them, and that these acini receive and filter the humor driven thither." (n. 197).

(*z*) The hepatic veins do not seem to be continuations of the branches of the vena portæ; for these branches alone are invested by the capsule of Glisson, or the production of the common membrane of the liver: but they appear to arise in the glands themselves, from the little threads or villi of which the pulpy substance of the gland was said to consist internally. Thus perhaps they spring, by a continuity the most subtile, from the extremities of the hepatic artery; in the same manner as the little veins in the body throughout. In these glands, then, vessels of a new kind arise, distinct from the branches of the vena portæ; from which, moreover, they recede. "Afterwards," says Winslow, "the first branches of these arteries, nerves and biliary ducts, leave the trunk of the great vein, and join in the same manner respectively, with the trunk of the small or hepatic vena portæ." (n. 196.) And Malpighi says, after Harvey, "This mass, by vessels which pass through the branch, receives and filters a certain humor, and by other different and new textures of vessels communicates it to the seed or embryo plant." (n. 197.)

(*a*) These roots seem to be continuations of the branches of the vena portæ, but by means of the glands, or the last and first termini. So that the branches of the vena portæ there cease, and are raised up anew in the form of biliary vessels; for these vessels come from the glands.

(*b*) The singular production or prolongation of the capsule cannot easily be comprehended by any merely general idea. It envelops the vena portæ, the hepatic artery, and the nerves, and in its progress to the glands, encloses them conjointly; and after its egress from the glands, it envelops the hepatic vein and pori bilarii, with which it is in a manner reflected towards the vena portæ, within the same common sheath. For the hepatic duct, composed of the pori bilarii, also proceeds to the concave region of the liver, beside the vena portæ and the gall-bladder, and thus as it were reascends. But the hepatic vein and the porus bilarius leave the branches of the vena portæ on the way, as we before pointed out (*z*); the former to go to the branches of the mence; and such as they are here, such do they continue through the whole passage; through vein, artery, biliary duct and capsule (*c*). In a word, the ducts dispense the bilific fluid (*d*); the hepatic arteries sprinkle into it the essentials of the blood (*e*); the hepatic veins imbibe from it the chyle (*f*); the capsule of

vena cava; the latter, to the hepatic duct. Wherefore the prolongations are said to return to the general capsule.

(*c*) See the particulars explained above, (*y*).

(*d*) The pori bilarii and the branches of the vena portæ, in their respective ascent from and descent to the glands, always hold a middle place between the arteries: which is in order that the porus bilarius may be enabled to attend fairly to the wants of all the surrounding parts. "The trunk of the vena portæ hepatica," says Winslow, "is in the middle of the bundle; the hepatic arteries lie on the right and left sides of this trunk." (n. 196.) See Glisson, Bidloo, and the Tabulæ of our authors.

(*e*) We may infer the manner in which the hepatic artery, through the whole of its course, sprinkles the fluid blood, the serum and the like, into the pori bilarii, from its perpetual ramifications over the capsule which invests the branches of the vena portæ and the pori bilarii: as Bidloo shewed by injection, and indicates in the following words in describing one of his Tabulæ: "The hepatic artery terminates on the common capsule of the hepatic vessels; also, on the coats of the pori bilarii, and on other parts, but principally on the coat of the vena portæ, and it is surrounded with a plexiform interlacement of nerves." (n. 199.) Shewing that the terminal discharges of this artery take place into the pori bilarii, as well as into the glands (see above), whose work the pori bilarii continue; and this, by the manifold forces, and the stimulating fibres with which it is so plentifully supplied.

(*f*) The perpetual absorption of the finer essences of the chyle by the hepatic vein, both throughout its course, and in the glands themselves, appears to be brought about by glandular means; for papillary and glandular forms abound in both the pori bilarii and hepatic duct, placed upon them, and growing to their parietes; just like those in the stomach and intestines; and in these glands the venous extremities must needs be inserted, and must take up the passing chyle, and transfer it into the hepatic vein, in the same way as in those organs. The nervous coat of the gall-bladder, according to Heister, "sometimes has minute glands upon it," and this "structure seems also to extend to the biliary ducts." (n. 192.) See also Winslow, n. 195. Ruysch represents the

Glisson transmits the serum, and caters for the lymphatics (*g*), in the same manner throughout as here in their very glands or

same in one of his figures. Authors have observed, moreover, that many glands are adherent to the hepatic duct, and that the structure of the pori bilarii is similar to that of the ductus communis itself. Reverhorst, in one of his Tabulæ, exhibits a large gland connected to the cystic duct. Add to this, that in the concave part of the liver of the ox, we generally discover a number of glands, as large as almonds, nuts, or small eggs. See Heister, *Comp. Anat., de Gland.*, n. 391. From the comparison instituted below (*h*) between the hepatic duct, and the stomach and intestines, it will be seen, that both have a similar *modus operandi*, and a similar method of elaborating and refining the chyle: so that the porus bilarius and hepatic duct continually separate and throw out the defecated and purified chyle, consequently, by means of similar venous extremities, into the hepatic vein.

(*g*) The permeability of the capsule was shewn above, n. 208 (*s*). It may be further proved by injecting the hepatic duct; for the very lymphatics are raised thereby, which would not be the case if there were no permeability. "The lymphatics," says Heister, "are shewn by the inflation of the artery, or of the hepatic duct." (n. 191.) The duct itself and the pori bilarii are surrounded by filamentary and cellular tissue in their whole extent; for the capsule ramifies in such a manner, that it is the common covering of all, and the particular covering of each. "Every branch of the vena portæ," says Winslow, "artery, nerve, and biliary duct, has its own capsule, and all the four have a common capsule, distinguished from the former by cellular septa." (n. 196.) That the lymphatics of the liver imbibe their fluid from these very sources, may be inferred by analogy from the lacteals of the intestines. For the chyle which is to be purified in the liver, is thrown out into both the veins and the lymphatics, just as in the intestines. Thus the lymph thrown out rises to the surface of the viscus, and there first is taken up by the lymphatics; according to our assertion in the chapter on the intestines, and in other parts of the work, namely, that the cellular tissue is the very emporium of the lymphatics; for none of their roots, so far as I am aware, have hitherto been found in the substance of the viscera. Winslow says, "We observe very distinctly numerous lymphatic vessels, on both the convex and concave sides of the liver: but it is more difficult to trace those which accompany the filamentary substance through that viscus." (n. 194.) Nevertheless, I cannot deny that the lymphatics of the liver may imbibe a part of their humor from the secretion of the arteries on the surface;

principles. Thus the liver is at work in every point, separating, discriminating, and purifying the chyle and the blood.

210. The pori bilarii in their whole extent, from their commencement to the hepatic duct, and the hepatic duct to the cystic, work, knead, grind, or treat the chyle entrusted to them, just as the stomach and intestines treat the food; for they rectify and purify the semi-digested alimentary juice or chyle (*h*). In like manner, they also correct, divide, and lay open to the innermost, the hard, heavy and resisting blood, which

as for instance, of the diaphragmatic artery: but since they have sometimes been observed by Nuck and others, to contain not a limpid, but a muddy and bilious fluid, hence I am persuaded that the greater part of their lymph is derived by transudation from the branches of the vena portæ, and from the pori bilarii and hepatic duct. What would be the use of this continuation of the cellular tissue to the very least parts, unless a secretion of bilious lymph were carried thither along the continuous cellular passages? "The substance of the liver," says Winslow, "appears to contain a membranous or filamentary tissue, which connects the ramifications and extremities of all the vessels to each other. This tissue seems to be a most multiple prolongation of the capsule, and of the external membrane of the liver." (n. 194.)

(*h*) This is principally inferred from the structure of the gall-bladder, which exactly resembles that of the hepatic duct, and this, that of the pori bilarii. For the hepatic duct has similar rugæ or valvulæ conniventes to those in the jejunum; also similar glands, &c.; so that this duct is constructed on the model of the intestines; and this, because it works, reduces, and digests the chyle itself, in the same manner as the intestines do, the food. Like them also, the hepatic duct sends out the properly chylified or purified essences, into the veins and lacteals; and excretes the fæces, which in this case are the biles, and which remain after the final correction. "In the cystic duct," says Heister, "I saw a number of transverse and oblique valves or membranes, which divided the duct as it were into cells: these, however, did not entirely close the duct in any part, but were disposed much like the valvulæ conniventes in the jejunum or colon." (n. 191.) The structure of the pori bilarii answers to that of the hepatic duct. "The internal surface of all the biliary ducts," says Winslow, "that is, of the ductus hepaticus, cysticus, and cholidochus, when examined in water, appears to be of nearly the same structure through their whole extent." (n. 195.)

the spleen, the pancreas, and the other viscera transmit to them (*i*); and they mingle it with, and introduce and marry it to, the chyle. The residue, which has not been absorbed by the veins and lymphatics, is the bile. Hence the liver is the laboratory for the defecation of the chyle, for the lustration and regeneration of the blood; and also for the preparation of the bile.

211. The hepatic vein pours this fresh chyle, married to the blood, and now become as it were new blood, into the open, eager and thirsty branches of the vena cava; not transporting it thither continuously, but passing it in obliquely and at right angles (*k*); for here the sphere of the liver ends, and that of

(*i*) That the hepatic artery, arising from the cœliac, brings with it the blood intended for further correction, particularly the bilious blood, will be shewn presently when we come to speak of the gall-bladder: also, that the other branches bring a similar, but purified blood, adapted for refining the chyle, from the spleen and the pancreas into the portal sinus, see the analyses of those viscera. For as the emulgent arteries constantly draw off the urinous and obsolete serosity towards the kidneys, so the cœliac artery draws off the unclean blood; and as the pori bilarii and hepatic duct wring, sunder, and reduce the chyle, so do they also the hard, concreted, and resisting blood-globules themselves: those which do not permit of being laid open elsewhere, are here submitted to as it were the last torture and turn of the screw, and absolutely forced to yield. This laying open of the blood in the pori bilarii and hepatic duct, has also the secondary use, of mingling its elements after it has been divided, with the fresh chyle, and so making them serve as kindly menstrua; just as we see in numberless mixtures effected by chemical means,—of oils, spirits, syrups, brine, which combine with homogeneous materials. This we before termed, the inauguration, impregnation, marriage, and copulation of the chyle.

(*k*) See the authors cited at the beginning of the chapter, in confirmation of the assertion, that the branches of the hepatic vein pass obliquely and crosswise into the branches put forth by the vena cava; but whether or not the two are continuous in their minute terminal offsets, has not yet, I think, been decided experimentally; although branches of the vena cava have been observed to penetrate to a considerable depth in the liver, and according to the opinion of some anatomists, all the way to the very glands. But the form of the large seems to warrant us in concluding to a similar form in the small. Bidloo has observed, that the branches of the vena portæ never pass the vena cava and of the right side of the heart, begins. By this provision, the vena cava and the right auricle are prevented from seizing the undigested chyle and impure blood, the gifts of the liver, before the proper time; and from taking any larger or other quantity, than what the liver, after having duly concluded its labors, consents to offer.

212. The lymphatics carry down the lymph which has been expressed through the pores of the capsule, and has worked its way to the common coat of the liver, and there been abundantly imbibed, to the receptaculum chyli. In this way the two venæ cavæ equally divide the chyliferous stream between them, and deliver it to the intermediate heart, precisely in such quantity and of such quality as the extremities of the body demand and require (*l*).

into those of the vena cava; evidently in order that none but purified and defecated blood may be supplied by the liver to the vena cava. The ductus or canalis venosus in the fœtus, into which the blood likewise enters obliquely, is a still plainer proof of the same thing. Its place is supplied in adults by a number of ramifications, which were they not inserted obliquely, the vena cava and the right auricle of the heart could not possibly help scraping up the unpurified chyle and blood at every draught, to the complete destruction of the kingdom, &c.

(*l*) The immense number of lymphatics which issue from the surface of the liver, particularly from its lower part, may be seen from the Tabulæ of Nuck, Reverhorst, Ruysch, Bidloo, Verheyen, and others. Respecting the equation of the fluids, see above, n. 203 (*z*). The mesentery and the liver divide the chyle between them; the cisterna mesenterii conveys it through the thoracic duct into the left subclavian vein, and from thence into the vena cava superior. But the liver pours its supply into the vena cava inferior. Both are nearly equidistant from the heart: both throw the chyle into the right auricle. Hence an EQUATION is requisite, in order that the chyle which is coming from both may be even with respect to quantity and quality, and that no conflicts or struggles may arise. For the spirit descending from the cerebrum to the end of the jugular vein, demands different quantities of fresh chyle at different times; sometimes more and sometimes less: so likewise the vena cava inferior. The equation appears to be brought about by the lymphatics, which go from the liver all the way to the receptaculum chyli; and this, by means of communicating fibres between the hepatic and mesenteric plexuses. "The trunk of the vena

213. The hepatic duct collects the refuse of the chyle, which now constitutes the hepatic bile (*m*), from the pori bilarii; and still vexes and as it were macerates it through a considerable length of passage; and at length expresses it into the cystic duct, and, with the aid of the gall-bladder, eructates it through the ductus cholidochus into the duodenum, close beneath the pylorus: not with a view to its being thrown away, but to its performing use, and serving as a menstruum, and returning again to the venæ portæ with fresh chyle, and thus even with interest; and like the rest of the humors, establishing a continual circuit, after the image of the circulation of the blood (*n*).

214. Powers and forces are almost redundantly luxuriant in the liver; hence the exceeding richness and variety of the efficient causes, which unanimously conspire, and are excited and invited, to one determinate course of action. *They conspire,* the whole with every part, the lobes with the glands, and the sinuses, veins, arteries, and pori bilarii, with both, from greatest to least, at every point (*o*). *They are excited,* each to its task, by alter-

portæ," says Winslow, "is in the middle; the hepatic arteries lie on the right and left sides of this trunk; the nerves surround it on all sides, and communicate with the superior mesenteric plexus" (n. 196).

(*m*) The quantity of this bile discharged into the duodenum, in a dog, was ascertained by Nuck and Reverhorst to be about six ounces in twenty-four hours: from which they conclude, that it is nine ounces in the human subject during the same time.

(*n*) That the cystic and hepatic bile, as well as the pancreatic juice, are menstrua, dissolving the particles of the food, and operating kindly upon all sorts of saline, sulphureous, oily and fatty substances, purifying and combining them, no one, so far as I am aware, denies. The fact is proved, indeed, by their ascertained chemical properties. Consequently they perform the same office as the salivæ; and like them, they always return with interest of fresh chyle. Respecting the circulation of the bile, see Nuck and others.

(*o*) This we before termed, the *unanimous conspiring of efficients* (n. 203), which exists, when all the causes flow determinately into their effects, in a constant order and series: also, when the multitude of conspiring causes is so immense, that the difference would be scarcely appreciable, if myriads were obstructed or collapsed, or proved to be lifeless and inoperative. Now in the liver there appears to be actually nations of expansion and contraction (*p*), which flow backwards and forwards from generals to individuals, and from outmost to inmost, synchronously with the animation of the lungs (*q*). And

a multiplicity of the kind, in the branches of the vena portæ, hepatic vein, and vena cava; likewise of the hepatic artery, biliary duct, glands, filamentary tissue, and cells. "Nowhere in the body," says Boerhaave, "do so many viscera, vessels, humors and causes concur, to produce any fluid, as in the liver, to produce the bile." (*Inst. Med.*, n. 350.)

(*p*) This we before termed, *incitation* (n. 203), which is the state wherein everything is excited to its labors and tasks, by a certain propulsive, active and living force, and thus continually incited.

(*q*) That the systole and diastole which go on in the liver, are always exactly synchronous with the momenta of the respiration of the lungs, is demonstrable by numberless circumstances. For the viscera surrounding and enclosing the liver, are actuated by these momenta, and no others; as shewn above in the course of our analyses. The stomach and intestines completely suspend the liver; so likewise do the peritonæum and the diaphragm, by means of their ligaments, particularly of the grand common or suspensory ligament, which so connects the general divisions of the liver with the individual ones, that there is nothing but stands related to it as a kind of centre of motion, centre of gravity, or balance: moreover the liver, by the same ligament, is connected to the xiphoid cartilage. The liver is also connected to the diaphragm still more closely by what is called the coronary ligament, which effects not so much connection as positive unition between the two; since the peritonæum is not there interposed between them. The liver also adheres sometimes to the false ribs, sometimes in part to the abdominal muscles on the right side. Consequently, if the peritonæum, the diaphragm, the sternum, the ribs, and the muscles of the abdomen, undergo alternations of expansion and contraction, synchronous with the respiratory motions of the lungs, then the liver, connected as it is to all these parts, must necessarily undergo the same alternations. And if the common coat of the liver, continued from the peritonæum and the diaphragm, following the direction of the ligaments, also encloses in particular the vena portæ and hepatic artery, and produces a capsule, which penetrates uninterruptedly to the minutest parts, and invests them all, one by one, then this respiration must necessarily penetrate to the minutest particulars of the parts, and excite them likewise. Especially since the very nerves of the hepatic plexus, sundered as they are excited, so *they are* also *invited* to the same alternations (*r*); for while the vena portæ and hepatic artery contract, and extrude the blood, the glands and the veins expand, and with open bosom imbibe, and so allure and invite it; and this alternately (*s*). No interruption mars the correspondence of these active forces. All things, without and within, in both the convex and concave regions of the liver, proceed tranquilly, mildly, naturally, and spontaneously; so that scarcely any motion at all is perceptible. This is brought about by various provisions,—by the ligaments, fissures, and lobulation which prevail

into their ultimate fibrillæ, keep up the same thing; and at the same instant that the lungs call forth and propagate this motion from outermost to innermost, the fibres of this nerve do the like from innermost to outermost. Of course, then, the mass made up of these parts, cannot possibly fail to move, at the nod, and to the impetus, of such numerous and commanding forces.

(*r*) This we before termed, *invitation* (n. 203): it is produced by a kind of attractive physical force, corresponding to the force of Incitation, as follows;—when the vena portæ throws in the blood through its branches, then the glands which receive the blood, expand, and thus invite it, driven forward as it is, into their chambers. Hence there is a corespondence between the outermost and the innermost, or between generals and particulars, in this respect,—that the excitation of the former, and the incitation of the latter, are coincident in their momenta.

(*s*) The expansion and contraction of the parts or glands of the liver seems to be almost of the same nature as the diastole and systole of the heart; namely, that when the blood of the portal branch acts, the nervous fibrillæ, and the arterial and venous extremities, composing the glands, yield; but when the portal branch ceases to act, the nervous and vascular fibre instantly begins, and *vice versa*. But during the contractions of the vena portæ and its branches, the nervous fibrillæ also, and thereby the capillary extremities of the artery, and thereby again the capillary extremities of the hepatic vein, seem to be contracted; and thus to cause a dilatation of the cavity of the gland. This is certainly deducible from the argument of continuity, and from the chain of causes. But lest our reasonings should become obscured by over-subtlety in these most subtle matters, (in the same manner as the parts themselves,) we therefore prefer dwelling on such particulars as are general and visible to the sight.

universally throughout the organ (*t*). Such balance of parts and motions is necessary, that all the offices may have their turns, be fitly distributed, and regularly discriminated, and put on a level; that those things which are disjoined may be associated, and the spirit, the chyle, and the blood work together in mystic union. Thus it is that chylification, commenced in the stomach, is continued in the intestines; and that the lustration of the blood, commenced in the spleen, and continued in the pancreas, is at last completed in the liver.

215. The blood which is impure, spurious, resisting, hard, dirty, dark, disagreeable, or melancholy; or covered with saline spicula, or grumous and variously clotted and coagulated; or

(*t*) Respecting the ligaments and fissures, see Heister, n. 191; Winslow, n. 194. There are three ligaments; four fissures; and also certain eminences or portæ that produce a division of the liver into lobes: which lobes are both more numerous and more distinct in dogs and some other animals, than in man. It would, however, be very prolix, not to say superfluous, to give the peculiar uses of all these parts; inasmuch as it would not tend directly to the end of our analyses. We shall be content, therefore, with this general statement, and with the conclusion,—that all viscera designed for the separation of concreted particles, require to be divided and separated into lobes, united by ligaments, and parted by fissures, of the kind observed in the liver; in order that nothing may be done violently, but all with the utmost peace, concord and unanimity. The extracts of the food, commixed and infarcted with impure blood in the vena portæ, cannot be distinctly separated into the purest native chyle, all the way to their very first elements, or into lymph, or bile; nor can the purer essences of the arterial blood be as it were amalgamated with the virgin chyle; nor can all, thus properly filtered, be intimately united and married, without the favoring presence of a vernal temperature, and of a halcyon and tranquil state. The least violence would confuse the portions already distinct, and break the compact. This is the reason of the frequent discrimination of the surface into lobes, and of the fissures between the lobes; there being similar but still more perfect divisions in the inmost recesses of the liver, where there is nothing but is surrounded by cells and cavities. Similar appearances occur in the cerebrum, in both the lungs, in the kidneys, in the conglomerate glands, and in all other members that are intended for dividing and uniting particular elements.

that is of no further use in the circle of life; or that cannot be laid open, digested, corrected or purified, either by the pori bilarii or the ducts, the liver dismisses and rejects into a cyst underneath it, called the GALL-BLADDER—there to be more forcibly digested, broken up, vexed and expurgated (*u*); and in

(*u*) We have experimental evidence to prove, that the blood always differs in different subjects, and runs through infinite changes of state. The eye alone is sufficient to inform us when the blood is healthy, quick, and full of life; for in this case it is purple and florid, and has a virgin, native, and truly living glow. But when it is moribund, it then straightway appears lifeless, lurid, dirty-yellow, green, dusky, or even pitchy; and the serum in which it floats corresponds to it, being sluggish, viscid, and ugly. Not seldom also we find the blood grumous, uneven, full of bile, melancholic and heavy. Thus too we judge of the health by the blood abstracted in venesection. Nevertheless, when the crisis of the disease has passed, and the state of the body or of the animus changes, and when health returns, the blood immediately regains its former appearances of vitality, and puts on its pristine glow. Does not this prove most plainly, that the blood assumes the same state as the life itself, and undergoes perpetual purification and regeneration? Such being the case, the question arises—where its regeneratory and excretory organs are situated? In order to discover them, we must follow the thread of anatomy, and, in short, pursue the aorta itself, through its trunks, branches and twigs, into the very substance of the viscera; and, by diligent investigation, bring to light the offices of each viscus; not alone of any one viscus individually, but also of all at once, conjointly and in society: in this way, it will at length become apparent, that the gall-bladder is the ultimate asylum of the unclean and obsolete blood. The fact that the gall, that is, the cystic bile, derives its origin from the blood thus rejected, is not in the slightest degree ascertainable from the chemical examination of the gall, nor from its smell or color, nor, indeed, from any sensible indications whatever. For the blood produces infinite species of humors, whereof, generally speaking, no two agree in even their common properties; all likeness being lost. Who could say of the phlegm, oil and spirit, extracted or distilled from the blood by chemical means, that any part of them was once blood? Who could divine that saliva, milk, fat, semen, lymph, and similar menstrua, originated from the blood? Who could divine this much of the pancreatic juice, or of the cystic bile? Still less is it possible to arrive at the truth, from color, or by the sense of some cases it also sprinkles and spews in a menstruum, namely, the hepatic bile, through the hepatico-cystic ducts (x). This

sight; inasmuch as color itself is only a modification of rays, according to the disposition of the subtler parts of substances. Nor from taste,—bitterness for example; since taste results from the figured composition of the grosser parts on the very surface, and which either titillate or twitch the papillæ of the tongue. All these things are susceptible of so many varieties, that any material may easily acquire any color or any taste, just as it may acquire any form. I am induced and persuaded by these considerations, and to prevent the fallacies of the senses from producing error and chaos in my analyses, to follow anatomy closely, and with respect to the changes of state, to follow anatomico-medical observations. But in the Chapter on the Spleen, it will be my business to shew, that the blood conveyed to the liver is not all impure, but that some of it has been corrected by the spleen and pancreas, to enable it to serve as a menstruum for refining the chyle.

(x) Respecting the arteries which run to, supply, and penetrate the gall-bladder, see our authors. They are, 1. The proper cystic arteries, the gemellæ cysticæ, which arise from the right branch of the cœliac; and, 2. Several twigs and little trunks which come off from the hepatic artery, and descend to the gall-bladder, furnishing it with many branches, and anastomosing with its proper vessels. These latter are distinctly shewn by Ruysch in one of his figures, inserted in Mangetus, *Theatr. Anat.*, tab. lxiii., fig. 8; in which GGG "represent the twigs of the hepatic arteries arising from the substance of the liver, and running to the gall-bladder;" E, "the last offsets of the cystic arteries passing through the membrane of the liver:" thus shewing that the arteries pass backwards and forwards between the liver and the gall-bladder. Now as this artery (according to our previous assertion respecting the liver) conveys the impure blood to be purified and resolved in the liver, therefore it follows (from the same assertion), that whatever impurities cannot be defecated in the liver, are sent away into the gall-bladder. That the liver also transmits a small quantity of bile through the hepatico-cystic ducts to the gall-bladder, is almost a general opinion, although still questioned by some. See Heister, n. 191; also the sentiments of Bohn, Cheselden, Bianchi, and Morgagni. But others have proved the fact by numberless experiments. See Verheyen, *Corp. Hum. Anat.*, tract. ii., cap. xvii.; Bidloo; and many other authors. "There are some ducts," says Winslow, "which form a communication between the pori bilarii and the gall-bladder. For a long time these ducts crude and undigested blood, the gall-bladder treats, or reduces and wrings, in the same way as the intestines treat the feculent remains of the food, which they consign to the colon and rectum, to receive the last torture and the final expurgation. So

could be found in brutes only, but lately they have been seen in the human body" (n. 195). It is very certain that numerous filaments pass between the body of the liver and the gall-bladder; and still more, between the hepatic and the cystic ducts, which are connected together by many little cords; thus there is no lack of communication. But these filaments, like the hepatic duct itself, are so twisted, and so constricted by rugæ and valvulæ, that they can scarcely admit colored injections. However, be this as it may, it is of no consequence whether there are any intercommunicating ducts of the kind, or not. For the gall-bladder derives its bile, not from the bile of the liver, (which latter bile is different from the former in nature and origin,) but from the recrementitious and excrementitious blood of its own arteries, particularly of the gemellæ cysticæ; which, (as every ramification of the cœliac artery through the spleen, pancreas and liver, and as the functions of these viscera clearly shew,) are the ultimate secretories and excretories of such blood. To say nothing of the properties of this blood, as ascertained by the removal of the gall-bladder; and by the use of the latter, and the tinge on the colon where that intestine is connected to it: and of injections, which are arrested by the minute vascular brushes, blocked up as they are with grumous blood, and like the visible ducts themselves—the hepatic, the cystic, and the ductus cholidochus—tortuous and narrowed to the last degree, in order that they may correct the blood to the utmost. Since the gall-bladder has such numerous arteries, pray where are the corresponding veins? Their place, it may be said, is supplied by lymphatics. Whither then does all the blood go? Anywhere but to the gall-bladder, or the liver? Since then the office of the gall-bladder consists in the ultimate defecation, grinding, and disruption of this blood, it must necessarily summon the hepatic bile to assist it; not indeed either to originate or increase the gall, but to serve as a kind of menstruum, as it also serves as a menstruum for digesting the food in the intestines, and as a vehicle for conveying parts in a state of resolution into the lacteals. From these considerations it follows, that the hepatico-cystic ducts may be either present or absent; that is to say, according to the requirement, quantity, and quality of the cystic bile. That the hepatic bile dilutes and softens the cystic bile, and tempers its acrimony, is a well-known fact.

the liver throws this blood into the gall-bladder, which in construction, membranes, rugæ, glands and lymphatics, consequently, in mode and violence of action, closely resembles the intestines; but with a fitting difference and relation to the quality and nature of the object to be accomplished (*y*). Thus the gall-bladder is the ultimate press and expurgatory of the spurious and obsolete blood.

(*y*) A comparison may not improperly be instituted between the intestines, (particularly the large intestines, or the colon,) and the gall-bladder. For the intestines and the gall-bladder have similar membranes: they both have a common membrane, continuous with the peritonæum; a similar cellular membrane; a similar muscular membrane, in which the direction of the motive fibres is the same in both cases; also a similar nervous membrane, with similar rugæ or valvulæ conniventes. "The nervous coat," says Heister, "has rugæ or reticulations on its inner surface, is covered with an unctuous moisture, and sometimes has minute glands upon it." (n. 192.) "In one subject, I saw a number of transverse and oblique valves or membranes, which divided the duct as it were into cells: these were disposed much like the valvulæ conniventes in the jejunum or colon." (n. 191.) These valves have also been observed by Bauhin, Bidloo, Vestus and Plancus, exactly resembling a screw. *(ibid.)* "The internal coat," says Winslow, "has on the inside a great number of reticular folds, covered with small lacunæ, like perforated papillæ. These lacunæ are regarded as glands." (n. 195.) Ruysch likewise in one of his Tabulæ, represents the inner surface of the human gall-bladder, furnished with glands, which, he says, are connected not merely to the lower part adjoining the liver, but also to the opposite side, and to the neck of the gall-bladder, the ductus cysticus, and the ductus hepaticus. He also exhibits the valvular and cellular surface of the cystic duct itself, which he compares to that of a reticulated melon (n. 199). Now since the gall-bladder is abundantly supplied with muscles, it therefore follows, that it expands and contracts: and since it has valvulæ conniventes, like the intestines, it therefore follows, that it rolls, works and reduces the bile: since it has glands similar to those of the intestines, that it imbibes this blood after the working, subduing and secreting is finished: and in the same manner as the intestines transmits it as it were into its lacteals or lymphatics: of which the immense number has been shewn by Ruysch, Nuck, Bidloo and others, in their plates. Wherefore, what we said above respecting the small intestines, may be said

216. This cyst, however, not only is a kind of expurgatory blood-intestine, and as it were, gall-colon; but it also resembles the urinary bladder, which carries off and excretes the impure serum of the blood, by the mediation of the ureters and kidneys, and by the emulgent arteries, from the trunk of the aorta. The gall-bladder similarly derives and invites the impure blood, by the mediation of the liver, pancreas and spleen, and by the cœliac artery, and particularly by its branches, the gemellæ cysticæ, from the trunk of the aorta (*z*); and this, as in the kidneys, by virtue of its construction, situation and excitation, and of its operation resulting from these, as causes: for whatever the principles, intermediates and extremes of the body demand, the universal mass of the blood is bound to supply (*a*). Thus the gall-bladder is the excretory vessel of the impure blood, as the urinary bladder is the excretory vessel of the impure serum.

217. The quantity of the cystic bile is affected by numberless causes; speaking generally, by external, intermediate, and internal causes. For the most part it increases just in proportion

analogically of the hepatic duct; and what we said of the large intestines, and of the colon particularly, may be said of the gall-bladder and the cystic duct. And indeed similitude of functions is the reason why the gall-bladder applies and connects itself to the colon.

(*z*) These points will be illustrated in our subsequent Analyses, of the Spleen, the Pancreas and the Kidneys; where we shall also shew the kind of invitation, whereby the impure blood is drawn off into the cœliac artery, and the urinous serum, into the kidneys.

(*a*) See n. 203 (*z*). The *primaries* of the body are the cortical substances of the cerebrum, which invite the purest blood and extract its spirit. The *mediates* are the numerous viscera of the body, as the heart, the lungs, the liver, the pancreas, the spleen, the testicles, the epididymes, &c. The *ultimates* are the kidneys, the ureters, the urinary bladder, and the gall-bladder; also the mammæ, the skin, &c. All this invitation is brought about by means of a correspondent excitation or propulsion, by which the organs themselves are adapted from the very beginning, and afterwards by education and habit: consequently they are formed organically for use and by use. Thus the use explains the structure, and the structure the use. See n. 32.

as the hepatic bile decreases (*b*). The *external causes* which affect it, or the causes in the body, are disease and wasting of the viscera, particularly of those related by position, office and consanguinity, to the gall-bladder—as the liver, the pancreas, the spleen, the mesentery, the omentum and the conglobate glands (*c*): also disease of the stomach, the intestines and the salivary glands, producing indigestion, capricious appetite and

(*b*) From a chemico-rational and an anatomico-rational investigation of the origin and nature of the two kinds of bile, carried on by means of observations, it is evident that the hepatic bile is principally composed of the outcast essences of the more crude and barren chyle, yet with an admixture of particles of genuine blood in a state of resolution or division; and that the cystic bile consists altogether of the obsolete and outcast blood. Now if there be such a communion between the liver and the gall-bladder, that the liver transmits to the gall-bladder the blood which it does not itself digest, resolve, and feed with chyle; and if the gall-bladder transfers to the liver as much as the liver wants, when it cannot obtain a supply from other sources, then it follows, that the quantity of the one increases, as that of the other diminishes: unless, perhaps, as appears to be the case, the quantity of the more crude chyle, consequently also of the hepatic bile, increases, in proportion as the blood of the hepatic artery is not expended: for, as we said above, this blood performs the office of a medium or bridesmaid. Wherefore it does not seem to be an invariable law, that one species of bile increases in precisely the same ratio as the other decreases.

(*c*) The viscera of the abdomen, like their operations, ought to be divided into distinct series. The fauces, the œsophagus, the stomach, the intestines, and the salivary glands, constitute one, and indeed the first series; because they all extract from the liquids and solids taken by the mouth, the nutrient or sanguigenous chyle. The mesentery, the liver, and the conglobate glands, constitute the second series, and not only convey but also refine this chyle. The spleen, the pancreas, and with the latter the omentum, also the liver, constitute the third series, and purify the blood already formed. Thus the liver is the ultimate and general member of both the second and third series, and as it were the conclusion of both. The kidneys, the ureters, and the bladder, constitute the last series, and draw off the useless serum of the blood. Consequently, the viscera of the abdomen must be considered as four series, with distinct offices, but which nevertheless are related to each other mutually, and conspire unanimously to a certain end and use, the nausea: the use of unnatural food; the rawness, poorness, excess or defect of natural food: the blocking up of excreting orifices, and the stoppage of perspiration: the burning of fever, and the heats, chills and sweats of any disease whatever: in a word, all the causes which thus immediately affect the blood (*d*). The *intermediate causes* which affect it, or the causes in the disposition, are irritability, anger, envy, sorrow; also anxiety, trouble, and immoderate grief of all kinds, which boil within, and devour themselves, and infect the blood, and thereby feed themselves: in a word, all the causes which thus immediately affect the spirit of the blood (*e*). The *internal causes* which affect it, or the causes in the mind, are too intense application, indolence, frenzy, anxiety, particularly with respect to the future; the sudden and dreaded extinction of high or ambitious hopes; also various misfortunes, hatred, the constant regard of evil without intermixture of good, and other similar things, which immediately affect the interior nature of the spirits (*f*). Affected by these causes, the spirit, and consequently the blood, becomes hot, cold, hard, soft, obscure or coagulated, in a word, adulterated and bilious.

218. And the treatment of the cystic as of the hepatic bile, never ceases; but continues without intermission, not only in

most general or the unity of all, which is, the perpetual existence, that is to say, the subsistence of the blood.

(*d*) These being ultimate conclusions from certain analyses, and respecting the corporeal predicates of the body, and being, moreover, of a pathological character, therefore it will not be proper to enlarge upon them in this place; although each particular thereof is capable of illustration in many ways. On this account I pass over much that is connected with them, without further comment. Generally speaking, these particulars are conclusions and consequences from the analysis of the blood; which, indeed, I have already given in my *Economy of the Animal Kingdom;* but with the intention of inserting it in this work in the proper place.

(*e*) The reader will find these points illustrated in the Part on the Affections of the Animus, and their fluxion into the body.

(*f*) And these, in the Part on the Affections of the Rational Mind, and the desires of ends, that are determined into act by means of the will.

the fundus of the gall-bladder, but also in the neck, in the cystic duct, and in the ductus cholidochus; and afterwards throughout the intestines, from the duodenum to the rectum (*g*). Consequently, it is not thrown out as excrementitious, but to perform the greatest uses in its passage. The hepatic bile and the pancreatic juice temper it to suit every want and requirement of the digesting organs and their contents, in fact, with infinite variety: they sharpen, dilute, inspissate and sheathe it, and nicely adapt it to objects in quality and quantity, whether such objects require sharp or gentle stimulation, cleansing, smoothing, anointing, impregnating, corroding, or conjoining. At last, its purer parts being disengaged, in the same way as those of the gastric saliva itself, return, either to the receptaculum chyli, or to the vena portæ, with a booty of fresh chyle, and so with interest (*h*). The incorrigible, vapid, dead and filthy portion,

(*g*) The treatment of the cystic bile, like that of the chyle in the pori bilarii and hepatic duct, or in the stomach and intestines, consists in its perpetual rubbing, resolution, expression, or reduction to its primitive component essences; in order that it, like all the rest of the blood, may be sent back again into circulation: wherefore, the organs of this series, with respect to power of action, constantly increase all the way to the gall-bladder, where they are the most violent of all. That these outcasts of the blood are continually wrung and, as it were, thrashed in this manner, is evident not only from the gall-bladder itself, but also from the cystic duct, in which the valvulæ conniventes or rugæ are more numerous, and even the channels become finer and narrower, as well as tortuous, and at last sometimes split into branches between the membranes of the duodenum, and wreathe through various windings towards that intestine; which are clear proofs that the treatment of the biles increases. In some animals, as horses, pigeons, &c., which are thought to have no bile because they have no gall-bladder, the place of these ducts is supplied by a number of little ducts or receptacles, (see Wepfer, *Cicut. Aquat. Hist.*, p. 176) [4to. Basil., 1679.]: perhaps in the same way as in a certain human subject, mentioned in the *Hist. de l'Acad. Roy. des Sciences*, an. 1701, p. 55, [Paris, 1719.] But the torments of the bile do not yet cease: they continue to accompany it even in the intestines, and indeed all the way to the colon and rectum, in which there are similar but more savage instruments of torture.

(*h*) So far as I am aware, all our authors are agreed in the use they assign to both the cystic and hepatic bile. It will be sufficient to cite green and foul, tinges the alvine fæces, and is discharged with them. The blood, thus purified by the liver and gall-bladder, remains clean, fresh, lively, joyous, busy, purple and florid, burnished with living splendor; and unweariedly pursues its circle of life, ever born again like the phœnix, in newness and perennial infancy.

their words. "The use of the bile," says Heister, "is, to attenuate the chyle; to mix oleaginous with aqueous parts; to stimulate the intestines; and in part to alter the acid of the chyle" (n. 192). "The cystic bile," says Boerhaave, "resists acescent matters; and by its admixture with other things, gives them a similar power of resistance: it is saponaceous and detergent; it makes oils miscible with water; it dissolves and thins resins, gums, and viscid substances, and renders them uniform if it be rubbed with them: it is neither alkaline nor acid; but consists principally of oily, saline, and spirituous parts, diluted with water. It is incombustible, except when dried previously: the sharpest of the circulating fluids; the most susceptible of putrefaction; readily penetrating in all directions by transudation. The use of it, when mixed with the chyle and fæces, is, to attenuate, resolve and cleanse; to stimulate the muscular fibres; to mix all sorts of different things; to blunt saline particles; to divide coagula; to shorten the way for the chyle; to sharpen the appetite; to act as a ferment; to assimilate crude to concocted materials. The bile of the gall-bladder performs these offices more powerfully than that of the liver." (*Inst. Med.*, n. 99.) By distillation over a gentle heat, the bile yields a liquid almost like water in appearance, whitish, of an oily taste and a rank smell. When the residue is treated with water, and redistilled, it yields a lixivial salt, black and impure, and a whitish caput mortuum. From which facts, although only in the most obscure manner, we may conclude, that the bile is indurated and compacted blood, hardened by the presence of alkaline and urinous particles, yet being of the family of the blood, fitted for all the above-mentioned offices. But what light do we obtain from chemical analysis, without we know previously the nature of blood, oil, spirit, alkaline and lixivial salt, (of all of which there are indefinite genera, species and individual varieties;) also the cause of color, of smell, and of taste. Without these be thoroughly explored, we pursue shadows, and disseminate inanities, almost as intelligible to the stupid as to the wisest.

CHAPTER X.

THE PANCREAS.

219. HEISTER. "The pancreas is a large, flat gland, for the most part of a fleshy color, situated behind the stomach, and reaching from the duodenum transversely towards the spleen. It is connected with the duodenum, with the mesentery, splenic vessels, and spleen. In man it is a single organ, but in the dog and cat it is divided, as it were, into two parts. It is eight or nine inches long; about two finger-breadths, or two and a half, broad; about one finger-breadth thick; and about three ounces in weight. In man, it has something of the shape of a dog's tongue; it is broadest near the duodenum, and gets gradually narrower towards the spleen. It is surrounded with a membrane which is continuous with the peritonæum. Its substance is glandular; formed by a conglomeration of many lesser parts. Its arteries arise from the cœliac and splenic arteries, and its veins from the splenic vein: its nerves are from the par vagum and intercostal nerve. Whether it has lymphatics or not, is uncertain.

"The pancreas has an excretory duct, composed of a number of lesser ducts. This duct was first discovered in a turkey, by Maurice Hoffmann, at Padua, in 1641; and afterwards it was found by Wirsungus, a Bavarian, in the human body; as T. Bartholin, who was present, informs us. It is commonly single in the human body; sometimes, however, there are two ducts; this is always the case in the goose, the duck, the African fowl, and the pheasant; and there are three pancreatic ducts in the common fowl, the pigeon, the eagle, &c. It is situated in the middle of the pancreas, where it resembles an empty vein, of about the calibre of a thin straw. It terminates in the duodenum by an oblique aperture, four or five finger-breadths below the pylorus, usually at the same orifice with the ductus cholidochus; but sometimes it has a double aperture. In many animals it is inserted into the duodenum by a particular orifice, at a considerable distance below the cholidochus. The use of the pancreas is, to secrete a peculiar fluid, called pancreatic juice; which is of a salivary nature, and serves to attenuate the chyle." *(Comp. Anat.,* n. 216.)

220. WINSLOW. "The pancreas is a long, flat gland, of the conglomerate kind, situated under the stomach, between the liver and the spleen. Its figure resembles that of a dog's tongue. It is divided into two sides, a superior and an inferior; into two edges, an anterior and a posterior; and into two extremities, one large, which resembles the base of the tongue; the other small and rounded, like the apex of the same organ. The pancreas is situated transversely under the stomach, and enclosed in the duplicature of the posterior portion of the mesocolon. The large extremity is connected to the concavity of the first curvature of the duodenum; so that a great part of that intestine lies between the pancreas and the dorsal vertebræ. The small extremity is fixed to the omentum, near the spleen. The pancreas is composed of a great number of soft glandular molecules, so combined as to present the appearance exteriorly of one mass, the surface of which is rendered slightly uneven by numerous small convexities, more or less flattened. When these molecules are separated a little from each other, we find along the middle of the breadth of the pancreas a particular duct, in which several smaller ducts terminate laterally, almost as the branches of a tree terminate in the stem. This canal, named ductus pancreaticus, or ductus Wirsungi, is very thin, white, and almost transparent, and the extremity of the trunk opens commonly into the extremity of the ductus cholidochus. From thence it diminishes gradually, and terminates in a point, next the spleen. The small lateral branches are likewise pretty large near the trunk, and small towards the edges of the pancreas. In some subjects there are two pancreatic ducts, one lying above the other. The duct does not always pass longitudinally, but sometimes takes a serpentine course, winding from side to side, always, however, in the same plane. It is nearer the lower than the upper side of the pancreas. It pierces the coats of the duodenum, and opens into the ductus cholidochus, commonly a little above the prominent point of the orifice of that canal; and sometimes it opens immediately into the duodenum.

221. "Several years ago I observed in the human subject, that where the great extremity of the pancreas is connected to the curvature of the duodenum, it sends down an appendage, which adheres very closely to the following portion of the intestine; and upon a careful examination, I found a particular pancreatic duct, ramified like the

large one, which ran toward and intersected this great duct, into the extremity of which it opened, after having perforated the duodenum. This portion I call the small pancreas: it sometimes opens separately into the duodenum, in which we likewise occasionally see several small holes round the cholidochus, which answer to the pancreas.

222. "The arteries of the pancreas arise from the pyloric, duodenal, and chiefly from the splenic artery, which adheres very closely to the whole lower side of the pancreas, near the posterior margin. In its passage, it gives off many branches, termed pancreatic arteries, which run out more or less transversely from each side. The pancreas also receives some small ramifications from the great gastric and superior mesenteric arteries. The pancreatic veins are branches of the splenic vein, which is one of the principal branches of the great or ventral vena portæ. The splenic vein, like the artery, runs along the lower side of the pancreas, near the edge, in a shallow depression in the substance of the viscus. These veins answer to the arteries of the same name; and there are likewise other small veins corresponding to the small arteries, and which are productions of the great mesenteric vein, &c. The pancreas receives nerves partly from the hepatic, partly from the splenic, and partly from the superior mesenteric plexus: and it likewise receives a nerve, the chorda transversalis, from a flat ganglion lying between the two semilunar ganglia, &c. The pancreatic duct is not only double in some subjects, but the little collateral branches have communications in form of islands in several places in the body of the viscus." *(Exp. Anat., Tr. du Bas-Vent.*, n. 319—328.)

223. Boerhaave. "Under the back, the right side, and the bottom of the stomach, chiefly in the posterior layer of the omentum, close to the duodenum, is a large, pendulous, conglomerate gland, called pancreas; which, by virtue of its glandular structure, made up of innumerable arteries, secretes a humor from the cœliac arteries, and discharges it all into one common duct which terminates in the duodenum. This humor is almost insipid, slightly saltish, limpid, abundant, secreted continually, forced out by the motion, pressure, warmth, and close vicinity of the heart, and especially by the stomach when full and digesting. It is neither acid nor alkaline, but very much resembles saliva, both with respect to origin, to the vessels which secrete it, and to its properties. When mixed with the bile in the living subject, and digested with it in the same gut, it produces no sensible fermentation, but either mingles uniformly, or else passes on alone into the empty intestines; so that when blended with the chyle, fæces, bile, or mucus, its use must be to attenuate and dilute; to mingle all; to render the chyle capable of mixture with the blood; to fit it for passing through

the lacteals; to sheathe acrid particles; to abate the viscidity and bitterness of the bile, to alter its color, to mix it thoroughly with the chyle; to serve as a menstruum and vehicle; and so to alter the peculiar taste, smell, and other qualities of the food, that it may assume a kind of uniformity; and to repeat all these processes with the utmost rapidity." (*Inst. Med.*, n. 100, 101.)

224. See Maur. Hoffmann, *Diss.* to Horne's *Microcosmus*, and his *Idea Mach. Human.* Wharton, *Adenog.* Graaf, *De Succo Pancreat.* Vesalius, *De Human. Corp. Fabric.*, lib. v., fig. iv. Ruysch, *Thes. Anat.* iv., n. 94, not. 1, 2. Verheyen, *Corp. Hum. Anat.*, tract. ii., cap. xv. Mangetus, *Theatr. Anat.*, lib. ii., pars. i., cap. vii.; where he details experiments, in which dogs, from which the pancreas had been excised, lived in health for several months. Morgagni, *Advers. Anat.* iii., *Anim.* 19, where he relates that in the pancreas of the hedgehog he found bile, or at least a greenish yellow fluid. See also the plates of authors. Eustachius, *Tabul. Anat.*, tab. ii., fig. 4. Graaf, *De Succo Pancreat.*, tab. i.; where he exhibits the pancreas with its common and lateral ducts, the conjunction of the new vessel with the biliary duct, and some of the viscera connected to the pancreas. Bidloo, *Anat. Human. Corp.*, tab. xxxvi.., fig. 2; the pancreatic duct and its ramifications, with a portion of the duodenum, exhibiting an orifice with annular fibres, which form something like a sphincter. Verheyen, *Corp. Hum. Anat.*, tab. xi., fig. 1.

225. In order to know the office of the pancreas, it is absolutely necessary to know the offices of all the viscera among which it lies concealed, and to which it is connected; thus it must be regarded in society and series with many other organs, and not with one, but with both our eyes (*a*). If the pancreas be looked at singly, apart from the series of coördination and subordination of the other viscera, it appears to be only a large gland, a remarkable specimen of the conglomerate kind, which prepares a juice for tempering the bile, for anointing and cleansing the intestines, for digesting the food, and for diluting the chyle; but what further services it performs, for the blood, and for its fellow-members, remains in profound concealment. The following then is a summary of the offices of the viscera (*b*).

(*a*) See the Chapter on the Liver, n. 200.

(*b*) Men commonly judge of the use by the visible effect, and even think it right so to judge: for at what certain conclusion, say they, can the understanding arrive, without its masters and messengers, the senses? I know and believe, because I see. But this reasoning proceeds from our judging of a viscus by itself alone, and sometimes, indeed, of many by one, not of one by many: whereby we obtain no comprehensive view, but only avoid labor and fatigue; and in the end generate uncertainty and fallacies, which are manifested more and more, the more particulars we regard. Ultimate effects are all that we perceive by sense, for the senses—of sight, hearing, taste and smell—are in the ultimates of the world, and indicate nothing beyond the forms of the last compounds of things. Whence these last compounds obtain their form, essence and nature, the senses know not, but refer the question to the mind. Thus the senses shew only that the pancreas pours forth a certain juice, of a particular color, smell and taste; and that the

226. The offices of all the abdominal viscera are, preparing the chyle: purifying and refining it: introducing it into the blood: purifying the serum and the blood itself: secreting the better parts of both: and circulating them: excreting the worthless parts: and throwing them out. *Preparing the chyle,* is the office of the stomach and intestines: *purifying it,* is the office of the liver, the mesenteric and the conglobate glands: *introducing it into the blood,* is the office of the thoracic duct, the veins of the liver, and certain lymphatics (*c*): *purifying the blood,* is the office of the spleen, and in the fœtus, of the succenturiate kidneys: *secreting the better parts of the blood and serum,* is the office of the pancreas, the omentum, and lastly of the liver (*d*): *circulating them* is the office of the pancreatic and hepatic ducts,

pancreas pours this juice into the duodenum. But if we rest here, we shall never discover what lies in this juice; what properties it possesses; where it comes from, or what purpose it is intended to serve. For whatever the juice has in it, is derived from causes prior to the effect, or from its more universal offices; and if these be unknown, then reason comprehends nothing more than sensation. Therefore the causes, that is, the efficients, or higher effects, must be explored. (Every cause is an efficient relatively to the things below it, and an effect relatively to those above it.) These causes, as I before said, cannot possibly be discovered, without taking into account the office of those which act in the same society, and which constitute the more universal series of causes or effects. For the more universal the cause, the more members concur to it. Thus by exploring several, and embracing them together in the mind, we discover what any one member of the series contributes to the common object, and what the visible product, as in the present case, the pancreatic juice, contains internally.

(*c*) Certain of the lymphatics do not go to the thoracic duct, but pour the lymph of their particular viscus, gland or muscle, immediately into some adjoining vein.

(*d*) To secrete the better parts is really the same thing as to separate them from the worthless parts, which latter are sent away towards the kidneys: and, indeed, certain of the grosser parts of the blood and serum, are secreted by the pancreas and liver, and sent away by their ducts; but inasmuch as these are circulated, and again purified, I rather prefer to call them secretions than excretions. Besides, it is also the office of the spleen and pancreas, to prepare a universal men-

and of the intestines: *excreting the worthless parts of the serum,* is the office of the kidneys and the ureters: *throwing them out,* is the office of the bladder and the urethra: *excreting the worthless parts of the blood,* is the office of the gall-bladder: *throwing them out,* is the office of the ductus cysticus, ductus cholidochus, and ultimately of the colon and rectum.

227. Thus all the abdominal viscera form one series, which may be called, the superior universal series; for they all respect a common effect, end and use, namely, the blood. But this series is divided into three inferior universal series; into one which primarily respects the *chyle;* into another which respects the *serum;* and into a third which separately respects the *blood* already formed (*e*). Each of these series, however, is further subdivided into several other inferior series; there being series which prepare the *chyle;* refine it; introduce it into the blood. Next, series which purify the *serum;* secrete it; excrete it: and series which do the like for the *blood.* Each of these is again divided and ramified into several others: for example, of those which *prepare the chyle,* there are some that grind the food; some that digest it, or extract its essential juices; some that besprinkle it with saliva; some that unite the saliva with the extracts, so producing the *chyle.* (*f*).

struum for refining the chyle, and marrying it to the blood; of which office we shall treat in the Chapter on the Spleen.

(*e*) The chyle, the serum, and the blood, are distinct things: one is the source of the other. The chyle itself in the arteries and veins is called serum; and the purest serum, united to the spirit, produces the blood. The blood thus completed, requires to be distinctly purified, in order to become the universal subject, parent and seminary of all things of bodily life. The purificatories of the blood, are, primarily the spleen, secondarily the pancreas, and lastly the liver:—these in the abdomen: there are others in the chest, and even in the head itself, namely, the lungs and the cerebrum, of which we shall speak elsewhere.

(*f*) There are many other both subdivisions and superpartitions, which terminate only in the very simplest divisions of this series, or in the capillary extremities of the vessels: which is the reason why no two of even the minutest veins or arteries carry blood or serum of a similar nature; each being dedicated to some most particular office, distinct from that

228. We may now see the close affinity that subsists between these series or viscera, and between their offices; and in what order and with what connectedness they are mutually subordinated to each other. And this, particularly with respect to the pancreas, which, according to a common opinion, may be either present or absent, indifferently (*g*). The truth, however, is, that the pancreas is a kind of link between the spleen and the liver; it purifies the blood for the spleen, and draws off the serum; the parts secreted, it sends away by three paths; one part to the omentum, mesentery, and mesocolon (*h*); another to

of the vessel next to it. Such differences occur not only between the vessels of different viscera, but also between those of one and the same viscus; as here in the pancreas, where no two glands invite or secern an absolutely similar blood or serum. Yet still, from differences in a continual series, results a compound most thoroughly adapted to nature's end. See n. 188.

(*g*) It having been proved by many experiments on dogs, that the pancreas and spleen might be removed, and the animals still survive, vigorous, voracious, and truly canine, it has therefore been supposed, that these viscera are inserted, merely to fill a vacant space in the abdomen; but that this is not true will be seen below (*l*).

(*h*) Respecting the adhesion of the pancreas to the omentum and mesentery, Heister says, "[The pancreas] is connected with the mesentery, splenic vessels, and spleen." (n. 219.) And Winslow says "The small extremity is fixed to the omentum, near the spleen:" and again, "The pancreas is enclosed in the duplicature of the posterior portion of the mesocolon." (n. 220.) That the arteries and veins also pass to and fro between these bodies, may be seen in anatomical tables. "The pancreas," says Winslow, "also receives some small ramifications from the great gastric [artery]," (which artery lies upon the omentum in a large part of its course, as shewn in one of Ruysch's Tabulæ,) "and superior mesenteric artery. . . . There are small veins corresponding to the small arteries, and which are productions of the great mesenteric vein." (n. 222.) That the fat derives its origin from the grosser elements of the divided blood, and from the purer elements of the serum, see the Chapter on the Omentum. Thus, when the pancreas purifies the blood, and sunders it into its primitive elements, it seems to reject a part into the omentum, to send forth a part into the vena portæ, and to excrete a part into its middle duct. See also what is said of these subjects below, n. 231 (*x*).

the porta hepatis (*i*); a third to its own duct: the latter furnishes the pancreatic juice, which is thrown into the duodenum, and then circulated, in union with the hepatic and cystic bile (*k*). Thus the pancreas, as the intermediant member, causes the action and sequence of effects, ends and uses to exist and subsist in its fulness; and consequently the subordination of efficient causes likewise to exist and subsist in the greatest state of perfection (*l*).

(*i*) Namely, by the veins, which like the arteries, are connected to the inferior edge of the pancreas, and which go to the vena portæ. "The pancreatic veins," says Winslow, "are branches of the splenic vein, which is one of the principal branches of the vena portæ. The splenic vein, like the artery, runs along the lower side of the pancreas, near the edge," &c. (n. 222.)

(*k*) See the similar statement respecting the hepatic and cystic bile, in the last chapter. Boerhaave also pronounces for the same opinion, n. 223, *ad. fin.*

(*l*) This then is the cause of the existence of the pancreas and spleen, and of the actuality of their offices. For nature, in all her methods and processes, ever intends and attempts the greatest degree of perfection. Therefore, when she is proceeding, or about to proceed to an end, she calls forth to her assistance all the causes which can possibly promote it—proximate, intermediate, remote and ultimate—that is to say, an entire causal series of subordinations, in order that effects, ends and uses may never defeat her principle: and these causes she multiplies in proportion to the nature, eminence and necessity of the end. More or less causes may be subordinated; but the more there are, the more sufficient and full is the effect, and the more perfect the state: the fewer, the more imperfect the state; and the defect is generally supplied in the ultimate viscus; as in this case, in the liver. This is the reason why two pancreases have been found in some subjects; see Winslow, n. 221; and two pancreatic ducts in others; see Heister, n. 219; and Verheyen, *Corp. Hum. Anat.*, tract. ii., cap. xv.; as indeed "is always the case in the goose, the duck, the African fowl, and the pheasant; there are three pancreatic ducts in the common fowl, the pigeon, the eagle," &c., n. 219. In creatures of simpler structure, as insects, we find neither spleen nor pancreas, but a large liver. When, therefore, either is deficient, there is always some defect of the natural functions. But the pancreas is so constructed, that in the absence of the spleen, and the liver is so constructed, that in the absence of the

229. No series can be complete or effective without involving at least a trine; that is, a first, a middle, and a last (*m*). These three must be so ordered, that the first term disposes the second, and disposes the ultimate both mediately and immediately. Thus there is a trine that purifies the blood, namely, the spleen, the pancreas, and the liver. A trine that secretes the blood and

pancreas, it may still discharge its natural functions. And it may even be inferred from the vascular connexion, that in the absence of one, its office is transferred to the next member of the series, this being furnished with additional organs, fibres and vessels, to make it competent to sustaining the office by itself: as might, I feel convinced, be observed in dogs, after excision of the spleen or pancreas.

(*m*) Nothing can be bounded, completed, or perfect, that is not a trine. Sometimes even a quadrine is necessary, or a still more multiple series or sequence, exactly according to the ratio between the first and last term, that is, to their distance from each other, and the nearness or remoteness of their relationship. Meanwhile, whatever be the relation, there must be at least a trine, to procure harmony. Otherwise, no termination or conclusion is possible. To instance only geometry, arithmetic, physics, rationals, and logic. In *geometry*, two linear extensions alone, take in nothing and conclude nothing: a third thing is respected as the concluding agent, and therewith as the conclusion; whether in a triangle, a body of trine dimension, an algebraic equation, or any other thing of this class. In *arithmetic*, two numbers form only a ratio, but when a third term is added, or generated by the two first, we then have an analogy, either continuous, or harmonic, or of some other kind. In *physics*, two powers or forces, regarded as causes, always likewise respect some third, whereby an effect is produced, and in this, a fourth, or fifth, and so on; which is to be the ultimate effect of the series. In *rationals*, nothing which deserves to be called a judgment, such as ought to exist in all the conclusions or actions determined by the will, can possibly be formed from two reasons,—there must always necessarily be some third. In *logic*, two premises are requisite, to constitute a full syllogistic form, or a full argument; more than two in a sorites. What is at last concluded from two, becomes the property of the conclusion itself, but this it derives from the premises. So in every science and art: the binary is ever the imperfect; hence some third thing is always involved either tacitly or openly. This is universally the case in the anatomy of the body, which is the mirror, prototype, and complex of all arts and sciences.

serum, namely, the pancreas, the omentum, and the liver. A trine that circulates the secretions, namely, the pancreatic, the hepatic, and the cystic ducts. A trine that prepares the chyle, namely, the stomach, the small intestines, and the large intestines. A trine also that secretes and excretes the worthless parts of the serum, to wit, the kidneys, the ureters, and the bladder.

230. As all the viscera of the abdomen perform a common series of operations, so also every viscus, organ and member represents in itself a kind of series; wherefore each member separately should be contemplated as a kind of series, complex, society, and complement of the operations which are peculiar to it (*n*). This is particularly apparent in the pancreas; which, to constitute it such a series, is divided into masses, congeries, and tuberosities, and these into lesser ones, and into least, which

(*n*) Series embraces in it everything which exists and coëxists; for there are both successive and simultaneous series, but the latter always arise from the former, and the one contemplates the other in itself, as the cause contemplates the effect, and the effect the cause; as children, their parents; or as a father contemplates his wife in his offspring: and there are as many series as there are analogies, conclusions, and effects. But it is my intention to deliver the Doctrine of Series in a separate form: I will here explain only so much as respects the series of substances and operations consequent thereupon in the pancreas. The pancreas itself is most distinctly divided into lobes, tuberosities, glomes or balls; and these into their glands or unities: so also are its ducts, vessels, &c. "The pancreas," says Winslow, "is composed of a great number of soft glandular molecules, so combined as to present the appearance exteriorly of one mass, the surface of which is rendered simply uneven by numerous small convexities, more or less flattened" (n. 220). With respect to the ducts, the same author says, "We find along the middle of the breadth of the pancreas a particular duct, in which several smaller ducts terminate, almost as the branches of a tree terminate in the stem" (*ibid*). And the same thing holds good of the blood-vessels, which proceed from greatest to least, and from least to greatest, in a kind of continual series. We also find the same in all the other members and viscera, and particularly in the liver, the lungs, and the cerebrum. Since these points appear to be most distinctly exemplified in the pancreas, therefore the pancreas may be regarded not undeservedly as the model of the conglomerate glands of the body.

latter are the very glands, wherein the progression of its series ends, and from which it begins. Every thing is a series, and in a series. The very knowledge of effects depends upon a distinct notion of the subordination of causes in any given series.

231. When the pancreas is examined carefully, in series with its fellow-members, it instructs us respecting the nature of the subordination of efficient causes, and of the influx, communication, and conclusion of operations, that is, of the effect (*o*). I. The SPLEEN flows into the pancreas, and the pancreas into the spleen, by means of nervous fibres and blood-vessels (*p*);

This was termed above (n. 203), *continuously successive progression.* But we need not dwell longer on these particulars, since the same thing occurs in all the other organs.

(*o*) The animal body is as it were the temple of all the sciences, both natural or physical, and rational or philosophical. In the body, we learn the things which are indeterminate in the sciences; namely, how they are determined, and reduced to act or practice. Here, therefore, we are instructed to the fullest degree, respecting the manner in which causes are subordinated, and in which, when subordinated, they flow mutually into each other, and communicate their actions. Lastly, how the effect becomes continuous with its cause or causes. In the pancreas all these things are represented more perfectly than in a bulky volume of doctrines. Every one must understand them if he attends to the several particulars.

(*p*) To avoid prolixity, I pass over the influx of nerves from the spleen into the pancreas, from the pancreas into the liver, and from the liver into the gall-bladder; and I dwell only upon the influx which takes place by the arteries and veins, which are the proximate causes. The nerves flow in nearly in the same way as the vessels, only in a more universal manner; for the spleen derives its proper fibres from its own splenic plexus, and its common fibres from the mesenteric plexus. So likewise the pancreas, the liver, and the gall-bladder. By means of these fibres, these viscera are inaugurated into a communion and society of operations, which are visibly represented by the blood-vessels. "The pancreas," says Winslow, "receives nerves partly from the hepatic, partly from the splenic, and partly from the superior mesenteric plexus: and it likewise receives a nerve, the chorda transversalis, from a flat ganglion lying between the two semilunar ganglia" (n. 222). But of this higher and universal influx, which is the cause of the lower influx by the vessels, (thus the principle of the cause, or, in the lan-the *pancreas*, likewise, into the liver, and the *liver* into the gall-bladder. II. The SPLEEN derives its blood from the trunk of the aorta, by the cœliac artery (*q*): the *pancreas* derives its blood from the same artery; and also immediately from the aorta, by a branch of the cœliac, and from the superior mesenteric artery (*r*): but the *liver* derives its venous blood from the

guage of the schools, the causality,) we shall speak in the Part on the Organism of Animal Motion.

(*q*) See the next Chapter, on the Spleen.

(*r*) Since the pancreas is the mediate purificatory of the blood, as the liver is the ultimate purificatory of both the blood and the chyle, therefore it is necessary that we examine thoroughly the influx of vessels, or of blood; that is to say, the source whence this member derives the supply which is the material and subject of its operations. That the pancreas derives its blood not merely from the spleen, but also, like the spleen itself, from the aorta, as well as from other sources, is sufficiently evident from anatomical descriptions. First, with respect to the derivation of blood by means of the spleen, Winslow says, "The arteries of the pancreas arise chiefly from the splenic artery, which adheres very closely to the whole lower side of the pancreas, near the posterior margin." (n. 222.) Next with respect to the blood which comes immediately from the aorta. "The right branch of the cœliac artery," says Heister, "gives off the right gastric and epiploic arteries, and the pancreatic artery: the left branch gives off [among others] the great splenic artery, which gives branches to the pancreas," &c. (*Comp. Anat.*, n. 293.) Or even mediately from the aorta by the gastric artery, which also arises from the right branch of the cœliac artery. "The pancreas," says Winslow, "receives some small ramifications from the great gastric artery." (n. 222.) See also his *Exp. Anat.*, *Traité des Arteres*, n. 192. That the pancreas likewise receives a certain other branch from the aorta, and indeed, from its trunk, below the origin of the cœliac artery, that is, from the superior mesenteric artery, see above, n. 222. This artery arises in front from the descending aorta, a little below the cœliac, and is distributed to the mesentery and intestines, particularly the small intestines. Consequently, the pancreas is supplied from three sources with the blood that it is to purify: thus in part, it acts independently of the spleen, and is even an efficient cause, by itself; so that if the action of the spleen be diminished, or fail, the pancreas and the liver may be able to supply its place.

pancreas, and thereby from the spleen; also from the stomach, intestines, and mesentery: its arterial blood likewise from the aorta, by the right branch of the cœliac, and by a branch of the diaphragmatic artery (*s*). III. The SPLEEN sends the purified blood into the splenic vein, and into the lymphatics; it receives the blood intended for purification, by means of the pancreas (*t*): and transmits it when purified, also by means of the pancreas, to the porta hepatis (*u*). The *pancreas* sends forth the blood that has been purified by its peculiar method, partly into the omentum (*x*); partly into the pancreatic duct; partly into the great

(*s*) Respecting the ramifications distributed to the liver and gall-bladder, see the preceding chapter.

(*t*) See note (*r*), just above. "The arteries of the pancreas," says Winslow, "arise chiefly from the splenic artery. [This artery] in its passage, gives off many branches, termed pancreatic arteries, which run out more or less transversely from each side." (n. 222.) From these the pancreas derives a part of its blood, which, inviting into its glands, it secerns and purifies, almost in the same manner as the liver secerns and purifies the blood of the vena portæ.

(*u*) "The pancreatic veins," says Winslow, "are branches of the splenic vein, which is one of the principal branches of the great vena portæ. The splenic vein, like the artery, runs along the lower side of the pancreas." (n. 222.) The branches to the great mesaraic vein, take the same direction towards the vena portæ; for the vena portæ in this situation, is termed the great mesaraic vein. See Winslow, *Exp. Anat.*, *Tr. des Veins*, n. 241.

(*x*) See n. 228 (*h*), and the Chapter on the Omentum. The pancreas, as is well known, is destitute of lymphatics, although the spleen, liver and gall-bladder are abundantly furnished with them. "In dogs," says Verheyen, "the lacteals run from the duodenum, through the pancreas, to the cisterna mesenterii; but they do not belong to the pancreas: yet it may be doubted whether, at the same time, they do not serve as lymphatics to it; although it is not to be questioned that there are also lymphatics proper to the pancreas." (*Corp. Hum. Anat.*, tract. ii., cap. xiii. xv.) But when we understand the methods whereby the pancreas purifies the serum and blood, the properties of its secretions, and of the fat, and the construction of the coats of the omentum and mesentery, (which are so perfectly cellular, that they resemble a continuous lymphatic thrown down into a plane, respecting which see above,) it will be clearly evident, that the pancreas, like the liver, splenic vein, and thence to the porta hepatis (*y*). The *liver* sends its purified blood into the vena cava, and into its own lymphatics; but the residue that is not purified, into the hepatic duct; and the blood that is to suffer the last expurgation, into the gall-bladder. IV. At last the PANCREAS, the LIVER, and the GALL-BLADDER mingle their secretions; this being the ultimate effect of the efficient causes, and the conclusion of the opera-

throws out its first secretion towards the surface, where it is not taken up by the lacteals, but by the cellular coats of the omentum, and succeeding viscera, and there partly changed into fat, and partly absorbed by the lymphatics of those viscera. For in order that the omentum may serve as a kind of soft pillow, placed beneath and between the stomach and intestines, at the same time that it exhales an unctuous humor for anointing them, it is absolutely necessary that the pancreas be destitute of lymphatics, and transmit its fatty and unctuous humor to those viscera, principally by means of the omentum: thus not by way of the arteries or veins, but by the usual lymphatic channel. That the glands of the pancreas, like those of the liver, are surrounded by exquisitely fine cellular membranes, will be seen presently; consequently, that they likewise throw out a certain lymph, but of a fatter kind, towards the surface. Such being the case, we have here another, and indeed the most noble office of the pancreas.

(*y*) See note (*u*). But the question occurs, In what manner does the pancreas purify the blood, and what blood does it purify? From the influx of vessels, and the parallel case of the liver, we may collect and conclude, that first it submits to rigid examination the arterial blood which is delivered to the spleen; and separates it, as well as its own proper blood, into three species; for the great splenic artery runs along the lower border of the pancreas, dipping into it, and thence passes on to the spleen, and returns as a vein, (which is also called the splenic,) and into which it then pours the pure portion of its blood. This is the reason why that vein dips more deeply into the substance of the pancreas; and why the pancreatic juice is given forth in such large quantities. "[The splenic artery]," says Winslow, "in its passage gives off many branches, which run out more or less transversely from each side. The splenic vein, like the artery, runs along the lower side of the pancreas, in a shallow depression in the substance of the viscus" (n. 222). Wherefore "the duct is nearer the lower than the upper side of the pancreas" (Winslow, n. 220). See the comparison instituted between the pancreas and the liver, n. 233.

tions (*z*). These particulars show in what way the pancreas is a mediate cause, subordinated between the spleen and the liver; namely, that it acts both mediately and immediately; or that it is an operative and efficient cause, both through the spleen, and by itself: so likewise the liver (*a*): consequently, when either is deficient, or not sufficient, the other takes its place, and assumes the reins, though in a comparatively indistinct, insufficient, and imperfect manner (*b*). The same considerations also shew, that the pancreas in one way makes common cause with the spleen;

(*z*) Namely, in the cholidochus, according to the observations of our authors. "[This duct]," says Heister, "terminates usually at the same orifice with the ductus cholidochus, but sometimes it has a double aperture" (n. 219). See also Winslow, n. 220. Something similar to this obtains in every effect, conclusion, and result; namely, that the causes flow into the effect, not by a single mediation or subordination, but by several.

(*a*) The modes of subordination of causes [in their progress] to effects, are various: and something of their nature may be seen by thoroughly examining the effect, supposing always that the signs of the causative agents be given. This progression or subordination may, indeed, be illustrated by comparison with the analogies of progressive ratios in geometry, arithmetic, algebra, and physics, and by series of causes in metaphysics. But to explain it more adequately, and more suitably to common apprehension, it may be likened to rivulets which unite to form a river, and this river again to form a third, in which still more numerous rivulets join. Or it may be compared to a number of paths, which unite at a crossway, and by a kind of royal road proceed to a second crossway, where still more numerous paths meet.

(*b*) See n. 228 (*l*). From the use here explained, it must now be conspicuously evident, what the pancreas contributes to the common good of the kingdom, and that the purificatory apparatus of the blood would be crippled and imperfect, if the pancreas were absent: also, that a more full and perfect state results from additional associates, whether pancreases or ducts; and that by this means provision is occasionally made for the kingdom, if haply the spleen or pancreas lose their vigor, which is no uncommon case; or if a too thick, slow, acrid, or bilious blood disturbs them. Nature is ever obviating the perilous contingencies that befal the body, and from the very commencement she guards against everything that may wrong or destroy it.

in another with the liver; and that the liver in another way again, makes common cause with the gall-bladder; and notwithstanding, that there is a continual series, and an analogous progression of causes and operations.

232. In consequence of the coördination and subordination of the viscera and their offices, no portion of blood can possibly circulate anywhere in the trunk, without being lustrated, purified, and expurgated by means of these members: the blood belonging to the middle region, by the spleen; that of the middle, and of the lower or abdominal region, by the pancreas; and that of the middle and lower, and at the same time of the higher or thoracic region, by the liver. For the *spleen* derives its blood from the trunk of the aorta by the cœliac artery (*c*); the *pancreas* from the same trunk, and by a branch of the same artery, and also below, by a branch of the superior mesenteric artery (*d*). The *liver*, likewise from the aorta by the cœliac, and also from the diaphragmatic arteries (*e*). So also the *gall-*

(*c*) It arises a little below the diaphragm, and above the emulgent arteries. Thus, as the emulgent arteries draw off the valueless serosity of the blood, so the cœliac artery draws off its sanguineous impurities, before the aorta descends or distributes itself further.

(*d*) See above, n. 231 (*r*). The superior mesenteric artery, immediately after its origin below the cœliac, sends a small branch towards the pancreas, and to the duodenum, and after having passed through the mesentery, is reflected to the ileum; then dividing into numerous twigs, it ramifies over the small intestines, by which it is distributed to the vermiform appendage, and even to the colon and mesocolon. It also anastomoses with the inferior mesenteric artery, which supplies the large intestines, and gives off the internal hæmorrhoidal artery. The pancreas, therefore, which claims the first branch, at the very entrance and threshold to these arteries, immediately draws off the whole of the impure blood of this region, and prevents the natural functions of these viscera from being stifled or destroyed by noxious blood, or by obstructions or aneurisms of the vessels.

(*e*) The diaphragmatic or phrenic artery, arising beside the inferior muscle of the diaphragm, anastomoses with the arteries of the succenturiate kidneys; with the adipose, intercostal, internal mammary, mediastinal, pericardial, and cœliac arteries; consequently, with all the arteries of the superior or thoracic region. Moreover, the liver derives *bladder*. Such is the extent of operations, when these four viscera are yoked and associated, and unitedly constitute the wheel of the machine.

233. The glandular nature of the pancreas is illustrated by that of the liver; the two being related in form, operation, and use: a comparison may therefore be instituted between them. The LIVER is covered with a common coat derived from the peritonæum; so is the *pancreas* (*f*). The common coat of the LIVER grows finer and finer until it reaches the glands; so does the common coat of the *pancreas* (*g*). The LIVER is divided into lobes; the *pancreas* into masses, which are its lobes. (*h*). Each

its blood from the abdominal region by the same mesenteric artery as the pancreas, and also by the cystic arteries. The liver, according to Heister, is supplied with arteries, by the cœliac, cystic, diaphragmatic, and sometimes also by the superior mesenteric arteries (n. 191). Thus, agreeably to my proposition, no portion of blood circulates in any part of the trunk, but is lustrated, purified, and expurgated by means of these members. In other words, the provinces are so divided among these three viscera, (the spleen, the pancreas, and the liver,) that there is not the minutest drop of impure blood in the system, but is derived and invited to one or the other, as the state of the body demands: just as there is no portion of the impure serum, but is derived and invited by the emulgent arteries to the kidneys.

(*f*) That the common coat of the pancreas, like the common coat of the liver, is a production of the peritonæum, is stated by Heister. "[The pancreas]," says he, "is surrounded with a membrane which is continuous with the peritonæum" (n. 219).

(*g*) That the common coat, growing constantly thinner and thinner, penetrates at length to the glands, is thus declared by Verheyen: "The glands of the pancreas," says he, "are connected together, both by vessels, and, according to De Graaf, by a membrane proper to each; and all together are surrounded by a coat of considerable strength, arising from the peritonæum, by virtue of which the whole pancreas is kept firmly in its place." (*Corp. Hum, Anat.*, tract. ii., cap, xv.)

(*h*) Respecting the division of the liver into lobes, (and which lobes are more numerous in dogs and other animals than in man,) see the Chapter on the Liver. The pancreas is not, indeed, divided into similar lobes, but it is distinctly divided into tuberosities, little masses, and beds, amounting in number, as it appears, to forty or fifty. These lobe of the LIVER is subdivided, and ultimately into glands; so is each mass of the *pancreas*. Every gland of the LIVER is composed of a last production of the common membrane, of arterial and venous threads, and of fibrillæ; so is every gland of the *pancreas* (*i*). A branch of the vena portæ approaches, enters, and perforates each gland of the LIVER; a branch of the splenic artery, each gland of the *pancreas* (*k*). Whatever the branch of the vena portæ conveys to the gland of the LIVER, is separated into three kinds; so likewise is what the branch of the splenic artery conveys to the gland of the *pancreas*. The gland of the LIVER conveys away one kind of secretion, by the hepatic veins, to the great vena cava; the gland of the *pancreas*, one kind, by the pancreatic veins, to the splenic vein. The gland of the LIVER throws out a second kind by fine cells to the surface, and ultimately into the lymphatics; so likewise does the gland of the *pancreas*, but into the cellular and adipose coats of the omentum, mesentery, and mesocolon (*l*). The gland of the LIVER expresses its third secretion through the pori bilarii into the common or hepatic duct; and the gland of the *pancreas*, its third secretion, through its biliferous passages or vessels into the great pancreatic duct. The LIVER transmits its bile into the ductus cholidochus; the *pancreas* transmits its juice into the same duct. All these things the LIVER performs silently, tranquilly, mildly; so also does the *pancreas* (*m*). When the gland of the LIVER ex-

partitions even shew themselves in the unevenness of the surface; and are of the kind usually called lobes.

(*i*) See note (*e*), and the Chapter on the Liver, n. 209. The arteries ramify perpetually, dividing at last into the minutest threads, in order to reach their ultimate goals or little stations; which are none other than the very glands. The veins likewise flow back from least to greatest, and of course from the same ends or beginnings.

(*k*) See the Chapter on the Liver, n. 209. You will also find this illustrated in n. 231 (*y*).

(*l*) You will find these particulars explained above, n. 228 (*h*), and n. 231 (*x*).

(*m*) Respecting the liver, see n. 214. The pancreas, like the liver, has no muscular coat to excite it to manifest motion. The nature of the functions of both these members, renders it requisite that this should be the case. For to allow various species of fluid to be dispands and invites the blood, then the porta contracts and impels it; interchangeably, reciprocally, and alternately. The *pancreas* acts by a similar invitation, incitation, alternation, and reciprocation: and *both*, indeed, by movements synchronous with the respiration of the lungs (*n*).

234. The pancreatic juice can still less be said to be excreted and thrown away, than the hepatic and cystic bile: like them it is circulated, and like them it is of use in its circle (*o*). By

tinctly secerned, separated from each other, and thrown out, the most perfect concord of motions is absolutely necessary: so that when the gland invites the blood, the splenic artery shall intrude it, or express it from itself, and *vice versa*. Consequently, the pancreas, like the liver, enjoys an alternate reciprocation of expansion and contraction. See n. 214. Such *alternate reciprocation* and *reciprocal alternation* of expansion and contraction, (for so this mode of action may not improperly be designated,) produces apparent rest, wherein no motion of either kind is discernible; for the general contraction of the mass, and the synchronous expansion of the glands, and *vice versa*, puts on all the appearance of a state of absolute tranquillity; although there is not even the least part, or the least thing in a part, but is leaping and pulsating, or opening and contracting alternately.

(*n*) These matters need no further illustration than was given above, in the Chapter on the Liver, n. 214. The parallelism of causes, operations, effects, and uses, abundantly confirms this.

(*o*) Respecting these subjects, see the Chapter on the Liver, n. 213 and 218. The pancreatic juice has a nobler nature and use than the hepatic bile, and a nobler still than the cystic bile. For this juice is secreted before the hepatic bile; and moreover, the hepatic bile contains not only the impurities of the blood, but also the feculencies and crudities of the chyle. The nexus of communicating vessels, and the tenor or series of operations and consequently of causes in the pancreas, seem sufficient to prove, that the glands of the pancreas do not invite, receive, or admit all the blood of their artery, the splenic, but only that portion which they are competent to secern, and that the rest is sent away, by the anastomosing branches of the little arteries, into the little veins, and so into their vein, also called the splenic, to go ultimately towards the liver. This seems a possible inference from the frequency of the anastomoses, from the softness of the substance of the pancreas relatively to that of the liver, and from a number of other circumstances which we must now pass over. The softness of the panthe mixture of the pancreatic juice and the two kinds of bile, a universal salivary menstruum is prepared (*p*), which is tempered and adapted, with infinite variety (*q*), to every necessity and demand of the intestines digesting, and of the food to be digested. The pancreatic juice is the first cause (*r*), the hepatic

creas results, indeed, from the thin and loose connection between the glands, which themselves are somewhat hard, as well as from the quantity of vessels which go to and from it; consequently also from the frequency of the anastomoses.

(*p*) Every humor which nature produces is of so perfect and consummate a character, as to be a universal menstruum; for whatever can be in it, is placed there; namely, all the requisites not only for continuing the life of the body, but even for perpetuating it. If defect or fault there be, it arises from ourselves,—from the intemperance of our appetites, the unbridled excess of our passions, and the collision thereof with the affections of the mind and soul, and from very numerous other causes.

(*q*) Were we to classify and compute all the possible varieties of the pancreatic juice, and of the hepatic and cystic biles, we should have a sum exceeding the power of numeration: for a kind of infinity may be predicated of each variety, with respect to quality and quantity. This variety, which to our conception is infinite, is further multiplied by the commixtion of these three mutually-corresponding humors; a menstruum being thus prepared, capable of exact adaptation to every possible use for which it is designed. But as the pancreatic juice, and the two kinds of bile, collectively taken and united, constitute a universal menstruum for digesting the food detruded by the stomach into the intestines, and which for the most part has been already seethed and exhausted; so the very blood of the splenic vein, after it has passed through the pancreas, and been there a second time lustrated or purified, is a universal menstruum for refining the chyle in the liver. So, again, the lymph of the lymphatics, of both the liver and many other viscera, is a universal menstruum for refining the chyle that escapes by way of the mesentery. Thus, there are as many ministering menstrua and appliances, as there are departments in the elaboration and refinement of the chyle: respecting these, therefore, we shall treat in the following Chapter, on the Spleen.

(*r*) Respecting the use of the pancreatic juice, see Boerhaave, n. 223, above: and respecting the use of the hepatic and cystic biles, n. 218 (*h*); and of the saliva, *passim*. It would be superfluous to add bile is the second, and the cystic bile is the third, according to the order of the viscera themselves.

more in this place. For this juice may rightly be considered as both a salivary and a biliary menstruum. That it is a *salivary* menstruum, no one denies: that it is also a *biliary* menstruum, is shewn by the pancreatic juice of the hedgehog, as observed by Morgagni, who says, "In the pancreas of the hedgehog I found bile, or at least a greenish yellow fluid" (n. 224). Respecting the pancreatic juice, the reader may also consult the descriptions of numerous authors, and, I may add, their contentions, as to whether it partakes of acid, salt, or alkali; of which contentions, indeed, as the pancreatic juice, admits of infinite varieties (*q*), there seems to be little hope of an amicable adjustment.

CHAPTER XI.

THE SPLEEN.

235. HEISTER. "The spleen is a viscus of a blackish red color, placed on the left side of the stomach, under the diaphragm, and close to the ribs. It is usually single, but sometimes several spleens have been found instead of one. Its figure varies; not seldom it is something like a tongue, concave towards the stomach, and convex towards the diaphragm and ribs; often, however, it is irregular, with fissures in various parts. It is connected with the stomach by the vasa brevia; and with the pancreas, omentum, diaphragm and left kidney, by membranes. Its size is various. Its length is usually five or six inches in the human body; but in pigs, dogs, &c., it is much longer and thinner. It is about three inches broad, and one thick, and weighs about twelve ounces. In man, as well as in the pig and dog, the spleen has but one membrane; but the spleen of the calf has two, of which the external or common membrane is robust, and adheres but loosely by blood-vessels to the internal or proper membrane, which, when the external is removed, transmits flatus. The vessels of the spleen, considering its size, are remarkably large. Its artery is from the cœliac, and is called the splenic artery: in the human subject it readily transmits water, air or mercury, when injected, into the veins. In the calf, the splenic vein is transformed into cells, soon after it enters the spleen; but in the human subject, it ramifies very much through the entire spleen, as the veins do in the other viscera. In the calf, both vessels enter at one extremity; but in the human body they enter by various distinct branches, all over the concave or internal surface. The nerves of the spleen are from the splenic plexus. The spleen has no excretory duct: but it has lymphatics, which go to the receptaculum chyli. Its substance is stated to be cellular and glandular. In the calf, indeed, it

is cellular; but in the human subject it is vascular and fibrous. What authors describe as glands in the spleen, Ruysch has proved to be vessels merely. One or two lymphatic glands, about the size of a bean, are frequently found on the outside of the spleen, near where the vessels enter it. *(Comp. Anat.*, n. 218.) There is no one of the viscera the use of which is more obscure than that of the spleen, and hence none has given rise to greater differences of opinion. Many who have found that it may be excised from animals, without causing death, have held, with Erisistratus, that it is a useless and superfluous part of the body, or an error of nature. Others suppose it to be made only for the sake of preserving the equilibrium of the body. Some have maintained, with Hippocrates and Aristotle, that it attracts watery materials from the stomach: and some of the ancients, with Galen, have thought that it was the sewer of the humor called atra bilis, or melancholia. Some have thought that a kind of ferment or menstruum, necessary for the operations of the stomach in digestion, was secreted in the spleen, and conveyed to the stomach. Some have supposed the spleen the origin of laughter, according to the adage,

'Splen ridere facit, cogit amare Hepar.'

Others declare that the spleen inspissates the blood. Havers was of opinion, that it prepares the mucus which is secreted in the mucous glands of the joints. Schelhammer, and with him some of the recent English anatomists, as Lister and Purcell, suppose it to be a diverticulum for the blood in the more violent motions: and to this Purcell adds, that it inspissates the blood; and by means of I know not what acid, produces a precipitation, which serves for the better secretion of the bile. Other persons have produced other uses, but it would be tedious to recount them all. My own opinion is, that by virtue of its situation and structure, the spleen supplies a thin and very fluid blood to the liver, in order that the other thick blood conveyed to it from various parts, and from which the bile is to be secreted in this dense viscus, may be rendered more fluid, obstruction of the liver be thus prevented, and the secretion of the bile promoted. For if the splenic vein be opened in a living dog, or other animal, both I and others have found that a thin, florid and subtile blood flows from it;—never a thicker blood than that in the rest of the veins." *(Comp. Anat.*, not. 21.)

236. WINSLOW. "The spleen is naturally divided into surfaces, extremities and edges. It has two surfaces, one external and slightly convex, and one internal, which is irregularly concave: two extremities, one posterior and large, one anterior, smaller and somewhat depressed

two edges, a superior and an inferior, on both of which in some subjects, there are several little inequalities. The inner or concave side is divided by a longitudinal groove or fissure, at which the vessels and nerves enter in human subjects. The spleen is connected to the stomach by the vessels called vasa brevia; to the extremity of the pancreas by branches of the splenic artery and vein; and to the omentum by twigs of the branches which the same artery and vein send to the spleen, and which run in the longitudinal groove. The spleen is connected to the edge of the diaphragm by a particular membranous ligament; and in some subjects to the stomach and colon by other ligaments. The figure of the spleen is not always regular; sometimes it has considerable fissures both in its circumference and on its surfaces; and sometimes it has appendages. I have sometimes found small, distinct spleens, more or less round, and connected separately to the omentum, at some distance from the anterior extremity of the spleen. Its covering adheres to it so closely in the human subject, that it is difficult to distinguish the common from the proper coat; whereas in some animals, as oxen, sheep, &c., we may clearly see two coats, separated by cellular substance. This covering seems to be no otherwise a continuation of the peritonæum, than by the intervention of the omentum and mesocolon. Even in the human spleen, the two coats may be distinctly observed at the place where the vessels enter by the longitudinal fissure. Its substance is almost wholly vascular, that is, composed of the ramifications of all kinds of vessels: in oxen, the reticular tissue prevails; in sheep, the cellular. In oxen and sheep there are no venous ramifications, but only open and branching sinuses, except a small portion of a venous trunk perforated on all sides, at the extremity of the spleen. In the human spleen we see glandular granules, as in those of other animals; and through its whole extent, there are numerous venous ramifications: and between them, everywhere, an extravasation of blood, retained in a kind of downy, transparent and very delicate tissue, which extends through the whole spleen. This downy tissue, after having surrounded all the ramifications, terminates in almost imperceptible cells, which communicate with each other; for if we blow through a small hole made in the membranous covering, the whole spleen will be immediately inflated. The surface of the spleen of oxen and calves is visibly full of a great number of lymphatic vessels, but it is a difficult matter to discover them in the human subject. The splenic artery, which is one of the principal branches of the cœliac, runs along the lower side of the pancreas, in a winding course, to the spleen. The splenic vein, which is much larger than the artery, is but little inflected in its course. When the artery and vein have got beyond the extremity of the pancreas, they give off at once several branches, which immediately afterwards divaricate in the same plane, run in the duplicature of the neighboring part of the omentum, and lastly intersect each other on both sides, in their common plane, all the way to the fissure on the inner or concave side of the spleen. These arterial and venous branches enter the substance of the spleen together, by the same fissure; and the cellular substance of the membranous duplicature of the omentum, accompanies them. We may likewise observe, that at this place the coat of the spleen sends from its concave side a portion of a lamina, which is reflected into the fissure, and penetrates into the substance of the spleen. The nerves of the spleen are numerous, and come from the splenic plexus. They send forth at different distances, a number of filaments in form of an irregular network, around the arterial ramifications. The arteries, veins and nerves, after entering the spleen, divide and subdivide into a great number of ramifications, and accompany each other to their very last divisions. They are enclosed in a kind of common cellular sheath or capsule, which both surrounds all the three, and sends off particular septa between them. This capsule seems to be formed by a continuation of the cellular tissue of the omentum, and by the before-mentioned lamina of the coat of the spleen. The capillary extremities of all these ramifications, both arterial and venous, end in the small downy cells, already described. Malpighi regarded these cells as distinct capsules or follicles, each containing a small gland. They all communicate together, so that we may inflate the spleen through an opening in any part of its surface. In oxen and sheep there are no venous ramifications; but as soon as the splenic vein enters the great extremity of the organ, after a course of half an inch or an inch, it disappears, and leaves nothing but a canal perforated on all sides. The beginning of this canal still exhibits some remains of venous coats, but the canal-form is soon lost, and then we find nothing but sinuses, hollowed out in the reticular substance in oxen, in the cellular substance in sheep. The splenic artery and nerves there ramify in a particular sheath, somewhat as in the human subject. The extremities of the capillary ramifications seem to swim or float in the cells, and to fill their downy tissue with blood. At the ends of several of these capillaries, I have observed small corpuscules, disposed like bunches of grapes; and I have seen two little tubes going out from each corpuscule, one short and open, the other, long and small, which latter was lost in the sides of the spleen. I imagine that the long tube may be the origin of a lymphatic vessel, especially since we find these vessels so very numerous in the spleens of oxen. These small corpuscules may readily be discovered in these spleens, when boiled, and the coat taken off. They are indeed much larger before than after boiling, but they are not so solid, and disappear when cut. The same sort of corpuscles are found in the human spleen, but they are so extremely small as not to be visible without the microscope." (*Exp. Anat.*, *Traité du Bas-Vent.*, n. 330—349.)

237. Malpighi. "The spleen is surrounded by two membranes. The external membrane is of considerable strength, and covers the entire spleen in such a manner, as to contain it in a kind of purse or bag: this is particularly apparent in sheep. In some cases, however, a small portion of the spleen in contact with the stomach is left uncovered. This membrane has nerves, veins, and arteries, which run [over it] lengthwise; and it is supplied by the arterial extremities which ramify through the interior substance of the spleen: consequently, when the membrane is taken off, a number of reddish points are seen. . . . We also observe veins, interlacing to form a kind of loose rete, and falling into considerable branches, which ultimately proceed to the trunk of the splenic vein, where it enters the spleen, or else are continued onwards to the omentum. . . . Great numbers of lymphatics also run under the external membrane. They are remarkable at once for their large size, and for forming a singular rete with unequal meshes. The humor which they contain is sometimes of a yellow, and sometimes of a reddish color; and is supported by numerous valves, and pours at last into the receptaculum by considerable ducts which pass through the omentum. After taking off the external coat, we come to a second coat, which is smooth and firm, covers the spleen completely, and is nowhere perforated, excepting at the ingress and egress of the vessels. It is made up of a regular, elegant, and singular interlacement of fibres. . . . This membrane has been frequently found ossified. . . . Not seldom, in sheep particularly, I have observed in it calculous concretions of a substance like gypsum, as well as melicerides and other tumors. . . . This membrane is supplied by the ends of the arteries; for if the splenic artery, pervading the interiors of the spleen, be filled with ink or air, the extreme branches, towards the whole of the circumference, immediately exhibit a further prolongation and subdivision, and three or four little branches are seen rising from them. The same thing is observed when the splenic vein is filled with air or ink; large branches ramifying here and there on the lining membrane, are filled and raised, and when the spleen is cut through the middle, we see considerable branches like roots of trees, and of the same color as the injected fluid. . . . The spleen occasionally acquires a preternatural shape and size, from the channels in its substance becoming relaxed by the copious afflux, and deficient efflux, of impure

humors; of this the records of medicine contain numberless instances. *(De Liene*, cap. i., *De Lienis Membranis.)*

"The whole of the spleen is full of fibres. These proceed from the internal membrane, and are prolonged transversely to the opposite portion thereof, or to a kind of capsule or common sheath of the vessels, which runs through the middle of the spleen. They do not keep in the same plane, but repeatedly meeting, and the peculiar substance of the fibrillary, or capillary parts, subdividing and halving, they inosculate with other similar portions or divisions, so as to form a wonderful interlacement and network; and when they approach the membrane, they bifurcate, and are inserted into it by multitudes of branching productions. *(ibid.*, cap. ii., *De Fibris per Lienem dispersis.)*

"With respect to the vessels of the spleen, the artery, first discovered by Arantius, arises from the aorta. It enters the spleen in different ways in different animals: in oxen and sheep, by a single branch, which divides into numerous twigs inside the organ; but in man, and in dogs, horses, and other animals, it enters by three, four, or even more branches. These, accompanying the extended tubuli of the veins, (which are larger in the spleen than elsewhere,) and giving off twigs here and there, terminate at last, by their extreme ramifications, in certain newly-discovered corpuscules, and in the cavities or spaces of the spleen; the rest being expended on the external coat. All this may be very readily seen in the spleens of oxen, pigs, and sheep. . . . The extremities of the artery are very numerous, and together form a kind of cauda equina, which moreover, (by removing a small portion of the substance of the spleen,) is better seen in horses than in other animals. These numerous extremities closely embrace certain little corpuscules scattered through the spleen, and which we shall consider presently: but by other offsets, further prolonged, they terminate on the membrane by which the spaces or cavities of the spleen are formed. The rest of the arterial extremities are prolonged to the investing membrane, as may be seen by examining the spleen when its arteries are filled with ink or air. It also [sometimes] happens that the branches of the arteries which pass to the circumference of the spleen, are reflected towards the surface of the same [membrane], and interweave with the offsets near them. . . . The entrance of the veins into the spleen is well known to all; for the splenic vein is a very remarkable trunk of the vena portæ: it enters the cavity of the spleen by one or more productions. . . . In oxen, the vessels of the spleen are surrounded with a kind of membranous production; that is to say, the vein and artery and the two branches of nerves unitedly enter the spleen; so that before they come to it, they may be easily separated from each other. By slitting open the venous canal with a pair of scissors, (if the section be made in the thinner part, where but little of the substance of the spleen is interposed between the canal and the exterior membrane,) we immediately bring into view a small artery, accompanied by the nerves. Over this is extended a certain thin and transparent venous coat, or what answers to such. We observe, but with difficulty, a fine continuation of this coat or membrane, passing lengthwise, and easily separable from the artery and nerves underneath it. One thing is remarkable, namely, that this vein, near its entrance into the spleen, (when on the outside,) consists of two coats; but as it enters, the thicker and external of these is taken up by the internal coat of the spleen, which forms the sheath of the vessels, to which it is closely connected: but the other coat passes onwards. . . . This large venous duct exhibits a number of foramina, as it follows the ramifying offsets of the arteries and nerves: of necessity, therefore, the orifices of the branches are observed in the open trunk in determinate spaces, directly corresponding to the arteries, and when the orifices open into two branches, the venous tubuli exhibit a further progression. This venous coat is also perforated with other foramina, for among the open orifices of the branches, there is a multiplicity of stigmata or points, not in that part under which the arteries and nerves run, but in the opposite part, and at its sides. Although the coat of the venous duct has these openings, yet the accompanying branches of the artery and nerve have no corresponding offsets. . . . When the nerves and vessels are removed, we come to a considerable membrane, which encases or wraps round the bundle of vessels, and, therefore, may be termed their common sheath or capsule. It arises from the internal and proper membrane of the spleen, which membrane is reflected at the ingress of the vessels, and enters the cavity of the spleen, and forming a tube, accompanies the divisions of the vessels, which are gathered into a bundle inside it. It is conspicuous, not only in oxen, but also in man; as well as in sheep and similar animals. Its structure is not the same in all parts: under the artery and nerves, where the body of the spleen is thicker and deeper, it is very dense and thick, so as even to form tubuli: but in the opposite part of the tube, or of the splenic vein [ramus splenicus], it seems gradually to lose its membranous nature and structure, and to become like a net. This capsule is perforated with the same orifices as the vein just described; for it follows the divisions of the venous sinus, and of the artery, through the spleen. In it terminate, or from it arise vast numbers of the extreme ends of the fibres of the spleen, deep in the substance of the organ, which running transversely through the spleen, are fixed by one end in the exterior coat, and by the other end, in this capsule, as a centre. In the latter case, they terminate by an expansion of their proper fibres, and by an implantation similar to that described above as happening in the surrounding membrane: and it seems extremely probable, that great numbers of fibres arise from the capsule itself, as branches from a trunk, inasmuch as the ends of the capsule are expended in these fibres. *(ibid.,* cap. iii., *De Vasis Lienem percurrentibus, eorumque Capsula.)*

"We begin to have some knowledge of the structure of the spleen. If the splenic artery be tied, and the vein well inflated, either by means of a syringe, or with the mouth, the spleen will swell to a great size; and the same will happen, only to a less extent, if, after tying the venous duct, the inflation be made through the artery. If the spleen, in this swollen state, be dried, and afterwards cut open, you will find the whole mass of it made up of membranes, forming cavities and cells, like a honeycomb. . . . In the spleens of calves and sheep, the cells are connected by fibres and vessels running transversely, and as walls are strengthened by cross-beams, so these cells are supported by this singular contrivance of nature. . . . The cells communicate with each other mutually by open orifices; whereby they open into the splenic duct or vein, not only into its extreme branches, but also into the sides of the great trunk, by means of the already-described stigmata or points, in which the branches of the arteries and nerves are not inserted. . . . The membranes of the cells are supplied with arteries; for sometimes when ink was thrown in through the artery, I have observed reticular plexuses; and when mercury was used, it filled still smaller branches, pervading the membranes of the cells. *(ibid.,* cap. iv., *De Lienis Substantia.)*

"In the spleen we observe clusters of glands, or rather of vesicles or sacculi, in immense numbers, dispersed throughout the organ, and closely resembling bunches of grapes. These minute glands are of an oval figure, and about the size of the glands of the kidneys. I have always found them of a white color, and although the blood-vessels of the spleen may be fully injected with ink, and play around the glands, yet the latter still preserve their whiteness. They appear to consist of a membranous, but soft and very friable substance; the cavity in which is invisible by reason of its smallness, but still may be conjectured to exist, because these bodies collapse when cut. They are extremely numerous; indeed, almost innumerable; and are placed in a singular manner in the cells already described, all over the spleen, and hang from productions of the capsule, or from fibres arising therefrom; consequently from the extremities of the arteries, and from the ends of the nerves: moreover, the ends of the arteries wind about them like tendrils, or ivy. They usually hang in bunches, each cluster consisting of seven or eight glands. It is not so easy to see them in man as in oxen, sheep, and goats. *(ibid.,* cap. v., *De quibusdam Corporibus per Lienem dispersis.)*

"In a dog from which the spleen had been excised, no symptom of ill-health was observed, excepting that it became voracious, taking its food very greedily. The excretions were natural. It was, however, remarkable that the animal passed large quantities of urine, and very frequently. In body, it was well-conditioned and fat; but I noticed one external peculiarity, namely, a swelling in the right hypochondrium, which caused the last ribs to protrude beyond the others." *(ibid.,* cap. vi., *Cogitata quædam circa Lienis Usum.)*

238. See Boerhaave, on the action of the spleen, *Inst. Med.,* n. 312—328. Glisson, *De Hepate.* G. Hoffmann, *De Liene.* C. Drelincourt, *De Lienosis.* Vesalius, lib. v., and Tabulæ. Schelhammer, *Anal. Anat.,*[x] epist. x., where he states, that the glands of Malpighi, connected to the interiors of the spleen in clusters, are a mere fiction. Ruysch, *Thes. Anat.* ii., vii., x., &c.; and his various Tabulæ: 1. *Ep. Anat.* iv., tab. iv., fig. 1; where he represents the splenic vein and its wonderful ramifications, and reminds the reader, that the minute offsets there delineated as simple, are compound, consisting of mere fasciculi like little brushes. 2. *ibid.,* fig. 2; where he shews the trunk of the splenic artery, and its serpentine course in the human subject; here likewise remarking, that the minute offsets, exhibited as simple, are really fasciculi or brushes of vessels. 3. *ibid.,* fig. 3; a portion of a calf's spleen, so dissected, that its interstices or cells, and transverse fibres are brought into view. 4. *ibid.,* fig. 4; where he shews two little branches of the splenic vessels and the fasciculi attached to them, and remarks concerning the latter, that they have been separated to make them more distinct; but that in the natural state they lie close, one upon another. 5. *Thes. Anat.* vii., tab. i., fig. 1; the spleen of a giantess, prepared with great care, and shewing the extremities of the blood-vessels, unfolded throughout, like fine down or cotton; but which, before this unfolding took place, appeared under the form of glands. Bidloo, 1. *Anat. Hum. Corp.,* tab. xxxvi., fig. 3; the spleen, with its arteries and veins and their various windings; also the depression in the capsule. 2. the membrane of the calf's spleen, with its lymphatics[x]. 3. *ibid.,* fig. 5; the inside of the proper membrane. 4. *ibid.,* fig. 6; the spleen, freed from its coats and glands, shewing its blood-vessels, lymphatics, cells, and fibrillæ. 5. *ibid.,* fig. 4; a portion of the spleen excised transversely. Verheyen, *Corp. Hum. Anat.,* tract. ii., cap. xvi., tab. xi., fig. 1, 2.

239. WHATEVER the members of the body desire or demand from the universal mass of the blood, is cheerfully accorded to them, even if it has to come from the extreme boundaries of the kingdom. The *heart* and *lungs*, the members of the thorax, receive the whole of the sanguineous fluid (*a*); in order that they may call together and survey the supplies, and afterwards distribute them through every province of the kingdom, and dispense them among the viscera, according to their offices. The *cerebrum* and *cerebellum*, and the *medulla oblongata* and *spinalis*, are presented with the firstlings of the blood, breathing fresh life, and capable of subserving their elevated functions (*b*). The

(*a*) It is well known, that all the blood of the body flows not only through the heart, but also through the lungs; from the right ventricle of the heart into the lungs, and from the lungs back into the left auricle. These two integral members wholly occupy the superior region of the body, being appointed to that station, in order that they may provide all the other members of the body with blood, and the lungs for the further reason of providing the members with universal motion.

(*b*) The blood carried up into the cerebrum is the most pliant of all, delicate, florid, and ruddy, and more full of life and spirit than the rest of the blood; such, indeed, as the most pure organs of the cerebrum, devoted to the very inmost senses, require. The carotids are the arteries of the cerebrum; the vertebrals, of the cerebellum. The carotids arise as branches from the arch of the aorta, just as it issues from the heart; and after having supplied the higher region of the body, that is to say, the region of the head, they enter the cerebrum. Thus these arteries take up the firstling blood, or what first comes from the heart, and seize the choicest and most suitable portion numberless *glands* planted in the head and thorax, receive the serum intended for secretion, that is, the lixivial and salivary serum (*c*). The *spleen*, the *pancreas*, the *liver*, and the *gall-bladder* have assigned to them the sluggish, livid, and impure blood, freed from the serum, and adapted for inaugurating the new chyle (*d*). The *stomach* and *intestines* have a similar blood, and which is capable of marriage with the chyle (*e*). The *kidneys* receive the useless, effete, concreted, or urinous serum, thrown out by the former organs. The generative members external to the abdomen, as the *testes*, *epididymes*, and *vesiculæ seminales*, receive a blood full of spirit and life. The *uterus*, the *placenta*, and the *fœtus*, receive a similar blood, and which is tempered to every want of the growing embryo. The *mammæ* take a mild and fatty blood, distilled from all the corporeal elements, and from which, consequently, a humor may be prepared, that will readily again relapse into chyle, serum and blood. All the other members receive such blood as serves their purpose and suits their office. Furthermore, whatever a part of any member requires from the mass of its blood, that likewise the member is bound to supply, whether the part be a gland, a vesicle, or any organ ever so minute: this again is bound to supply the requirements of every vessel; this again, of every

of the whole mass; for whatever kind is expended on use, that kind is proportionately supplied.

(*c*) Namely, the parotid, maxillary, sublingual, palatine, and pharyngeal glands, &c.; respecting which see the Chapters on the Tongue, and on the Lips and Palate.

(*d*) The aorta, emerging from the heart, delivers the firstlings of its blood, as we before said, to the head and cerebrum. As it descends, it imparts several little streams to the thorax, by the intercostal branches. It next gives off from its trunk the diaphragmatic arteries, and the cœliac, whose blood it distributes to all the abdominal viscera; but the principal part thereof to the nearest; namely, to the spleen, the pancreas and the liver. That these viscera demand the blood alone, separated from the serum, and that may serve as a menstruum for impregnating the chyle, will be seen presently.

(*e*) That is to say, both mediately through the spleen by the vasa brevia, and also immediately from the same cœliac artery which supplies both the stomach and intestines.

nervous fibre; and this lastly, of every simple fibre (*f*). Hence we see the nature of the economy of this kingdom, namely, that it is replete with order, stupendous, divine.

(*f*) As an entire member demands and receives from the whole mass a quantity and quality of blood exactly proportioned to its expenditure, so does each part from its own viscus, and each individuum or singular, from its part: so that at last the simplest fibre is the prime cause of the whole effect. This fibre, moreover, it is,—solely and singly,—that rules in the whole corporeal system; for besides this fibre, and the spirit of which it is the vehicle, there is nothing that acts and lives in the universal body. And if this be the only fibre that acts and lives, and if it also be determined according to every representation and intuition which constitutes the nature and essence of the soul, it follows, that all the principles of the operations of which we are now treating, proceed absolutely from the very soul; that is to say, from the nature which is her essence. For the state or animus existing in the cerebrum, that is, in its inmost organs which live by the soul, determines a corresponding state in all the fibres, which flow from the cerebrum, as direct radii. Hence it is no member of the body, regarded in its compound, that causes the mass to supply such a quantity and quality of blood, as the offices of the several members require; but this is caused by the simple, purest, innermost, veriest fibre, which is the essential form from which all other forms are derived. Inasmuch as this fibre is a determinate representation of the soul, therefore it follows, that such as is the nature of the soul, such must inevitably be the series of causes to the ultimate effect. This, as above, may be shewn by a single example. The representations of the rational faculty are determined into correspondent actions by means of the will, and this, through the muscular fibres, which although muscular, are composed nevertheless of vessels and nervous fibres. If then the rational mind itself produces such effects, and every motive fibre is most obedient to it, what must not the simple fibre accomplish, according to the whole representation or nature of the superior mind, or soul, of whose actions we are ignorant. The single object of this mind or soul, while it lives in the body, is, to preserve all the powers of the body in their primitive integrity, or to make them subsist as they at first existed; wherefore it applies all its forces, and strains every nerve, to enable its system to enjoy established and lasting soundness, and every individuum or singular, to perform its office, according to the whole order and state of its nature. From this, the first cause of all effects must

240. Each organ derives its power of action from the very nature of its employment or office; for its office determines its construction, situation, connexion, potency and force; in a word, the organic disposition of its parts, and its motion, or its mode, proportion, and correspondency of expansion and contraction (*g*). Hence it follows, that it also determines an *incitation* of each member, part, and particle, to the performance of its offices; and an *invitation* of such blood as suits the office (*h*). Thus

be derived; but of these subjects we shall speak further in our Psychology.

(*g*) It is not enough that the principle or first cause excites the second, and the second, the third, to act exclusively in one particular manner; but it is also necessary that the second and subsequent causes be so disposed, as to comply in all things with the first causes: which cannot possibly be effected without a correspondence of all things. This is the reason of the stupendous organism that exists in the living body; where such, indeed, is the *fabric*, that the last things, by a wonderful subordination, conspire with the first, so that the end of the cause is apparent from the very effect. With respect to the motions of expansion and contraction, there is, in both, such a consent and concurrence of prior and posterior causes, that when the fibres expand by virtue of inmost causes, that is, by virtue of the causes which are constantly impressed upon them by means of the cerebrum, composites also likewise expand by virtue of ultimate causes, communicated by the motive fibres and membranous tissues. Hence it is that the respirations of the lungs are coincident with the animations of the brains; of which coincidence we have treated already in a particular work, and are about to demonstrate it more fully in our Analysis of the Cerebrum. But in the present essays, the nature of structures, and the organic disposition of parts, are our subjects of enquiry; wherefore, at this outset, we must not dwell longer on these particulars.

(*h*) Respecting incitation and invitation, see above, n. 153, 163, 203, and other places. *Invitation* is a kind of attraction, that may be likened to the physical attraction of a syringe, air-pump, or syphon; for while the minutest organs, as the glands and vesicles, expand by means of the fibres, they then invite the blood that is waiting in the nearest and adjoining branches; just as the vesicles of the lungs invite the air. But *incitation* is produced by the motion of contraction, when these same little organs again compress themselves, and express the fluid they have lately invited. That a wonderful correspondence between whatever the members of the body desire or demand from the universal mass of the blood, is cheerfully accorded to them, even if it has to come from the extreme boundaries of the kingdom.

241. The use both determines and unfolds the reason of the structure, but the structure, apart from the use, does not give a reason for itself, except as interpretable by examining numerous effects and corresponding causes in series (*i*). In the spleen the eye sees nothing more than a structure and conflux of fibres and vessels: nay even when excised, its absence is scarcely to be recognized excepting by its vacant space (*k*). Nevertheless, it is

these motions, that is, between expansion and contraction, prevails throughout the system, so that when the glands, by their expansion, invite the blood, then the branch, the trunk, and the body of the viscus, contract and impel the blood, may be seen in the Chapter on the Liver. The same thing obtains also in the pancreas, in the spleen, and in all the other organs which secern liquids. Having premised thus much, we must now devote our attention to the spleen.

(*i*) See n. 32. The very organic form resembles the end inscribed upon it; and thereby produces effects or uses suitable to the end. This organic structure corresponds exactly to the representation, which, as we before said, constitutes the essence or nature of the soul; and which representation may be likened to an idea of thought in our minds. Thought itself is the form of many ideas; and action, which results from the determination of many motive fibres,—all, as it were, so many forces that collectively represent the corresponding organic fabric,—corresponds exactly to the form of thought. In like manner the form of the ideas which constitute the nature of the soul, is exactly represented in the organic fabric of the body, and of all its viscera. Wherefore given the effect, and the end is also given; and in the fabric we comprehend what the end intends, or in what manner the effect acts: as by action, or the form of the muscles acting, we know what the mind is contemplating. But to develop the end from the fabric alone—from such a fabric as that of the viscera—is a task worthy of an Apollo. To what extent we can advance in this direction with respect to the spleen, is now to be shewn.

(*k*) The spleen has no visible excretory duct, such as distinguishes the pancreas, the liver, and other glandular members: and besides the spleen, there are several glands, as the thymus, and certain of the conglomerate class, which do not put forth a duct; and consequently offer no fluid, and present no result, from which the mind can infer certain that it has a use. Nature never produces even the smallest point, without having an end or use in view. But before we interpret the use from the structure, we will give a concise description of the organ.

242. The spleen, lying in a narrow section of the body, and connected by vessels or membranes to the aorta, the pancreas, the omentum, the stomach, the colon, the diaphragm, and the left kidney (*l*), is surrounded by a covering, or rather by two coverings (*m*). Interiorly, it is entirely made up of little compartments and follicles of different shapes, connected together by strong fibres; that is to say, it is divided into innumerable cells, which are continuous productions of the common internal membrane of the viscus (*n*). The venous ramifications appear to

their use. This circumstance, coupled with the fact, that animals and even men after excision of the spleen, have still been found to enjoy good health, and to continue in that state for a long series of years, has given rise to a number of random conjectures respecting the use of the organ, which you may see recounted above by Heister, n. 235, and still more copiously by other authors, and by Malpighi himself at the end of his description of the spleen. (*De Liene,* cap. vi.)

(*l*) Namely, to the *aorta,* by the left branch of the cœliac artery, which furnishes the splenic artery. To the *pancreas,* by the branches of the same artery, as well as of the splenic vein. To the *omentum,* by offsets from the same artery and vein, and also by lymphatics and membranes. To the *stomach,* by the vasa brevia, and occasionally by particular ligaments. To the *colon* sometimes by a similar ligament. To the *diaphragm,* by a membranous ligament. Consequently, not immediately to the *peritonæum,* like the other viscera, but mediately by the coats of the viscera connected to it.

(*m*) In some animals, the spleen is manifestly surrounded by two membranes; but in man it is thought to have only one, on account of the very close connexion between the two. Nevertheless, a duplicature is found at the place where the vessels enter. "The covering [of the spleen]," says Winslow, "adheres to it so closely in the human subject, that it is difficult to distinguish the common from the proper coat. [Nevertheless] the two coats may be distinctly observed at the place where the vessels enter by the longitudinal fissure" (n. 236). According to Malpighi, "a small portion of the spleen in contact with the stomach, is left uncovered" (n. 237).

(*n*) This cellulation is sufficiently demonstrable by means of injecbe surrounded by the same membrane. The cells intercommunicate by common orifices; causing the spleen to represent one continuous cavity, with lesser and least divisions. Nerves and arteries pass into this compages, and veins and lymphatics pass out of it. The common capsule conducts and encloses them as soon as they enter. Numerous *nerves*, both from the common and proper plexus, accompany the vessels, particularly the arteries. The splenic *artery*, arising from the trunk of the aorta by the left branch of the cœliac, runs along the lower border of the pancreas, and then dividing into several branches, sinks into the spleen (*o*). As soon as these branches enter, they split into offsets, capillaries, and fine threads; one part of which ramifies on the cellular parietes, and there terminating, opens into the cavities; another part ends in certain spheroidal granules or vesicles; and another part again is reflected to the surface of the viscus. The *veins*, similarly dividing and ramifying, and perforated all over with little foramina, increase from small to large, and at last all at once rise into a great vein, termed the splenic vein; which running along the lower border of the pancreas, discharges itself into the vena portæ hepatis (*p*). The *lymphatics*,

tions. See Winslow, Malpighi, and the Tabulæ of various authors, of Ruysch particularly. Respecting the ramification and threefold distribution of the arteries, see Malpighi, above, n. 237.

(*o*) Respecting the passage of the splenic artery through the lower border of the pancreas, Winslow says, "Immediately after the origin of the splenic artery from the cœliac, it runs towards the left, under the stomach and pancreas, to the spleen. It adheres closely to the posterior part of the lower side of the pancreas, to which it gives several branches, termed pancreatic arteries." (*Exp. Anat., Tr. des Arteres*, n. 190.)

(*p*) It is of great importance to explore carefully the course of this vein to the liver, inasmuch as it is the principal excretory duct of the spleen, and shews its use. We shall here again adopt Winslow's description. "The splenic vein," says he, "is one of the three great branches of the vena portæ, and is indeed in some sense a subordinate trunk of that vein, It runs transversely from right to left, first under the duodenum, and then along the lower side of the pancreas, near its posterior edge. In this course it gives off several veins, namely, the vena coronaria ventriculi, the venæ pancreaticæ, and the venæ gastricæ, mounting to the surface along the delicate membranes of the cells, of the veins, and of the capsule, run like drains through the omentum and mesocolon to the receptaculum chyli (*q*). The spleens of animals are similar in structure, excepting that the flux of their vessels has a different determination.

243. From this description of the structure of the spleen, it is evident, that the spleen pours out the whole of its arterial blood, into its cellular compartments (*r*); and there works and

or the gastro-epiploica sinistra, and epiploica sinistra. It likewise often gives origin to the hæmorrhoidalis interna, one of the three principal branches of the great vena portæ. It terminates afterwards by a winding course, and divides into several branches that go to the spleen, one of which produces the small veins called by the ancients vasa brevia." (*Exp. Anat.*, *Tr. des Veines*, n. 251—253.) But, begging pardon of so illustrious an author, we may here remark without disparagement, that it would be better to describe the splenic vein, in the true and natural order of its fluxion, from the beginning to the end in the vena portæ, and not in inverse order, from the end to the beginning. For the veins proceed from their branches to their trunks and receptacles, and not from the latter to the branches. I am well aware that this inverse way of describing the veins is usual with authors, yet it is one which greatly misleads the student. It is as though we should describe the receptaculum chyli as dividing into the numerous vessels of the mesentery, and these again into the numberless vessels of the intestines; or rivers as running from the sea; or, by way of simile, the wife as marrying the husband, and the husband as being married to the wife. This order, then, must be inverted, to enable us to understand the true nature of the defluxion of the splenic vein towards the porta hepatis.

(*q*) Respecting the abundance of lymphatics in the spleen, Malpighi (besides what he asserts in the description prefixed to this chapter) also says, "So great is the abundance, and such is the minute division of the lymphatics in the spleen, as to admit of neither description nor representation. Their trunks pass out about the thicker portion of the spleen, where the blood-vessels have their ingress and egress." See the Chapter on the Glands, n. 173.

(*r*) The method adopted in the spleen of purifying the blood, is peculiar; not in use anywhere else in the body, so far as I am aware, excepting in the penis. This peculiarity consists in the circumstance, that the spleen extrudes all the arterial blood out of its vessels or nareduces it (*s*): that the veins absorb a part of this blood by their numerous orifices (*t*); and that the lymphatics claim a part (*u*);

tural channels, into certain little membranous crypts; producing what in other parts is called *extravasation*. For none of the arterial blood escapes from the viscus, (excepting perhaps by the branches reflected to the surface,) but after the before-mentioned extravasation, it is forced into the veins, and thus converted into blood of a different kind, not properly venous. That the blood conveyed by the splenic artery is all eliminated out of its vessels, is a fact so well-established by the common consent of anatomists, that we have no need to attempt to prove it. Besides, it is perfectly conspicuous from the termination of the arteries, which takes place on either the little parietes of the cells, of the sheath, and of the veins themselves, or else in certain bullar forms and spheroidal corpuscules, or else by reflexion to the surface, according to the observations of Malpighi. With respect to the first mode in which the arteries terminate, namely, on the little parietes of the cells, where they open by their extremities into the cellular cavities, Winslow says, "The extremities of the capillary ramifications seem to swim or float in the cells, and to fill their downy tissue with blood" (n. 236). With respect to the second mode, namely, the termination of the arterial threads in certain glanduloid corpuscules, the same author says, "At the ends of several of these capillaries, I have observed small corpuscules, disposed like bunches of grapes" (n. 236). Malpighi took great pains to demonstrate these clusters, and he terms the corpuscules, "vesicles and sacculi," and also likens them to bunches of grapes. He further describes them as of an oval figure, and about the size of the glands of the kidneys, and of a white color; as collapsing when cut; as hanging in the cells, either from productions of the capsule, or from fibres arising therefrom, as bunches of grapes hang from the vine; and he says that the ends of the arteries play around them, like tendrils or ivy. See n. 237. I am perfectly aware that Schelhammer and other writers do not consider these corpuscules as glands, and even do not admit their existence; but on the other hand, Ruysch expressly affirms them, and has delineated them in one of his figures. In the spleen of a giantess, (a portion of which is represented, *ad vivum,* in his *Thes. Anat.* vii., tab. i., fig. 1,) he shews "the extremities of the blood-vessels, unfolded throughout, like fine down or cotton; but which, [as he says,] before the unfolding took place, appeared under the form of glands" (n. 238). In his *Ep. Anat.* iv., tab. iv., fig. 4, he represents them as complete spherules, with the remark, "that in the natural and it also seems probable that the cellular compages itself throws out a part into the omentum. Thus the spleen has all

state they lie close, one upon another" (n. 238). Thus they are acknowledged to be in a manner glandular forms, by this great anatomist; only that being so abundantly furnished with vessels, we are to take care not to call them glands. But to return: that these orbicular, or, according to Malpighi, oval forms, also eructate the blood, or sanguineous juice, into the same cells, derives further proof from an observation of Winslow. "I have seen," says he, "two little tubes going out from each corpuscule, one short and open, the other long and small" (n. 236): and again he says, "The same sort of corpuscules are found in the human spleen, but they are not visible without the microscope" *(ibid., ad fin.)* What may be the destination of the vessels which are reflected to the surface, has not, I believe, been yet determined experimentally. It is, however, plain from all the foregoing considerations, that the arterial blood of the spleen is extruded from its vessels, and into its cellular down, or lanuginous tissue.

(*s*) That the cellular compages of the spleen works and reduces the blood thrown out by its arteries,—this is a necessary consequence of its very texture or fabric. For certain filaments are continued from the proper coat of the spleen to the parietes of these cells; and there are also a kind of sinewy fibres that run transversely from wall to wall; and there are also vessels and fibres, which constantly dipping into the parietes, contract and expand alternately: hence the very structure of each cell, made up, as it is, of similar components, must necessarily contract and expand in like manner. So likewise must the vascular extremities which are conglomerated into spherules, and the sheaths which grow from and are implanted in those cells and tissues. For if we admit the expansion and contraction of the spleen itself, and its vessels and fibres, we can have no good right to deny the same to the cells, which are composed entirely of a production of the exterior coat, and of vessels and fibres. If then there be an alternate expansion and contraction, it follows that there is a working and kneading of the enclosed blood. But with respect to the fibres of the surface, which are supposed to be continued to the parietes of the cells, Malpighi, (who doubts whether they are motive fibres[x],) says, "[The fibres] proceed from the internal membrane, and are prolonged transversely to the opposite portion thereof, or to a kind of capsule or common sheath of the vessels. . . . and are inserted into it by multitudes of branching productions" (n. 237). He also mentions certain transverse [trabecular] fibres, and likens their crossings between the parietes of the cells, to the these excretory ducts; namely, the great splenic vein, and a host of lymphatics.

244. The uses of the spleen as manifested by this structure, are, to draw off the impure blood from the trunk of the aorta; and so to break it up and prepare it, that in the first place, it may serve the liver for its purificatory office, and as a menstruum for refining the chyle: also, that it may serve the stomach as a similar menstruum; and likewise the mesentery. In the second place, that it may serve the omentum, and in the last place, the kidneys: not to mention other numerous subaltern uses subordinated to these primary and secondary uses, and which are commonly of wide diffusion and extensive ramification. But let us treat separately of each of these uses.

girders and beams under arches and between the walls of rooms (*De Liene*, cap. ii.); which fibres likewise prove the actual existence of such motion of the cells. Meanwhile, if we admit the presence of such multitudes of these two kinds of fibres, and that each has a power, at once inherent and constantly communicated, of alternate motion, it must then follow incontestably, apart from all further compilation of authorities, that these cells contract and dilate their spaces, and by reciprocal contorsion work and teaze the enclosed blood to and fro within their parietes, and at last throw it out, by the little orifices, into the veins.

(*t*) Authors treat largely of the perforation of the veins of the spleen; shewing that the veins, particularly the venous trunks, are perforated on all sides with open foramina; in like manner the capsule, according to an observation of Malpighi; and that the blood, expressed or extravasated from the arteries, cannot be driven anywhere but it finds an asylum or place of reception, in some vein allotted to it. This is still more plainly visible in the trunk of the splenic vein of oxen.

(*u*) The roots, or rather orifices of the lymphatics, by which the humor is here taken up and pumped out towards the surface, seem to be more numerous than can ever be detected by the microscope. The escape or passage of this lymph appears to take place by the continuous membranes of the common capsule which surrounds the fasciculi of vessels and nerves; also by the coats of the veins, which are continuous with the common coat of the viscus. In Ruysch's *Ep. Anat.* iv., tab. iv., fig. 1, several lymphatics are exhibited, issuing from the viscus along the coat of the splenic vein. And in Malpighi's description,

245. *That the spleen draws off the impure blood from the trunk of the aorta.* This is proved by the position of the cœliac artery, close under the diaphragm, and between the thorax and abdomen (*x*): also by the nature of the blood which is carried upwards by the carotid arteries to the organs of the head, and to the cerebrum (*y*); also of that which is carried down to the

certain stigmata or dots are mentioned, different from the foramina with which the vein is perforated, and which appear to be little entrances for the lymph separated from the blood, and determined hither. "The venous coat," says Malpighi, "is also perforated with other foramina, for among the open orifices of the branches there is a multiplicity of stigmata or points" (n. 237). And Winslow has observed that two kinds of vessels proceed from the glanduloid corpuscules. "I have seen," he remarks, "two little tubes going out from each corpuscule, one short and open, the other long and small, which latter was lost in the sides of the spleen. I imagine that the long tube may be the origin of a lymphatic vessel" (n. 236).

(*x*) The office of the heart consists in calling together the forces of the blood, and transmitting them to every corner of the body; but not in assigning to the respective viscera the quantity or quality of blood which their offices require. This, as we before observed, is the privilege and duty of the viscera themselves, bestowed and implanted by the assistance and mediation of their fabric, situation, connexion, and modus operandi, and by the nature of their activity. Hence the blood is not intruded on these members by the aorta, but is invited, that is, allured and attracted, by the members themselves. Respecting the forces which invite it, I intend to treat separately in another place. It is not, however, sufficient that the member be furnished with these forces; a particular situation is also necessary as a secondary cause. With respect to the situation of the cœliac artery, (by which the spleen and many organs associated with it in function, derive their blood from the aorta,) it occupies the boundary or crossway mediate between the thorax and abdomen; for it arises immediately after the intercostal and diaphragmatic arteries, which latter is sometimes connected to the cœliac. Thus this artery guards and opens the first door to the viscera of the abdominal region.

(*y*) Every tyro is aware that the carotid is the artery of the head, and ultimately of the cerebrum; and that this artery carries off the purer blood, which is fresh, virgin, and rich in spirit; such as is required by the sensoria of the head,—as the eyes, the ears, the nostrils, kidneys (*z*); and lastly of that which is carried down to the organs of generation (*a*); and to the muscles of the lower re-

and the tongue,—but particularly by the general sensoria; those, we mean, of the cerebrum. And when these organs have subtracted the pure and infantine blood, the consequence is, that the residue, which flows down through the aorta, is comparatively unclean, old and polluted, but nevertheless mingled with the purer blood. So that when not only that supreme region, but also the thoracic region and its arms and levers, with the spinal marrow, which demands a similar blood, have subtracted their due, the rest is comparatively worthless: particularly what goes to the first gate of the abdomen—the cœliac artery—and to the emulgent arteries and the kidneys. This blood, close as it is to the middle septum, may properly be termed, mediastinal blood; and classed among effete, servile, and mediastinal things. Wherefore, in consequence of this loss of the purer blood, it is indispensable to the animal economy, that certain purificatory organs be placed in this situation, in order to carry off the worthless part.

(*z*) In the line of the descending aorta, a little below the cœliac artery, we find the emulgent or renal arteries, which emulge and draw off from the red blood all the feculent portion of the serum. From these arteries it is again evident, that organs purificatory of both the serum and the blood, are applied, and that the superior province is assigned to the purificatories of the blood; in order, namely, that they may draw off the unclean blood, and thus set free the serum, and so prevent a quantity of impure blood also from flowing into the kidneys, and disturbing and confusing their secerning operation. This seems to be the reason why dogs, after excision of the spleen, are constantly desiring to micturate. "Wherefore," says Boerhaave, "does a continual desire to micturate follow excision of the spleen? The situation of the renal artery explains this phenomenon." (*Inst. Med.*, n. 327.) For in this case, the urine is more acrid; that portion of it, namely, which is not expurgated by means of the spleen. But this expurgation is not the work of the spleen alone, but also of its associates in office, that is to say, of the pancreas and the liver.

(*a*) The spermatic arteries also come off not far from this point, in some cases from the renal arteries themselves; and they supply the testicles, epididymes, vesiculæ seminales and prostate gland in males, and the ovaries and the other organs corresponding to those of men, in females, with a blood turgid with life and spirit. Without a previous fining of the blood, and derivation of the more barren and unyielding portion thereof, to the spleen, the pancreas and the liver, the use of gion (*b*). This necessity, indeed, is enjoined by the very end itself; namely, that every member be provided with the blood which it desires, and that the blood, if impure, be prevented from de-

these members would soon cease, and their finest ducts be obliterated: so that the laboratories of nature's ultimate work would perish, her fire be extinguished, and her end defeated. The excision of the spleen does indeed produce the effect of a greater quantity of blood being diverted, and sent to the emulgent and spermatic vessels; and of the urine becoming more acrid, and the venereal excitement, stronger: but whether these consequences would be lasting, I much doubt. "Animals," says Boerhaave, "are more salacious after the removal of the spleen, but how long this continues, is a question.". (*Inst. Med.*, n. 327.) Morgagni observes, "That a dog, after excision of the spleen, was still prolific; but that the cystic bile lost its yellow color, and became of a dirty red." (*Advers. Anat.* iii., Anim. 19.) A plain proof this, that the blood in that case lies and presses in greater quantities at the entrances of the spermatic vessels, and excites the members of this artery to increased salacity, sometimes indeed attended with success or prolification, but variously according to the state of the pancreas and liver, (which varies in each individual,) to which the offices of the spleen are transferred. The same experience distinctly indicates, that the blood, without previous correction by the spleen, proceeds to the laboratory of the liver and gall-bladder, not in a clean, but in a comparatively foul state, and makes the cystic bile duskier than natural. See Chapter IX., where we gave an explication of the office of the Gall-bladder.

(*b*) All muscles subject to the government of the will require more clean and agile blood than the viscera of the body: wherefore a purer blood is sent to the muscles of the thorax and arms, all the way to the fingers; being supplied by the subclavian artery, which, with the carotid and vertebral arteries, takes up the heart's first blood. The abdominal muscles also demand a similar blood; as do all the other muscles of that region, or of the inferior portion of the trunk, down to the soles of the feet. The reason is, because their powers of expansion and contraction depend entirely upon the reciprocal action between the spirit in the nerves and the blood in the vessels: and particularly, because the blood is generated and exists solely as a means to the motion of the muscles and the sense of the organs,—that they may be made use of by man to produce rational actions, for the good of society. This is the end of our corporeal life: this is the end of the blood of filing the works (*c*). In order that the aorta may not pour in impure blood, and in order that the purificatory viscera—the spleen, the pancreas, and the liver—may draw it off, and invite a blood suitable to their offices, the cœliac artery comes off from the trunk, not obliquely, but at right angles; as also do the branches of the splenic artery in the interior of the spleen (*d*).

246. *That the spleen discriminates or separates this blood.*

that life; these, therefore, are the very reasons why the blood is to be prepared from the chyle, and purified; wherefore this is the end also of the viscera of the body. Hence it follows, that the whole of this inferior organico-motory machine would soon stop, if the impurities of the blood in its descent to these motive forces, were not first drawn off by the pancreas, the spleen and the liver, and the residue which flows down to these forces, thus rendered fit and suitable.

(*c*) This is a consequence of what we have just premised.

(*d*) The cœliac artery comes off from the aorta almost at right angles; unlike the other branches which go to the muscles, particularly to the voluntary muscles. "The trunk of the cœliac artery," says Winslow, "comes off from the aorta almost in a straight line." (*Exp. Anat., Tr. des Arteres*, n. 178.) This is also the case with the branches of the cœliac artery inside the spleen, as may be seen in Ruysch's Tabulæ; likewise with the intercostals, the external and internal carotids, the vertebral and many other arteries; (the rest, and those particularly which are determined to the muscles of the will, come off obliquely;) evidently in order to prevent the aorta from injecting the sanguineous stream into the viscera, with the force of its own rapid torrent, and intruding it uninvited; and to allow it to be drawn out and invited by the viscera themselves, that is to say, such a quantity and quality thereof, as the nature of the use on which it is expended requires; so that the viscera may have the power of demanding it, and the heart, only that of supplying it. This accounts for the circumstance, that the cerebrum subtracts the purest portion of the blood; the kidneys, the stale part of the serum; the spleen, the pancreas, the liver and gall-bladder, the impure portion of the blood; and the other viscera, exactly what the nature of their business requires. Nor should it be forgotten, that the splenic artery goes to the spleen in a tortuous and serpentine manner, as the internal carotid artery goes to the cerebrum, and that this breaks the impetus of the torrent of blood, and gives the pancreas and spleen the opportunity of selecting and attracting it.

This is effected first by means of the pancreas (*e*), which clears away the viscid and grosser matters of that blood, and pours them into, and sends them away by, the pancreatic duct (*f*); also by means of the omentum, which swallows the fats of the blood (*g*). The spleen does the rest: it takes up the blood, now separated from the serum, left to itself, alone and companionless (*h*); this it again amends, corrects, extrudes from its vessels

(*e*) The pancreas is the uniting medium between the spleen and the aorta, and between the spleen and the liver. The splenic artery passes under the posterior border of the pancreas, and gives off to it the pancreatic arteries: the splenic vein also takes the same course, and applying itself to the same border of the pancreas, takes up the pancreatic veins. Thus the way of the arteries leads to the spleen, and that of the veins leads from the spleen, through the pancreas. See the Chapter on the Pancreas, n. 234, and the scholia.

(*f*) Inasmuch as the splenic artery first applies itself to the pancreas, and inasmuch as such numerous branches proceed from it, and penetrate to the very glands; and the glands not merely secern from the blood the lixivial pancreatic juice, but also send back many ducts or veins into the great splenic vein, therefore it follows as a necessary consequence, that the pancreas first absterges the splenic blood; for the portion absterged produces the abundant pancreatic juice: whence it appears, that it takes off the salivary and redundant part of the blood, and circulates it, as the liver circulates its bile, and as the numerous salivary glands circulate the saliva.

(*g*) The splenic artery, on its way to the spleen, is connected not only to the pancreas, but also to the omentum; and hence it may be concluded, that it there deposits its fattiness; inasmuch as the branches which it puts forth penetrate even into the omentum. We shall have to shew presently, that there is an intimate connexion between the spleen and the omentum; such a connexion, in truth, that sometimes, for the sake of a kind of common and reciprocal use, certain little succenturiate spleens are formed in the omentum. "I have sometimes," says Winslow, "found small, distinct spleens, more or less round, and connected separately to the omentum, at some distance from the anterior extremity of the spleen" (n. 236). But this is the arterial communication—of the omentum with the spleen; that there is also a reciprocal communication of a closer kind—of the spleen with the omentum, by the veins, will be shewn presently.

(*h*) It is to be observed as a peculiarity in the spleen, that it reinto its cells or loculi, there works it, rolls it about, shakes it up, bursts its bonds, discusses accretions and removes all impediments; so as to leave no clots, pieces, threads, or tangles, but a free and unfettered blood, consisting of pure globules (*i*): the

ceives no serous blood, but only purely globular or red blood; for which end, these two purificatories—the pancreas and omentum—are prefixed to it. Its fabric and modus operandi do not admit of any other fluid. The serum, rough, as it frequently is, with saline spicula, readily becoming viscid, and being of a cold nature, would unduly irritate, and even break up its cellular down. And moreover, the serous humor would be of no use to the liver as a menstruum for impregnating and inaugurating the new chyle; inasmuch as the serum consists for the most part of rejected and antiquated chyle; and is, therefore, neither sanguigenous nor copulative.

(*i*) Respecting this peculiar modus operandi of the spleen, namely, the extrusion of the blood from its vessels, see n. 243 (*r*). "Through its whole extent," says Winslow, "there are numerous venous ramifications: and between them, everywhere, an extravasation of blood, retained in a kind of downy, transparent and very delicate tissue, which extends through the whole spleen" (n. 236). But what the blood is, and the manner in which the spleen acts upon it, must be explained, before we can shew the reason why the spleen eliminates the blood, and what share such elimination has in its purification. With respect to *what the blood is:* it is well known that the blood in the arteries and veins, is a mixture of serum and blood; and that the genuine, ruddy, globular blood, has not always the same properties, or appearance: also that the blood may be either of a yellow, livid, dusky, or pitchy hue, or of a comparatively bright and florid red; consequently either fresh, infantine, juvenile, or decrepit;—either pliant, soft, viscid, or hard; or spurious; or grumous; or concreted and coagulated into gritty, dusty or stringy pieces, or into lamellæ, and all sorts of threads; in which case it is commonly termed bilious, or atrabiliary; producing obstructions, varices, steotomata, aneurisms, and many other effects, which are the symptoms and causes of disease: these particulars are to be unfolded more fully in our Analysis of the Blood. *The manner in which the spleen acts upon the blood,* is shewn by the structure of the spleen; namely, that after throwing the blood out into its cells, all so many little gymnasia and fragibula, it rolls it about, works and reduces it; dashes it from wall to wall; thrashes it, but only gently; throws it out from cell to cell, and in every cell subjects it to a similar exercilegitimate, however, mixed with the spurious, the soft with the

tation; thus separates the globules from each other, combs down and wipes away accretions and projections or protuberances, tears asunder connected portions, shakes off their bonds, and thus breaks up, loosens and sets free all clots, lamellæ and irregular pieces, leaving the globules naked and distinct. *The reason why it eliminates the blood*, and throws it, as a thing of no value, out of its paternal vessels and from its natal soil, is, because no similar place or opportunity for unbinding it occurs in either the arteries, the veins, or the glands; for in the vessels, if it be grumous and conglutinated, it is poured around, and remains the same in the narrow passages, relaxing their coats, and weakening their powers of action: still less is there any such place in the glands, into which such blood is not admitted. From the foregoing considerations it may now be seen, *what share this elimination has in the purification of the blood;* namely, that when the above-mentioned sanguineous clots or pieces are sent out from the full orifices, through the ends of the arteries, into more spacious chambers, by simple exercitation and conglomeration they are easily rendered divisible, and separated from each other; for the particles of the blood are spherical, and hence on any slight motion, the clots burst their shackles, and the globules claim their freedom. That blood of this nature is poured into the spleen, is plain, not only from the structure of the spleen itself, from the office of the pancreas and omentum, and from the function of the liver, but also from an infinity of phenomena and experimental facts. For the records of medicine furnish abundant proofs, of the spleen having been clogged with similar blood: and the same thing is shewn in acute affections of the spleen, which sometimes impede the respiration, when the muscles, called into motion, would force the spleen to submit its blood to exercitation: for the spleen is subject to the power of many viscera, and unlike the other organs, it does not depend immediately upon the peritonæum. Meanwhile, respecting obstructions of the spleen, see the works of anatomical authors, where you will find that cases have occurred in which the spleen was blocked up with obnoxious blood of the kind, not in its cellular down only, but also particularly on its surface, (for many arteries are reflected from the internal compages towards the surface, in order to draw off any pituitary matter which still adheres to the blood,) thus producing conglutination of the vessels, schirrus, callus, and various other diseases. "The membrane [of the spleen]," says Malpighi, "has been frequently found ossified. . . . Not seldom, in sheep particularly, I have observed in it calculous concretions of a substance like gypsum, as well hard, and the young with the old (*k*). The blood thus prepared is a menstruum for refining, impregnating, inaugurating, and copulating the new chyle (*l*). This menstruum the spleen sends

as melicerides and other tumors. . . . The spleen occasionally acquires a preternatural shape and size, from the channels in its substance becoming relaxed by the copious afflux, and deficient efflux, of impure humors; of this the records of medicine contain numberless instances" (n. 237).

(*k*) The fabric of the spleen is sufficient to shew, that it cannot break up the blood-globules, but only separate those which are sticking together: for its cellular spaces are too large to act upon the globules individually; besides which, the cellular parietes are not furnished with motive fibres, but are somewhat soft and yielding. The breaking up therefore must be accomplished in the extreme arterial capillaries, in the glands, and particularly in the pori bilarii of the liver, which are furnished with trituratory apparatus, like little intestines. See n. 210. Otherwise the blood would be of no use to the liver in its two-fold office. That blood-globules of various kinds and ages, are commixed in the spleen, may be concluded also from the circumstance, that none but hard, antiquated and effete blood coagulates into clots and pieces. The softer and more recent blood refuses all such concatenation; and for this reason it is divided in the arteries, possibly by means of collision, and goes away into the lymphatics, in order that the older blood may be given to the liver. But nevertheless it follows, that the blood, whatever be its quality, supposing only that it is divided into parts, and separate from the serum, appears perfectly pure, so far as chemical examination is concerned; for the serum and the concretion of the globules, are what defile the blood. Whether the blood-globules be hard or soft, whether in the flower of youth or the rust of age, they are still voluble or mobile, fluid, and carry nothing but blood; for whatever be the quality of the blood, it still contains the purest of the serum, and the spirit. This is the reason why the blood of the splenic vein appears to be more clear and pure than the rest of the blood. See Heister, n. 235. "This blood," says Boerhaave, "is fluid, free, full of spirits, abounding with lymph, coagulates slowly, is in a state of intimate mixture, does not easily change into heterogeneous matters, is ruddy and purple. Such is the blood sent out from this viscus through the great splenic vein." (*Inst. Med.*, n. 324).

(*l*) In the Chapter on the Liver I pointed out, that that organ has two functions: 1. That it corrects the hard blood, and breaks it up back by the great vein to the pancreas, which pours in a similar menstruum, and propels it to the porta hepatis; a part thereof also it insinuates, for a similar use, by the vasa brevia, into the stomach. In this way the spleen *assists the liver, and assists the stomach* (*m*).

into its parts, for the purpose of renovating it. 2. That it refines the chyle, and inaugurates it into the blood. The spleen subserves both these functions, by supplying the liver, through the splenic vein, with a blood separated into globules: for the liver then transmits the hard, antiquated, and stale blood, through the branches of the vena portæ, into the glands, where, were it grumous, it could not enter. Next in these, and in the pori bilarii, it tears the globules asunder, and where it cannot itself accomplish this, it sends them away into the gall-bladder, to suffer the extremity of torture. Moreover, it mingles the broken up globules, that is, the purely sanguineous elements, with the chyle, which is thus initiated by them into the blood. With respect to the means and appliances for refining the chyle, the only available menstruum for this purpose is the blood which has been deprived of all its serum; for the kindliest and most kindred menstruum is the blood itself, into whose family the chyle is about to be adopted: for according to all the laws and infinite facts of experience and chemistry, the chyle cannot be united to anything nearer to the blood than a product of the blood-stock. It would be easy to adduce numbers of analogous cases from artificial chemistry; but they would only illustrate, not prove. Meanwhile, it is worthy of observation, that as the stomach and intestines require to be furnished with peculiar salivary menstrua for preparing the chyle, and these, gradually thicker and sharper, and at length with the pancreatic juice, and the two kinds of bile; so also the liver must of course be furnished with its peculiar menstruum, for refining the same chyle, and inaugurating it into the blood. The pancreas affords its menstruum to the intestines; both the pancreas and the spleen afford theirs to the liver; so that the more worthless parts of the blood, and the very serum, subserve the office of the intestines; and the blood itself serves the liver as a menstruum. Hence we may see what share these viscera take in the generation of the blood.

(*m*) That not only the splenic artery, but also the splenic vein, gives branches to the stomach, (these branches being termed vasa brevia,) see Heister, Winslow, and particularly Ruysch, in Mangetus, *Theatr. Anat.*, tab. lxi., fig. 7. B, where he gives a delineation of that vein, singularly unfolded. This distribution is evidently for the same reason,

247. The more vital and colorless blood, that is to say, the lymph, the spleen dispenses in a different manner: bringing it to the surface through the pores in the capsule and in the coats of the veins (*n*), and transporting it, either by the lymphatic tissue, or else by visible channels, to the omentum (*o*), and

namely, that the chyle of the stomach, in like manner as in the liver, in its first conception and nativity, may be inaugurated by the pure blood into the blood.

(*n*) Respecting the lymphatics, see n. 243 (*u*). Their common course appears to be, through the pores of the capsule and of the coat of the veins; for besides the reasons adduced in the passage just cited, the coat of both is continuous with the common coat which covers the spleen, and leads to the omentum. This lymph always first tends to the surface, that is, rises into the light of day, before it is taken up by its vessels, and conducted further. That it passes through the capsule, is evident from the connexions and prolongation of the latter. The capsule, according to Malpighi, "arises from the internal and proper membrane of the spleen, which membrane is reflected at the ingress of the vessels, and enters the cavity of the spleen, and forming a tube, accompanies the divisions of the vessels, which are gathered into a bundle inside it" (n. 237). The lymph passes likewise through the coat of the vein. "The vein," says Malpighi, "near its entrance into the spleen, (when on the outside), consists of two coats; but as it enters, the thicker and external of these is taken up by the internal coat of the spleen, which forms the sheath of the vessels, to which it is closely connected" (*ibid*).

(*o*) "Great numbers of lymphatics," says Malpighi, "run under the external membrane [of the spleen]. They are remarkable at once for their large size, and for forming a singular rete with unequal meshes. The humor they contain, pours at last into the receptaculum by considerable ducts which pass through the omentum" (n. 237). In the human subject, however, there is as it were a perpetual lymphatic tissue or duct, not discriminated into distinct vessels, as in animals; which, indeed, is the case, not in one organ only, but everywhere, and particularly in the cerebrum; for the spirit of human blood wherewith this lymph abounds, in kind, nobility, purity, and volatility, immensely excels the lymph of the lower animals: which seems to be the reason why the covering of the spleen, as Winslow says, "adheres to it so closely in the human subject, that it is difficult to distinguish the common from the proper coat; whereas in some ani-thereby to the receptaculum chyli. Thus, in its cells, in its capillary vessels, and even in its glandular clusters, the spleen separates and prepares a menstruum for refining the chyle, not only in the liver but also in the mesentery (*p*). It disposes and shares its gifts between the two, equally and justly, according to their means of requirement, and to the measure of their wants.

mals, as oxen, sheep, &c., we may clearly see two coats, separated by cellular substance. [Nevertheless], even in the human spleen, the two coats may be distinctly observed at the place where the vessels enter by the longitudinal fissure. . . . The arterial and venous branches enter the substance of the spleen together; and the cellular substance of the membranous duplicature of the omentum, accompanies them. . . . The capsule seems to be formed by a continuation of the cellular tissue of the omentum, and by the lamina of the coat of the spleen" (n. 236). Moreover, in the spleens of oxen and other animals, the same mode of secretion, and of purification of the blood, obtains, as in the human spleen, but the mode of discharge is different: that is to say, the blood received, rolled, and reduced therein, is extruded through the continuous cellular partitions, all the way to the surface, and there first taken up by the trunk of the great vein, and by the lymphatics. Whereas in the human spleen it is taken up by infinite radicles of veins and lymphatics in the very compages and substance of the organ; and this, on account of the excellency of this lymph in man, as we said before.

(*p*) The economic dispensation of blood in the spleen demonstrates, that the older and harder blood, which has circulated many times, and is defunct, is presented to the liver; for such blood is suitable for both its functions; that is to say, for both the lustration and castigation of the blood, and the inauguration of the chyle into the blood. Thus the blood which has either served already, or which cannot serve, has the office and province of introducing the new chyle into the bed-chamber of the veins, and thus of ministering in a hymeneal condition. But the fresher, younger, and softer blood, with its purer lymph, is derived into the glands of the mesentery, partly that it may be taken up by them and return immediately into the blood, partly also that it may serve as a menstruum to the chyle which is about to pass through the lacteals and the mesentery; for it is neither pure lymph nor pure blood. "The humor which they contain," says Malpighi, speaking of the lymphatics, "is sometimes of a yellow, and sometimes of a reddish color" (n. 237).

248. The blood which is thus separated and prepared, by means of its structure, situation, connexion, modus operandi, and peculiar activity (*q*), the pancreas invites from the common stream; the spleen from the pancreas; the liver from both; the vena cava from the liver; the right side of the heart from the vena cava; the lungs from the right side of the heart; the left side of the heart from the lungs; and the whole animated system from the left side of the heart. So likewise by way of the lacteals: the tongue, palate, fauces, and œsophagus demand, absorb, and invite the food; the stomach invites it from them; the intestines from the stomach; the mesentery invites the purest food, that is, the chyle, from the intestines, and the lymph from the other members; the thoracic duct, from the mesentery; the vena cava superior from the thoracic duct; and from this again, the right side of the heart, the lungs, and the left side of the heart, and from the latter the whole animated system.

(*q*) We have already treated of the structure of the spleen, and of its situation, connexion, and modus operandi, but not yet of the nature of its activity, which seems to be very similar to the activity of the pancreas and of the liver; namely, that like those organs, the spleen contracts, and constringes its artery and vein, and at the same instant, the cells and glandular acini or vesicles open, and invite the advancing blood to come into their recesses; during which time, the foramina leading into the veins appear to be closed: but again, when the spleen and vessels expand, that the cells and glandular clusters contract. That there is such a reciprocation and alternation of motions of expansion and contraction, is proved by the connexion of the membrane at the surface with the blood-vessels; and by the continuation and dissemination of that membrane, and of the vessels and nerves, throughout the interior compages of the spleen: for the proper or internal common coat of this viscus penetrates all the way to the so-called glands, and as in the pancreas and liver, invests them; the capsules being the offspring and production of the ends of both the arteries and the fibres. These glands, according to Malpighi, "hang from productions of the capsule, or from fibres arising therefrom; consequently from the extremities of the arteries, and from the ends of the nerves" (n. 237). The consequence is, that here also, as in the liver, all things are carried on tranquilly, so that no movement of any part is apparent; and the incitation is exactly correspondent to the invitation.

But the invitation is met by a corresponding incitation (*r*); hence there is an equation and perpetual circle of all things, and each has its task, power, place, and life allotted to it by the community.

249. There are then these three viscera, the liver, the pancreas, and the spleen, whose especial province it is to purify the blood; but the main duties devolve upon the liver. From their association in function, their succession of operations, their connexion with each other and with the aorta, and from the nature of each, and the power it possesses of demanding its tribute at will from the system, we may now see clearly, that when the spleen is excised, destroyed, or diseased, the pancreas undertakes and sustains a part of its office; the liver a part; the glands a part; and several members of both the abdomen and the thorax, another part (*s*). Thus the office of one member is

(*r*) Respecting invitation and incitation, see n. 153, 163, 203, 240 (*h*). To which we may add, that by these means, not only a particular quantity, but also a particular quality of blood, is demanded by every viscus. The *quantity* is various—great or small—according to the degree of activity; but the *quality* varies according to the state of the fibres, particularly of the simple fibres, which give the veins the power of exercising a choice, and of refusing and rejecting, or selecting and greedily imbibing, what comes near them. Thus whatever the extremities desire, demand, or consume, is supplied from the whole mass of the blood. But such invitation is not enough;—there must be also a corresponding incitation, or an activity to advance and accomplish this effect.

(*s*) Although the office of the spleen has been so described, that it may seem as if the liver acted entirely under its direction, and were subject to its rule, yet experience shews, that the purification of the blood may be performed without the spleen, inasmuch as it is not suspended after that organ is excised. What really happens in this case, we are principally instructed by the connexion of the vessels and fibres; namely, that the pancreas first takes the place of the spleen, next the liver itself, and afterwards several of both the abdominal and thoracic viscera; so that the office of one organ is distributed among many; although never with complete success. With respect to the *pancreas*, it first takes up the blood which should be purified by the spleen, and cleanses it of its heterogeneous parts, and of its serum; in order that indeed distributed and extended; yet never without danger and loss, and disturbance of that state and order, which the nature

the blood sent out from its laboratory may be able still to serve as a menstruum for the liver, and meet the urgent necessities of the case. The pancreas also, like the spleen, remits its purified blood into the splenic vein, and transmits the vein onwards from the spleen to the liver: so that the arteries and veins still proceed to the liver by the same way, although the office of the spleen has ceased; and there is no doubt that the pancreas, in this case, strains every nerve, and works with increased energy and activity. The *liver* likewise, (whose vena portæ is formed not only of the splenic but of many other veins,) at the same time attracts its blood immediately from the aorta by the right branch of the cœliac artery; and that this blood also serves as a menstruum for purifying the chyle, was shewn in the Chapter on the Liver. But all the blood that is coagulated into clots and pieces, and which the liver cannot transmit through the branches of the vena portæ to the glands, it either does not invite or attract, or if attracted, it sends off into the gall-bladder. That notwithstanding, the liver is distressed, its arteries and veins turgid, its passages opened, its narrow ways widened, and the whole organ laboring with some new and unwonted effort, is evident from its intumescence, or swollen state. "In a dog," says Malpighi, "from which the spleen had been excised, I noticed one external peculiarity, namely, a swelling in the right hypochondrium, which caused the last ribs to protrude beyond the others" (n. 237). And Boerhaave asks, "Why is the removal of the spleen followed by a tumor of the right hypochondrium, and an enlargement of the liver?" (*Inst. Med.*, n. 327.) Moreover, a quantity of grumous and uneven blood is rejected both by the liver, and immediately by the aorta, through the gemellæ cysticæ, into the gall-bladder; and hence the dark color of the bile, after excision of the spleen, as observed by Morgagni. But when this tetrarchate of viscera which purify the blood, has been destroyed, nature, fearful of the destruction of her commonwealth, does not bestow the power of this office upon this pair of viscera only, but also distributes and transfers it to several others; indeed, to all which either are furnished, or can be furnished, with similar laboratories; particularly to succenturiate or additional pancreases, and to many glands in various situations; nay more, she sometimes builds and inserts new spleens, and perhaps several of them, in the omentum; respecting which, see Winslow, n. 236; or even supplies a new succenturiate pancreas in addition to the large pancreas; respecting which, see again the same author, n. 221, in the Chapter on the of the supreme mind that governs the body, has conferred and established, as being in the highest degree perfect, and harmonious with itself (*t*).

Pancreas. Not to mention, that in process of time, she increases the powers of the rest of the viscera, as of the stomach, the kidneys, and the lungs, (which also purify the blood in a peculiar manner,) and perhaps even excites the genital members, (and the penis itself, which in such case is more lascivious, and which expels the blood from the vessels in the same manner as the spleen,) and gives these parts a not dissimilar power of purification: so that the office of the spleen is distributed. But who can enumerate nature's ways and contrivances, or describe her stupendous resources, for restoring order, and repairing ruin and mischief?

(*t*) I will not dwell on the inconveniences and ailments which the animal suffers during the first days after the extirpation of the spleen, (and which consist, according to Boerhaave, in borborygmus, vomiting, and nausea), but only on those which are permanent, and do not cease but with life; and which are, indigestion, and crudity of the chyle, as if it rushed into the veins in undue quantity; causing voracity and constant hunger. "A dog," says Malpighi, "from which the spleen had been excised, became voracious, taking its food very greedily" (n. 237). Respecting the constant desire of micturition, see above, n. 245 (*z*) and (*a*). Besides which, the blood, growing thicker and thicker, infarcted with heterogeneous matter, and concreted, must be continually on the increase, until at last the quantity is overwhelming; and until, by obstructing both the capillaries and the larger vessels, it causes anxiety at the heart and in the chest, followed by sadness, melancholy, hypochondriasis, and a change from a happy and cheerful disposition, to grief, lamentation, or corroding sorrow. This is the reason why the ancients considered the spleen as the cause of laughter. See Heister above, n. 235. But to pursue all these subjects would exceed our space; we must now, therefore, pass on to the Omentum.

CHAPTER XII.

THE OMENTUM.

250. Heister. "The omentum is a membranous part, usually furnished with a large quantity of fat, and lies under the peritonæum, immediately over the intestines. It is called by some authors rete and reticulum, from the number of little foramina that appear in it when it is raised; but these foramina are not natural. The Greeks called it epiploon. It usually occupies only the upper part of the abdomen; but sometimes it extends to the lower part also. Inferiorly it is floating and free; superiorly it is connected in front with the fundus of the stomach, and with the duodenum and the spleen; behind with the colon and the pancreas. It is composed of a very fine double membrane, forming a kind of hunter's bag, and furnished with a bursal cavity; of fat contained in fatty cells, which form a kind of ducts and surround the blood-vessels, and have between them areolæ or membranous spaces; also of arteries, the arteriæ epiploicæ, which come from the cœliac, and are very numerous; of veins, from the vena portæ, and principally from the splenic branch; of nerves, from the intercostal and par vagum; and of lymphatics. It has a remarkable natural foramen, discovered by Winslow: this foramen is situated under the great lobe of the liver, near the neck of the gall-bladder; and the omentum may be beautifully inflated through it. *(Comp. Anat.* n. 208.*)* It is to be sought in this situation between a certain membranous ligament, which joins the beginning of the duodenum and the neck of the gall-bladder to the liver, (on the side of the eminence, which is the root of the lobulus Spigelii,) and another ligament which connects the colon to the pancreas. These two ligaments, where they unite, leave between them an interstice or aperture, as Winslow says, of four or five lines in diameter, in a child of four or five years old, through which aperture, the eminence just

mentioned, passes. If a large tube be let in at the aperture and the rest of it closed by the fingers, the whole cavity of the omentum may be inflated; and in this state it resembles a kind of sack or bag, or as the French term it, a bursa, but of an irregular figure. The same author observes, however, that this cavity is not formed solely by the omentum; but also by the upper surface of the mesocolon, and by part of the colon and the stomach; and finally, by the membrane which occupies the space between the two orifices of the stomach, and which he calls the lesser omentum. The use of the foramen he supposes to be, that if any humors chance to be collected in the cavity of the omentum, they may have a way out by this means, especially when the person lies on his back. *(Comp. Anat.*, not. 7.*)* The adipose ducts of Malpighi are nothing more than the lateral ramifications of the blood-vessels, which deposit the fat in the cells surrounding the blood-vessels. These cells, as in the rest of the fat, communicate with one another, and with the veins. *(Comp. Anat.*, n. 208.*)* Malpighi himself, in his posthumous works, acknowledges that he doubts whether these adipose ducts are distinct vessels or not; and Morgagni has justly hinted, that they are not necessary as such; since it appears that the secretion of fat may be carried on here, as well as in other parts, from the arteries, into the adipose cells; and that the fat may afterwards be absorbed again from the latter by the veins, without the intervention of a third sort of vessels, such as Malpighi supposed to exist. Rivinus also denies the existence of these ducts. *(Comp. Anat.*, not. 8*)*.

251. "In considering the fat or membrana adiposa, we are to observe its situation, under the cutis almost all over the body; and besides this, in the interstices between the muscles, in the orbit of the eye, in the omentum and mesentery, about the kidneys, and in various other places. It is composed of a thin transparent membrane, furnished with innumerable cells or loculi, that communicate with each other; and of the fatty, oily, or butyraceous matter of the blood, filling the cells; and which is secreted, not from glands, but from the little arteries of the membrane. In bodies greatly emaciated, we find only this cellular membrane, with the cells empty. Authors make a distinction between adeps or sevum, and pinguedo; according to this distinction there is no such thing as true adeps in the human body. The blood-vessels of the membrana adiposa are numerous; they come from the neighboring parts, and enter and as it were surround the membranous cells. The nerves are few, and hence the membrane has but little sensibility. Authors mention certain peculiar vessels, which they call vasa adiposa; but these do not appear to be distinct from the lateral branches of the blood-vessels, and from the cells above mentioned. That there is a

circulation, motion, or return of the fat to the veins, is very evident from the consumption of it in certain diseases, and in animals when violently exercised; in both which cases the disappearance of it is often extremely sudden. The uses of the fat are, to serve as a kind of covering to the body, defending it from cold, and from the acrimony of saline particles; to preserve the flexibility of the skin, of the muscles, and of the other parts between which it is placed; to facilitate the motions of certain parts, as the eyes, the maxilla; to fill up empty interstitial spaces, and by that means to add greatly to the beauty of the parts, as in the face, the neck, the eyes; also to facilitate the distension or dilatation of the parts, the spaces between which it fills up, as about the anus, the vagina, and ossa ischii, which give way in the passage of the fæces, and in the birth of the fœtus; to prevent painful attrition, particularly in the soles of the feet, in the nates, and in the pubic region, in which it forms a kind of pillow or cushion; perhaps also, in time of need, to serve in some way as nourishment for the body, by reëntering the veins." *(Comp. Anat.*, n. 201.*)*

252. Winslow. "The omentum is a large, thin and fine membranous bag, surrounded on all sides by numerous portions of fat, which accompany and even invest the same number of arteries and veins, adhering closely to each other. It extends more or less over the small intestines, from the stomach to the umbilical region. The superior portion is in a manner divided into two borders; one of which is fixed along the great curvature or convex side of the arch of the colon, and the other along the great curvature of the stomach. The commissure or union of these two borders on the right side, is fixed to the common ligament of the duodenum and colon, and to the contiguous parts of these two intestines. That on the left side is fixed to the longitudinal fissure of the spleen, to the extremity of the pancreas, and to the convex side of the great extremity of the stomach. It is likewise attached to the membranous ligament which sustains the ductus cholidochus, and connects it to the trunk of the vena portæ ventralis. With the exception of these adhesions, the omentum lies loose between the anterior wall of the abdomen, and the intestines. The anterior and posterior portions are generally called the laminæ of the omentum, but it would be more proper to call them folia, alæ, or some such name. The membrane of the omentum is, through its whole extent, made up of two extremely thin laminæ, joined together by a cellular substance; the quantity of which is very considerable, accompanying the blood-vessels in broad bands, proportioned to their ramifications. These cellular bands are more or less filled with fat. Besides this membranous bag, which I name, the great omentum, there is another much

smaller, which differs from the large, not only in size, but also in figure, situation and connexion: this I name, the small omentum. This small bag is fixed by its whole circumference, partly to the small curvature of the stomach, and partly to the concavity of the liver, in front of the sinus of the vena portæ, so as to surround and contain the prominent portion of the lobulus Spigelii. The small omentum is thinner and more transparent than the other. Its cavity diminishes gradually from the circumference to the bottom, which in some subjects terminates in several small cavities or fossulæ, more or less pointed. Its structure is pretty much the same as that of the great omentum; it being composed of two laminæ; with a mixture of the same cellular and fatty bands, which are considerably finer than in the other. We see from this description of the two omenta, that in the space left between the lower side of the stomach and upper side of the mesocolon, they have a very broad communication with each other; so that if either of them contained in its cavity any fluid, it might readily pass between the stomach and mesocolon, and so into the other bag, especially when the stomach is empty. Therefore by means of this interstice between the stomach and mesocolon, the two omenta form one cavity, which opens into the cavity of the abdomen by one common orifice, situated near the commissure on the right side of the great omentum. This orifice is semilunar or semicircular, and formed by the union of two membranous ligaments, whereof one connects the beginning of the duodenum and the neck of the gall-bladder to the liver; the other connects the contiguous portion of the colon to the same viscus, and extends to the pancreas. This produces an incurvated border, which surrounds the root of the lobulus, leaving an opening wide enough to admit the end of the finger. To discover this orifice, we need only raise a little the great lobe of the liver, and find out the root of the lobulus; and when we have found it, insert into it a tube, wrapped round with cotton, wool or tow; then, if we blow gradually, the air will inflate the sides of the great omentum, and give it the appearance of a large bladder, irregularly divided into several lobulations by bands of fat, which appear in this state like so many fræna between the lobulations. This experiment succeeds better in young, lean subjects, than in old or fat subjects. When we handle these membranes with dry fingers, the membranes stick to them so closely as hardly to be separated without being torn; as we see by the reticular holes that appear in those portions of the membranes that have been thus handled. In that case it is to no purpose to blow through the orifice already mentioned; and it is owing to these small accidental holes that the membranes of the omentum have been supposed to be naturally reticular. The membranous laminæ of the small omentum are continuous partly with the external membrane of the liver, partly with that of the stomach, and partly with the membrane that lines the neighboring portion of the diaphragm. Those of the great omentum are continuous partly with the same coat of the stomach, and partly with the external covering of the colon, and consequently, with the mesocolon: and they likewise communicate with the covering of the spleen. We may satisfy ourselves concerning these continuations, by making an orifice in one of the laminæ of the omentum, near the stomach or colon, and by blowing into the orifice, through a tube well fitted to it, when we shall find the air insinuate itself gradually under the common coats of the stomach and colon. The fatty appendices of the colon and rectum have always appeared to me to be small or supplementary omenta or epiploa. They are situated at different distances along these intestines, being particular elongations of their common or external coat. They are of the same structure as the large omentum, and there is a cellular tissue contained in their duplicature, more or less filled with fat. Near the intestines, each of them forms a broad, thin basis, and they terminate by very irregular mammillary processes, thicker than their bases. These bases are at first disposed longitudinally, then obliquely, and lastly, more or less transversely, especially near the rectum. The appendices are for the most part separated from each other, but some of them communicate slightly. When one of them is inflated, it forms a small irregular bladder, and the air penetrates under the external coat of the colon or rectum. Besides these appendices epiploicæ, there are several adipose strata, near the mesocolon, which may likewise be looked on in the same light as the former. The arteries and veins of the great omentum are branches of the gastric arteries and veins, and for that reason, go by the name of gastro-epiploicæ, dextræ and sinistræ. The arteries on the right side answer to the hepatic artery, and those on the left side, to the splenic: and both communicate with the arteria coronaria ventriculi, and respectively, with the mesenteric arteries. The gastro-epiploic veins on both sides, answer in the same manner to the distribution of those of the vena portæ. The vessels of the small omentum come chiefly from the coronariæ ventriculi; and those of the appendices and fatty strata, are ramifications of the arterial and venous network of the colon and rectum." *(Exp. Anat., Tr. du Bas-Vent.*, n. 351—370.)

253. Malpighi. "The omentum is composed of a thin extended membrane, excavated and produced into sacculi and striæ, or rather into a capacious bag; and this membrane follows and embraces the branches of the arteries and veins, so that where at last they terminate

reticularly at the sides, they are still accompanied and invested by it uninterruptedly. These sacculi also contain fat; and each sacculus holds together and invests minute lobuli of it, of different shapes, supplied by vessels of their own, and covered with their own membrane; as we have observed with respect to the ramifications of the lungs, in that they also are formed of lobuli adapted to each other. . . . In sheep, oxen, and other animals, the omentum consists undoubtedly of a double membrane, in the middle of which the vessels proceed and ramify, with an abundance of fat lying at their sides, pretty much in the same manner and of the same figure as we see in the human omentum. The membranes are extended continuously in the intermediate spaces, and in some portions they are so far obliterated, as to produce the appearance of a rete, or fibrous brush of vessels. In the centre of striæ of this kind are contained the arteries and veins which pass from the fundus of the stomach, and particularly from the spleen, to the liver. These vessels ramify all over the omentum, and their branches uniting reciprocally, enclose conspicuous spaces, almost like those formed in the leaves of trees by the fibres of the leaf: the multiplicity of these spaces produces the appearance of a network. Other offsets or twigs likewise come from the veins and arteries in the very middle of the striæ, and proceed to the lobuli of fat, of which the whole sacculus or stria seems to be composed. In the areolæ or meshes surrounded by the striæ, where we see the reticular distribution of the blood-vessels, an immense number of bodies are observable, which running out in the manner of a rete, with an uninterrupted and sometimes tortuous course, proceed from stria to stria, and occasionally passing over the nearest striæ, go to others more remote. Sometimes these bodies, (as I lately observed by the microscope in an owl,) are not only filled with fat, and evidently distinct from the network under them, but where they terminate, one would say they discharged fatty globules. In portions of the omentum of dogs, where there was still some trace of membrane remaining, I have often seen a number of these ducts, running over the network; and where the omentum is completely reticular, without any membrane covering it, (if it be examined carefully while the viscera are warm,) these adipose bodies are observed, running transversely over the striæ and adipose sacculi, and at length terminating, by a singular interlacement, in the vascular network. . . . Whether these adipose bodies be hollow, like veins and arteries, so that the fluid fat passes as it were through tubuli; or whether they are as it were filaments through which the adipose particles run, it is hard to say; for they are so small and transparent, that we cannot determine their exact minute structure by sight. To me it seems most probable that they are vessels, because in animals just killed they appear uniformly turgid, and look like little intestines full of globules. . . . In some fishes there is at the fundus of the stomach a remarkable membrane, which contains vessels different from the blood-vessels. . . . The fat may be observed in all parts, particularly about the skin and in the interstices of the muscles, where we find a number of membranes resembling sacculi, and excavated lobuli, which membranes are adherent to some other membrane lying under them, and which is thicker, and serves them as a support and basis. Through this proceed the veins and arteries, at the sides of the trunk of which is accumulated a mass of fat, resembling as it were a heap of millet-seed. The blood-vessels expand like trees into branches, and to the extremities of the branches are appended the membranous sacculi or lobuli, full of fatty globules, which growing on the branches like leaves, complete the resemblance to a tree. These membranous sacculi are of different shapes; some of them are slightly depressed, and almost oval; and others are like the lobuli of which the lungs, or rather the conglomerate glands, are composed. They are connected together by membranes, which give them consistence and form, and by a network of vessels. The larger portions of fat are made up of several lobuli placed one upon another. When the globules of fat are pulled out of the containing membrane, they leave a cavity or membranous covering; so that if the interior of the pellicle be examined by a microscope, it has the appearance of a honeycomb made up of cells. But whether all the adipose globules be invested again with a pellicle of their own, or not, the eye is unable to decide; this much, however, is visible, that the globules are connected to the sides of the vessels, as the stones of the grape are appended to the stalks. Through these membranes run very minute vessels, forming a net-like expansion, and which represent a little omentum: these penetrate the lobuli intimately, and are connected to the fatty globules: but sometimes they are covered with a fine membrane. Through this same membrane divide and ramify adipose vessels like those found in the omentum." *(De Omento, Pinguedine, et Adiposis Ductibus.)*

254. Swammerdam. "In the cossus or vermis scarabæi, the fat consists of very small, and as it were sandy globules; which, when viewed with a microscope, seem to be again composed of innumerable and yet smaller particles; all of which are supported by very thin and transparent membranes or coats. The pulmonary pipes run up and down through the fat. Its figure is not regular, but various; generally, however, spherical. In the silkworm, it is very irregular. When the fat of the cossus is further examined by the microscope, it appears of unspotted whiteness, and covered with little transparent bubbles,

almost like hydatids. The fat of larger animals is of the same nature." *(Bib. Nat.*, p. 311, 312, tab. xxvii., fig. 9, 10.)

255. See Glisson, *Tract. de Ventriculo,—Omenti Hist.*, p. 111, (Amstel., 1677); where he asserts, that the membranes of the omentum are so very thin and pellucid, and so full of foramina, that in many animals a candle may be blown out through them, even when doubled. Ruysch, *Thes. Anat.* ii., tab. v., fig. 1; where he exhibits the omentum, "furnished with myriads of blood-vessels, but with no foramina." Boerhaave, on the action of the omentum, *Inst. Med.*, n. 329—338. Eustachius, tab. ii., fig. 1; the omentum with numerous vessels; and fig. 3, a portion of it, hanging from the fundus of the stomach, with exceedingly minute adipose glands. Winslow, *Hist. de l'Acad. Royale des Sciences*, an. 1715, (*Mem.*, p. 234, 235); where he treats of the foramen of which we have cited a description from Heister. Verheyen, *Corp. Hum. Anat.*, tract. ii., cap. ix. T. Bartholin, *Anatomia Reformata,—de Omento*, p. 45, (Lugd. Bat., 1651); where he states that Galen removed the omentum from a gladiator, who afterwards easily took cold, being obliged to keep his belly covered with wool; and that catarrh, alvine flux, lientery and phthisis arise from defect of the omentum. See also Schenckius, *Exercit. Anat.,—De Pinguedine:* cap. xxix., p. 141-152, (Jenæ, 1662). Fabricius ab Aquapendente, Vesalius, &c.

256. It is impossible to know either the nature, the origin, or the use of the fat, without we know what the blood carries in its globule, and what it carries in the serum, extraneous to the globule; for the fat, in the omentum, and in the other coverings and diverticula where it is condensed, derives its origin, its nature, and its use, from the blood (*a*). *What the blood carries in its globule,* is determinable by examining variously its generation, offices, condition and structure; and the result is, that the blood is compounded of two natures; one, spiritual, whereby it acts in the fibres, and serves, and communicates with, the soul; the other, material, whereby it subsists in the vessels, and inclines and belongs to the body. The blood derives the former nature from the spirit of the cerebrum; the latter, from the chyle of the body: thus the chyle gives the blood its essential embodiment (*b*). Hence we may see *what the blood carries in its*

(*a*) It is notorious how many opinions there are respecting the origin and nature of the fat, and respecting the use of the omentum. The bare enumeration of them occupies many pages in the works of different authors. But what others think is not our concern, but only, what inference is deducible from experience, or what conclusion the series of experimental facts demands.

(*b*) This embodiment or body comprises everything which derives its existence from the elements of the world, or from the elemental world. That elements of this kind reside in the blood, and form the blood-globule, will be explained at large in Part III., where we treat of the Heart and the Blood. For the sake of forming the body of the blood, all the chyle is prepared, that is to say, all the inmost essences to which the food can possibly be laid open by the elaborate mechanism of so

BB 2

serum, namely, a material that is to be united to the spirit, in order that these two natures associating their forces in the unanimous body, may perform aright whatever is requisite to be done. From these considerations it is evident, that in proportion as the serum is pure and simple, it is near and akin to the blood; and that in proportion as it is impure and compound, it is disconnected and remote from the blood. In the former case it is adopted, as homogeneous, kindred, and consanguineous; but in the latter case it is rejected, as heterogeneous, foreign, and disinherited (*c*).

257. To explain the nature of the spirit of the blood, belongs to the province of psychology; to explain the nature of the body thereof, to angiology. In this place we cannot speak in detail of either; nor can we mutilate the subject. We can only recapitulate (*d*), that the blood-globule contains, besides the spirit which inhabits it, the first elements, simples, and unities of all kinds of compounds; and that these are kept so arranged in it, that it admits of being readily resolved into each: and such being the hidden contents of the globules, there is, therefore, no possible fluid, and no possible formation from fluids, either hard

many organs and viscera, are allured and extracted from the ingesta. But of these subjects we have treated already.

(*c*) The serum appears to admit of being divided into three general classes. *Firstly,* into the purest serum, which is proximate to the blood, or which consists of the same elements as the blood itself, with respect to its embodiment, and which, as soon as the spirit or prior essence is present, passes and changes into blood. *Secondly,* the serum which is not yet reduced to such simplicity, but which nevertheless comprises similar elements, and admits of being readily resolved into them: this serum also is closely related and allied to the blood. *Thirdly,* the serum which cannot be divided into similar elements, being so concreted and clogged, as to be perfectly unsuitable: this is said above to be disconnected and remote from the blood; and it is actually repelled, and thrown off by the various outlets. But these three species require to be further subdivided into several other inferior species.

(*d*) See what was said above on these subjects, in the Chapter on the Glands, n. 179, from which the particulars immediately following are taken.

or soft, whereof the blood cannot furnish the principles from its own bosom, and which it cannot compound therefrom by means of the organs. Whence it follows, that the blood has in it nothing of those things which are expressed by the received chemical terms,—nothing alcaline, or acid, or lixivial, or urinous, or sulphureous, or oleaginous, or consequently, fatty,—in short, nothing that is expressed by any formula proceeding from, or reducible to, our sensual conceptions (*e*); yet still that the blood has in it infinite things; in short, that like a seed, it involves, in virtue and potency, all those things that can possibly be compounded of simples.

258. Every time the blood is resolved into its primitives, simples, and generative elements (*f*), a wise caution is exercised, to prevent any portion of these its two essences and natures, from being lost, or escaping from the microcosm. The soul, which is a kind of tutelar divinity to the body, gathers up the spirit at such times with especial care, and by the lymphatics, arteries and veins, recalls it either to the cerebrum, or into the blood. In like manner its embodiment or bodily part (*g*). But should the supply of the latter be over-abundant, it deposits the exuberant portion in the cells of the omentum, or in the analogous loculi or compartments of some of the other adipose tissues (*h*), and there puts it by for use, somewhat as the bee lays

(*e*) No one, I suppose, can be ignorant, that names have been given to those things only, which have reached the intellect by the path of sensation, or sensual intuition; but that of things which are above or beyond that path, we have procured no distinct notion, and no idea; wherefore they are not reducible to terms. This is the reason why those things which exist in the blood, have betaken themselves so completely into the interiors of the sphere of the senses, that they can be expressed by only the most general formulæ, that is to say, in the most obscure manner; indeed, by general symbols only,—by simples, unities, elements, principles, and the like; albeit, in even these there are various distinctions, and various subordinations, degrees, harmonies, &c.

(*f*) See n. 180, and the Analysis of the Blood.

(*g*) That is to say, the elements and simples, of which the body of the blood-globule is composed.

(*h*) Respecting which, see Heister, n. 251. These loculi for colup its honey in cells. That this portion constitutes the milky and virgin fat, such particularly as exists in the epiploic repositories, is clearly proved by the conception, birth, clothing, and nursing of the fat; by its sympathy and correspondency with the blood; by its increase and decrease; by the fineness of its vessels, and of the coat which encloses it; and to sum up all, by the circle of uses (*i*).

lecting the fat are found nearly all over the body; to wit, in the cellular membranes.

(*i*) That the fat consists of disintegrated blood, or of the elements which constitute the body of its globules, or of the serum which most nearly approximates in nature to such blood, (respecting which serum, see one of the preceding paragraphs,)—the lymph, however, that renders it fluid being driven off,—is evident from the *conception* of the fat in the minutest arteries. For before these interior elements of the blood take refuge in the loculi of the omentum, the blood is purged of all its grosser serum; namely, both in the pancreas, which derives the impure and lixivial portions of the blood into the pancreatic salivary juice, and in the liver, which sends away similar portions, in union with the chyle, into the hepatic bile; so that nothing but the pure blood and the corresponding serum remain, which latter is conveyed into the cells by the epiploic arteries. The arteries, throughout the whole of their course, reject first the heterogeneous and grosser matters; then, as their branches become narrower, they reject those which come next in grossness or in purity. There now remains blood, either pure, or else surrounded by kindred and as it were consanguineous serum. Lastly, in the minutest capillaries, even the blood-globule is broken up, and its members or parts are sent away, either immediately into some vein, or into some lymphatic, or into some other reservoir. See my *Economy of the Animal Kingdom*, and Part III. of the present work, where I intend to treat of the Blood. The same conclusion is also deducible from the coming forth or *birth* of the fat; for the fat hangs from the ultimate apertures of the capillary arteries, in the form of a dewy or vapory exudation. "The blood-vessels," says Malpighi, "expand like trees into branches, and to the extremities of the branches are appended the membranous sacculi or lobuli, full of fatty globules. . . . Through these membranes [of the globules] run very minute vessels, forming a net-like expansion, and which represent a little omentum: these penetrate the lobuli intimately, and are connected to the fatty globules. Through this same membrane divide and ramify adipose

259. To enable the omentum to act as such a depository, storehouse, and dispensatory of the better parts of the blood, it

vessels" (n. 253). The latter vessels cannot be any other than the minute terminal appendices of the ramifications of the artery, filled with this whitish juice; consequently, than continuations of the blood-vessels all the way into the globules of fat, where, as in places of reception and entertainment, they deposit the precious lading of their blood. With respect to *nursing*, this fatty progeny is surrounded with coverings, as it were swaddling clothes, formed of convoluted productions of the vessels, and is not only fostered therein, but little breasts and teats are there as it were held out to it by the parent arteries and veins, so that no new-born babe can be nursed more sedulously. Malpighi has described admirably, not only how this globule is covered with a most delicate membrane, but also how its interior particles are supplied by the extremities of the vessels. See n. 253. "The fat [of the cossus]," says Swammerdam, "consists of very small globules; which, when viewed with a microscope, seem to be again composed of smaller particles; all of which are supported by very thin and transparent membranes or coats" (n. 254). *From the sympathy of nature between the fat and the blood*, in that the fat is so frequently demanded back by the blood; whence its sudden decrease and increase, and this, from hour to hour, in some species of animals: shewing that the blood is perpetually refreshing itself from this its store, and relying on a power of self-renovation in the omentum. "This humor," says Boerhaave, "must be considered as most subtile; whether we regard its origin and nature, or its perpetual consumption and replacement." (*Inst. Med.*, n. 331.) Consequently it seems, that the blood, when pressed by necessity, supplies its wants exclusively from this its own melliferous food. *From the fineness of its vessels, and of the coat which encloses it:* for the coat of the omentum itself—the common coat of all—is extremely delicate and attenuated. "The membrane of the omentum," says Winslow, "is made up of two extremely thin laminæ" (n. 252). Still thinner is the membrane of each cell; for it is split off from the common membranes; and still thinner the membranule which surrounds the fatty globule; and perhaps there is yet a thinnest, which is invisibly thin: which membranes are so many proofs, that the contained parts or the fat are of similar purity; for the content corresponds to the continent: we may judge of the one by the other. Respecting *the circle of uses*, see a subsequent portion of the present chapter. To these considerations we may add, that similar elements, which are the primitives of nature's compounds, when brought together, never unite is necessary, *Firstly,* That it be expanded in the form of a veil over all those viscera which prepare and eliminate the chyle, the serum, and the blood (*k*). *Secondly,* That it take up, in the first instance, and diffuse throughout its membranes and fringes, all the arterial blood which passes through the cœliac porta, and its three entrances, and which is distributed through the before-mentioned viscera (*l*). *Thirdly,* That it transmit the whole collection of its veins into certain common channels, which lead to

into any other species of humor, than some very fine oil, resembling a sort of spirit. The last result of artificial distillation is a similar oil.

(*k*) That is to say, the expanse in which such material is stored up, must needs communicate with all those viscera, which in any way contribute to the existence of the blood; consequently, with the stomach, intestines, mesocolon, mesentery, liver, pancreas, and spleen. That the omentum is connected to all these viscera, is fully confirmed by our authors. "The omentum extends," says Winslow, "more or less over the small intestines, from the stomach to the umbilical region. The superior portion is in a manner divided into two borders; one of which is fixed along the great curvature of the colon, and the other along the great curvature of the stomach. The commissure of these two borders on the right side, is fixed to the common ligament of the duodenum and colon, and to the contiguous parts of these two intestines. That on the left side is fixed to the longitudinal fissure of the spleen, to the extremity of the pancreas, and to the convex side of the great extremity of the stomach. It is likewise attached to the ligament which sustains the ductus cholidochus, and connects it to the trunk of the vena portæ.... The small omentum is fixed partly to the concavity of the liver, so as to surround and contain the prominent portion of the lobulus Spigelii" (n. 252). It is divided into alæ, small omenta and appendices epiploicæ, each of which has its peculiar province.

(*l*) The whole of the cœliac artery, which is distributed to all those viscera that are preparatory and purificatory of the chyle, the serum, and the blood, passes through the network of the omentum, as through a common plane or court; and this, in order that it may there deposit and reserve so much of the better essences of the blood, as will require to be expended; and thus consult use and necessity. For the right branch of the cœliac artery, which gives off the gastric, pyloric, hepatic, cystic and intestinal arteries, skirts the border of the omentum, and gives off the epiploica and gastro-epiploica dextra. So in like manner does the left or splenic branch, which gives off the porta hepatis (*m*). And this, in order that it may constantly maintain a store, consisting of the genuine material of the blood, as a provision for the conveniences, expenses, changes of state, wants, and other prospective contingencies of life;—in order that the superfluity of the blood may be lopped away; its poverty relieved; and that the blood itself, as well as the serum and the chyle, may be elaborately tempered to every use.

260. In the animal kingdom, the series, chain, progression, and circle of causes, involve a corresponding series, chain, progression, and circle of uses; for the principle of the cause, which is eminently a living principle in this kingdom, regards nothing but perpetual ends (*n*). The effects which the cause produces from this principle, are the effects of an end, consequently of a use (*o*). Hence there is a similar progression of uses, as of ef-

the epiploica sinistra: and it also communicates with the mesenteric arteries.

(*m*) The veins correspond to the arteries, and springing from numberless roots, speedily form channels, which discharge themselves at length into the vena portæ. All these particulars may be best obtained from the Tabulæ and elaborate descriptions of anatomical writers.

(*n*) I have judged it proper to excerpt this short statement from my Psychology, in order that we may comprehend, albeit obscurely, how the body is a representation and image of the soul; namely, that all the effects which belong to the body, involve uses, which are ends, and which the soul comprehends, and inasmuch as it is living, respects or regards: so that the organic body is only a mechanism of effects, which are all represented in the soul, as ends. Also, that we may comprehend what the word nature signifies in the animal kingdom; namely, all that principle which acts in the cause, or which is the prime activity of the cause: but in the animal kingdom this principle is living, and represents to itself the series of those ends, and determines [the body] into action, conformably to the representation. None but an intelligent being can represent to itself ends, nor can any but a living being have intelligence, for living consists in exercising intelligence; consequently, there can be no intelligence in the body, except as intelligence in the soul.

(*o*) A cause, in order to be efficient, must necessarily involve an active principle. This principle is commonly called nature by philosophers, inasmuch as nature is the complex of all such principles or active forces. There can be no cause, without it exist as such from such fects; a similar progression of effects as of causes; and a similar progression of causes, as of ends, the series of which ends is in the very soul itself. Such is the nature by which the living body is said to be governed. This nature does not proceed one hair-breadth, that is, does not design the minutest fibre, or the smallest vessel, still less the entire fabric of an organ, without stamping upon it her series of ends, and respecting a use; and, indeed, in the primary use, respecting a mediate use; in the mediate, an ulterior use; and never in anything does she respect an ultimate use, without at the same time respecting in it the primary use. This is what we mean by the circle of uses. Such a circle is represented to the life in the omentum. Its *first use* is, to reap the superfluous part of the blood, and lay it up in its cells, as grain is laid up in barns; for the sake, or with the end, of providing for a coming winter or period of dearth; also of consulting the welfare of all the contiguous viscera which prepare the chyle and lustrate the blood; that each may have its just quantity, and its proper quality (*p*). The *second use* is, that the whole of this repository, or, as we may now term it, reticulum or membrane—the omentum—be employed in covering all these members; and in such a manner, that they may be severally provided for by this their seminary, according to the measure of their need; consequently, in conjoining them into a society of coöperation; and moreover, in defending them from the injurious effects of heat, cold, halitus, and various impulses (*q*). The *third use* is, that the omentum, while it purges

a principle, and no effect without a cause: therefore, when a cause is supposed, a principle and an effect are also supposed; the cause being taken as the mean of both. In the series of what follows, it will be shewn, that whatever belongs to the class of causes, must be formed organically, in order that it may serve as a cause: also, that there is not a single effect in the living body, which may not be likewise a cause, and in order respect subsequent effects perpetually, as *de novo* efficient causes; in short, that nature never designs any ultimate or last effect, without so involving in it the first, as to imitate a perpetual circle.

(*p*) This will be illustrated in what follows.

(*q*) These are the principal uses which are attributed to the omentum by anatomists; but they are not the primary, but only the seconitself of gross, stale and rancid substances, exhales rank animal vapors through its open foramina, and therewith besmears and anoints the viscera, particularly the stomach and intestines, (which are continually wreathing through an everlasting spiral within the cavity of the abdomen,) and prevents their axes from creaking, and their wheels from drying by mutual friction, and growing sluggish and stopping in their gyre (*r*). The *fourth use* is, that this blood-seed—this pabulum with which the granaries of the omentum are distended—be prolific and fertile; that is, be distributed for the common good of all, and for the particular good of each; for it is taken out from the omentum in such quantity and quality, as the hunger and thirst of the blood require; and in such quantity and quality as the stomach, intestines, liver, pancreas, spleen and mesentery demand, to prevent. them from losing their labors and their pains. This then is the reason why the pellicle of the omentum is formed with so tender and fine a care, and its cells distended with sanguineous nectar; and why the blood again ruminates and feeds upon the latter: to wit, that the omentum may not only be a support to the stomach, but by laying on its fat, may pour fruitful fertility over every field. This last effect or use is also absolutely the first in the cause; in other words, it is the end; and thus it completes the circle of uses (*s*).

dary and general uses; for they are not the uses of single parts, but of the whole network or compound.

(*r*) Respecting the foramina through which the effete fatty humor transpires, not only to anoint the viscera and intestines, but also to perform the office of a menstruum, we refer the reader to the following pages.

(*s*) All things which involve an end, constitute a circle; for the end is in the first, the middle, and the last, and regards itself as effected and followed out in the last, which thus return to it. But the word, circle, is of very wide signification. There is the SIMPLE CIRCLE, such as that of the blood, which after having completed a circle through the arteries, returns back to the heart, and into the arteries, through the larger veins: there is a similar circle of all the humors in the body. There is also the PROGRESSIVE CIRCLE, or the circle which proceeds from greatest to least, or from a compound to its unities, and back from the unities to the compound: this also is termed a circle. Such

261. FIRSTLY; *That the epiploic reticulum deposits in its cells the exuberant portion of the blood, under the form of fat.*

is the circle of the blood, when it is distributed from the largest trunks to the branches, and from these to the minute twigs, or their leaves, and after arriving and being divided to its unities in the smallest capillaries, it flows back again by corresponding venous twigs into corresponding trunks, and thus returns to the place from which it started, or the heart. There is a similar circle of all the fluids of the body, as well as of the vessels containing them. Besides this, there is also the TRANSCENDENT CIRCLE, or the circle that ascends and descends from one series to another, or from one degree to another. Such is the circle between causes and effects, or between prior and posterior things: such is the circle by which prior causes produce posterior causes, and ultimately, postreme causes or effects, and by which these effects return, with a corresponding passage, to the prior causes. Such is the circle of the resolution and regeneration of the blood. Thus, when the volume of blood is divided to its unities, or globules, and these again are resolved into their purer or prior globules, and these again into their least globules; and when these least globules again unite into their [larger] globules, and ultimately into the globules of the red blood,—this is termed a transcendent circle. The circles of substances, whether they be fluids or solids, involve corresponding circles of all their accidents; for substances are the subjects of accidents: and therefore, corresponding circles of uses. To take the case of the omentum. Here, the first thing is, that the genuine sanguineous elements are laid up, for the manifold uses of life: also, that the whole expanse, or the compound of all the cells, performs uses in like manner,—confined, however, to a narrower sphere. Next, that the exhalations from this fat are of still further profit and advantage to the commonwealth, by anointing all the viscera that have a gyratory motion. Lastly, that the whole of the provision there collected, returns back into the blood, and thus constantly begins the circle where it ceases. But if it be the character of an intelligent being alone, to contemplate ends, and to do nothing without an end, surely that being who can complete a circle of ends or uses, in which every the least point involves a use,—who can thus produce a connected chain of perpetual utilities relative to life,—must indeed have consummate intelligence. If any one, with the mind alone, has either the disposition or the power to contemplate the government of the universe, the order of nature, and the fluxion of things, he will most assuredly discover that there are innumerable such circles, (and not, indeed, circles, but gyres higher than the circle as

This is proved by the cœliac arteries, nearly the whole of which, as they pass down to the chylopoietic and sanguigenous viscera, apply themselves to the upper border of the omentum, and skirt its fatty fringes, as rivers run along their banks (*t*); and then by straight offsets (*u*) go forth into its expanded web, where

commonly understood,) in which there is not the least point but flows from some use, and tends to some use, and thus gyrates through perpetual revolutions. To establish these perpetual circles, is altogether beyond the power of nature, and can be the act of only a most intelligent being, which disposes nature in this manner: for producing perpetual circles of uses, is identical with pure intelligence, wisdom, and more peculiarly with Providence. No person who reflects upon these things as occurring in the world, can reasonably doubt that the nature of the universe is governed by an all-wise and provident God. And no one who attends to the corresponding things in the animal body, can reasonably question the existence of a soul, which is constantly ordaining similar things in her particular kingdom. Moreover, from the very uses or ends which the human rational mind represents and proposes to itself, and from the series and nexus in which it determines or produces them into act, but especially from the quality of the ends themselves, we may conclude with certainty respecting the characters of individual men, or their intellectual and mental endowments.

(*t*) That the inferior gastric arteries skirt the very boundary which forms a common border to both the stomach and omentum, see Ruysch, *Thes. Anat.* ii., tab. v., fig. 1: and that the same holds good of the splenic arteries, see the Chapters on the Pancreas and the Spleen. "The arteries and veins of the great omentum," says Winslow, speaking of them generally, "are branches of the gastric arteries and veins, and for that reason go by the name of gastro-epiploicæ, dextræ, and sinistræ. The arteries on the right side answer to the hepatic artery, and those on the left side, to the splenic" (n. 252).

(*u*) That the branches of the gastro-epiploic artery come off almost at right angles from their trunk into the omentum, see Ruysch's Tabula before mentioned. Hence, agreeably to an observation made above (n. 245, *d*), the blood is not poured into the epiploon in a violent torrent, but only so much of it is diverted thither, as the more voracious members—the stomach, the liver, the intestines, and the pancreas—do not carry away by their more rapid force. The omentum, in these respects, may be compared to a river, whose waters are caught and absorbed by numerous gulfs, but with a level plain adjacent, in they spread and ramify. Also, by the cellular textures, which environ every branch in the form of folds or fringes (*x*). Moreover, by the extent of this veil, which when excessive, sometimes reaches even to the pubes. Also, by its capacity, such as beseems a great repository, and which becomes manifest in fatty dropsy or anasarca. Lastly, by nature's provident circumspection, coupled with the actual necessity, that there be somewhere a receptacle and asylum for the superabundant blood; in order to prevent the least vessels from swelling to the size of the greatest, and sanguineous inundation, heat and redness, overwhelming and confounding the lowest things and the highest; as we see imaged on a small scale in the minor deluges which constitute plethora (*y*).

which, as a port, a large portion of the tired stream finds rest, and escapes from the rushing torrent.

(*x*) That the cellular beds in which the fat is collected and preserved lie close to the sides of the vessels, like fringed or fimbriated borders, is thus declared by Malpighi: "[When the veins and arteries]," says he, "at last terminate reticularly at the sides, they are still accompanied and invested uninterruptedly by the sacculus or membranous striæ. These sacculi also contain fat" (n. 253). "The membrane of the omentum," says Winslow, "is made up of two laminæ, joined together by a cellular substance; the quantity of which is very considerable, accompanying the blood-vessels in broad bands. These cellular bands are more or less filled with fat" (n. 252). And that there are "areolæ or membranous spaces" between the bands, see Heister, n. 250. This is best shewn by inflating and raising those cellular strata. Inasmuch as these receptacles lie close to the streams of blood, and the latter, by a singular process of diminution and evolution, terminate therein, it is therefore evident that they deposit there some rich and superfluous material, which they are unwilling to carry back into the blood, unless absolutely required for use.

(*y*) From plethora, which is a particular or limited inundation of it may be even the purer blood, we may infer what would be the consequences of a total inundation; namely, immense distension, disruption, and blocking up of the vessels; inflammation, uniform and unvarying redness, loss of shape, a confused pulp and structureless vascular mass; torpor of all determinations, actions, forces, and motive fibres; dullness of the senses,—of the internal senses particularly; the trunk, a mere carcase; distinct exercise of no viscus; rout and confusion of the func-

262. All the other coverings, tumuli and cavities of the body, which are distended with stuffing and fat, have the same object in view (z): but the office of the omentum is confined to the viscera contained in the abdomen,—to preventing an undue quantity, or an undue quality of blood, from being supplied and assigned to them. In consequence, the splenic artery glides under the omentum as it passes to the pancreas and spleen; in like manner also the gastric artery as it goes to the stomach; and the other arteries likewise—intestinal, cystic and hepatic (a). Thus the omentum culls its share, and abstracts its provision on the way; and when it has filled its wallet, it sends away the remainder, and dispenses and portions it among its fellow-members, in proportion to their offices, their wants, and their voracity (b).

tions of all the viscera; resulting in the most terrible and dangerous diseases. Plethora produces similar effects, only in a lesser degree. The blood, which is being constantly generated, and constantly nourished with rich chyle, would rapidly increase and cause a deluge of this kind, unless various asylums and places of reception were provided, such as the adipose tissues, containing repositories for superfluous materials. For it is evident that the fat is either the recrement, or the deposit of the blood; whichever it be, the blood must lose precisely as much in quantity as the fat gains; so that in either case the fat takes off the superabundance of the blood. But as the fat also decreases, and is consumed, this is a sign, that in substance it is kindred to, and homogeneous with, the blood.

(z) Namely, all over the circumference of the body, where it constitutes a peculiar covering, termed tunica adiposa; in all the interstices; around the muscles; in all protuberant parts; most of all about the pubes, the kidneys, and the mammæ: the fat is also abundant in the mesentery and mesocolon; similar repositories, cells, and loculi exist throughout; similar increase and decrease: whence it may be inferred, that the use of the omentum in this respect is confined to the abdominal viscera.

(a) See what we stated just above, note (t).

(b) See the explanation given above respecting the equation of the fluids in the animal body (n. 203, z). The mass of blood that is brought in by the cœliac artery, and all its branches, and by the mesenteric artery, (which usually anastomoses with the cœliac,) and that is afterwards distributed among all the chylopoietic and sanguigenous

263. SECONDLY; *That the epiploon covers and connects these its viscera.* This is proved by its expansion over the broad fundus of the stomach, in the manner of a bag or corslet; also, over the volume of the intestines (*c*): by the attachment and connexion of the epiploon itself and its appendages, to the duodenum and colon (*d*); and its free and floating movement in

viscera, must consist of various portions, widely differing from each other in nature and constitution. For the stomach draws from the mass a peculiar quantity and quality, suited to its actions and wants, and to the quality of the chyle; the liver, another peculiar kind and quantity; the pancreas and the spleen, another; thus each of the viscera, a different sort. This artery, at its commencement, carries all these species at once, and commixed into one; but to be separated and drawn out distinctly by the viscera themselves, which demand and invite their own quantity and quality. The purer portion which remains after the partition, cannot flow back into the aorta, and it does not seem consistent with nature's plan that it should be expended on the chyle; inasmuch as it is already new blood, in its first vernal estate; wherefore the omentum seems to be placed near these viscera, in order that this portion may be derived into it, and that thus all things may be allotted properly and distributed accurately. Such then is the equation of quantity and quality in the blood, and which may be compared to a kind of equilibration, of which the omentum seems here to act as the balance.

(*c*) The omentum "extends," says Winslow, "more or less over the small intestines, from the stomach to the umbilical region. The superior portion is divided into two borders, one of which is fixed along the great curvature or convex side of the arch of the colon, and the other along the great curvature of the stomach. . . . The small omentum is fixed partly to the small curvature of the stomach," &c. (n. 252). This is the reason why the omentum is spoken of as a covering, and compared to a pouch or bursa.

(*d*) Respecting its connexion with the duodenum, Winslow says, "The commissure or union of the two borders on the right side, is fixed to the common ligament of the duodenum and colon, and to the contiguous parts of these two intestines" (n. 252). Respecting its connexion with the colon and rectum by means of the appendices adiposæ, or little omenta, he says, "The fatty appendices of the colon and rectum have always appeared to me to be small or supplementary omenta or epiploa. They are situated at different distances along these other parts (*e*). *That it defends them from the injurious effects of heat and cold:* this it does by means of the fat, which breaks, moderates and tempers both. *From the injurious effects of the halitus or vapors* with which the abdominal cavity abounds: for its rich oil softens acrid, blunts saline, and repels aqueous substances; it also exhales a vapor, which seizes, absorbs, and sheathes them; and in order that it may drive off and keep away these things from the laboratories, and prevent them from spoiling the works, it is formed of a complicated double membrane, and anointed with grease; and moreover is imperforate (*f*). In order that it may protect *from assaults,* it is extensile, contractile, elastic, and when it is acted upon, it yields like a tendon, and reacts (*g*).

264. THIRDLY; *That the omentum exhales a rank moisture, wherewith it anoints the viscera of this region, and prevents them from sinking into atrophy and lethargy.* This is proved by the offensive vapor which constantly fills the cavity of the abdomen, and anoints and moistens the viscera thereof. A necessary condition, to prevent these members—which must be always constantly rolling their lubricous surfaces—from drying, shrinking, and becoming inactive (*h*). On which account, this fatty cover-

intestines" (*ibid*). That there is a connexion and union between them and the intestines, is proved by the fact, that when the former are inflated, the air passes all the way into the cellular coats of the latter.

(*e*) "With the exception of these adhesions," says Winslow, "the omentum lies loose between the anterior wall of the abdomen, and the intestines" (n. 252): by virtue of which condition, it is accommodated to every viscus,—to the stomach which lies above it, to the intestines which lie beneath it, and to the liver, the spleen, and the pancreas, which are connected to it; and this, in whatever way these viscera rise, expand or contract, and with whatever motion, ordinary or extraordinary.

(*f*) That the epiploic membrane, through the whole of its extent, is formed of two very thin layers, connected by cellular tissue; and that it is perforated by no foramina, (inasmuch as the cellular tissue may be inflated into the form of a large bladder,) see n. 252. The omentum, says Ruysch, has "no foramina" (n. 255).

(*g*) See what we said just above, note (*e*).

(*h*) All the members of the body that are covered by any mem-ing (*i*) has a large opening, [the foramen of Winslow,] communicating with the whole extent of its cavity, and situated precisely at the general hinge, on which the small and large intestines, the stomach, the liver, the gall-bladder, and the pancreas, hang and turn (*k*). And when the serous humor, thrown out by the spleen particularly, and polluted with effete fat, is eructated from this opening, it is diffused round, readily and uninterruptedly,

branous investment, or loose sac, are surrounded by a peculiar humor. The heart, for example, has its humor—the liquor pericardii; the cerebrum has its humor, as observations testify; the lungs also have theirs, which likewise fills up the cavity of the chest. The tongue has its saliva; the eye and the other sensorial organs have their fluid; the joints and articulations of the bones and cartilages have theirs: the muscles are surrounded by the fat. The viscera at present under consideration, have theirs, which when the abdomen is opened, reeks and exhales, and gives forth a rancid, porraceous smell. If members which enjoy continual mobility, were once to become dry, their motion and office would be at an end.

(*i*) That is to say, the omentum.

(*k*) But on these subjects let us hear Winslow, who first discovered and described this foramen, and who, moreover, attributes to it the same use as that mentioned in our text. (See *Mem. de l' Acad. Roy. des Sciences*, 1715 (Paris); and Heister, n. 250, above.) "This orifice," says Winslow, "is formed by the union of two membranous ligaments, whereof one connects the beginning of the duodenum and the neck of the gall-bladder to the liver; the other connects the contiguous portion of the colon to the same viscus, and extends to the pancreas," &c. (n. 252). From the meeting as well as from the course and extension of the ligaments, it is evident, that this spot is the post and common hinge of all the viscera; consequently that the exudation given out here, is diffused in all directions, and creeps and insinuates itself continuously along the coats of all the viscera: also, that this spot is the centre of the motions of the viscera, to which (as in the body throughout) the humors are derived from their surfaces. There is a passage along the hepatic ligament, all the way from the vena portæ, by the lobulus Spigelii; likewise, from the gall-bladder; and along another ligament, from the colon and pancreas, &c. Hither also, for the same reason, pass the cystic, hepatic, and pancreatic ducts, or the ductus cholidochus communis; they pass, that is to say, from a region of motions and as it were storms, to a haven of rest.

over all the gyres, fastenings and points of the visceral surfaces: the very motion being sufficient to besmear them all (*l*). Thus the omentum and the spleen together, prepare a certain unctuous humor (*m*), wherewith they smear, lubricate, cleanse and open

(*l*) It is a constant law in inanimate or purely mechanical machines, that when sebaceous, bituminous, or greasy substances are smeared over any of their axes, such substances diffuse themselves in all directions by virtue of the mere continuity of motion. How much more must this be the case in living machines, which are under the government of a superior motion, and describe a perpetual circle, and where there is a circumference wherever there is an axis, and *vice versa;* agreeably to the description of such motion in the Chapter on the Intestines. Now, in order that not even the least point of the intestines may be destitute of this unguent, it is poured on exactly at the beginning of the small intestines, and nearly close to the end of the large intestines; and thus glides inwards both along the colon, and along the duodenum; and the more certainly so, inasmuch as the motions of these intestines are reciprocal, and centre in the cœcum or at the valvula coli, during every revolution. See n. 128.

(*m*) The nature of this unctuous humor may in some measure be inferred or conjectured, from its source and composition; that is to say, from its being the serosity thrown out by the liver and pancreas, and particularly by the spleen. For the spleen, agreeably to what we stated in the last chapter, separates all the serosity from its blood, and throws it out to its surface; from the surface there is a continuous passage into the omentum. Hippocrates seems to have intimated as much, in his *Liber de Locis in Homine*, where he teaches, that in dropsy, when the omentum, as well as the whole body, are dissolved and attenuated, serum escapes from the vessels instead of fat; adding, that such serum is derived from an enlarged spleen. See Mangetus, *Theatr. Anat., de Omento.* This also is the reason why certain of the moderns, as Vesalius, Schenckius, and others, have considered the spleen to be the source of the omental fat; which appears to be borne out by those little spleens which are sometimes found inserted in the omentum. But as there is a straight and ready passage by means of the cellular tissue, from the spleen into the omentum, and as the office of the spleen also consists in separating the serum from the blood, it therefore results as a necessary consequence, that the serous humor, which is thrown out principally through the before-mentioned foramen, springs chiefly from this source; but also from the pancreas and liver, following the directhe outside of the intestines, as the liver and the pancreas each prepare a peculiar juice, wherewith they perform the same offices for the inside of the intestines. It also seems probable that this cellular and organic membrane supplies the large intestines with a menstruum for macerating the exhausted food; for there is a communication between it and the cellular coat of the duodenum and colon (*n*), and a continuous passage thence to the vermiform appendix, which pours a similar menstruum into the fundus of the cœcum, and so into the gorge of the colon (*o*). What the omentum supplies to the peritonæum, will not, I hope, always escape the sagacity and penetration of scientific investigators.

265. FOURTHLY; *That the blood demands back this deposit, and with these its elements, which constitute its proper food, at*

tion of the ligaments. But this humor, in its passage through the omentum, cannot but carry away with it that portion of the fat, which by a long stay in the latter, is becoming decomposed and rancid, and which is thrown out, to prevent the fresh portion from being tainted: so that there is also a recrementitious part in this membrane, and which here, as elsewhere, is converted to use. When such serosity, clogged or infarcted with recrementitious fat, does not exhale, it constitutes fatty dropsy or fatty anasarca. In some animals, this appears to be the origin of the adeps, from which the suet or sevum is produced.

(*n*) That there is an uninterrupted continuity of passage between the omentum and the cellular coats of the intestines, has been demonstrated by injection. "We may satisfy ourselves," says Winslow, "concerning these continuations, by making an orifice in one of the laminæ of the omentum, near the stomach or colon, and by blowing into the orifice, through a tube well fitted to it, when we shall find the air insinuate itself gradually under the common coats of the stomach and colon," &c. (n. 252).

(*o*) That the appendix cœci vermiformis pours a new liquid, adapted for anointing and lubricating the wavy folds of the colon, and particularly for macerating the fæces, into the fundus of the cœcum and the gorge of the colon; and that this liquid is proximately obtained from the cellular coat of the intestines, remotely, from the cellular coats of the peritonæum and of the abdominal viscera; and that the appendix draws off and discharges the useless and harmful portion of it, just as the cœcum itself draws off and discharges the alvine fæces; see n. 138 (*b*) and (*c*).

once satisfies its own hunger, and appeases that of the stomach and the other viscera. This is proved by the equation between the increase and the expenditure, or by the circulation of this fat, which is an appendage to the circulation of the blood (*p*): by the implantation of the venous and arterial extremities into the very inmost or recondite parts of the globule, where the vital oil is floating (*q*); (for therefrom the fat-absorbing vessels,

(*p*) See n. 258 (*i*). "That there is a circulation, motion, or return of the fat to the veins," says Heister, "is very evident from the consumption of it in certain diseases, and in animals when violently exercised; in both which cases the disappearance of it is often extremely sudden. . . . Perhaps, in time of need, it serves as nourishment for the body, by reëntering the veins" (n. 251).

(*q*) The microscopic examination of the fat seems to prove, that it increases in purity by degrees, proceeding to the inmost, and that what we see by this artificial means, is many times compounded. For the fat appears to consist of spheroidal, oval corpuscules, covered with a fine membrane, which is channelled by little arteries and veins; so that according to Malpighi's description, the pellicle of a globule of fat, resembles a miniature omentum. This globule appears to be made up of other prior and smaller globules, each with similar vessels, but here unfolded into their ultimate twigs and leaves. In the innermost, therefore, must dwell this fat in its earliest infancy,—delicate, fluid, milky, lying at the little venous teats, perfectly obedient to the smallest suction exerted by its particular vein. From this description it is evident, that the apparent tenacity or cohesiveness of the fat, is entirely owing to the coverings and little bands in which it is enclosed, as it lies like an infant in its swaddling clothes. And inasmuch as it cannot be freed from either these bands, or from the grasp of the vessels, it necessarily simulates the appearance of a viscid and sluggish moisture; although it is far from having any inherent sluggishness, inasmuch as it is being constantly licked up, like the thinnest milk, by the very smallest veins. "The blood-vessels," says Malpighi, "expand like trees into branches, and to the extremities of the branches are appended the membranous sacculi, full of fatty globules. . . . Whether all the adipose globules be invested again with a pellicle of their own, or not, the eye is unable to decide; this much, however, is visible, that the globules are connected to the sides of the vessels, as the stones of the grape are appended to the stalks" (n. 253). From the foregoing considerations it appears, that the blood-vessels in this part do not stop till they reach like bees, take out, and as it were ruminate the treasured store): by the sudden and frequent efflux of the same vessels from these inmost parts, their change into visible capillaries, and their confluence into branches (*r*); and at length of the branches into the venous channels, the splenic and gastric (*s*), which convey these necessaries and commodities to the vena portæ (*t*). Some little is also delivered to the lymphatics, in order that it may be conveyed to the cisterna mesenterii (*u*). Thus the liver, in order to

their ultimates, and that the globules are germinations as it were of the membrane, as we shall also find to be the case with respect to the cortical substances of the cerebrum. Thus, of course, the arteries must deposit the purest corpuscules of the blood, and the veins demand them back again.

(*r*) What myriads of vessels there are, is evident from the manifold divisions and subdivisions of the little arteries and veins; for even those which are large enough to be visible, are reckoned by myriads. Ruysch, in his *Thes. Anat.* ii., tab. v., fig. 1, exhibits the omentum, which he says, "is furnished with myriads of blood-vessels" (n. 255).

(*s*) That the veins correspond to the arteries, and in great numbers discharge the blood into other veins which go direct to the vena portæ, is evident from the Tabulæ and descriptions of our authors. The veins of the omentum, according to Heister, are "from the vena portæ, and principally from the splenic branch" (n. 250). "The gastro-epiploic veins," says Winslow, "on both sides, answer to the distribution of those of the vena portæ" (n. 252).

(*t*) On these subjects, Boerhaave says, "Inasmuch as the human omentum is not observed to have any excretory duct, excepting the venæ epiploicæ, dextræ and sinistræ, it is therefore probable that the venous blood of the omentum, rich in lymph and oil, is all poured into, and mixed with, the blood which is flowing to the liver." (*Inst. Med.*, n. 332.)

(*u*) Lymphatics also proceed from the omentum; yet it seems doubtful whether or not they carry away any portion of the genuine oil of the arteries. The cells are certainly permeable from one to another, both for the passage of the splenic serosity, and of the fat which has escaped from its coverings; and since the beginnings of the lymphatics open here in all directions, it seems very probable that a small portion of this fat is absorbed by them. Perhaps the adipose ducts of Malpighi carry away something of a similar nature; see the sate the hungry chyle, and occasionally to repair sudden losses, calls forth these rich preserves and dainties from the cellular treasury and storehouse of the omentum, repealing the right by which the latter holds them; and bestowing them upon the blood, brings them into the market, that every member may draw upon them, according to its requirements.

description given of them in n. 253; although, according to Heister, Malpighi retracted his opinion about them, in his posthumous works (n. 250). Moreover, the lymphatics of the omentum seem to collect similar constituent and primitive elements of the blood, and at the same time the spirit of the blood, contained in the humor thrown out by the spleen, pancreas and liver, and which, not being put into the fat, would otherwise be lost. See n. 258.

CHAPTER XIII.

THE SUCCENTURIATE KIDNEYS.

266. Heister. "The succenturiate kidneys, called also capsulæ atrabilariæ, or glandulæ renales, were first described by Eustachius. They are two yellowish, compressed glands, lying on the upper part of each kidney. They have a very narrow cavity, moistened with a brownish liquor of a sweetish taste. Their figure is irregular—square, triangular, oval, &c. Their size is various, but in adults they are about the size of a large nux vomica. In the fœtus they are larger, and sometimes exceed the kidneys themselves in magnitude. The membrane that surrounds them is very thin; it closely invests their glandular substance, and connects it with the kidneys. Their blood-vessels come sometimes from the aorta and vena cava, but more frequently from the emulgent vessels. Their nerves are from the renal plexus. Their lymphatic vessels are numerous. They have no known excretory duct, and their use therefore is not certainly ascertained; but they seem to be of more service to the fœtus than to the adult. *(Comp. Anat.*, n. 220.) The Academy of Bordeaux feeling that the use of these glands was so very obscure and uncertain, several years since offered a prize to any person who could discover it; but no one, so far as we yet know, has appeared who could claim the reward. Some little time after this, however, Valsalva published, and also wrote (privately) to many people, that he had found out their real use, and had discovered certain ducts, going from them to the epididymes in males, and to the ovaria in females; and that thus these glands contribute to generation. Valsalva promised a distinct treatise on the subject; but although he lived for some years after this promise, the treatise never appeared; nor has any one been able since to find the ducts. Not only then are the ducts themselves doubtful, but the size of these glands in the fœtus, and

their diminution in the adult, make greatly against the theory of their being of use in generation." *(Comp. Anat.,* not. 23.)

267. WINSLOW. "The capsulæ atrabilariæ, renes succenturiati, capsulæ renales, glandulæ renales, or, as we should incline to call them, the glandulæ suprarenales, are situated on the upper end of each kidney, a little obliquely, that is, nearer to the inner edge and concavity of the kidney, than to the outer and concave margin. Each of these glands is an oblong body with three sides, three margins, and three points. Its length is about two-thirds of the breadth of the kidney, and the breadth of its middle portion is about a third of its extent between the two extremities. Its color is a dark yellow. It has an anterior, a posterior, and a lower side, which last may be termed the basis; and it has, consequently, as many edges, the superior of which may be called the crista. One of its extremities is internal, or turned inwards, towards the sinus of the kidney; the other is external, or turned outwards towards the gibbous part of the kidney. The figure of this gland may also be likened to a cock's-comb, or to the top of a helmet. The surface of these glands is uneven; the anterior surface is the broadest, and the inferior side, or basis, is the narrowest. Along the middle of the broad or anterior side, a groove runs from the edge of the inner extremity, a little above the basis, to the point of the other extremity, and divides this side into two parts, like the middle rib of the leaf of a tree; and on the lower side, under the basis, there is a kind of suture. The blood-vessels of these glands come from the emulgent and diaphragmatic arteries and veins, from the aorta and vena cava, from the cœliac artery, &c. They are termed the capsular arteries and veins; and they seem to be invested by a sheath, that penetrates into the glands. They are not always derived from the same sources, nor is their number the same in all subjects; there is generally one pretty large vein which runs along in the groove. The nerves, on each side, are supplied by the neighboring semilunar ganglion, and by the renal plexus which depends on it. In the inside of these capsules, there is a narrow triangular cavity, the surface of which consists of short, strong villi, of a yellowish color; but in children it is reddish, and of a dark brown in old subjects. The sides of this cavity are connected to each other by a number of filaments; and they appear to be wholly glandular, and covered with very fine, small follicular corpuscules: at the top of the gland, they are in contact. Inside the cavity, we find a granulated and follicular substance, which fills it almost entirely; and the blood-vessels are distributed on this substance, as well as on the sides of the cavity. If the section be begun at the great extremity of the capsule, and be continued through the upper edge, and if the

lateral portions be afterwards separated, the glandular substance comes in view, in the form of a kind of crista, raised from the middle of the bottom of the cavity. This glandular body or nucleus adheres more closely to the bottom or basis of the cavity, than to the sides, especially near the great extremity; but nevertheless, it is distinct, and may be separated, both from the basis and sides, to which it is connected by a number of filaments. It adheres least to the basis near the small extremity. The capsular vein generally comes from the renal vein, and is large compared to the arteries, which are very small. This vein communicates with the inside of the capsule, much in the same way as the splenic vein communicates with the cells of the spleen; for it may be inflated by blowing into any part of the capsular cavity, and the air likewise passes into the renal vein. The cavity contains an unctuous fluid, more or less viscid, of a yellowish red color, which with age changes gradually into a yellowish purple, a dark yellow, and a black yellow; and sometimes into a black, but even then if it be spread thin on a large surface, it appears yellow. I have sometimes found it not only reddish, but mixed with real blood. These capsules are very large in the fœtus, and diminish as age advances.

268. "The renal glands lie sometimes directly on the top of the kidneys, but I never found them on the gibbous part. The gland on the right side is partly connected to the diaphragm, under and very near the adhesion of the great lobe of the liver to that muscle. That on the left side adheres to the diaphragm immediately below the spleen. The connexion between the capsules and the diaphragm is confined to the neighboring portions of the inferior muscle. The capsules are attached to the kidneys by the cellular tissue of the adipose membrane, of which a thin portion insinuates itself between them and the kidneys, and also between them and the diaphragm; so that they are connected to both, entirely by the cellular tissue: and for this reason the connexion appears to be effected in many subjects by a layer of fat. The venous groove already mentioned sometimes sinks so deep in the anterior side, that the upper part of this side seems to be separated from the lower; but this is seen most distinctly when the capsule is examined in clear water. When the capsular vein is opened lengthwise with the point of a lancet, we find in it several small openings, many of which are only the orifices of branches of the veins, while others are simple holes; and it is perhaps through these that the air passes into the vein, as already mentioned. On the outer surface of the capsules we find a very thin, distinct coat, separate from the cellular tissue that surrounds them. Sometimes this coat is raised by an uneven layer of fat, which gives it a granular appearance, and makes the capsules look pale, like fatty

bodies. The liquid in the cavity appears sometimes in the fœtus, and in young children, of a blueish red. To discover the uses of these capsules, we must also attend to their external conformation, which is commonly more regular in the fœtus and in young children than in adults and old people. We must likewise take into account the consistence and solidity of their substance, which is greater before birth, during childhood, and in youth, than in advanced and old age. They are sometimes very soft, and very much wasted, so that it is not remarkable that when they are removed from their cellular and fatty coverings, they should resist more in the latter, on one side, than in the former. This perhaps may be the reason why the figures given of them are so very irregular, and so different from what I have demonstrated for above twenty years past." (*Exp. Anat., Tr. du Bas-Vent.*, n. 431—447.)

269. MALPIGHI. "When one of the succenturiate kidneys is divided lengthwise, (for these glands are solid and firm,) its substance is seen to consist of fibrous bodies, passing from the circumference towards the centre, just as in the kidneys. These bodies are yellowish, and supplied with blood-vessels: [whence] they have [also] a reddish tinge. Probably they are tubular and excretory vessels. They arise from certain yellow bodies, placed between their extremities and the surrounding membrane. These yellow bodies are generally oval, not seldom depressed, and probably are the loculi of glands, the humor separated by means of which is sent through the tubular bodies, as excretory vessels, towards the centre. The excretory vessels open into, or at least are continuous with, a kind of cinereous, mucous substance, the nature and minute composition of which I have not been able to trace with certainty, although sometimes it has seemed to be made up of minute roundish particles. I have, however, clearly made out that it is supplied with a beautiful network of vessels, and particularly with vast numbers of white nerves, reticularly interwoven; whence it seems probable, that a further separation is carried on by the nerves in this substance, or rather that it is an appendage of the excretory vessels and their extremities: inasmuch as it is immediately connected to a broad and capacious duct, which passing lengthwise, opens externally, and discharges itself into the emulgent veins: while at its other extremity it gives off copious branches, whereby it receives the humor secreted by the glands and tubuli. This duct or cavity is lined by a fine membrane, which is studded with innumerable foramina of irregular figure; whence it is probable that great numbers of excretories open into the cavity, and that there is a passage from it into the before-mentioned foramina; as we see exemplified in the kidneys. I

have often noticed an analogous structure in certain glands situated in the intestine that adheres to the fleshy stomach in hens and chickens. These latter glands are exceedingly numerous, and surrounded with beautiful blood-vessels, and they are supported externally by the muscular membrane of the intestine. They are something of the shape and size of a pea, and open by a free orifice, or excretory vessel, into the cavity of the intestine, discharging a mucous fluid like barley gruel. Glands of this kind are made up, as usual, externally of a membrane or loculus, and of a principal substance or organ. Internally they have a cavity, which exhibits great numbers of orifices of unequal shape and size. Between the external membrane and the central cavity, there is a thick substance, which on a diligent examination, appears to be resolvable into filaments and tubular bodies. Wherefore, it is likely, that the juice secreted in the surrounding follicle, is conveyed into the cavity through the tubular bodies, as excretory vessels; an extended network being formed of their extremities; its unequal meshes representing as it were so many stigmata or points, by which the mucous juice is first received, and then derived into the cavity." (*Opera Posthuma,—De Structurâ Glandularum Conglobatarum*, &c., p. 144, 145. Amstelod., 1598.)

270. See Eustachius, *Opuscula Anatomica,—De Renibus*, p. 39, (Ven. 1564); where he states, that this gland is so connected to the external membrane of the kidneys, which is continuous with the peritonæum of the transverse septum, that without great care be used when the kidneys are taken out, one is apt to overlook it altogether, on account of its close adhesion to the transverse septum, &c. Eustachius, however, says nothing respecting either the internal structure or the vessels of these glands. Morgagni, *Advers. Anat., Anim.* 50., where he remarks, that in the fœtus, these glands seem to constitute one body with the kidneys: but that in adults there is a greater distinction between the two. T. Bartholin, *Anatomia Reformata*, p. 120, 121, (Lugd. Bat., 1651.); where he relates, that in one subject he saw as many as four succenturiate kidneys, and the same number of smaller ones below the emulgent vessels. Boerhaave, *Inst. Med.*, n. 364. Verheyen, *Corp. Hum. Anat.*, tract. ii., cap. xix., tab. xii., fig. 1, E. See also Casp. Bartholin, Wharton, Veslingius, Bauhin, and Riolan.

271. The first and obscure ages through which we pass, that is to say, while we are concealed in the ovaries, afterwards in the Fallopian tubes, and lastly in the mother's gravid womb, glide by as though they did not belong to us, but to some other being beyond us or before us (*a*) : for at this time nature reigns supreme, and spins and weaves the threads, and designs the states and fates of the coming life (*b*). We live then only as

(*a*) Nothing is properly ours that is not determined by our will, consequently by the rational mind as an antecedent. The will alone is ours; therefore the actions proceeding from it are ours, and are imputed to us, and we are remunerated, punished, or estimated, according to them. Thus the quality of an individual, when predicated of himself, is predicated really of his will; consequently of the rational conclusion which precedes the will, and creates its essence. The rest belongs to nature, and we merely make use of it, and attribute it to ourselves; but it is no more ours, than the actions of animals are theirs, which result from blind instinct of nature, and are determined without their consciousness. But whatever we predicate of nature, we mean to predicate of the soul. Although the soul be ours, yet the same will cannot be predicated of it as of the rational mind; for the soul is attracted by its nature, and while it lives united to the body, it transfers the power of voluntary arbitrament to its inferior mind, which constitutes the human rational mind. But these perhaps will be regarded as dark sayings and paradoxes, until the soul's nature has been discovered, and we therefore leave the further explication of them until that time.

(*b*) That the soul has in it this nature, and that the soul therefore is the fashioner of its body,—that is to say, is what clothes itself with a body, has already been indicated several times, and will be explained stamina and unfinished webs, possessing latent powers, but the exercise of which is deferred, until all things have been perfectly completed, and such preparation made, that we ourselves are enabled to enter upon our lives (*c*).

272. During this primeval time, while we are living as embryos in the womb, our *intuitive mind,*—which is one day to constitute the rational mind, and to make the life of the body our property and possession,—destitute of the light of the senses, lies quite buried in the deepest obscurity of night and ignorance. The *will* likewise, having no sense to determine it, and no inclinations to excite it to activity,—with its subject motive fibres or living forces,—remains inactive, and folded up in a state of bare potency (*d*). The *lungs,* which open the scene, and commence the drama of this life (*e*), now constricted and

in the proper place. Here I will merely apprize the reader, that whenever I mention nature, I mean the soul, whose nature it is that produces this web, and builds these frames, exactly in its own image, or in representation of its own nature, and in the course of its development into effects; that is, in the development of its ends into uses.

(*c*) *That* life, as we have just stated, is ours, which is determined into action, or an appearance of action, by some foregone intuition, consideration, and conclusion of the mind, by means of the faculty termed the will. Upon this life we do not enter until after birth, or until we leave the mother's womb; nor fully then, but by degrees only, as we advance in age and increase in understanding: thus it increases gradually with our power of action, until at length we may be said to be in possession of ourselves. But in proportion as the dominion of the inferior mind, which we term the animus, (and to which belong the affections of the body,) increases at the same time, in such proportion the intellect may [be said to] be deprived of its rights, self-possession, and dominion. But these subjects will be discussed in another place.

(*d*) Respecting these subjects, we must refer the reader to our Psychology, where we intend to shew what share the senses and the soul have respectively in the formation of the intellect. Sense flows in with the whole of the instrumental cause; but the soul gives the power of making use of these instrumental appliances, that is, of revolving them into rational forms, of combining them analytically, and forming them into conclusions.

(*e*) The manner in which the lungs occasion the body to live, that is, to feel and act, will be shewn in the Chapter on Motion and Sensation.

closed, neither emit nor admit the vital breath of the body. The *stomach* makes no use of its folds, and digests no food into chyle; for the blood is not now to be generated from chyle, but is all introduced from the maternal store, in prodigal abundance, as use demands it (*f*). The *intestines* do not creep through their rhythmic movements; for they crave no after-banquets from the poor and empty stomach. The *pancreas* is void of office,—it emulges no juice from the blood to carry to the vacant intestines (*g*). The *spleen* likewise is exempt from duty; for why should it lustrate the blood, and separate it from the serum, to serve in the liver as a menstruum for the chyle, when not a drop of chyle is supplied thither by the stomach (*h*)? Nor,

Without the use of the lungs, our life would be without action, and our actions without life. Were it not for the continual respiration of the lungs, even the voluntary muscles could not be called forth into their alternate motions; consequently, neither could the sensorial organs apply themselves to objects, or receive their modes distinctly, and convey them to the general sensorium. In a word, all the functions peculiar to the body, commence, and cease, with the respiration of the lungs. Wherefore, until these living bellows are opened, we live in the body as though we were not in the body.

(*f*) Namely, through the placenta and umbilical vessels, into the liver.

(*g*) Since the pancreas, (according to the description of its offices in Chapter X.,) purifies the blood, and prepares a salivary menstruum, which, when mixed with the two kinds of bile, is of use and service to the intestines, and also to the food, therefore the exercise of its offices must evidently be suspended, so long as the intestines contain no materials for digestion into chyle. When uses and offices are suspended, operations also are suspended.

(*h*) The offices of the spleen, (described in Chapter XI.,) consist in attracting the old, hard and grumous blood, and cleansing it entirely from serum, so as to enable it to be of use to the liver, both in refining the chyle, and in inaugurating it into the new blood: also in transmitting the serous portion into the omentum, in order that it may be thrown out in conjunction with the rank fatty exhalation of the omentum itself, into the abdominal cavity, and yield an unguent, for anointing the lubricous backs of the intestines, as well as the surfaces of the other viscera: lastly, also in conjunction with the omentum, in supplying the large intestines with a menstruum for maccrating the refuse of the food.

during this period, do *the kidneys, the ureters, and the bladder* rise, and expand their tubuli, pelvis, and calices, to receive any urine from the blood or the serum. These, and many of the other *ministries* of corporeal life, withheld from their offices, continue in the deepest repose, and simply allow themselves to be fashioned and informed for coming uses. All things now proceed according to the tenor of nature's ends: no desire of the mind, no lust of the will, and no allurement of sense interrupts them.

273. Inasmuch then as there are so many members not employed on the public service, the supreme power of the kingdom devolves upon a few; merely, in fact, upon the cerebrum, the heart, the liver, and certain subordinate organs, among which latter are the suprarenal capsules. The members among which the empire is divided in this simple form and tranquil state of government, sustain all the offices upon their own shoulders; and prevent those members which have not yet been admitted into society, from hastening or rushing untimely into their offices; but provide in the meantime that they grow up into the powers of the same, and that each is inaugurated by a secret guidance into the uses resulting from its office.

274. The SUPRARENAL GLANDS, before the other abdominal viscera have been initiated into their offices, administer a kind of sovereignty; but afterwards, without absolutely intermitting the peculiar operations which they have undertaken from the beginning, they yet restrict them to a few particulars. In both

All these offices pause, so long as the stomach and intestines are not digesting, and the vena portæ is taking up no chyle from them: the office is suspended with the use, and the action with the office. Besides, it does not appear that the embryonic blood needs these purificatory organs, since the purer portion only is attracted from the mother's store; and which is sufficiently lustrated, in both the uterine placenta, in the whole passage through the umbilical vessels, and lastly, in the liver; and no emotions of the animus afterwards arise, to disturb or defile it: for as long as nature reigns supreme, the blood undergoes none but the most natural changes, which do not adulterate it. Add to this, that were the spleen admitted to a participation in the work, all the other members, its partners in office, which perform with it one series of operations, would also be summoned into play along with it.

lives, they render valuable service to their fellow-kidneys, and also to the spleen, and, speaking generally, to all the members of this region,—both to those surrounded by the peritonæum, and to the genital members which are bound up with it. But first I will give a succinct description of the province which they administer, and of their *modus operandi*.

275. That these glands have a kind of systole, is proved by their close attachment to the peritonæum extended under the diaphragm (*i*), all the way to its inferior muscle (*k*): by the insulcation and fissure in which their vein is lodged (*l*): by the numerous vessels on their convex and concave surfaces (*m*):

(*i*) That the renal glands adhere very closely to the lower part of the diaphragm, where the latter is invested by a coat of peritonæum, is declared by Eustachius. "This gland," says he, "is so connected to the external membrane of the kidneys, which is continuous with the peritonæum of the transverse septum, that without great care be used when the kidneys are taken out, one is apt to overlook it altogether, on account of its close adhesion to the transverse septum" (n. 270).

(*k*) The capsule "on the left side," according to Winslow, "adheres to the diaphragm immediately below the spleen. The connexion between the capsules and the diaphragm is confined to the neighboring portions of the inferior muscle" (n. 268). Thus any vibration which reaches this extreme part of the diaphragm, is communicated to the capsules, which are adherent to it. They also draw certain vessels from this source.

(*l*) All bodies which expand and contract alternately, have the signs of their motion imprinted on their surfaces. This is the case with both the liver and the spleen. The marks or impressions are generally occupied by the blood-vessels, which are collocated for the most part in the very stream of the motion. The same thing is observable throughout the body. From this it is evident, that the constrictile and expansile motion of this viscus, converges and is determined to its insulcation or fissure. Respecting the extension of the suprarenal vein along this furrowed channel, Winslow says, "Along the middle of the broad or anterior side, a groove runs from the edge of the inner extremity, a little above the basis, to the point of the other extremity, and divides this side into two parts. . . . There is generally one pretty large vein which runs along in the groove" (n. 267).

(*m*) Wherever there is a large number of vessels, particularly on the surface, there is also a multitude of motions, which simulate one by their abundant supply of nervous fibres: by the glandular follicles on both their surfaces: by their purely tubular intermediate substance (*n*): by their triangular cavity, and the connecting bands therein (*o*): by the juice which is conveyed into it, and expressed through the foramina, and which juice is far

general pulse; and that this is continued through the whole compages [of the capsules] to the very cavity, is evident from the passage of those vessels through the little membranes and parietes, even of the tubuli, all the way to their other end, in the cavity itself; where in like manner they are succeeded by other vessels, reticularly interwoven. The sanguineous color of the glandular substance in this situation, proves what a great number of vessels it contains. "The [tubular] bodies," says Malpighi, "are yellowish, and supplied with blood-vessels; [whence] they have [also] a reddish tinge. They arise from certain yellow bodies, placed between their extremities and the surrounding membrane" (n. 269).

(*n*) That there are glandular forms on both the outside and inside, and that the intermediate substance is tubular, is plainly declared by Malpighi in the passage just quoted; to which he adds, that the "substance [of the succenturiate kidneys] is seen to consist of fibrous bodies, passing from the circumference towards the centre," and that "these bodies are generally oval, not seldom depressed, and probably are loculi of glands" &c. (n. 269). Respecting the glands connected to the concave surface, Winslow speaks at considerable length; see n. 267, 268. Consequently, if the follicles be glandular, that is to say, if they possess a cavity, and alternately receive and throw out a humor, this must necessarily be accomplished by a power of compression and expansion, as in other parts of the body: equally so, if any secretion, or mixture of secretions, be sent through the hollow tubuli.

(*o*) The fact of the parietes being held together by connecting cords or ligamentary sinews, is a clear proof of the existence of an expansion and constriction, whose modes and bounds are regulated by the cords; as in the longitudinal sinus of the dura mater. If this corpuscule were at rest, it would have no need of bonds. The truth is, these cords are so many limitations, not only of its space, but also of its motion; and they serve to prevent its disruption, and to reduce or oblige the expanded cavity to return to its pristine dimensions. "The sides of this cavity," says Winslow, "are connected to each other by a number of filaments" (n. 267).

from limpid (*p*) :—from such systolic motion results an active force upon which their operation depends (*q*). The renal glands, by virtue of this action, intercept and attract a part of the arterial blood which is descending to the inferior region of the body, either from the aorta, or from the emulgent, diaphragmatic, or cœliac arteries,—from one, or from many together (*r*).

(*p*) The juice, not limpid but yellowish, which is enclosed and sealed up in these glands, is another proof of their alternate contractions; which must necessarily be strong, and proportioned to the resistance of the humor, in order to drive the latter through the foramina which we shall mention presently: otherwise, it would undoubtedly stick in the cavity, and by its accumulation, would extend the parietes, until it caused the destruction and disruption of the viscus. But this fluid increases and decreases; which is a proof of the presence of a corresponding active force.

(*q*) The various circumstances which we have adduced, clearly shew, that these glands have a certain alternate action, or systole and diastole; but the question occurs, Whether or not this action be coincident with the pulse of the heart? This is easily answered: for in the primeval state, before the pulmonic bellows are opened, or breathe, there is no reciprocation of motion excepting only that of the heart; so that at this time these capsules cannot possibly vibrate any otherwise than synchronously with the heart. But after the primeval state, that is, after birth, when the lungs have been opened, and installed into their reciprocal movements, it is evident from many considerations, that these glands follow the same alternate movements as all the other abdominal viscera; and thus undergo changes of state even as respects their motion, and come over to the pulmonic movements; nor does it appear that they can possibly pass into any other, inasmuch as they are connected and fixed to the diaphragm and peritonæum, as well as to the kidneys. The case of these glands seems to be similar to that of all the other members;—even of the cerebrum itself, which in this primeval state draws its breath in exact coincidence with the systole and diastole of the heart. But immediately after birth, and the opening of the lungs, it alters its motion, and comes down to the pulmonic respiration; as I shall endeavor to shew further in my Analysis of the Cerebrum.

(*r*) Respecting the derivation of arterial blood into these glands, Winslow says, "The blood-vessels of these glands come from the emulgent and diaphragmatic arteries and veins, from the aorta and

As they attract this blood from the arteries, so also they attract a large quantity of serum from the cellular coat of the peritonæum, and of the neighboring viscera, above all, immediately from that of the kidneys. This is proved by their place of insertion under the diaphragm; by their uninterrupted communication with the kidneys; by the permeability of the membranes; and likewise of the coats which surround the vessels at this spot (*s*); by the fulness of the same in the fœtus, in which

vena cava, from the cœliac artery, &c. They are not always derived from the same sources, nor is their number the same in all subjects" (n. 267).

(*s*) That the succenturiate kidneys are as it were two corcula or little hearts, for circulating, not the blood, but the serum which flows through the cellular tissues of the peritonæum and its viscera, may be inferred from many considerations; that is to say, by following the course of the arguments brought forward in this chapter. From their *insertion* into the peritonæum, where it is connected to the lower surface of the diaphragm, (see above, *i* and *k*,) at the very spot to which the humor contained in this tissue, is naturally derived. That this spot is a kind of centre of confluence, may be seen in the Chapter on the Mesentery. "The capsules," says Winslow, "are attached to the kidneys by the cellular tissue of the adipose membrane, of which a thin portion insinuates itself between them and the kidneys, and also between them and the diaphragm: so that they are connected to both entirely by the cellular tissue" (n. 268). Nor can it be questioned that a sero-aqueous stream is constantly permeating this tissue, inasmuch as the tissue is made up of continual cells, which intercommunicate; and inasmuch as any fluid injection distends large portions of it. And a similar tissue also accompanies the peritonæum under the diaphragm. The fact is rendered still plainer in dropsy, tympanitis, and similar diseases. The sources of this humor are as many in number as the viscera; the whole expanse of the peritonæum is its field. If it has sources and springs, it must of course have a circulation, and it must have places of discharge; and if the latter, surely the primary place thereof must be, at the centre of the various motions; consequently at the place where the renal capsules are situated. These glands have a similar *communication* with the kidneys; as shewn by Winslow's observation just quoted. That *the veins and arteries are likewise tunicated* and cellulated, is proved by the experience of many authors. "The capsular arteries and veins," says Winslow, "seem to these membranes abound in serum, which in the fœtal state serves instead of chyle (*t*). This serum is commixed with the arterial blood, in the glandular follicles which occupy the surface, and in which the extremities of the cellular texture also terminate (*u*). The mixture is conveyed through tubuli of a similar nature, and which have a similar operation, into the common cavity (*x*), where it is rolled about in a mass. A part

be invested by a sheath, that penetrates into the glands" (n. 267). Respecting the sheath of the vessels we shall have much to say in the Chapter on the Kidneys. And as this vaginal covering is derived from the peritonæum, and penetrates into the very glands which occupy the surface, it therefore follows, that the humor attracted from the peritonæum by the activity of the glands, is constantly poured into the glandular compages itself, and thus carried into circulation.

(*t*) The large quantity of humor that permeates these membranes in the fœtus, may be inferred from the softness, delicacy and tumidity of the membranes and cuticles, and from the infinite number of arteries that go to the latter, and towards the circumferences or surfaces, during embryonic life; from the copious secretion for which there is as yet no passage by the kidneys into the bladder; from the liquor amnii, which exudes entirely from the cuticle; and lastly, from the re-absorption of the same liquor amnii, which is like true chyle. These circumstances go to prove, that this humor, which is the serum secreted from the maternal blood, can have no other proper receptacle assigned to it, than the extensive tract of the peritonæum and pleura. If so, there must then be certain glandular organs, placed at the very centre, for the purpose of maintaining the circulation of this humor.

(*u*) We are still more fully confirmed in these views, when we bear in mind, that the peritoneal coat is prolonged into the very glands or minutest follicles of this viscus, and there terminates; as we also observed before of the glands of the liver and pancreas. And the tissue of the peritonæum is prolonged thither, not only immediately, but also mediately, by the coats of the vessels, which are rooted in the very glands. Thus of necessity, the serum extracted from the cellular coat, is thoroughly mingled with arterial blood, in the very glands or follicles. That the cellular coat terminates in these glands, see above (*s*): and likewise the vessels (*m*).

(*x*) Respecting which, Malpighi says, "These yellow bodies are probably the loculi of glands, the humor separated by means of which is sent through the tubular bodies, as excretory vessels, towards the of this mass is absorbed by the veins which reticularly overspread the cavity (*y*); a part, by the glands which in great numbers occupy its parietes (*z*); a part is expressed through the foramina by the renal gland itself (*a*); and from the foramina into considerable veins, perforated as in the spleen (*b*). The quantity of serum commixed with the arterial blood, so increases

centre" (n. 269). That the operation, as commenced in the gland itself,—that is to say, the mingling of the serum with the blood,—continues throughout the tubuli, may be inferred for the same reason, as adduced above in treating of the liver, respecting the pori bilarii (n. 209). For the gland is only the head of its duct or tubulus, and therefore, its principle of action, and causes in the parts which succeed it, a similar action; which is also shewn by the dark color wherewith the tubuli are tinged, this being produced by the ramification of arteries through them here and there, in the same manner as through the gland itself. "When one of the succenturiate kidneys," says Malpighi, "is divided lengthwise, its substance is seen to consist of fibrous bodies, passing from the circumference towards the centre, just as in the kidneys. These bodies are yellowish, and supplied with blood-vessels; [whence] they have [also] a reddish tinge" (n. 269). This substance wherewith the concave surface is covered, and which seems to be produced by the extremities of the tubuli, is compared by Malpighi to "mucous substance" *(ibid.)*, and by Winslow is said to "consist of short, strong villi" (n. 267).

(*y*) Respecting the vessels in the cavity, Malpighi says, "An extended network is formed of the extremities [of the excretory vessels]; its unequal meshes representing as it were so many stigmata or points" (n. 269). And Winslow says, "The blood-vessels are distributed on the sides of the cavity. . . . The capsular vein communicates with the inside of the capsule" (n. 267).

(*z*) "The sides of the cavity," says Winslow, "appear to be wholly glandular, and covered with very fine, small follicular corpuscules" (n. 267).

(*a*) Respecting these foramina in the cavity, and the corresponding foramina in the veins, see Malpighi, n. 269.

(*b*) Respecting these, Malpighi says, "This [venous] duct or cavity is lined by a fine membrane, which is studded with innumerable foramina of irregular figure" (n. 269). And Winslow says, "When the capsular vein is opened with a lancet, we find in it several small openings (n. 268). This vein communicates with the inside of the capsule, the vein, that it greatly exceeds the parent artery in size (*c*). This vein is itself the real excretory vessel which is said to be wanting in these glands (*d*). The part left in the cavity, and which resembles bile, seems to be an extract, or quintessence, whereof a small quantity will convert a large quantity of serum into blood (*e*). Thus the suprarenal gland is constantly maintaining a circulation of the chyloid serum through the cellulated coats of the peritonæum and its viscera; just as the embryonic heart maintains the circulation of blood through the vessels; and it unites this nutritious and richly spirituous liquid to the blood, and remits the product into the sanguineous circulation. These fruits of its labors it pours forth into the emulgent vein, and into the vena cava, close under the stream which the liver

much in the same way as the splenic vein communicates with the cells of the spleen" (n. 267).

(*c*) That the serum of the cellular tissue of the peritonæum, in these glands is mingled with the blood, is clearly evident from the large size and tumid appearance of the vein as compared to the artery, and which proves that a greater quantity of liquid is carried out, than is brought in by the artery. "The capsular vein," says Winslow, "is large compared to the arteries, which are very small" (n. 267).

(*d*) Malpighi describes this vein as a "broad and capacious duct, which opens externally" (n. 269); and also acknowledges it as an excretory vessel.

(*e*) That the juice of the cavity of these glands, which at one time is of a yellowish, at another time of a brownish color, is a kind of extremely pure extract of the blood, and has a power of imparting a certain sanguineous tincture and nature to the serum, is here stated as a conclusion from what has gone before: inasmuch as it is the office of these glands to commix the embryonic serum or chyle with the blood. This much at least is certain, that immense numbers of lymphatics permeate these capsules after birth, and in adult subjects, and that no more serum is admixed, than is wanted for the blood. Hence it follows, that this juice is an extract most rich in blood, one drop of which will ensanguinate a little volume of serum. The bile, which it resembles, is also a product of the blood, but of the stale blood. Whether this juice has the like property in adults, I leave for further enquiry; also, whether experience be in agreement with what I have previously stated.

brings out through the canalis venosus (*f*); and this, in order that the fresh maternal blood, which has been driven through the umbilical vessels, and through the glands of the liver, at this first passage into the vena cava, may be mingled thoroughly with the embryonic blood which has been just tempered in the renal glands, and cohobated with the chylific essence; so that the quality of the blood that is returning to the auricle of the heart, and which must be sent thence into the arteries without passing through the filters of the lungs, may be accommodated to all the necessary uses: also, that the quantity demanded by the heart and the little viscera, may be instantly supplied by this short cut, from the proper store, and that no larger quantity of the maternal blood may be extorted by the liver, than is just calculated to supply the want (*g*). Moreover, these glands

(*f*) This is shewn by its very situation, for it eructates the venous blood into the cava, close under the ductus venosus of the liver. "The gland on the right side," says Winslow, "is partly connected to the diaphragm, under and very near the adhesion of the great lobe of the liver to that muscle" (n. 268). For the vena cava, from the liver to its bifurcation, receives the greatest supply of blood by the emulgent veins, which are the largest of all the veins sent to it from the trunk in this part. The blood of the capsules is poured into the emulgent veins from above, and passing through them, speedily rises in the vena cava, to lose its stream in that of the liver: not to mention other little streams which communicate occasionally through the diaphragmatic veins. Thus, the renal capsules appear to perform a somewhat similar office in embryos, to the liver in adults; namely, to mingle the chyle with the blood. (I call the embryonic serum, chyle, because it is not unlike chyle, in its purity and other properties, as appears from the liquor amnii, which is serum of the kind, and yet serves for chyle.) But these chylo-serous corcula—the renal capsules—are deprived of their office, as soon as the liver begins to refine the chyle, and to inaugurate it into the blood.

(*g*) That embryos suck out the maternal blood through the umbilical vessels; and that the blood is by no means poured in without invitation by the womb, will be shewn in Part V. Were it not for the insinuation of this blood into the cava, close under the ductus venosus of the liver, more than is wanted would certainly be emulged from the maternal store, and the security of the embryo would be destroyed.

regulate the quantity of blood that passes down to the inferior viscera, and thus prevent them from rushing precipitately into their offices before their appointed time. The suprarenal glands administer this high office, when nature alone holds the rod of empire; and therefore, at this period, they are tumid, large and agile: but when other noble organs engage in the direction of affairs, and also distribute certain provinces among the rest of the viscera, then the power of these glands is greatly circumscribed: hence, as they advance in age, they contract and waste, and at length are consumed as of but little account.

276. These glands render especial and valuable service to the kidneys, to which they are closely bound. *In the brief life before birth,* they divert the abundant stream of serum from the voracious gorge and swallow of those organs: they draw it into themselves through the cellular recesses underneath their membranes, and remit it into the sanguineous circulation (*h*). For it must be of very great importance, to prevent the kidneys from seizing this innocuous and nourishing serum of the blood, and banishing it from the little circle and gyre of embryonic life. On this account, these glands are closely connected to and implanted on the kidneys (*i*); and separated from them by only

(*h*) We shall shew in the following Chapter, when treating of the Kidneys, that the greatest care is exercised to prevent the finer part of the serum of the blood from being thrown out along with the urinous part, and to separate the former, and throw it towards the surface: and inasmuch as this part is intended to be remitted into the blood, the renal capsules are placed ready, to draw it off from the cellular tissue. Thus these glands are receptacles and absorptories of the finer lymph, to prevent it from escaping in combination with the urine, and to restore it to the blood. This is the case after birth. But in embryos, all the serum which passes to the kidneys is drawn into the capsular vortices, for this serum is not yet polluted with urinous impurities, but is like well-defecated chyle, nourishing and genuine: wherefore its loss is anxiously prevented. For the human embryo has no allantoid coat for receiving the worthless serosity ejected through the urachus, as the embryos of animals have. Respecting the circulation of this serum, we shall treat more fully in the Part on the Membranes of the Embryo, the Amnion and Chorion.

(*i*) The suprarenal glands lie closer upon the kidneys in the fœtus a filamentary or cellular tissue (*k*): and for the same reason, the glands and kidneys have common and intercommunicating vessels, surrounded with a similar capsule or continuous sheath (*l*): and moreover, a large number of fibres are sent to these glands by the renal plexus (*m*). Thus they are diverticula, and at this time, true succenturiate kidneys. *In the life after birth,* they still keep this appointment of embryonic life; for they prevent the kidneys with their now gaping jaws from devouring the pure and nourishing part of the serum (*n*); they draw off the lymph rich in chyle and spirit, by the same porous channel, and restore it to the blood through the twofold avenues of the veins and lymphatics (*o*): they also tincture it, as it appears, with the essence of their juice, and after the bond is broken, they again espouse it to the blood.

277. But although these succenturiate kidneys stand in a

than after birth; and this, in order that they may transfer to themselves the whole of the serum. "In the fœtus," says Morgagni, "these glands seem to constitute one body with the kidneys, but in adults there is a greater distinction between the two" (n. 270); and Winslow says, "These capsules are very large in the fœtus, and diminish as age advances" (n. 267). Which shews that there is a more remarkable mutual intercourse between them and the kidneys in the former period than in the latter.

(*k*) See the opinions expressed and the citations made just above, in note (*s*).

(*l*) See note (*s*), and the following Chapter, on the Kidneys.

(*m*) The immense quantity of fibres given off first to these capsules from the common and renal plexuses, is evident from the number of fibres which enter them. "The nerves," says Winslow, "on each side, are supplied by the neighboring semilunar ganglion, and by the renal plexus which depends on it" (n. 267): and Malpighi says, "I have clearly made out that it [the mucous substance] is supplied with a beautiful network of vessels, and particularly with vast numbers of white nerves, reticularly interwoven" (n. 269). That during the obscure or uterine period of life, the office of the kidneys is transferred to the renal glands by the action and through the influx of these fibres, may be easily demonstrated.

(*n*) See note (*h*), just above.

(*o*) That they abound in lymphatics, see Heister, n. 266.

conjugal and social relation, as respects their office, with the kidneys, yet inasmuch as they include the ordinances of two distinct lives, they always, by their aid, contribute in some way to the purpose and use of the other viscera; and while they extract from the peritonæum and the common lake of the blood the purer portion of the serum, as the kidneys extract the more feculent portion, they also claim for the spleen the blood deprived of serum, and which will be serviceable to it in performing its work (*p*): and while they also carry off the very flower of the blood, and marry it to the virgin serum, they claim the senile and comparatively worthless portion for the spleen and pancreas, and for the liver and gall-bladder (*q*). They also snatch it away from the spermatic syphons and the testicles, and

(*p*) We shewed in the Chapter on the Spleen, that the spleen demands a blood deprived of serum, and that on this account the splenic artery passes under the border of the pancreas, in order that the pancreas may take away the serum, and throw it into the pancreatic juice; so that the spleen may not only separate and divide the blood which has coagulated into clots and lumps, but also submit to the liver the pure blood that may be a menstruum for the chyle, which the liver refines; and at the same time, that the liver may reduce and break it up more easily, and cast it into the bile, and thus renovate it. In this business, the renal glands perform a valuable service for the spleen, inasmuch as they carry off the purer portion of the serum, in the same way as the kidneys carry off the stale portion. But with respect to the blood's mode of purification, it is very different in the spleen from what it is in the renal glands: for in the spleen the blood is purged of all serosity, whereas in these glands the blood is intimately mingled with the serum; that is to say, with the finer serum which is proximate in essence to the blood. It would be very desirable to state the particular operation of each in a single word, did language admit of our expressing distinctions which are at present so little known or recognized. The operation of the succenturiate kidneys seems to be a kind of *initiation* of the serum into the blood: that of the spleen, a *lustration* of the serum from the blood.

(*q*) For the reason given in the analysis of these bodies, namely, that the old and hard blood may be resolved into its constituent parts, and thus regenerated, and that at the same time it may serve as a menstruum for the chyle.

thus hinder and prohibit the immoderate influx, downpour, and seizure of the flower of the blood, into those wanton and voracious organs (*r*). This the renal glands transmit from the aorta, by the shortest circle and passage, into the middle of the vena cava (*s*), and submit to the hepatic blood, swollen with new

(*r*). The spermatic vessels arise sometimes from the emulgent arteries, sometimes immediately from the aorta; consequently, in the very vicinity of the capsular or atrabiliary vessels. That the purest blood, rich in spirit, prolific and genial, flows into the seminal vessels, is sufficiently evident from the office of the viscera to which they convey it. Not only the kidneys but also the biliary capsules are placed close to these vessels, and they all drink their beverage as it were out of one patine or goblet; but the true renal arteries imbibe a widely different blood to the capsular arteries, and the spermatic, to both. The viscera themselves, by virtue of their construction, make their choice, and allure, invite, and demand their blood, as we have frequently stated before. Inasmuch then as the testicles, epididymes, vesiculæ seminales, and prostate gland, seize so large a quantity of the pure, new, and regenerated blood, and with such avidity and ardor, it seems, therefore, to be necessary, that some organ should be appointed over them, to temper and moderate their fire, and to rescue a portion from their grasp, and transmit it by a short cut into the vena cava—the middle venous lake—so that the sanguineous mass may not be despoiled of its noblest part by the flames of the animal inclinations. Hence it seems probable, that when we abstain longer than usual from venery, this richly endowed and prolific blood makes a short gyre and circle through these capsules, as diverticula.

(*s*) In no part of the body is there a shorter passage or transit from the aorta into the vena cava, than through these capsules, and through the kidneys themselves; the latter being, however, somewhat longer than the former. The blood drawn up through the carotid arteries, has to pass through the whole of the superior province—through the province of the head—before it is returned by the jugular vein. The blood poured into the subclavian arteries, passes all along the arms, to the tips of the fingers. The blood of the intercostal arteries is diffused through the whole region of the chest, and through all its muscles. The blood of the diaphragmatic artery circulates through the plane of the diaphragm. The blood of the cœliac artery does not arrive in the vena cava, until it has passed through the spleen, the pancreas, the stomach, the intestines and the omentum. So likewise the blood of chyle, and commix the two, and deliver them to the heart. Thus all things proceed auspiciously, according to nature's decrees. But the case is altered as age advances; for in process of time these organs exchange their once powerful and florid estate for a mean and wasted condition; their function is alienated and cut off; with this, their use is confined and lessened; and their texture is restricted and unwoven as their use diminishes: from manifold causes (*t*).

the mesenteric and iliac arteries, &c. Wherefore it appears, that the shortest passage for the blood from the aorta into the cava, is through these glands, and it is still shorter when the [capsular] branches arise from the aorta itself, as represented by Eustachius, *Tabul. Anat.*, tab. i., fig. 2, 4, 5. Thus no other members transmit the blood in so short a time as the capsules, or are so competent as they to supply the venous blood from the abundant arterial blood which passes to them, when the heart requires this to be done, and when the viscera, superior and inferior, require it from the heart. Provision is made in this way for the hepatic blood particularly, which, through this short passage, is qualified for every use, according to every want, purpose, and necessity of state. So all things proceed auspiciously, according to the decrees of nature.

(*t*) There are many causes why the renal glands gradually decrease, waste, lose their shape, are removed from their fellow-kidneys, and as it were grow old and die. Speaking generally, these causes may be divided into natural and accidental. The *natural causes* appear to be, that the serosity of the peritonæum, as well as the nobler blood, are sent in another direction; namely, that one portion [of the former] is carried away by the salivary glands, and converted into saliva; that another portion is taken by the pancreas, and forms the pancreatic juice; another by the kidneys themselves; that a part is taken by the omentum and the other adipose coats or planes, and made into fat, to be laid up therein for use; a part also by the genital members, which require a large quantity of fine serum; as will be seen in Part IV. Besides which, the lymphatics of nearly all the viscera, convey it in large streams to the receptaculum chyli, and to the cup at the head of the thoracic duct, as well as immediately into the veins. In a word, quite a different form of government has succeeded to the former simple form: thus the function of these glands is alienated, and their use diminished. The *accidental causes* are the numerous affections or passions of the animus, and diseases of the body, which too often interrupt, and even pervert, these natural circles and states. Luxury, and the immoderate lust of indulging the palate and cramming the belly, stimulates the salivary glands, and the other viscera, to pour forth salivary juices and digestive menstrua in immense abundance. Immoderate venery also draws off the purest serum and blood from their natural circles, and dries the very membranes. All the other storms of the animal mind are similarly hurtful, and confound and defile the blood, striking it with heat or cold, or filling and clogging it with phlegm. Not to mention an infinity of other causes, and old age itself among them, which closes up the minute vessels, blocks the passages, undermines and demolishes the strength of manhood, and induces torpor and sterility upon the very blood and its serum. By all these things, the function of the suprarenal glands is interrupted, and their use diminished. "To discover the uses of these capsules," says Winslow, "we must attend to their external conformation, which is commonly more regular in the fœtus and in young children than in adults and old people. We must likewise take into account the consistence and solidity of their substance, which is greater before birth, during childhood, and in youth, than in advanced and old age," &c. (n. 268).

CHAPTER XIV.

THE KIDNEYS AND THE URETERS.

278. HEISTER. "The kidneys are two red viscera, of something the shape of a kidney bean. They are situated in the loins, one on each side, with their concave part turned inward, their convex part outward. They are placed near the two last false ribs; and one of them is not unfrequently found a little above the other. Sometimes the left kidney is the highest, sometimes the right, contrary to the common opinion. The kidneys are connected with the loins, the lower ribs, the colon, the succenturiate kidneys, the renal vessels, and the ureters. They have two membranes, an external or common membrane, termed the adipose membrane, which is strong, surrounds the kidneys loosely, and is furnished with its own proper vessels. The other membrane is proper, and is very thin, and everywhere adheres closely to the substance of the kidneys. The kidneys are five or six finger-breadths long, three broad, and about a finger-breadth and a half thick. The surface is, in adults, smooth and even; but in the fœtus, often irregularly divided into a number of lobes, as indeed is always the case in calves, oxen, bears, &c. The vessels of the kidney, like those of the liver, are inclosed in a membranous or capsular production of the peritonæum. The arteries and veins are large, and are called emulgent or renal vessels; sometimes there is only one of each, sometimes there are several: the veins come from the vena cava; the arteries from the aorta. The nerves are from the renal plexus. The excretory duct is called the ureter. The lymphatics pass to the receptaculum chyli. The substance of the kidneys is firm and hard, and is of two kinds. The exterior or cortical, according to Malpighi, is glandular; but according to Ruysch, is throughout elegantly vascular. (*Comp. Anat.*, n. 219.) On account of the tubercles sometimes found in the kidneys,

many writers have endeavored to shew that their substance is glandular; but the same may be said of these tubercles as of those in the liver. *(Comp. Anat.*, not. 22.) The interior, which is tubular, consisting of the tubuli urinarii Bellini, terminates in eight, ten, or twelve papillæ, which open by a multitude of little foramina into the pelvis: but these papillæ are not found in all subjects. The pelvis is a membranous cavity, sending out several processes, called the tubuli of the pelvis, which surround the renal papillæ. *(Comp. Anat.*, n. 219.)

279. "The ureters are two membranous canals, of about the thickness of a quill, but of varying diameter. One ureter usually extends from each kidney to the bladder. At their origin from the kidneys, they form an enlarged infundibular pelvis. They terminate in the posterior inferior part of the bladder, passing obliquely in between its membranes, and opening into the bladder in a valvular manner, by narrow orifices, which admit nothing into them from the bladder. They are not straight, but somewhat bent, so as to resemble the letter *ſ*. Their substance is membranous, and made up of three coats; 1. a common coat from the peritonæum; 2. a thin muscular coat; 3. a nervous coat, covered with a lubricous humor, and sometimes exhibiting glands. The blood-vessels and nerves come from the adjacent parts. The use of the ureters is, to receive from the pelvis the urine secreted in the kidneys, and to convey it to the bladder; for when the ureters are obstructed, the urine is suppressed, there being no passage for it into the bladder. They are often found of an unnatural size, from the passage through them of renal calculi." *(Comp. Anat.*, n. 221.)

280. Winslow. "The posterior surface of the kidneys is much broader than the anterior; and the upper extremity is likewise broader and a little more curved than the lower. The depression in the small curvature is oblong and uneven, resembling a sinus, surrounded by several tubercles. Vessels:—the descending aorta and inferior vena cava lie between the kidneys, close to the bodies of the vertebræ, and to each other; the artery being to the left, the vein to the right. Each of these vessels sends out transversely to each side, a considerable branch, which goes to the kidney, and enters its sinus or depression by a number of branches. Sometimes there are more than one of each kind, which is oftenest the case with the arteries; and this, sometimes on one side only, sometimes on both sides. The veins are anterior to the arteries, because the aorta lies close to the spine. Nerves:—each renal artery is surrounded by a nervous network, termed the renal plexus, which supplies a great number of filaments to the kidneys: these filaments come partly from the semilunar ganglia of the two great

sympathetic nerves, and partly from the hepatic and splenic plexuses. The renal plexus sends likewise some filaments round the renal veins. Coats:—the kidneys are surrounded by a very loose membranous and cellular covering, called membrana adiposa, because in fat persons the cells of this substance are filled with fat. This was for a long time considered as a duplicature of the peritonæum, the true membranous lamina of which, however, covers only the anterior surface of the kidneys, and consequently they lie external to the sac of the peritonæum; because the portion of peritonæum which covers them, cannot be regarded as an entire coat; so that the only common coat they have is the cellular tissue, which likewise invests the renal arteries and veins in the form of a cellular sheath. The proper coat of the kidneys is composed of two laminæ, having between them a very fine cellular substance, which may be made visible by blowing in between the laminæ. The external lamina is very thin, and adheres very closely to the internal lamina, by means of the cellular substance. The internal lamina sends a number of processes into the substance of the kidneys, from which it cannot be separated without tearing. The surface of the external lamina is smooth, polished, and glistening; and it renders the whole convex surface of the kidney very even and uniform in adults: in children, this convex surface is in a manner divided into lobes or tuberosities, almost as in oxen and calves; and in grown persons we sometimes find the same inequalities. The blood-vessels, after entering the kidneys, ramify in all directions, and these ramifications give out capillary branches, which go all the way to the surface, where they appear like irregular stars, and supply the proper membrane. The proper membrane of the kidney passes to the sinuosity where the renal vessels enter, and in the form of a sheath or capsule accompanies all their ramifications into the intimate substance of the organ, and contributes in part to form the pelvis and calices or infundibula. Sometimes a considerable vessel is observed to go into, or come out of, the middle of the convex surface of the kidney; but this is not common. In case this occurs, there is a depression at which the proper membrane enters, and communicates with that portion which goes in by the sinus. The tunica adiposa or common coat, which invests the great vessels until they enter the kidneys, does not seem to accompany them any further; but terminates at the sinus, in the interstices between the ramifications. Structure:—we may distinguish three different substances in the kidneys: an exterior, dense, granular, and in a manner cortical substance; a middle and more internal substance, medullary and radiated, and which is termed the substantia striata and tubulosa, because it appears to be made up of small radiated tubes or canaliculi; and an inner or third substance, which is only a continuation of the second, and terminates on the inside by papillæ; for which reason I have named it, substantia papillaris. These three substances may be seen distinctly in a kidney cut into two equal parts, through the great curvature. The cortical substance will then be observed round the whole circumference; and by the microscope we perceive it to be of a spongy, granular, and slightly waving texture; all its parts adhering together closely in radiated lines. Its color is a bright, whitish grey. By fine injections, and during inflammation, we discover an infinity of small capillaries, which run in various directions between and round the different portions of this texture: and by the microscope we see likewise a number of small red corpuscules, more or less round, and disposed almost like bunches of currants. These corpuscules are perhaps only the ends of the vessels, cut more or less obliquely, filled either with blood or with colored injection. The other two substances, that is, the medullary or striated, and papillary, are really but one and the same mass, of a redder color, the convex side of which rises at several places into largish tubercles, lodged in the same number of cavities or depressions. The radiated striæ are afterwards continued to the papillary portion; and the papillæ form so many centres of these radii, opposite to the tubercles. The medullary substance is likewise distinguished from the cortical, by the arterial and venous arches, which send branches and ramifications on all sides; and its color is more or less red. The papillæ, which are only a continuation of the medullary substance, are often a little paler than that substance. They are ten or twelve in number, perfectly distinct from each other, and form each a cone, with a broad base and an obtuse apex. At the end of each papilla we see, even without a microscope, in a small depression, several very minute orifices, through which little drops may be observed to issue when the papillæ are compressed. These are drops of urine, which being filtered, partly in the cortical, partly in the medullary substance, afterwards pass through the tubuli of the papillæ, and are discharged by these orifices. The pelvis:—each papilla lies in a kind of membranous calyx or infundibulum, which opens into a common cavity, called the pelvis: all the calices or infundibula open into it separately. This pelvis is membranous, being of the same structure as the calices, of which it is a continuation; and its cavity in man is not uniform, but divided into three common fundi or gulæ, each of which contains a number of infundibula, or calices and papillæ. Sometimes we find two or three papillæ in the same infundibulum. At the place where these infundibula surround the bases of the papillæ, they send productions into the medullary substance, which accompany the blood-vessels, and serve for capsules or sheaths to all the vascular arches, both arterial and venous, and to their different ramifications, quite through the cortical substance, and as far as the surface of the kidney.

281. "The ureters:—after the infundibula have contracted in a conical form round the apices of the papillæ, each of them forms a small short tube or gullet. These various tubes, uniting together at different distances along the bottom of the sinus of the kidney, form three large tubes which go out from the sinus, in an oblique direction from above downwards, and immediately afterwards unite into one trunk. This trunk becomes a very long canal, called the ureter. In man, the three tubes supply the place of the pelvis in animals, and may more properly be termed the roots or branches of the ureter, than the pelvis; which name would best suit the trunk, as being larger than the rest of the ureter. The ureters are commonly two in number, one for each kidney; but sometimes there are more than two. The artery is in the upper part of the sinus, and partly in front of the vein: the vein is about the middle, and between the artery and ureter; the ureter is in the lower part, a little behind the vein, and partly surrounded by a branch of the artery. The superior branch of the ureter is longer than the inferior, because of their oblique direction from above downwards. From this description, we see that in the human kidney there is no other common or uniform pelvis but the trunk or head of the ureter, and the three great branches. To have a true idea of their disposition, we must imagine that the ureter enters the kidney at the lower part of the oblong sinus, and that it increases in size as it advances, and divides into three branches before it enters the substance of the kidney. One of these branches may be reckoned a direct continuation of the ureter, and it is longer than the rest. The angles between these branches at their bases, or at the head of the ureter, are not pointed, as those of other ramifications, but consist of a round curve, which is generally surrounded by fat. The first branches of the ureters produce other small branches at the bottom of the sinus, which are disposed in pairs. These collateral branches enlarge, and form the calices or infundibula in which the papillæ are lodged: the great circumference of which produces in the substance of the kidney the different sheaths of the vascular arches, and of their ramifications. The internal lamina of the coat of the kidney is continued round these sheaths, and the external lamina expands round the first branches, round the trunk, and round all the rest of the ureter. If the trunk of the ureter be split on that side which is next the vertebræ, and the section be continued to the extremity of the superior branch, we may

observe immediately above the trunk two holes, lying near each other, which are the orifices of the small collateral branches or gullets of the infundibula: in each of which we may observe the apex of at least one papilla. The ureters run down obliquely, and with a slight curve, from the kidneys to the front of the lateral parts of the internal or anterior surface of the sacrum, and passing between the rectum and bladder, they terminate and open into the latter. The ureters are very elastic canals, which yield in every direction, and again directly regain their pristine calibre, unless they have been long and unnaturally stretched. They are composed of three proper coats; the first or external of which is whitish, of a very compact fibrous texture, and yet very extensile, and appearing like ordinary cellular tissue degenerated. The middle coat is of a reddish color, stronger than the first, and made up of different layers of fibres, which intersect each other, and it is difficult to distinguish whether they are muscular or simply membranous. The innermost coat is in some measure ligamentary, and lined by a very fine membrane, which covers an exceedingly delicate vascular network. It is slightly granulated like shorn velvet, and moistened all over by a mucilaginous liquor. It has several longitudinal rugæ, which are intersected and in a measure interrupted by a great number of small transverse rugæ. Besides these proper coats, the ureters are invested also by the cellular tissue of the peritonæum, the membranous lamina of which covers about two-thirds of their circumference, sometimes more, sometimes less, but never surrounds them entirely: so that when they are examined in their natural situation, they look like cords lying behind the peritonæum, and standing out more or less into the cavity of the abdomen, together with that portion of the peritonæum which covers them." *(Exp. Anat., Traité du Bas-Vent.*, n. 394—429.)

282. MALPIGHI. "In oxen, tortoises and birds, the exterior surface of the kidneys is not uniformly smooth and continuous, but divided by grooves and depressions into a number of large polygonal, and for the most part hexagonal masses. In the bear, the kidney has the appearance of a bunch of cherries, the interstices being intimately penetrated and filled up with fat. The same kind of composition is sometimes clearly observable in human adults, and frequently in the human fœtus; the kidney being made up of small portions separated by deep fissures; and even in adults, traces of this division are always present internally. . . . . These subdivisions are not confined to the surface of the kidneys, but penetrate intimately, and form in some measure a distinction in their substance. In oxen, and even in human subjects, the large masses that in some animals are bounded by conspicuous fissures or interstices, have at any rate internally a peculiar

and well-defined shape, and seem to resemble polygonal pyramids. In these masses, then, you will find precisely the same divisions in the human subject, the dog, and cat, and that they intimately pervade the substance of the kidneys. They are aided and maintained by the branches of the blood-vessels, which ramify in part over the external surface: the internal branches likewise, are reflected when they reach the surface, and bound the same spaces. . . . Nor does this discontinuity of the external surface terminate here, for when the membrane has been recently removed, while the substance of the kidneys is yet soft, a number of very short, round corpuscules present themselves, rolled up like little worms, and not differing greatly from those which are found in the substance of the testes, when the coverings are removed, or the testes themselves divided through the middle. The whole of this may be seen, by pouring ink on the outside of the kidney, then gently wiping it off, and applying a good microscope; by the use of which we may sometimes see also remarkable branches of vessels, with certain globules, (of which we shall speak presently,) appended to them; the branches lying under the external surface, and sometimes comprehending interstices between them. In dogs and other animals, where the kidneys have a groove in their concave portion, to allow the ingress of the vessels, we have a clear view of these vermicular productions, which after making the above-mentioned short reflexions and circumvolutions, go straight along the dorsum of the kidney, following the external surface, towards the pelvis. This is a proof, that there is a connexion between these vermicular vessels composing the external surface of the kidneys, and the descending vessels, all the way to the pelvis. (*De Renibus,*—cap. i.; *De Renum extimæ Superficiei Divisione.*)

"The color of the human kidney differs in different parts; exteriorly, its substance is somewhat red, and becomes gradually paler towards the pelvis. . . . Certain it is, that on the external part of the kidneys, there are fine canals in immense numbers, and which are not muscular [carnei] but membranous, and seem to be analogous, both in structure and use, to other excretory vessels, as the salivary ducts and the pori bilarii. Although these canals, by their aggregation, principally make up this portion of the substance of the kidneys, and by their softness, and their intimate connexion with each other, present a decidedly fleshy appearance, yet if a deep fissure be made in a macerated kidney, by gently tearing it open with the fingers, we shall find that almost all the portions which seem fleshy, fibrous or parenchymatous, are excretory and membranous vessels. . . . I have at length determined with certainty, that all the urinary vessels do not go to the surface, but that some of them, passing through the arched branches of the blood-vessels, end here and there at the sides, and that a large portion of them is reflected inwards, and there terminates. . . . In all the kidneys that I have hitherto been enabled to examine, I have found a multitude of minute glands; and this, uniformly, both in quadrupeds, tortoises, and human subjects. The best way to see these glands, is, to inject the emulgent artery with some black liquid mixed with spirits of wine, so as to fill the kidney and blacken its surface; for then, even with the naked eye, (when the membrane of the kidney is removed,) the glands come into view immediately, appended here and there to the bifurcations of the arteries, and dyed with the black color of the injection: and when the same kidney is divided longitudinally, between the fasciculi of the urinary vessels and in the interstices you will find almost innumerable glands of the same kind, appended, as apples are appended to the branches of the apple-tree, to the blood-vessels filled with the black injection, and which moreover have a beautiful arborescent appearance. *(Ibid.,*—cap. ii., *De exteriori Renum Substantia,* &c.)

"These glands, placed in this external part of the kidneys, are almost innumerable, and as I think, probably correspond to the urinary vessels, of which the mass of the kidneys is composed, and of which there are more than forty in every one of the fasciculi that constitute the minute divisions already mentioned in all kidneys. The glands seem to be round, like the spawn of fish, and when a dark liquid is injected through the arteries, they become black, so that one would pronounce them to be surrounded by the extremities of the vessels, twining about them like ivy, and crowning them with wreaths; but with this limitation, that only the principal part, which is appended to the branch of the artery, is rendered black, but the remainder preserves its natural color. They are so connected with the branches of the arteries, that they grow from, and are appended to, the internal branches, and sometimes to the external branches, reflected inwards, and divided into numberless twigs. This may be clearly demonstrated, by injecting any colored liquid through the emulgent artery, in which case the glands will become of the same color as the continuous arteries, and ocular demonstration will thus be afforded readily, of the connexion of the former with the latter. The glands are also in relation [societatem habent] with the veins that accompany the arteries; for when the veins are filled with dark injection, although the glands continuous with them are not filled, yet the color penetrates so far, that no space intervenes between the glands and the extremities of the colored veins. Add to this, that the glands are sometimes white and nearly transparent, and sometimes red. . . . Although liquids injected through the arteries, fill the glands, yet I have never been able to find them penetrating the urinary vessels; and the same remark applies to injections through the veins: and again, when I injected through the ureters, the liquid colored only a few elongations of the pelvis, but never reached the excretory urinary vessels, called by some, fibres, so as to blacken the glands. Wherefore, after vainly using many contrivances, I have not been able to see the connexion between the glands and urinary vessels. In the kidney of a living animal, I thought I saw some kind of connexion and continuity, yet it was not so decided as to amount to sensible demonstration. (*Ibid.*,—cap. iii., *De internis Glandulis Renalibus*, &c.)

"In the more perfect animals, we observe a continued elongation of the urinary tubes and blood-vessels, all the way to the centre or pelvis of the kidneys; and in birds, to the bifurcation and dilatation of the ureter: we shall now, therefore, speak of this part. The urinary vessels then, arising from the gibbous portion and extremity of the kidney, and passing through the vascular arches, go straight towards the centre, and in some instances composing a kind of fasciculus, terminate in the form of a papilla; so that the whole compages of these fibres or vessels, seems to resemble a polygonal pyramid: in other cases, inasmuch as they unite only in one continued tract, or as it were extended papilla, the urinary vessels arising from the whole circumference of the kidney, terminate towards the middle, where the urine exudes into the pelvis through certain oblong foramina, from all the little mouths of the vessels opening internally, as I have observed in sheep, dogs, cats, and other animals. This part, produced by the meeting and ends of the vessels, used anciently to be compared to a new moon, or to one of the natural longitudinal divisions in a gourd. . . . Certain it is, that [in such kidneys], all these vessels go to the papillæ only, and never open on portions of the superextended pelvis; and therefore the urine is not excreted through the pores of the pelvis, but through the papillæ. I know for a certainty that this is also the case in man; for by careful dissections often repeated, I found that the urinary vessels of the human kidney, which resemble solid and compact fleshy fibres, terminate in evident and distinct papillæ, that stand out and open into the pelvis, and are lodged in an equal number of tubuli thereof, and for the most part amount to as many as twelve. . . . And although when the pelvis is removed, there appear certain continued protuberances produced by the urinary vessels, and closely connected to the pelvis, yet they are certainly not the extremities of vessels, but only their sides, going to the papillæ. . . . The urinary vessels which arise from the glands, and proceed to the papillary body, have not all exactly the same termination, but the urine flows from a fissure in the apex of the papilla, so that when this is laid open, we find the extremities of the different urinary vessels, of different lengths: but sometimes the urinary tubuli and blood-vessels go first to the apex, and are then reflected to the sides, where, not far from the extremity of the papilla, we find a kind of cavernous foramen, in which both the reflected and straight vessels terminate; so that when the papilla is compressed, the whole of the urine escapes through this foramen. . . . The whole of that portion of the kidney of which we have been speaking hitherto, has its own blood-vessels, which proceeding from the incurvated branches, and passing downwards, embrace the urinary vessels in the manner of ivy: and in living animals, when the emulgent vein has just been tied, these blood-vessels are completely turgid; shewing that a larger quantity of the obstructed blood collects in them, than in the other vessels. But these vessels are obliterated near the end, towards the papillary substance; whence, on account of their fineness, and the closeness of the [other] vessels, the appearance is presented of a membranous or nervous substance; and I have often doubted even whether it is not lined by a coat or reflexion from the pelvis, inasmuch as it is closely connected to and embraced by the tubuli thereof: but as in goats I cannot find any [coat] between the pelvis and the collection of urinary vessels, therefore I argue that this cannot be the case uniformly. *(Ibid.,—cap. iv.; De Residuo Substantiæ Renum usque ad Pelvim.)*

"The emulgent arteries and veins, on entering the cavity of the kidney, divide into several considerable branches, and running out on the continuation of the ureter, that is, on the pelvis, among the fat, at length, (beside certain appendages named tubuli [calices], which are elongations from the pelvis and surround the papillæ, and still contained in the same sac,) curve into an arch towards the gibbous part of the kidney, and go to meet, and are united by anastomosis to, other branches, and from this arch again, towards the gibbosity of the kidney, give off numbers of branches, which themselves form a kind of reticular interlacement, and run out between the little lobes into which the congeries of urinary tubuli is subdivided, all the way to the external surface, from which they are reflected a little to the sides, and at length terminate internally; their last twigs ending at the glands before mentioned. And inasmuch as a kind of round space is bounded by the tubuli of the extended pelvis and the arch of vessels, which space is filled by the urinary vessels put together to form the papillary substance, therefore here and there branches arise, forming an arch under the gibbosity, and uniting by mutual anastomosis, and the spaces defined by them, according to some observations, resemble the cells of a honeycomb. Offsets of these vessels also pass from the arch below, around the urinary vessels, and are distinguished by containing blood. In birds, the form of the ureter or pelvis is extremely beautiful: a sort of whitish vessel courses along the whole length of the concavity of the kidney, and here and there gives off branches, which gradually widening, after the manner of trumpets, admit innumerable collections of white vessels proceeding from the circumference; so as to make one doubt whether these white vessels arise from the division of the pelvis, as the fingers proceed from the hand, or whether these are broad portions of the pelvis, resembling funnels, into which the white urinary vessels discharge the serosities." *(Ibid.,*—cap. v.; *De Vasorum Propagine, et Pelvi.)*

283. See Eustachius, *De Renibus,* and *Tabul. Anat.,* tab. i., fig. 1, 2, 3, 4, 5, 6, 7, 8, 9; where he shews the different modes of ingress and egress of the vessels to and from both the kidneys and their capsules. Fig. 13; the kidney of the bear. Fig. 14; three kidneys, one on the right side, two on the left, the superior left kidney, however, being very small. Fig. 15; the interiors of the kidneys, with the carunculi or papillæ of the mamillæ closing the open extremities of the branches of the urinary vessel. Fig. 16; the branches of the artery, vein and urinary vessel, distributed through the substance of the kidneys. Fig. 17, 18, 19; the substance of the kidneys between the branches of the vessels. See also tab. iii., fig. 10, 12, 14, 15, 18, 20, 21, 23; tab. iv., fig. 1. Bellini, *De Structura Renum.* Littre, Obs. sur les Reins d'un Fœtus humain de neuf mois, in *Hist. de l'Acad. Royale des Sciences,* —*Hist.,* p. 45;—*Mem.,* p. 111—118; an. 1705, (Paris, 1730:) and in Mangetus, *Theatr. Anat.,* tab. lxiv., fig. 10, 11, 12, 13; where he exhibits the external surface of the kidneys with the vessels as very vesicular and ampullated; an appearance ascribed by Heister to disease, in his *Comp. Anat.,* not. 78, and n. 365, seq. Ruysch, *Thes. Anat.* iii., tab. iv., fig. 2, 3, 6; *Thes.* iv., tab. i., fig. 1: or in Mangetus, *Theatr. Anat.,* tab. lxiv.; where fig. 2, shews the surface of the kidneys, with innumerable vermiform blood-vessels passing over it. Fig. 3; the kidney, cut through the middle; the vessels, by a serpentine course, degenerating into the tubuli Belliniani, and becoming continuous with them; the renal papillæ, perforated by vast numbers of little foramina or mouths. Fig. β; a papilla is exhibited separately. Fig. 4; the kidney so divided as to shew the vermicular course of the blood-vessels, through the interiors. Fig. 7; the kidney of a sheep, with the papillary substance, where it is to be remarked, that before the small arteries degenerate into the urinary ducts they undergo a change; although it is indeed true, that by filling the arteries the ducts are also filled, so that the two appear to be continuous; and that the margin of the pa-

pillary or semilunar substance is perforated by innumerable openings. Fig. 9; the vermicular course of the vessels. Nuck, *Adenographia:* and in Mangetus, *Theatr. Anat.*, tab. lxiv., fig. 1; shewing the kidneys and ureters, the emulgent vessels, their branches, and the lymphatics. Boerhaave, the action of the kidneys, *Inst. Med.*, n. 351—365. Verheyen, *Corp. Hum. Anat.*, tract. ii., cap. xviii., tab. xii., fig. 1, 3. T. Bartholin, *Anat.*, p. 116, (Lugd. Bat., 1651); where he states, that in a dog's kidney he once found a worm resembling a snail, and nothing but the external coat of the kidney remained. Morgagni, *Advers.* iii., Anim. 35; that glands are sometimes found in the ureters. Coschwitz, *Diss. de Valvulis in Ureteribus.* Ruysch, *Obs. Anat.* xv., where he speaks of preternatural enlargement of the ureters.

284. The aorta, after leaving its starting-place in the heart, and describing the arch whence the carotids arise, and passing directly downwards to the middle septum of the body, opens into various streams,—into the phrenic, cœliac, mesenteric, and renal arteries; but not impeded thereby, it still pursues its way, until it reaches the extreme boundaries of the course (*a*). And although it impels its blood in a downward stream, yet it plucks and draws forth from the very middle of the torrent, the exact quantity and quality which the viscera below the septum demand and solicit (*b*). The aorta only detrudes the stream,

(*a*) We learn sufficiently from angiology, the manner in which the aorta divaricates into branches; namely, that soon after it issues from the heart, and immediately that it has given off the subclavian artery, it curves downwards; and that the carotid and vertebral arteries arise from the bend of the arch, not as continuations of the trunk, but as branches, (according to an observation of Ruysch); and that a little beneath the incurvation, the descending trunk gives origin to the intercostal arteries; and as soon as it reaches the middle septum, or diaphragm, it opens its bosom, and gives origin to the diaphragmatic artery; next to the cœliac and mesenteric, and at length to the emulgent arteries: and that it bifurcates afterwards into the iliacs, and thenceforth into various branches. Hence it appears, that this great and common artery deposits its freight about the middle septum, and near the threshold of the abdomen, and that this freight is distributed among the viscera of this region.

(*b*) It is perfectly evident, from the power possessed by the aorta, that it simply conveys the blood along its canal: for its muscular coat drives the blood onwards, but by no means allots the quantity or quality thereof which the viscera require; this being a part of the office of the opens its bosom, and brings its commodities into the public store: but from this common bequest, the viscera respectively select, seize and absorb what is most available to their purposes. All the chylopoietic or abdominal viscera require, and stipulate for, one genus of blood (*c*), and which is also brought to them

viscera themselves, which invite and attract into them the passing current, as we have often mentioned before. It is an incontestable fact, that the kidneys require one species of blood; the spermatic vessels and the testicles, another; the liver, the pancreas, and the spleen, another; the stomach and intestines, another. And the viscera require, not only different species of blood, but also different quantities—more or less—according to all their different wants and uses. Inasmuch, therefore, as the aorta only detrudes the blood along its pipe, it can have no power of assigning the proper quantity and quality to the viscera; but this must evidently be done by the viscera themselves, which, as we have explained and proved above, require, demand, and summon it. Thus the kidneys demand the blood which is fullest of stale serum; the testicles and genital members elect the fine and spirituated blood; the spleen, pancreas, and liver take the impure blood, which may be serviceable as a menstruum, and as a kind of hymen, to introduce the chyle to the bed of its spouse. Unless we inquire into the causes of this physical invitation or attraction, we shall never be led to understand the causes of these manifold effects. This power which the viscera possess, of summoning the blood from the trunk of the aorta, must be very great indeed, inasmuch as it draws forth from a stream that is descending with the force and rapidity of a torrent, such a quantity and quality as the office of each viscus requires. It seems to be a constant law, that the blood driven down by the action of the muscular coat, never turns off laterally, still less at right angles, unless a stronger force overcome a relatively weaker one; whence the attractive power of the viscera seems to exceed the detrusive power of the aorta upon the blood; particularly since these arteries,—for instance, the cœliac and emulgent arteries,—come off almost at right angles.

(*c*) We term that a genus of blood which is demanded by all the viscera at once, and with it as it were one common mouth; and this, in order that in the sanguineous mass which is derived from the trunk of the aorta into any common artery, there may be such a mixed quality, that they can all be provided for by it, according to the nature of their uses. Thus, the blood which is introduced into the emulgent arteries, must be of such a nature and condition, that the kidneys may obtain through a single entrance; but from this genus, each viscus chooses, brings out, and takes, its own species, quality, and quantity. The viscera which do not belong to this province, or to this department of business,—the kidneys, the ureters, the bladder, the suprarenal glands, the testicles, epididymes and vasa deferentia,—likewise, as it were by agreement, seize their peculiar genus through the common entrances and currents of the emulgent or renal arteries, from the rapid stream of the passing aorta, and appropriate it as a common stock, from which each separates, extracts, and summons, its particular species, share, quantity and necessaries. Thus all the different members drink different fluids from a single goblet (*d*): and so likewise does every part, from every branch allotted to itself and its associates.

285. These corporations of viscera (*e*), all the members thereof, and the parts of the members, command and take, first from

from it their proper quantity and quality, the renal capsules, theirs, and the spermatic vessels, (which frequently descend from the emulgent arteries,) theirs. This is entirely owing to the fabric, vessels, and fibres, which contract, relax, crave, seize, loathe and reject, according to the state of the body and the animus of the cerebrum. The effect which comes at last to our knowledge, or senses, is only general and obscure, but that of each fibre is distinct.

(*d*) If we attend to general uses, or the uses of all the viscera, and to the specific uses of single viscera, and the particular uses of their parts, we arrive at a knowledge of the species of blood which each demands; for all things take place according to use: wherefore the investigation of the uses of the viscera is the primary object of our analyses. In this way we shall learn why the genital members derive their blood from the same channel as the kidneys, although the former require the purest blood, and the latter, blood defiled with worthless serum: also, in what manner the kidneys purge this blood for the genital members, or draw off its impurities. See above, n. 239, 240.

(*e*) Those viscera that are employed generally in a similar office, constitute corporations; for instance, all the chylopoietic or abdominal viscera, which belong to one society; likewise, the excretories,—the kidneys, the ureters, and the bladder; also the genitals,—as the testicles, epididymes, vesiculæ seminales, and the prostate gland, in male subjects, —to all of which the spermatic vessel is prefixed, and they divide its blood among them distinctly, according to requirement and use.

their steward, the aorta, and next from its subordinates, that is, from every artery proceeding from the aorta, whether cœliac or emulgent, not only a peculiar species, but also a particular quantity: the *Species* which the fibres covet, which the veins gladly imbibe, and the lymphatics sip, as being the most desirable: a *Quantity*, according to the hunger and thirst of each, and the increased action excited thereby. It is to be observed, however, that they take different and unequal species and quantities, each according to its varying conditions and changes of state (*f*).

286. The principle, as we said above, generates the cause: the principle in conjunction with the cause constructs the body and all its parts; after which both together, by setting the structure in motion, produce the effect, whence comes the use, after the likeness of the end represented in the principle. If the use be unknown, as in the case of the more internal viscera of the abdominal cavity (*g*), then the structure must be opened,

(*f*) The mutability of state of the viscera, its causes, and the effect resulting from it, are subjects which must be discussed presently, as constituting a department of vast extent, and without a comprehension of which, it will be fruitless to attempt to obtain a knowledge of the varieties which exist in the animal kingdom; still less, of the causes why the urine, or excrementitious serum, as thrown out from the mass of the blood, presents such different characters, varying, in fact, with every state of the body and of the animus; and sometimes so obviously, that the condition of the body may be inferred from it, often with great correctness.

(*g*) That is to say, of the liver, the pancreas, the spleen, and the suprarenal capsules, which lie internally, and put forth no excretory duct to the outward parts, which conveys a liquid particularly distinguishable from the blood. The use of the other viscera is shewn by their ultimate effects only; as of the intestines, by the rectum, which is the last intestine, and discharges the alvine fæces. The use of the stomach is known from that of the intestines: not indeed from its structure, but from its continuity with the œsophagus, and thereby with the fauces, and inferiorly, with the intestines all the way to the rectum. The use of the kidneys is revealed in the same manner;—not at all by their structure, but by the bladder which is their ultimate, and by the urine which it excretes. The viscera placed in the extremes and examined in all its aspects, and the use must be interpreted, and in fine unfolded therefrom, by the connexion of causes, and the concatenation of phenomena or signs (*h*). But if the use be known, as the use of the kidneys is known by that of the bladder, we have then to enquire what is proximate, what continuous, what remote, and what supreme, in the cause which generates the effect and communicates to it the use. From many particulars presented in the structure of the kidneys, it is evident, that they expunge the stale serum from the arterial blood; and also draw off the sluggish and noxious phlegm from the peritonæum. That they detrude these materials through their tubuli, papillæ, calices, infundibuli, pelvis, and ureters, by a motive force resembling muscular action. That from these outcasts, throughout the whole of the passage, the veins recover and absorb the finer elements of the blood. That the lymphatics redeem the spirituous and essential portions, and restore them to the blood. But these results are brought about with infinite variety; differently according to every state which is induced, either naturally or accidentally, by internal, intermediate, or external causes. They are, however, all referable to a certain alternate motion, from which, as being the principle of the series, we shall now begin.

287. *That the kidneys are most busily actuated by alternate expansions and contractions, in the whole and in every part.* This is proved by their connexion with the peritonæum, the colon, the lower ribs, and the emulgent vessels in both sexes; with the diaphragm by means of the renal capsules, and with the ureters (*i*): by the depressed, curved, and almost concave

immediately manifest their use; but of those placed in the intermediates, for instance, of those completely enclosed in the abdomen, and not visibly continuous with the extremes, the use is not thus conspicuous; wherefore, their offices must be inferred from their structure, and from the series of subordination of causes.

(*h*) See the Chapter on the Spleen, n. 241.

(*i*) Respecting these connexions, Heister says, "The kidneys are connected with the loins, the lower ribs, the colon, the succenturiate kidneys, the renal vessels, and the ureters" (n. 278). From this we may conclude, not only to the existence of the alternate motion of these bodies, but also to their times of alternation; namely, that like form of their internal border (*k*): by the luxuriant and wonderful distribution of vessels over their whole circumference, and to

the other viscera of the abdomen, their movements are exactly synchronous with the respirations of the lungs; the same as the peritonæum, the abdominal muscles, the diaphragm, the lower ribs, and the colon. The peritonæum itself, tied up, as it is, to the lower ribs, and covering a large portion of the kidneys, cannot but communicate its momenta of motion to the kidneys, during every alternation: so likewise the other powers and levers,—the colon, the ribs, and the diaphragm. On these accounts, I am satisfied, that the momenta of the expansion and constriction of the kidneys, are synchronously alternate with the momenta of the respirations.

(*k*) "The concave part [of the kidneys]," says Heister, "is turned inward, their convex part outward" (n. 278). Their internal border is hollow or depressed, but their circumference is made up of protuberances. The blood-vessels, as well as the urinary vessels, comprising the pelvis and ureters, have their ingress and egress by this depression; which is still more evidently the case in birds, where the ureter is attached along the sinuosity. "In birds," says Malpighi, "the form of the ureter or pelvis is extremely beautiful: a sort of whitish vessel courses along the whole length of the concavity of the kidney, and here and there gives off branches, which gradually widening, after the manner of trumpets, admit innumerable collections of white vessels proceeding from the circumference" (n. 282). The sinuosity indicates, that this spot is as it were the central region of motion,—the place from which the motion of expansion and constriction begins, and into which it returns, and where it terminates. In the structure of every member intended for motion,—that is to say, for a motion in which all the parts enjoy mobility, and conspire to the same motion, and throw themselves into the general motion, and the general motions concentrate themselves in the similar little motions of the parts,—we may always discover throughout a certain relation of circumference, axes, and centre; for without this, the motion would have no regular form. This is the case in the kidneys, and in short, in all the viscera: in all, the last term is also the very first. Hence this depressed region of the kidneys gives to the other parts the power of perpetuating the motion, according to nature's appointment; that is to say, it gives the power of reciprocation from innermost to outermost, and from outermost to innermost; wherefore I have placed this sinuosity among the proofs of the motion of the viscus; particularly, as the motion is continued to their middle and central portions (*l*): by the never-failing accompaniment of the vessels by nervous fibres (*m*): by the con-

the ureters, which enter the kidney at the same part, and seem to be furnished with a kind of visible muscular coat.

(*l*) Between the common and proper membranes of the kidneys there is one uninterrupted sheet of vessels, so arched and inflected, as to represent stellar and lunar forms. Vessels also pass through the tubular substance all the way to the infundibula and pelvis, and some are reflected back to the surface, in order that they may in a manner connect the innermost motion continuously with the outermost. Of these vessels, and their course, we shall have an opportunity of speaking at greater length presently. Meanwhile, since the kidneys are scarcely anything more than vascular congeries, and every point in every vessel has its own pulse, (that pulse, namely, by which it is committed to the gyre,) there must therefore necessarily be as many divisions in the motion, as there are little arteries, and points thereof: thus, the confluence of infinite little motions produces a single motion, which is general, and renders the motion of the viscus conspicuous to the sense. Eustachius has admirably described the peculiarity of the first division of the arteries and veins, and Boerhaave has adopted his account in the following words, "The kidneys receive from the descending aorta either a single large branch, or several branches, and at the same time, a second membrane: from this first branch, or branches, they receive four or five considerable offsets, from these many smaller offsets, and from these again, an immense multitude of arched twigs, invisible to the naked eye, and which are distributed through all parts of the kidneys." (*Inst. Med.*, n. 352.)

(*m*) That the renal plexus, corroborated also by fibres from the other plexuses, embraces the renal vessels, (the arteries particularly,) and penetrates with them to the minutest parts of the mass, is thus stated by Winslow: "Each renal artery," says he, "is surrounded by a nervous network, termed the renal plexus, which supplies a great number of filaments to the kidneys: these filaments come partly from the semilunar ganglia, and partly from the hepatic and splenic plexuses" (n. 280). The nerves and all their fibres have in them perpetually the animation of the cerebrum, or the same reciprocation of motion as their principles; and inasmuch as these fibres weave round and enter all the arteries and veins, and excite them to the same action or motion, therefore there is of necessity a perpetual cause of motion from the innermost, and which coincides exactly with the causes in the outermost: consequently, the cerebellum, by means of the fibres, keeps up the intinuation of both the membranes to the most internal parts (*n*): by the division of the viscus itself into lobes, pyramids, papillæ, and finally into the urinary tubuli (*o*): by the abundant ivy-like

terior motion of the viscus; and the lungs, by means of the peritonæum and the diaphragm, and the other appendages, keep up its external or general motion. Such are the principles of these motions, and the causes of their coincidence with the pulmonic motion, which again is concurrent with the animatory motion of the cerebrum.

(*n*) If the cause of the common motion reside in the neighboring and connected parts, as here in the peritonæum, and consequently in the membrane, the proper membrane particularly, which also surrounds the vessels; and if this membrane be propagated towards the inmost, all the way to the centres, the cause of motion must necessarily proceed thither along with it. That the membranes of the kidneys, especially the proper or interior membrane, ramify to the inmost parts of the viscus, and cover all the vessels, both arterial and venous, and construct the papillæ, infundibula, &c.,—this is a fact fully discovered and admitted by anatomists. "The external lamina," says Winslow, "expands round the first branches, round the trunk, and round all the rest of the ureter" (n. 281). "The internal lamina sends a number of processes into the substance of the kidneys. . . . The proper membrane of the kidney passes to the sinuosity where the renal vessels enter, and in the form of a sheath or capsule accompanies all their ramifications into the intimate substance of the organ, and contributes in part to form the pelvis and calices or infundibula" (n. 280). See above, note (*i*).

(*o*) The partition of the viscus into lobes, and its consecutive divisions and subdivisions, are most evident signs of motion. The human kidneys, it is true, in adults particularly, do not present the same external appearance of division as the kidneys of some animals; but still traces of it remain evident enough in the substance of the organ, and shew that divisions did exist in the embryonic and infantine kidney: thus, although the traces are obliterated externally, yet they become conspicuous when the membranes are removed. The distinct partition of the viscus involves a corresponding distinctness in its motion: each congeries contributes to the power thereof. Respecting the division of the kidneys into lobes, Winslow says, "In children, this convex surface is in a manner divided into lobes or tuberosities, almost as in oxen and calves; and in grown persons we sometimes find the same inequalities" (n. 280). And Malpighi says, "These subdivisions are not confined to the surface of the kidneys, but penewreathing and twining of the vessels round these tubuli, exactly as round the motive fibres of muscles (*p*), whereby power and force are added to the motion: by the reservoirs beneath the

trate intimately, and form in some measure a distinction in their substance. In oxen, and even in human subjects, the large masses have at any rate internally a peculiar and well-defined shape, and seem to resemble polygonal pyramids. In these masses you will find precisely the same divisions in the human subject, the dog, and cat, and that they intimately pervade the substance of the kidneys" (n. 282). Furthermore, there is a similar division of the tubuli or uriniferous vessels in the interior of each mass or compages; and thus the kidney wears the imprint of motion universally.

(*p*) It is, I think, particularly worthy of observation, that the whole of the medullary, tubular, or striated substance, exactly resembles a fasciculus of motive fibres; the tubuli being constructed in the same manner as the motive fibres of muscles; that is, surrounded by luxuriant arterial and venous ramifications; shewing that the arteries, together with the nervous fibres which they carry with them to these ultimate goals, have the entire control of their expansion and constriction: notwithstanding which, they are so constructed, as to perform at the same time the office of tubuli, and to afford a passage to the urinous drainings. Hence we may, I think, infer without doubt, that these tubuli constrict and expand the whole of this compages by muscular action; and that the entire mass of the papillary or pyramidal substance, extends, retracts, contracts, and dilates, towards the infundibula; and by this means excusses its humor, both through the tubular passages, through the little mouths of the papillæ, and so forth. That the tubuli uriniferi are surrounded by vascular ramifications, is thus declared by Malpighi: "The blood-vessels proceeding from the incurvated branches, and passing downwards, embrace the urinary vessels in the manner of ivy: and in living animals, when the emulgent vein has just been tied, these blood-vessels are completely turgid; shewing that a larger quantity of the obstructed blood collects in them, than in the other vessels" (n. 282). Malpighi also likens these tubuli to muscular fibres, when he says, "The urinary vessels of the human kidney resemble solid and compact fleshy fibres" (*ibid*). Which is the reason why the old writers termed this substance, parenchymatous and carneous, and the papillæ themselves, carunculæ. "This substance," says Eustachius, "if we may judge from its appearance, is carneous, dense, and very solid and hard" (*De Renibus*, cap. ii.) That muscular substance answers to this description, will be shewn in my Rational Myology.

tubuli, subordinated to each other in a kind of series, and which yield when the tubuli are acting (*q*): by the ureters themselves, and their action on their contents, which is an image and representation of the action of the tubuli (*r*). When once we know the motion and structure of these organs, we can then understand what is their *modus operandi* in every state.

288. *That the kidneys expunge the stale serum from the arterial blood.* This is too well known to admit of a question: but the power and *modus operandi* whereby the blood of the kidneys removes and eliminates the serum alone from the middle of the torrent, and of this, only the worthless and redundant portion, without itself escaping either simultaneously or subsequently (*s*),

(*q*) Under the tubuli uriniferi, or at the end of the papillary substance, are the tubuli which proceed to the pelvis, and at the sides there are certain carunculæ which exercise compression. See n. 290. From this it may be seen how providentially and divinely all things are constructed,—in order that the freest exercise may be given to the expansile and constrictile motion, and a hollow field left underneath to enable the superior and more solid part to extend in every possible way. Something similar is also met with in other members, particularly in the cerebrum, where the ventricles give the power of motion, and determine its direction, as in the present case, towards those last boundaries, which are also the first; that is, towards the concave fundus.

(*r*) The only difference between the urinary tubuli and the ureters, is the difference between the whole and a part, or between great and small. The ureters are the urinary tubuli which convey the sero-urinary volume towards the bladder; but these least tubuli carry only little drops or minute streamlets of that ichor towards their bladder—the pelvis. The ureters are elastic, expansile, contractile canals, invested, moreover, with a kind of muscular coat, or with something very like one; this is also exactly the case with the tubuli of the kidneys, which are thus represented in the ureters. The ureters, according to Heister, are "made up of three coats," the middle of which is "a thin muscular coat" (n. 279). "The ureters," says Winslow, "are very elastic canals, which yield in every direction, and again directly regain their pristine calibre, unless they have been long and unnaturally stretched" (n. 281).

(*s*) It is indeed obvious and known generally, that the blood, in its progress to the minutest branches, frees and clears itself from serum, displacing it all round into lateral tubuli, in order that the blood may can be seen only by comparing the difference between the mode of fluxion of the genuine blood, and that of the worthless serum; although even after such comparison, unless we have previously explored the blood, these points will appear in but an obscure and glimmering light. The red or globular blood, considered separately from the accompanying serum, when driven by a slight impulse, tends to flow, and, where it has an opportunity, actually does flow, in gyres and spiral meanders; for its nature whirls and incites it into forms of the kind. Not so the serum, especially what is rough, uneven, sluggish, inert and gravitating, and lends no assistance of its own to any motive force: when such serum is acted upon, it neither endeavors, nor is able, to run forth in any but straight lines, in the direction of its tangent,—it is dragged unwillingly into every curve that it describes (*t*): hence it scrapes and presses heavily against the sides

flow by itself into the capillary offsets of the little twigs. But the ways and means are not so obvious, whereby one and the same stream, which is at once sanguineous and serous, and these two, as it is commonly thought, in the most intimate union, can be so separated in even the smallest vessels, that no portion of the blood shall escape in company with the serum; of which fact a most remarkable example is presented in the kidneys, through which the whole mass of the serum—if not the whole mass of the blood—passes, in the course of a few rounds of the circulation; for the serosity is separated from the whole mass of the blood, in these ultimate strainers particularly.

(*t*) This must always appear somewhat dubious or obscure to the senses; we mean, that the red blood, which is globular and pure, when left to itself, on any slight impulse, tends to go forth, consequently to be carried, not into a straight line, but into a sort of perpetual curve. The fact is proved in the first place by visible effects themselves; by the infinite circumflexions of the vessels into arches and circles, not only in the borders of the mesentery, but in all the extreme parts. Leeuwenhoek observed some thousands of such circles in a space a few lines in breadth: and they are visible to even the naked eye in all the extremes. Thus in the minute glands of both the body and the cerebrum, where the vessels are left in comparative freedom, and at their own disposal, they revolve into similar wreaths, belts, and windings, and institute their games more nimbly there than where the vessels are straight. It may be inferred from this, as a visible effect, that the blood, of its own nature and accord, desires and endeavors to run forth into similar of circumflexed vessels, and flies off through all the foramina and little mouths which open on the walls of the arteries, and through the little tubes proceeding therefrom, still following the direction of the impulse received (*u*). Thus while the blood

circles. This is also plain from the very nature of the blood itself. For the blood-globule, besides the elements which constitute its bodily part, contains also a spirit or prior blood, which is to the highest degree active, that is to say, in the perpetual enjoyment of active or living power; and by this principle the compound blood itself,—the blood-globule,—is actuated. That this principle is of such a nature that it aspires to nothing less than the perpetual-circular form of motion,—this will be shewn in our Analysis of the Spirit, and in our Doctrine of Forms. For in proportion as essences and substances are superior and prior, in the same proportion they are more perfect with respect to the form which they manifest by their fluxion, and which exactly resembles their substantive form itself. Consequently, such substances never tend into a straight line, but into a kind of perpetual and unlimited circular. From them the blood derives a similar nature, and thus it occupies the axes of the curved vessels. And as at the same time it is preëminently fluid, because globular, and as the serum is sluggish to the last degree, and delayed continually, because angular, therefore the latter must necessarily be displaced from the middle of the torrent of blood to the sides, and consequently be thrown out. But such is the strong impression on the mind, that all bodies driven by any force run out only in straight lines, that our proposition must necessarily appear paradoxical. But I will not dwell upon these prejudices of infancy, further than to remark, that they must be put aside if we really wish ever to arrive at the causes of things. Corpuscules which have no activity of their own, but require to be driven and pushed,—that is, which are not elastic, but act from gravity alone, and resist in proportion to their gravity,—when acted upon, are determined into straight lines, for they have no active principle which can bend or determine them into any form. Not so, elastic corpuscules: these react to the same degree that they are acted upon, and take from themselves even the direction of their motion. But this, as we said before, is still only obscure, and as it were dimly visible.

(*u*) If we lay it down as a rule, that the serum, in proportion as it is concreted, is in the same proportion shapeless, rude, and unfit for circular motion, and when impelled, tends into none but straight lines, it will follow as a matter of course, that if it be carried in a circumflexed and performs a gyre suitable to its nature, with increased celerity and force, in the arterial vessel, it always tends to the median axis, and displaces to the sides and circumferences everything discordant with it, or opposed to its peculiar fluxion; consequently the rude, inert, and obsolete serum; and the more discrepant the serum, the more the dissension, division, and rapidity of proscription increase. To effect this end, all the renal arteries fold and wreathe into circles and arches, as it were little stars and moons; and this, from the very threshold, or on the surface, close under the common coat, over the whole circumference of the kidneys (*x*). And the arteries or ramifications

gyrating vessel, it must impinge on every part of the vessel, and never comply with the stream of blood, but be dragged unwillingly from point to point; and as soon as it meets any little opening, it must find an asylum therein, and first by the force it has received, and next by the impulse of the vessel into which the opening leads, escape from the vortex. That all these orifices lead into little ducts, which unite to form one urinary tubulus, may be seen below, n. 289 (*a*).

(*x*) That the renal arteries, by their mediate and ultimate ramifications, form perpetual arches and circumvolutions of this kind, that is, describe gyres exactly conformable to the natural fluxion of the blood,—this is exhibited by Ruysch in several of his plates, (see Mangetus, *Theatr. Anat.*, tab. lxiv., fig. 2, 3, 4, 6,) where he also shews the particular appearance of those gyres; and further, that very similar circumflexions are in some places continued all the way into the interior parts. Respecting these arches, Malpighi says, "When the membrane has been recently removed, a number of very short, round corpuscules present themselves, rolled up like little worms, and not differing greatly from those which are found in the substance of the testes, when the coverings are removed, or the testes themselves divided through the middle. . . . In dogs and other animals, we have a clear view of these vermicular productions, which after making the above-mentioned short reflexions and circumvolutions, go straight along the dorsum of the kidney, following the external surface, towards the pelvis" (n. 282). And Winslow says, "By fine injections we discover an infinity of small capillaries, which run in various directions between and round the different portions of this texture," &c. (n. 280). Thus much appears certain, that in no other part of the body is there, on the surface, such a continual vascular circumflexion and gyration, as in the kidneys.

which enter their intimate substance, do not terminate there, but are presently reflected, and form similar glomeruli, or balls of convolutions (*y*); in order that they may thus come into their natural play: moreover, they encircle the fistulæ Bellinianæ also in a similar manner; and this, at the beginning, the middle, and the end, where at last they are obliterated and escape from the gyre (*z*). Thus throughout, by instituting these natural games, they drive out, expel, and send away the hostile bands of the stinking and urinous serum.

289. From a careful examination of the interiors and little intestines of the kidneys, the following appears to be their mode of excreting and eminging the urinous ichor; namely, that the arterial blood circulating at full speed through the vessels, throws the sluggish and tardigrade serum to the circumferences; constantly agitates it, as streams agitate the sand; and squeezes

(*y*) That the vessels are reflected from the interior substance, by a process of revolution, to the surface, and there form little balls and arches, is thus declared by Winslow, "The blood-vessels, after entering the kidneys, ramify in all directions, and these ramifications give out capillary branches, which go all the way to the surface, where they appear like irregular stars" (n. 280). Malpighi says, "The emulgent arteries and veins, on entering the cavity of the kidney, divide into several considerable branches, and . . . . curve into an arch towards the gibbous part of the kidney, and go to meet, and are united by anastomosis to, other branches, and from this arch again, towards the gibbosity of the kidney, give off numbers of branches, which themselves form a kind of reticular interlacement, and run out between the little lobes into which the congeries of urinary tubuli is subdivided, all the way to the external surface, from which they are reflected a little to the sides, and terminate at the glands" (n. 282).

(*z*) That the vessels surround the tubuli uriniferi or Belliniani, in the manner of ivy, see above, note (*p*), where Malpighi graphically describes their circumflexion. They form indeed similar circles in all directions. But Malpighi has also observed, that the vessels seem to be obliterated at the end of the tubuli. "These vessels [which surround the urinary vessels after the manner of ivy] are obliterated," says he, "near the end, towards the papillary substance" (n. 282). In other words, when they have performed their office, and thrown out the hostile serum, they change into filaments, which contribute to form the membranes of the papillæ and infundibula.

it through the foramina into the fine, straight ducts proceeding from them (*a*). This urinous excretion is driven down through these ducts or tubuli, (which resemble tufts of vessels,) by a constrictile action, into either the glandular corpuscules (*b*), and

(*a*) The origin and nature of the abundant cortical substance of the kidneys, which appears to be only partially vascular, are unknown. For this substance admits none but the finest colored injection, like that employed by Ruysch; and in the natural state, it is not red, but white, and clearly proceeds from the blood-vessels. "The cortical substance," says Winslow, "will be observed round the whole circumference; and by the microscope we perceive it to be of a spongy, granular, and slightly waving texture; all its parts adhering together closely in radiated lines. . . . By fine injections, and during inflammation, we discover an infinity of small capillaries, which run in various directions between and round the different portions of this texture" (n. 280). "Certain it is," says Malpighi, "that on the external part of the kidneys, there are fine canals in immense numbers, and which are not muscular [carnei] but membranous, and seem to be analogous, both in structure and use, to other excretory vessels, as the salivary ducts and the pori bilarii" (n. 282). If we may be allowed to judge from the connexion, use and origin of similar structures in other parts of the body, these vessels seem to be prolongations from the foramina of the blood-vessels, of the arteries particularly; from which source very many of the membranes in the body derive their filaments. Thus they are the first and simplest of all the urinary ducts, and take up the serum immediately from the arteries, and convey it to the tubuli uriniferi; for that they are continuous with those tubuli will be seen presently. Wherever in the universal body any secretion takes place from arterial blood, particularly in the extremes, (as for instance, under the skin,) similar offsets exist in immense numbers, resembling white, glistening vessels, and which draw off the serosities, and do not admit blood, unless it be violently intruded, and unusually heated, as in inflammation: in the same manner exactly as in the cortical substance of the kidneys, which is filled by Ruysch's injections, and therefore is classed among vascular structures.

(*b*) Respecting the renal glands, Winslow says, "By the microscope we see a number of small red corpuscules, more or less round, and disposed almost like bunches of currants" (n. 280). Littre and Malpighi treat of these corpuscules at some length. "In all the kidneys," says Malpighi, "that I have hitherto been enabled to examine, thence into the tubuli of the striated, medullary or parenchymatous substance, or else immediately into the tubuli themselves, by which it is taken up and collected from all the little confluent rills (*c*). From these tubuli uriniferi or Belliniani, it is again

I have found a multitude of minute glands; and this, uniformly, both in quadrupeds, tortoises, and human subjects" (n. 282). And among other particulars, he adds, that they may be shewn very distinctly by injections, both immediately under the membrane, and "between the fasciculi of the urinary vessels, and in the interstices, appended, as apples are appended to the branches of the apple-tree, to the blood-vessels filled with the black injection, and which moreover have a beautiful arborescent appearance" . . . . and that "they seem to be round, like the spawn of fish:" he also describes their connexion with the blood-vessels *(ibid.)* Eustachius and Bellini mention and describe these glands. But in any case, the finest or first ducts before spoken of, are continued into these glands, and also into the tubuli uriniferi; and even appear to generate and form them; consequently, deposit their humor in both the glands and tubuli. According to the order of nature in other parts, corpuscules of this kind are usually prefixed in the manner of little heads to tubular fibres proceeding from them. But inasmuch as there is also an immediate communication with the urinary vessels, therefore the quantity of such glands appears to vary in different subjects.

(*c*) All these particulars are clearly deducible from our knowledge of the connexion of the fine ducts already-mentioned with the glands, and of both the ducts and glands with the tubuli uriniferi: they rest upon experimental evidence. With respect to the connexion of these minute ducts with the glands, Malpighi says, "They [the glands] are so connected with the branches of the arteries, that they grow from, and are appended to, the internal branches, and sometimes to the external branches, reflected inwards, and divided into numberless twigs" (n. 282). They are, moreover, somewhat whitish, and colored injection does not enter them, although the little arteries themselves wherewith they are surrounded, are colored thereby. With respect to the continuation of these little ducts into the urinary tubuli, Malpighi again remarks, that certain circumstances afford "a proof, that there is a connexion between the vermicular vessels composing the external surface of the kidneys, and the descending vessels, all the way to the pelvis," &c. *(ibid.)* Ruysch, in his Tabulæ mentioned above, (see n. 283,) states several times, that the serpentine vessels of the cortical thrust downward by a similar action, and driven out, indeed, into a certain little cavity,—through the papillary orifices into the calices or infundibula, which are a kind of vessels narrowed inferiorly (*d*): in this way it is expressed through the pervious tubuli or pipes, along the lateral surfaces, into more capacious reservoirs, that is to say, into the pelvis. Finally, the drainings collected in this higher bladder are thrown out with a similar but more compound constrictile power, through the ureters, and by these again, as above, through narrow channels, into the grand cavity and receptacle of the bladder (*e*). Thus the same

substance degenerate into urinary ducts. "It is indeed true," says he, "that by filling the arteries the ducts are also filled, so that the two appear to be continuous" *(ibid.)* Respecting the continuation of the glands into the tubuli uriniferi, Malpighi says, "The glands, as I think, probably correspond to the urinary vessels of which the mass of the kidneys is composed" (n. 282). In another place he says more confidently, "The urinary vessels arise from the glands, and proceed to the papillary body," &c. *(ibid.)*

(*d*) We gather from the unanimous voice of all our authors—of Eustachius, Ruysch, Bellini, Malpighi, Winslow, and Heister—that the tubuli uriniferi, taking up the droppings of urine from the numberless little ducts, at last eminge them at the end of the papillæ, and in fact, along the membranous parietes wherewith the pyramidal fasciculi are invested. This is clearly proved by the extremities of the papillæ being perforated by many little holes, as pointed out by Ruysch (n. 283), and as described by Winslow in the following words, "At the end of each papilla we see, even without a microscope, in a small depression, several very minute orifices, through which little drops may be observed to issue, when the papillæ are compressed. These are drops of urine, which being filtered, partly in the cortical, partly in the medullary substance, afterwards pass through the tubuli of the papillæ, and are discharged by these orifices" (n. 280). Malpighi speaks constantly of the termination of the same tubuli on the membranous parietes; the case being here exactly analogous and parallel to that of the ducts which proceed from the infundibula to the pelvis, and from the ureters to the bladder: shewing that the excretions in the smallest and largest parts agree in this particular—that they follow the course of the membranes.

(*e*) Respecting the termination of the ureters in the bladder, Heister says, "They [the ureters] terminate in the posterior inferior action is present and persistent from the first and highest springs and streams, to the last and lowest. The most minute and the *primitive ureters* are the single threads of the cortical substance,—those which proceed from the arterial vessels, and pass into either the miliary glands, or immediately into the tubuli uriniferi. The *middle ureters,* originating from and constructed by the primitive ureters, are the fibres of the medullary substance,—the tubuli uriniferi,—which terminate in the papillæ and cavities. The grand and *ultimate ureters* are what extend from the pelvis to the bladder. By these last, therefore, we may be instructed respecting those which immediately precede them, and by both the last and the middle, respecting the first; for the one is the image of the other, with only such a difference between them as exists between what is prior and what is posterior (*f*).

290. Thus in no place does the recrementitious serum of the blood pass off spontaneously, nor is it here invited with the welcome that is usual in the other organic structures; but being sluggish, resisting, and obstinate, it is expelled as an enemy, by forcible motion and sheer violence, and *along the whole way, is detruded from tube to tube, and from cavity to cavity* (*g*). That

part of the bladder, passing obliquely in between its membranes, and opening into the bladder in a valvular manner, by narrow orifices" (n. 279). For a description of the short tubes extending from the infundibula to the pelvis, see above, n. 280.

(*f*) The discriminations and connexions between prior and posterior things, will be explained in the Doctrine of Order and Influx. They respect both quantity and quality. In proportion as things are prior, in the same proportion they are more universal, more multiple, and more perfect; resembling, or assimilating to, each other in general, not in particular. Hence a proportion may be obtained, not between operations only, but in some measure between dimensions also; the space which the superior things occupy being first given. As the tubes correspond to each other mutually, so also do the cavities prefixed to them. The same proportion that exists between the bladder and the pelvis, exists also between the pelvis and the calyx, and between the calyx and the gland; that is, with respect to cavities and modes of action; but with a difference of perfection. To enter fully into these proportions, would be tedious, inasmuch as they all follow one law.

(*g*) There is this difference between the *modus operandi* of the this is the case, is demonstrated by the serpentine wreathings and windings of the vessels of the cortical substance, from the surface, to the fasciculi formed by the union and straight prolongation of the medullary or tubular substance (*h*): by the conterminous and interjacent glands: by the urinary tubuli themselves being wide above, and constantly narrowing as they

liver, &c., and the *modus operandi* of the kidneys,—namely, that in the kidneys there is a violent detrusion into the underlying cavities or receivers, and not a kindly invitation; which detrusion is the cause of the motion treated of in n. 287. But this detrusion is not continuous from the first blood-vessels all the way to the pelvis or the bladder, but is broken up into certain determinate lengths, or stadia. The first stadium extends from the arterial vessels to the glands; the second, from the glands, through the tubuli, to the calices; the third, from the calices to the pelvis; and the fourth, from the pelvis to the bladder. It is, however, continuous within each stadium, from cavity to cavity: thus one receptacle succeeds another, and there is a diverticulum at the end of each. Such is the progression from prior to posterior, and from posterior to postreme. But when the urinary juice is expressed from the tubuli into the cavity under them, then, according to the laws of the before-mentioned Doctrine, the uppermost cavity, as the head, first compresses itself, or begins the motion of compression, which thus proceeds and creeps on in order, all the way to the end, or to the cavity beneath, which at this time is in its state of expansion; and so on. Moreover, whatever the little head does, whether that head be a gland or a cavity, the whole tubulus continued from it, does the like, as we before observed when speaking of the liver.

(*h*) We have already treated of the vermicular winding of the cortical substance, which is admirably shewn in Ruysch's Tabulæ. Inasmuch as this substance is composed of the very earliest ducts, which receive the first urinous drippings, therefore it is inflected and wavy, after the manner of the ureters. Its very inflexion indicates, that there is an alternate expansion and constriction of its little ducts; also, that the medullary substance under it, swells up, and rises out of its place, during the time of the diastole. Such being the case, therefore the cortical substance necessarily presents the appearance of vermicular fluxion: for the individual ducts of this substance are not fettered in their course in the same way as the tubuli under them; which latter, being congregated and compacted into fasciculi, act in common, even above themselves, by extension.

descend, giving rise to the polygonal pyramidal bodies, and to the mamillary form, or the papillæ (*i*): by the foramina with which these teats are perforated: also, by the infundibula, their form, their contiguity, their speedy contraction into tubes, the insertion of these tubes into the pelvis, the narrowing of the pelvis to form the ureters, the final contraction of the ureters into tubes, and the implantation of the tubes in the bladder: in a word, by the continuous progression from wide to narrow, and the sudden change from narrow to broad; thus from the minutest threads to the glandular follicle, from the follicle to the urinary tubulus, from this to the cavity and calyx, from this to the pelvis, and so on through the ureters to the bladder. Thus all things conspire to testify, that the urine is urged downwards

(*i*) We have three convincing proofs of the fact, that the tubuli uriniferi are wider above than below, in the same manner as the ureters themselves, which have a broad pelvis placed above them, and end in contracted tubuli. The *first* proof is, that glandular follicles are placed upon the tubuli, in the same manner as the pelvis is placed upon the ureters; for according to Malpighi, the follicles and tubuli are continuous bodies. The *second* is, that the tubuli uriniferi all together form a kind of fibrous fasciculus, of a pyramidal figure, reaching from the surface, where they swell to a considerable size in the manner of lobes, to the intimate or central parts, where they terminate in papilliform points: thus as the fasciculi themselves, in their descent, contract gradually, so also do the tubuli with them. The *third* proof is, that the capillary vessels, at the end of these tubuli, begin as it were to vanish away, and contract out of sight, according to Malpighi's account. If then the several tubuli constantly decrease in diameter as they descend to the teats of the mamillæ, then the urine injected into them must necessarily be detruded by violence, or force of compression. For the urinous humor, naturally sluggish, and sometimes viscid, grumous, and uneven, never passes down of its own accord; as we may be certain from the glairy matter which frequently lines the tubuli, and in some cases even distends the ureters themselves into a ventricular form. Moreover, at the sides of the papillæ there is a peculiar caruncular substance, not described by our authors, but admirably represented by Eustachius; (see his *Tabul. Anat.*, tab. i., fig. 15, *b b b;* tab. iii., fig. 10;) and which, if carneous [or muscular,] must greatly assist in compressing and opening the tubes of the papillæ.

by a kind of continual pressure, and this, from vessels into cavities, to the third and fourth degree of composition.

291. *That from these outcasts, throughout the whole of the passage, the veins recover and absorb the finer elements of the blood.* This is proved by the insertion of the veins into the glandular follicles; into the tubuli uriniferi; into the membranes bounding the lobular pyramids; into the parietes of the infundibula and pelvis; into the internal coat of the ureters (*k*), which are also dotted with glandular corpuscules, just like the intestines (*l*): by the application and implantation of the tubuli, as well as of the larger ducts, to and upon the coverings, that is, to and upon the coats investing the pyramidal bodies and the cavities themselves; whereby the humor, as it drips out, keeps close along the path of the membranes, where the little venous mouths open: moreover, by the action of the tubuli and ureters upon the urine which they contain; and which prevents any part thereof from passing by without being touched and saluted by the parietes, and by the little veins grouped upon them. Thus if anything that can still be of use, is interspersed among

(*k*) Our authors treat largely of the venous ramifications proceeding to both the glands, tubuli, calices, pelvis, &c. See Malpighi, n. 282. Thus there is scarcely a point where we do not find some little vein, or reabsorbent vessel, terminating. The vascular ramifications, according to Malpighi, "form a kind of reticular interlacement, and run out between the little lobes into which the congeries of urinary tubuli is subdivided, all the way to the external surface, from which they are reflected a little to the sides, and at length terminate internally; their last twigs ending at the glands," &c. (n. 282).

(*l*) Respecting the glands of the ureters, Heister says, "The nervous [or internal] coat [of the ureters] sometimes exhibits glands" (n. 279). Morgagni makes the same observation, n. 283. The ureters, in these respects, are almost like the intestines, inasmuch as they have valves similar to the valvulæ conniventes. See Coschwitz, *Dissert. de Valvulis in Ureteribus.* Winslow also remarks, that "the innermost coat [of the ureters] is lined by a very fine membrane, which covers an exceedingly delicate vascular network. It is slightly granulated like shorn velvet. It has several longitudinal rugæ, which are intersected and in a manner interrupted by a great number of small transverse rugæ" (n. 281).

the urine, it is absorbed, and returned into the blood; and nothing but the feculent and effete humor is rooted out or thrown away. But of the blood itself and the finer serum thereof, there is a continuous circulation within the vessels,—from the arteries into the veins, which are united by perpetual anastomoses both on the surface of the kidneys and in their intimate substance. The veins take up this blood, together with the richer and finer serum that is thrown among the urinous portion: they lead it away through the level plains between the greater and lesser divisions of the viscus, and at length convey it to the surface, into certain common branches; and thus export both, through increasing channels, into the emulgent veins, and from them into the vena cava. As the veins recover and absorb the corporeal parts of the blood, so (*m*).

(*m*) Inasmuch as absorbent vessels of two kinds, to wit, venous and lymphatic, are evidently brought together in all the cavities, tubes, members, organs and viscera, therefore it is of great importance to understand, what each kind of vessels absorbs respectively from the liquids applied to it. From a careful comparison and consideration of all the circumstances, I am fully convinced, that the *veins* imbibe only those things which belong to the bodily part of the red blood; as in the case of the infinite host of veins in the omentum, and the other fatty planes and beds, where they demand back only those matters which are to be incorporated with the blood: likewise in the skin, where they imbibe an abundant moisture, full of highly volatile salts and nitrous elements: in the intestines, stomach, œsophagus, fauces and tongue, where they absorb the chyle itself, of which the blood-globule is formed: and the same remark applies to the veins of other organs, and here, to those of the kidneys. For this humor is adapted to their comparatively large lips and orifices, and at the same time to the blood itself, which craves the serum, for the purposes of self-renovation, as ardently as a bride desires the favors of her husband. But the lymphatics—putting out of question those of the mesentery—are so fine and delicate, particularly their first lips, that they do not admit any other than the purest species,—anything, in short, but the lymph that is full of the spirit and life of the blood; that is to say, than the white, purer and prior blood, set free from the blood while in a state of resolution, and which has reached the summit or surface of the viscus. These views are supported by the nature of this lymph as explored chemically, by its passage into the receptaculum chyli, and at

292. *The lymphatics redeem the spirituous and essential parts, and restore them to the blood.* For the proper coat of the kidneys, or that which lies close under the common coat, consists of two thin laminæ, with a fine cellular tissue enclosed between them (*n*). This tissue accompanies the coat into the interiors of the organ, insinuating itself by the interstices that separate the lobes (*o*), and leads continuously to the papillæ, calices, and pelvis, where it meets another portion, and the two laminæ coming together, invest all the parts (*p*). This tunic, throughout its

length into the subclavian vein, where, as a bridegroom, it is to meet its future partner, the chyle; or where, as a soul, it is to meet its body. Respecting this lymph, we refer the reader to our Analysis, on the Glands generally.

(*n*) The surface of the kidneys is covered by two layers of cellular tissue; one, belonging to the common coat, and continuous with the peritonæum; the other, situated between the two laminæ of the proper coat, and very thin, like fine down or wool. Respecting the latter, Winslow says, "The proper coat of the kidneys is composed of two laminæ, having between them a very fine cellular substance, which may be made visible by blowing in between the laminæ" (n. 280).

(*o*) For the kidneys are divided into lobes, which proceed pyramidally towards the interiors. These lobes are their most general divisions, through which the fine cellular membrane insinuates itself, and penetrating the more internal parts, connects them with the most external; and thus diffuses, and most distinctly distributes and divides the common cause into the particular causes of the parts. Respecting this insinuation [through the lobes], Malpighi says, "In oxen, tortoises and birds, the exterior surface of the kidneys is divided by grooves and depressions into a number of large polygonal, and for the most part hexagonal masses. The same kind of composition is sometimes clearly observable in human adults, and frequently in the human fœtus; the kidney being made up of small portions separated by deep fissures; and even in adults, traces of this division are always present internally" (n. 282). And Winslow says, "The internal lamina sends a number of processes into the substance of the kidneys, from which it cannot be separated without tearing" (n. 280).

(*p*) This coat covers the surface, and passes in by the divisions, (which are very wide and deep in the fœtus,) towards the extremities of the papillæ, and towards the infundibula, and there meets another portion of itself, and thus coming from two different directions, it seems progress, digresses, and passes into the vessels, and overwraps and surrounds all their branches, interlacements, and reticulations (*q*); and it also appears to dip into the spaces between the tubuli uriniferi (*r*). Thus while the follicles, and the ureters, primitive, middle, and ultimate (*s*), carried away by their alternate systole, press the serum close against all points of their parietes, the purest lymphs and the winged spirit take swift flight through the little foramina and pores, and wherever they alight, they fly into the fine cellular tissue, and rise at all points through the continuous down, to the surface, into the coat surrounding either the kidneys, the arteries, or the veins. There the lymphatics joyfully receive their new guests, beset them with their little lips, load them with kisses, and congratulating them on their rescue from the unworthy serum, transport them to the receptaculum chyli, as it were linking together their wings for the purpose, in order to restore them to the blood (*t*). The suprarenal glands also, according to their ancient custom,

to subtend and enter the pelvis. For the lamina which penetrates the interiors, and expands over the bottom of the calices, is continued through the foramina and shorter ducts, to the inside of the pelvis, whose outside is covered by the same membrane coming round thither along the surface; so that this coat invests the hollow centre, within and without, and commands and subjugates it entirely, and causes it to be absolutely dependent upon both the general action of the kidneys, and the action of their veriest singulars, or minutest parts. "The proper membrane," says Winslow, "contributes to form the pelvis and calices or infundibula" (n. 280).

(*q*) According to Winslow, "The proper membrane of the kidney passes to the sinuosity where the renal vessels enter, and in the form of a sheath or capsule accompanies all their ramifications into the intimate substance of the organ" (n. 280).

(*r*) Eustachius mentions certain grooves in the cortical and medullary substance, and thinks that the urine passes down through these grooves: but they seem, from his description, to be nothing more than interstices between the tubuli, formed by the cellular tissue.

(*s*) Respecting the least, middle and ultimate ureters, and the corresponding cavities or follicles, see above, n. 289.

(*t*) Respecting the lymphatics of the kidneys, see Nuck, *Adenographia;* and Heister, n. 278, above.

draw forth for their own benefit and necessities, and rescue, whatever portion they desire. The veins and lymphatic vessels, when they have quaffed their new beverage,—the vintage being in a manner ended,—announce that the vines are exhausted, and that what now remains is mere urine.

293. As the blood is continually making its circle of life; that is to say, is in a constant revolution of birth and death; as it dies in its old age, and is regenerated or born anew; and as the veins solicitously gather together the whole of its corporeal part, and the lymphatics, of its spirituous part; and successively bring it back, refect it with new chyle, and restore it to the pure and youthful blood; and as the kidneys constantly purge it of impurities, and restore its pure parts to the blood;—so likewise Man, who lives at once in body and spirit while he lives in the blood, must undergo the same fortunes generally, and in the progress of his regeneration must daily do the like. Such a perpetual symbolical representation is there of spiritual life in corporeal life; as likewise a perpetual typical representation of the soul in the body (*u*). In this consists *the searching of the heart and reins,* which is a thing purely divine.

(*u*) In our Doctrine of Representations and Correspondences, we shall treat of both these symbolical and typical representations, and of the astonishing things which occur, I will not say in the living body only, but throughout nature, and which correspond so entirely to supreme and spiritual things, that one would swear that the physical world was purely symbolical of the spiritual world: insomuch that if we choose to express any natural truth in physical and definite vocal terms, and to convert these terms only into the corresponding spiritual terms, we shall by this means elicit a spiritual truth or theological dogma, in place of the physical truth or precept; although no mortal would have predicted that anything of the kind could possibly arise by bare literal transposition; inasmuch as the one precept, considered separately from the other, appears to have absolutely no relation to it. I intend hereafter to communicate a number of examples of such correspondences, together with a vocabulary containing the terms of spiritual things, as well as of the physical things for which they are to be substituted. This symbolism pervades the living body; and I have chosen simply to indicate it here, for the purpose of pointing out the spiritual meaning of *searching the reins.*

294. Sufficiently evident tokens exist, to enable us to infer with certainty, *that the kidneys also draw off the sluggish and noxious phlegm from the peritonæum, and perhaps some portion of the effete fat.* The reasons which indicate this, are, that the cellular texture of the peritonæum, (which is generally full of an aqueous and oily humor, more particularly after a few years of life are passed, and not, as in embryos, of a lacteal and chyliferous humor (*x*), is connected immediately—as well as by the mediation of the colon and cœcum—with the cellular texture of the kidneys and ureters, just under their common covering (*y*),

(*x*) It was shewn in the Chapter on the Succenturiate Kidneys or renal glands, that in embryos none but the purest serum permeates the cellular tissue of the peritonæum and of the other viscera. But after birth, when the form of government is completely changed, that is, when nature is given over to the dominion of the human will, and when innumerable intestine motions of the mind, the animus, and the body, tear the republic to pieces, and momently throw the blood into all sorts of different states, and inspissate, inflame and disorder both it and the serum, precipitating them into quicksands and whirlpools,—then the blood and serum change, and become turbulent, and after having circulated a certain number of times, they are in great part effete, and must be thrown off by the urinary vessels and the skin. Now serum of this quality frequently encumbers and clogs the tissues of the peritonæum.

(*y*) The general opinion is, that the common coat of the kidneys is continuous with the peritonæum; for it is similarly circumstanced with the common coat of the ureters. But the peritonæum covers only the anterior part of the kidneys. So far, therefore, the same cellular tissue accompanies both the common coat of the kidneys, and the peritonæum; and although not in the way of duplicature, yet there is no denying the continuation of the tissue itself; from which if the serum is to be excreted, there must necessarily be some general place of discharge, to avoid dropsical inundation. "This [tissue]," says Winslow, "was for a long time considered as a duplicature of the peritonæum, the true membranous lamina of which, however, covers only the anterior surface of the kidneys, and consequently they lie external to the sac of the peritonæum; because the portion of peritonæum which covers them, cannot be regarded as an entire coat; so that the only common coat they have is the cellular tissue" (n. 280). Furthermore, one of the kidneys is connected to the colon; the other, to the cœcum, where the and is likewise continued around the trunks and larger branches of the emulgent vessels, which follow the path of the fissures and interstices, towards the interiors of the organ (*z*). Furthermore, that as the renal capsules waste, and recede some little from the kidneys, this office, either wholly or in part, is transferred to the kidneys themselves, is rendered very probable by the circulation of this humor, by the necessity for its discharge, by the situation of these viscera (*a*), by the capacity of the emulgent vein relatively to the artery (*b*), by the necessity for

large and small intestines meet. That the common coat of the ureters is also continuous with the peritonæum, is stated by Winslow, as follows, "Besides their proper coats, the ureters are invested also by the cellular tissue of the peritonæum, the membranous lamina of which covers about two thirds of their circumference" (n. 281).

(*z*) That the [cellular] tissue of the same coat covers also the large vessels, is thus stated by Heister, "The vessels of the kidney, like those of the liver, are enclosed in a membranous or capsular production of the peritonæum" (n. 278). And Winslow says, "[This tissue] invests the renal arteries and veins in the form of a cellular sheath" (n. 280). Whither it accompanies them, see the same author, n. 281.

(*a*) Respecting the circulation of liquid through the peritonæum, the situation of the kidneys, (whereby they are enabled easily to derive into them ichor of this kind,) and the removal in some degree of the suprarenal glands from the kidneys, as mentioned by Morgagni, see the preceding chapter.

(*b*) That the emulgent veins are of greater capacity and diameter than the emulgent arteries, see Eustachius, *Tabul. Anat.*, tab. i.; and particularly Ruysch, in Mangetus, *Theatr. Anat.*, tab. lxiv., fig. 2 and 5, and Nuck, in *ibid.*, fig. 1, where the vein is delineated as very large in comparison with the artery. The same circumstance was remarked respecting the vessels of the renal glands in the last chapter. It shews, that a greater quantity flows back from the kidneys, than is seen to flow into them by way of the arteries, notwithstanding the quantity of stuff discharged into the ureters and bladder. The superabundant contents of the venous channel, appear to be derived from no other source than the peritonæum: unless we suppose the reflux from the emulgent vein into the cava, to be slower than the influx through the artery, and that hence the vein swells much more than the artery.

the dilution of the outcast and inactive urine (*c*), and by many other considerations (*d*).

295. *That these results are brought about with infinite variety, —differently according to every state which is induced, either naturally or accidentally, by internal, intermediate, or external causes.* This is evident from the actual effect, in other words, from the urine, which is continually varying. With respect to *color*, for instance, it may be either watery, transparent, opaline, greenish, grey, yellow, flame-colored, murrhine, lateritious, ferruginous, brownish, or blackish. With respect to *smell*, it may be either ambiguous, inodorous, ammoniacal, sulphureous, fragrant, or aromatic; or rank, putrid, and stinking. With respect to *taste*, it may be smooth, acrid, saline, mawkish, styptic, or nauseous. With respect to *temperature*, it may be either mild, moderate, febrile in various degrees, boiling, or fiery. With respect to *consistence*, it may be limpid, thin, turbid, thick, viscid, thready, pellicular, crusted, or cloudy, and it may have

(*c*) If we pursue the course of the larger vessels, and attend to the circumstance that the cellular tissue continuous with the peritonæum, ceases not far from the entrance of the branches, we shall find that this humor is poured into the calices, and at the same time mediately into the pelvis, and lastly into the ureters, but not into the tubuli uriniferi: thus, that the whole of the clear fluid is not reabsorbed by the veins and lymphatics, but that the lotium or sediment collected in the pelvis, seems first to be diluted with this humor, and prevented from forming glairy concretions, or becoming viscid and glutinous: so that the urine, by means of this vehicle, is more easily detruded through the ureters.

(*d*) That the kidneys draw off the effete and rancid fat by the same channel, may be inferred from the quantity of oleaginous matter with which the urine is sometimes impregnated; from the fat which fills the cellular spaces in their coats, and even the larger interstices; and from their connexion with the colon and cœcum, to which the obsolete portion of the fat of the omentum, mesocolon, and mesentery, is naturally derived, as was shewn in the Chapter on the Omentum. From these circumstances, I say, it may be inferred, that these fluid fæces, which do not exude by the skin, and are not effused into the cavity of the abdomen, or if effused flow back into the peritonæum, are also carried off and eliminated by these common excretory organs.

either a grumous, sabulous, or other deposit. As *tried by chemical tests,* it may exhibit the most various saline, phlogistic, phosphoric, oily, or earthy composition; it may be thick or thin, sanguineous, serous, or aqueous (*e*). In *quantity,* it may be either copious, sparing, or uncertain. It is different in every genus of animals, in every species, and in all individuals; according to their different ages, natures, temperaments, diseases, ailments, and affections: it sometimes varies within the day, and even within the hour. To these varieties conspicuous in the effect, correspond, as proximate causes, the same number of changes of state in the body of the kidneys (*f*); that is, either

(*e*) The urine has been the subject of innumerable chemical analyses, by which its properties have been explored generally: but specifically, such exploration must ever be impossible; for the properties of both the kinds of blood, arising from infinite causes, are as infinitely various as those causes; and the properties of their secretions and excretions are of course equally various. It may be sufficient to hear Boerhaave's account of the chemical constitution of the urine. "The urine," says he, "contains the watery part of the blood; its sharpest, most subtle, volatile, and alcaline salt; a very acrid, thin and volatile oil, almost in a state of putrefaction; also, a very fine, volatile, and impalpable earth," &c. (*Inst. Med.*, n. 375.) For further particulars respecting the water, salt, oil and earth, here mentioned, see the same work, n. 376—379.

(*f*) Every organ, viscus, and member of the body is so formed, that it may assume and undergo innumerable and, indeed, infinite changes of state; that is to say, with respect to *substance,* it may be altered, constringed and dilated into the most widely different forms, and put on shapes suitable to all the various impulses of the blood and spirits. And this is still more perfectly the case with the parts and their singulars or individua; for every substance is inclined and adapted to changes of state, in exact proportion to its priority and purity: in the power of change consists its highest perfection. Notwithstanding this, it must be of such a form or construction, that while it varies its states in infinite ways, it still subsists and remains constantly in integrity with respect to essence. With a view to this power, all the parts, particularly the interior and finer parts, are also comparatively freer, spontaneously active, perfectly mobile within their sphere of forces, and able and ready to contract and dilate, whenever the least sign of a cause of change appears. This is the case with the muscles, in the whole kidney, or in the graduated divisions and subdivisions of the parts, or in the vessels, the urinary tubuli, or the fibres: for the more numerous and indefinite the states into which any member, organ, gland, or muscle is capacitated for changing; and the greater the degree of subordination wherewith the general states correspond to the state of the individual parts, or the absolute singulars, the more perfect is such member or organ in its nature. Hence the embryonic, infantine, growing, and undiseased kidneys, are more perfect than the adult, senile, and diseased. But as these changes of state correspond, so they are referable, to the causes which alter the blood and its serum, and the spirit of the blood and its lymph; which causes are either external and belong immediately to the body, or internal and belong to the animus, or innermost and belong to the mind or proximately superior animus: and they are either natural, that is, necessary, or contingent. These very causes, however, are infinite; as numberless, indeed, as the causes of all the diseases of the body, of all the disorders of the animus, and of all the affections of the rational mind. Hence the vast diversities in the urine; and the prognostications, conjectures, and opinions to which these diversities give rise.

and hence every fibre thereof is capable of being compelled and excited to its own little motion, and actions of the most different kinds are produced, exactly according to the form of the essentials which determine the will. Each muscle in particular has such a power, and so have the muscles reciprocally and collectively, as we see in the countenance, which represents every state of the mind in so exquisite a manner. All the viscera have a similar power, and hence their changes of state are innumerable. As there are changes of state in substances, so there are changes of state in their forces and motions, (for the motion may be either quick, slow, strong, weak, obtuse, acute, even, or uneven, &c.,) and so likewise in all their other accidents; hence the infinite diversity of operations and effects. But to explain this vast subject distinctly would require volumes: we shall treat of it more at large in our Chapter on the Cerebrum, and its Cortical and Medullary Substances.

CHAPTER XV.

THE URINARY BLADDER.

296. HEISTER. "The urinary bladder is a hollow, membranous organ, something like a pear in shape, situated in the pelvis, and intended to collect and expel the urine. In adults it is capacious enough to hold about a pint of liquid. 1. In the human subject it is connected in a singular manner by the peritonæum to the os pubis, which is not the case in brutes. 2. It is connected with the parts of generation, by the urethra. 3. With the umbilicus, by the urachus and umbilical arteries. 4. In the male subject, with the rectum. 5. In the female, with the vagina. It is divided into a neck, a body, and a fundus. The coats of the bladder are thinner in the body and fundus than in the neck of the viscus. Its blood-vessels come from the hypogastric, umbilical, and hæmorrhoidal vessels in the male, and in the female from the uterine vessels also. Its nerves are from the intercostals, and from the sacral nerves particularly. Its lymphatics, so far as I know, have not been noticed excepting by Zeller. The bladder consists of three membranes. 1. The first, called the common membrane, is continuous with the peritonæum, and covers only the fundus: there is usually some fat found underneath it. *(Comp. Anat.,* n. 222.) The doctrine of a cellular coat in the bladder, may, I believe, be supported by weighty reasons; it is evident that between the external and muscular coats there is usually, if not universally, a quantity of fat, and that, often considerable; and that this is secreted, collected, and preserved for use in peculiar adipose cells: so that four membranes may be properly enumerated in the urinary bladder. On the same account, Morgagni has also rightly inferred the existence of a cellular coat of the same kind in both the stomach and the gall-bladder: he also stoutly defends and affirms the existence of the muscular coat of the bladder, in oppo-

sition to Bianchi, who attempts to deny it. The cellular coat of the bladder is scarcely to be seen in dogs. *(Ibid.*, not. 24.) 2. The muscular coat is composed of several layers of fibres, but which are principally longitudinal and transverse. 3. The nervous coat is covered with a mucous humor, secreted in the arteries or glands of this coat, which are sometimes conspicuous near the neck of the bladder. The sphincter is a series of transverse fibres running in the form of a circle under the straight fibres, and surrounding and closing the extremity of the neck of the bladder, to prevent the involuntary discharge of the urine: in the male it is connected to the fibres of the rectum, in the female, to those of the vagina. The openings of the bladder are three; two where the ureters enter, at which the urine comes in, and one much larger than these in the neck of the bladder, for the discharge of the urine into the urethra. *(Ibid.*, n. 222.) In the fœtus, the bladder is of a longer shape, and extends almost to the umbilicus." *(Ibid.*, n. 248.)

297. Winslow. "The bladder is a kind of membranous and fleshy pouch or bottle, capable of dilatation and contraction; situated in the lower part of the abdomen, immediately behind the symphysis pubis, and before the rectum. In shape it is a kind of short oval, broader before and behind than at the sides, rounder above than below when empty, and broader below than above when full. It is divided into a body, a neck, and a fundus; into an anterior, a posterior, and two lateral parts. The upper part is termed the fundus: the lower part is contracted to form the neck. The bladder is made up of several coats, almost like the stomach. That part of the external or common coat which covers the upper, posterior, and lateral portions of the bladder, consists of the true lamina or membrane of the peritonæum: the rest of the viscus is surrounded by cellular tissue, by the intervention of which the peritoneal membrane is connected to the muscular coat. The proper coats are three in number: a muscular, a nervous, and a villous, which latter is the innermost. The muscular coat is composed of several strata of fleshy fibres; the external fibres are mostly longitudinal, and the next to these are more inclined toward each side; and the internal, more or less oblique, and they become at length almost transverse. All these fibres decussate in various directions, and they are connected together by a fine cellular tissue, and may be separated by inflating that tissue. The nervous coat, so called, is nearly of the same structure as the nervous coat of the stomach. The internal coat is slightly granulated and glandular, and a mucilaginous humor continually oozes from it, which moistens the internal surface, and defends it against the acrimony of the urine. It appears sometimes altogether uneven on the inner side, being full of little projections and irregular

rugæ when empty and in its natural state of contraction: these inequalities disappear when the bladder is full, or distended by air. At the top of the bladder, above the symphysis pubis, we observe a ligamentary cord, which runs up between the peritonæum and the linea alba all the way to the umbilicus, diminishing gradually in thickness as it ascends. This cord had a particular use in the fœtus, being originally a production of the internal coats of the bladder, which production is termed, urachus. It is composed likewise of two other ligamentary elongations, which are the extremities of the umbilical arteries. These arteries come from the hypogastrics, run up by the sides of the bladder, and are hollow, and filled with blood even in adults, as high as the middle of the bladder, through all which space they likewise give off ramifications. Afterwards they lose their cavity, and become ligamentary as they ascend. At the upper part of the bladder they approach each other, and joining the urachus, form the cord, which may be termed the superior ligament of the bladder. The external fibres of the muscular coat are more numerous than the internal; and the anterior external, which are the longest, form a semicircle round the urachus at the top of the bladder, much like that of one of the fleshy portions which surround the upper orifice of the stomach and lower extremity of the œsophagus: this semicircle passes behind the urachus. The portion of the peritonæum which covers the posterior convexity of the bladder, forms a very prominent fold when the bladder is contracted, but which is obliterated when the bladder is extended. This fold surrounds the posterior half of the bladder, and its two extremities are elongated towards each side, forming a kind of lateral ligaments of the body of the bladder, which are more apparent in children than in adults. The lower part of the bladder, which deserves the name of fundus much better than the upper part, is perforated by three openings, one anterior, and two posterior. The anterior opening is formed by an elongation of all the proper coats of the bladder, in the form of a gullet, bent much in the same manner as the inner orifice of the rostrum of the head of an alembic: this elongation is the neck of the bladder. *(Exp. Anat., Tr. du Bas-Vent.,* n. 448—458.) The urethra is at first no more than a membranous canal, continued from the anterior opening of the bladder, at the place called the neck of the bladder, which is a name that would be more proper for this portion of the urethra. About a finger-breadth and a half from its origin, the canal joins a spongy substance, like that of the corpora cavernosa, only smaller. *(Ibid.* n. 531, 532.) In females, the membrane which lines the neck of the bladder, also forms rugæ, more or less regular, but that investing the cavity is full of irregular corrugations when the bladder is empty. (*Ibid.*, n. 644.) The other two openings in the true fundus of the bladder are formed by the ureters; which, in their descent, run behind the spermatic vessels, and then behind the lower part of the bladder, approaching each other. Each ureter lies between the umbilical artery and vas deferens of the same side: the artery lying on the outside of the ureter, and the vas deferens on the inside. Afterwards they get between the vasa deferentia and the bladder, crossing these canals, and then at about a finger-breadth from each other they begin to pierce the coats of the bladder. They run a little way between the muscular and nervous coats, and inclining nearer to each other, open into the bladder obliquely. The orifices of the ureters, in the bladder, are somewhat oval, and narrower than the cavity of the ureters immediately above the orifices. The edge of these orifices is very thin, and appears to be only a membranous fold, formed by the meeting of the internal coat of the bladder with the internal coat of the ureters. The arteries of the bladder, generally, come from the hypogastric or internal iliac arteries: in particular, they are branches of the sciatic, epigastric, and umbilical arteries on each side. The veins come from those of the same names as the arteries. The nerves of the bladder come from the crural, and also from the great sympathetic nerves, by means of the communication of the latter with the crural nerves. It has likewise a branch from the plexus mesentericus inferior. Besides the ligaments already mentioned, there are likewise two small ligaments, by which the anterior part of the true fundus of the bladder is connected to the ossa pubis." (*Ibid.*, n. 459—464.)

298. Morgagni. "The ureters terminate, by contracted orifices, in the bladder, and by the continuation of their fibres produce on each side a thickish, round and compact body, which usually projects posteriorly on the inside of the neck of the bladder. These two fleshy bodies [muscular cords] after a short course, unite, seldom in a straight line, namely, between the two ureters, seldom even in the manner of a curve, but passing obliquely downwards, they incline to form an angle, from which, in males, I have frequently found a kind of indefinable line prolonged downwards, and continued to that prominence of the urethra, termed caput gallinaceum. When this line or these bodies are drawn downwards with the fingers, the orifices of the ureters are seen to be contracted and closed. . . . The two bodies above described are found not only in human females, but in almost all the brutes, male and female, which I have dissected; having been placed there for some most wise reason by the supreme architect of living beings. (*Advers. Anat.*, i., n. 9.)

"By inspecting a great number of bladders in their natural position,

I have arrived at this result,—that nothing can be more various than the length, thickness, and prominence of these bodies, in males and females, indiscriminately." *(Ibid.*, iii., *Anim.* 42.)

299. See the descriptions and Tabulæ of the following authors. Eustachius, *Tabul. Anat.*, tab. iii., fig. 12; where he exhibits the urinary bladder with the urachus cut off, and the hypogastric arteries proceeding from the iliac arteries to the body of the bladder, and to the root of the penis. Fig. 18; a singular delineation of certain carneo-tendinous sinews weaving round the bladder, and of which he speaks in his book, *De Renibus.*, cap. xix., p. 75; although (according to Lancisi) he could not have shewn them without first taking off the external membrane, and perhaps also steeping in hot water the fibres which constitute the muscle of the bladder. Fig. 20; the blood-vessels which go to the back part of the bladder. Ruysch's *Thes. Anat.* x., n. 126; where he mentions that "innumerable blood-vessels run over the fundus of the bladder in a serpentine manner, very differently from those which are distributed through the skin, the kidney, the spleen, and the liver." Cowper, *Myotomia Reformata*, fig. 12, (8vo., London, 1694.) Graaf, *De Viror. Organ. Gener. inserv.*, tab. v. Drake, *Anthropologia*, tab. iii. John Browne, in Mangetus, *Theatr. Anat.*, tab. ix., fig. 1, 2; where he exhibits the bladder, and the muscle which he terms detrusor urinæ, together with the sphincter. Laur. Terraneus, in *ibid.*, tab. lxv., fig. 1; the external muscular surface of the bladder; the internal surface marked by certain white lines, one passing from above to below, and two others proceeding to it obliquely, and passing through the orifices of the ureters to the sphincter: also the neck of the bladder, and the beginning of the urethra. Boerhaave, on the action of the urinary bladder and the urine, *Inst. Med.*, n. 366—388. Verheyen, *Corp. Hum. Anat.*, tract. ii., tab. v.; tab. x., fig. 4; tab. xii., fig. 1, 4. T. Bartholin, *Anatomia Reformata*, where he mentions having seen two cavities in the bladder in some subjects; and refers to the case of the celebrated Casaubon; (respecting a calculus in the bladder of Casaubon, see Mangetus, *Theatr. Anat.*, lib. ii., cap. xii.) In this case, the adscititious or left bladder communicated by a round opening with the right, and great numbers of rugæ were observed on the internal surface, which was hollowed out as it were in burrows, the little cavities therein being full of small, round stones. Cheselden, *Treatise on the high Operation for the Stone.* J. B. Bianchi, observations on the mechanism in the neck of the bladder, in Mangetus, *Theatr. Anat.*, lib. ii., cap. xii., p. 414, tab. extra ord. iii., fig. 1, 2; and tab. iv., fig. 1, 2. (Genev., 1716.)

ANALYSIS.

300. All things in the living body are appointed to their respective places with the utmost fitness and propriety. Those which are superior in situation, are also superior in forces, in power, in dignity of office, and in use. The cerebrum and the sensoria, which belong to the head, hold the first rank: the heart, and the lungs adjoined to it, which belong to the thorax, hold the second rank: the stomach, the intestines, and their auxiliary members, which belong to the abdomen, hold the third; and the kidneys, the ureters, and the bladder, which are in the cis-abdominal region, hold the last and lowest rank (*a*). Such is the order of the whole; similar also is the

(*a*) Whoever attentively considers the order of things in the animal body, must be seized with admiration and amazement; for nothing can be more consummate. But all things should be examined not only with respect to their situation and connexions,—this is but an examination with the eye,—but also with respect to their particular offices and uses: the true admiration being reserved for those, who view all things with the rational or mental sight also. For those things which are superior, are also more universal, prior and more perfect,—the nobles of the kingdom, of which the rest are the servants and subjects. The cerebrum, therefore, is set in the highest place, in order that it may give laws to the body and its members, and govern them with absolute sway. The cerebrum has the cerebellum associated with it, in order that nature herself may have some voice in its counsels. These two brains have at once both the universal and the most particular administration of all things; so that nothing whatever can happen in the universal body, without their being made aware of it, and nothing can be done, without their either determining it, or consenting to it, or rejecting it. Under the brains are disposed the organs of the external senses.

order of the parts in all the organs, and of the individua [or unities] in the parts. The result is, perpetual relation of all things, mutual regard, everlasting subordination, perfection, finish, beauty; in a word, a form which maintains itself by virtue of the excellence of its order.

301. The URINARY BLADDER and the rectum lie under the feet and heels of all the other organs, in a secluded cavity

Sight holds the highest place; the rest come after it in order, according to the universality and dignity of their offices. This disposition or arrangement of nature, absolutely perfect in all its parts, generates a beauty which affects and delights the mind. Below this supreme or celestial region of the body, stand the inferior organs, inhabiting as it were the atmosphere of the earth: and first, the lungs and the heart, two viscera which like the former, exercise a kind of universal sway in the kingdom; but nevertheless a sway less universal than the cerebrum. For the lungs, by their respiratory motion, excite all things which belong to the body, into motion and life; and as the brains inspire all with a kind of spiritual life, so do the lungs inspire all with corporeal life. The heart operates similarly, by the derivation of the blood all over the body, and to the minutest parts. These viscera are appointed to the intermediate place between the cerebrum and abdomen. The viscera of the abdomen, beneath the middle septum, governed by their superiors, perform particular or private duties only; namely, of preparing the blood from the chyle; and thus provide, not for themselves alone, but also for their superiors and supremes, to which they carry their products as tributes. By the supplies thus brought, the superiors are enabled to administer the republic properly, and to maintain the connexion between the spiritual and corporeal lives. Thus the superiors give the inferiors the power to act, and the inferiors act, and execute their commands, accordingly. Below these again come the excretory viscera,—the kidneys, the ureters, and particularly the bladder, which dwell on the very threshold, partly within and partly without the cavity of the abdomen; for they are only partially surrounded by the peritonæum, their other part lying beyond it. Such then is the appointment of these viscera with respect to situation. But if we do not stop here, but keep our minds in the true series and order of all things, relatively to offices and uses, we shall see, that nothing can be more consummate or exquisite; and that even the mind itself can never sufficiently admire the infinite wisdom, perfection and beauty displayed therein.

termed the pelvis: and inasmuch as they are the lowest, therefore they suffer themselves to be pressed and in a manner trodden on, by all; for the whole of the body, and the parts, even to the most minute, engage and coöperate in compressing them, and in extruding the fæces carried down thither. Every time that either the liquid refuse, or the alvine fæces, are to be thrown out, the *stomach and intestines*, as well as the *other viscera of the abdomen*, particularly and specifically, and by means of the peritonæum, generally also, press in upon and impel them (*b*). The *lungs* also, by their inflation and expansion, fill the chest, and by means of the diaphragm, detrude the viscera of the abdomen, and superadd forces to forces, and afford the means, and determine the very mode, of acting upon the bladder and rectum (*c*). The *cerebrum* also, like the lungs, swells at the

(*b*) We all know perfectly well by common experience, that everytime the bowels are to be emptied, or the bladder evacuated, the abdominal muscles contract, and their action is determined to the contained parts, and ultimately, with a kind of constant direction, towards either the rectum, or the bladder. For whether the belly is pushed outwards, and expanded, or drawn backwards to the spine, and contracted, or thrust upwards or downwards, or in any other different way, the action is always according to the natural state, situation, pressure, continuity and action of the stomach, intestines, and bladder. So that when the fæces are about to be excreted, the common effort of the stomach seems to be directed towards the duodenum; of the duodenum, towards the succeeding small intestines, all the way to the cœcum; from the cœcum to the colon, upon which the abdominal muscles act at the same time from without, whereby the innermost effort of all concurs with the outermost. And when the urine is about to be excreted, these viscera seem to suspend their functions for a time, and to exert themselves, by a general force, to expel it.

(*c*) That at the time when the fæces are expressed from either the bladder or the intestines, the lungs hold their respirations, and extrude the diaphragm in various degrees, according to their expansion, and by this means press upon the viscera below the septum, upon the stomach particularly,—this is a point which we all know from personal experience. Whence it is evident that the lungs, by the mediation of the diaphragm, stomach and intestines, act upon even these lowest viscera. They likewise dispose the viscera of the abdomen, not by bare compression only, but also by a general action upon the nerves same time, and suspends its breath, stretching the nerves vigorously, and mediately by the superior, and immediately by the ultimate nerves, powerfully constringes these two sinks or reservoirs (*d*). The fibres of both the cerebrum and cerebellum, that is to say, nature and the will, unite together, and determine this action (*e*). The *skin,* and the *muscles* of both regions

which pass through the muscular portion of the diaphragm; in like manner also the muscles of this region, together with the ribs and vertebral column; and so order it, as that all things concur at once to one and the same effect.

(*d*) Every one must be conscious of this also. For at this time, all the voluntary muscles, that is to say, all the fibres which depend upon the cerebrum, or the will, are expanded, or kept in expansion;—the muscles, not only of the head, but also of the chest and abdomen, (I speak from the sensation itself;) and by the straining of their fibres, they give the viscera the power of disposing themselves for a common effort towards accomplishing these lowest uses. Thus the abdomen acts very differently from the chest, and the cerebrum from both; for the one flows into the other according to the manner whereby superior acts upon inferior; to wit, not in a continuous series, but in a series descending through degrees; acting upon the nervous fibres, and by them upon the motive fibres of the general muscles, and of the particular muscles of each viscus; more especially upon the lungs; and causing all things to coöperate, internally and externally, to this ultimate effect.

(*e*) That the two sympathetic nerves—the intercostal and the par vagum—arise from the medulla cerebelli; and that the crural nerves, coming from the bottom of the spinal cord, and from the cauda equina, are nerves of the cerebrum, and proceed not only to these extremes of the trunk of the body, but also to the very soles of the feet, in order to conjoin the lowest things with the highest, will be shewn in our Part on the Spinal Marrow, and on the Organism of Animal Motion. The bladder is supplied with both the sympathetic or cerebellar, and the crural or cerebral nerves, which indeed are joined therein by mutual anastomoses. And since the voluntary actions are referable to the cerebrum, and the natural actions to the cerebellum, and since the fibres bring with them all the animus of their parents, therefore it is evident that the ultimate of the will, and the ultimate of nature, concur in producing these ultimate effects. "The nerves of the bladder," says Winslow, "come from the crural, and also from the great sympaconcentrate themselves, the former to the umbilicus, the latter to the linea alba, and concur with the internal forces in throwing out these collected impurities (*f*). Lastly, the *bony levers* also,

thetic nerves, by means of the communication of the latter with the crural nerves. It has likewise a branch from the plexus mesentericus inferior" (n. 297). See Willis, *Cerebri Anatome,* and Vieussens, *Neurographia;* where the above communications are described and figured. A further consequence is, that the whole of the nervous system also concurs in producing this effect; for the inferior or third mesenteric plexus depends in a general manner upon the action of the two superior plexuses; and the crural nerves, upon the action of the whole of the spinal marrow, and consequently of the fibres which flow down to this point.

(*f*) Were we to describe particularly how all the common muscles of the thorax and abdomen, like the viscera themselves, concentrate their forces and coöperate, in order to exert a most general or single united effort upon the bladder and pelvis, during the discharge of the urine, our description would be tedious; and after all—taking into consideration that there are as many varieties of determinations, as varieties of the state of the evacuations—it would be of no great value. It is evident that the external muscles, although so numerous, all concur in direction with the viscera which they enclose. But in order that the individual forces of all the muscles may be concentrated to a few general actions, and these, to a single most general action, and that this may be determined to either the bladder or the rectum, it is necessary that the action of all should converge to the linea alba, where the abdominal muscles meet, and thereby to the lowest of the abdominal muscles; in short, to the triangularis, situated in the region of the pubes. That there is such a determination to this line, is evident from the grand common ligament of the bladder, which unites its vertex to the linea alba. In the fœtus, this ligament constituted the urachus; as may be seen in the plates of Eustachius, Bianchi, Terraneus, and other anatomists. "At the top of the bladder," says Winslow, "above the symphysis pubis, we observe a ligamentary cord, which runs up between the peritonæum and the linea alba all the way to the umbilicus, diminishing gradually in thickness as it ascends. This cord had a particular use in the fœtus, being originally a production of the internal coats of the bladder, which production is termed urachus" (n. 297). And as the tunics or integuments of the whole body stand related to the umbilicus, as their centre of gravity; and as these tunics, as we the *sternum*, the *scapula*, the *ribs* and the *vertebræ*, down to the ossa innominata—the *ileum*, the *pubes*, and the *ischium*—lend and apply their robust powers (*g*). Thus the whole, and the parts, even to the most minute, in order and in series, according to their situation and mode of action,—that is to say, according as the mode is more or less general or universal, near or remote,—engage in the work with one mind and aim, and fight for the common cause,—to drive away these enemies, that would otherwise spoil all their labors.

302. So strong and so united is the effort wherewith they all act upon the bladder, that it seems as though they intended

said above, together with the muscles underneath them, also conspire to compress the bladder, therefore the action must necessarily be determined to the umbilicus, and thence to the bladder; as this same ligament again demonstrates by its insertion in the umbilicus; to say nothing of the two other ligaments, which are the remains of what in the fœtus were the umbilical arteries. "This cord," says Winslow, "is composed of two other ligamentary elongations, which are the extremities of the umbilical arteries . . . . these arteries become ligamentary as they ascend. At the upper part of the bladder they approach each other, and joining the urachus, form the cord." *(Ibid.)*

(*g*) The manner in which all the bones of the trunk, like the muscles, conspire to produce this ultimate effect, is evident from the determination of the bones. The scapula, the sternum, and the ribs respect the vertebral column as a sort of axis, or as levers respect their centres of motion: this column consisting of mere centres or hypomochlia of such levers. This central column also respects its ultimate as its proper centre, and at the same time as its prop or pedestal: this consists of the ossa ischii, ilei, and pubis, which constitute the pelvis, in which the bladder is situated; wherefore also the bladder is connected again to the os pubis by ligaments. "There are likewise," says Winslow, "two small ligaments, by which the anterior part of the true fundus [or cervix] of the bladder, is connected to the ossa pubis" (n. 297). But the action of these ligaments is very obscure, and these bones appear rather to defend the bladder against the violent action of the viscera and muscles, so that it may sit safe in its seat, and suffer itself to be pressed and squeezed in the requisite manner. I say nothing of the concurrence of the other bones, as of the loins and arms; or of the arteries, which come thither from the iliacs, that is, immediately after the bifurcation of the aorta by the hypogastric arteries.

to drive it from its seat, and to throw the whole bag, as well as its contents, out of doors: notwithstanding which, it sits in its pelvis with the most perfect security, suspended from, and connected to, the very centres and foci of the bones, the muscles, the skin and the viscera, by the peritonæum. But the universal onslaught and action upon the bladder, are not only external, but also at the same time internal. The peritonæum and the muscle of the abdomen act upon it from without; the fibre of the cerebrum and cerebellum from within: yet, as the united and unanimous causes of but one effect, the external and internal concur, excite each other mutually, and coöperate. The harmonious relation between them, is proved by the insertion of the urachus with the two umbilical ligaments, into the muscular coat of the bladder (*h*): by the connexion of the peritonæum also with the muscular coat (*i*): by the passage and prolonged

(*h*) For the common ligament itself, composed of the urachus and umbilical arteries, and which passes between two laminæ of the peritonæum, is closely united with the muscle of the bladder, so that when the ligament is acted upon by the surrounding muscles and skin, the action passes by continuity to the very fibre which is the proximate cause of the action of the bladder; and the action of all the neighboring viscera results at last to this ligament, and from it to the true membranous structure of the bladder; as may be inferred from a single glance at this urinous bottle or reservoir. "This [ligament or] cord," says Winslow, "had a particular use in the fœtus, being originally a production of the internal coats of the bladder, which production is termed, urachus. It is composed likewise of two other ligamentary elongations, which are the extremities of the umbilical arteries. These arteries run up by the sides of the bladder, and are hollow, and filled with blood even in adults, as high as the middle of the bladder, through all which space they likewise give off ramifications. Afterwards they lose their cavity. At the upper part of the bladder they approach each other, and joining the urachus, form the cord, which may be termed the superior ligament of the bladder" (n. 297). These two arterial ligaments, according to this passage from Winslow, as well as according to Duverney, are united by all the ramifications which they put forth, to the muscular fibres of the bladder.

(*i*) That the peritonæum itself also is closely connected and attached to the muscular coat, becomes evident when we attempt to separate it. This is indicated by Winslow in the following words: "That part of course of the ureters between the muscular and nervous coats, and their intimate conjunction with both (*k*) : lastly, by the influxion of the nerves and nervous fibres.

the external or common coat which covers the upper, posterior, and lateral portions of the bladder, consists of the true lamina or membrane of the peritonæum: the rest of the viscus is surrounded by cellular tissue, by the intervention of which the peritoneal membrane is connected to the muscular coat" (n. 297). Thus, inasmuch as both the ligament and the peritonæum act immediately and proximately upon the muscle of the bladder, therefore they excite it to its natural action; consequently, to constringe the bladder according to the fluxion of its fibres. This appears to be the first thing which the external force contributes to the motion of the bladder.

(*k*) The ureters themselves are also inserted as ligaments into the sides of the bladder: and the motion which they bring with them is poured forth through the whole of this viscus, that is, through all its coats, and diffused from the points of insertion. The motion is primarily impressed upon the ureters by the kidneys, and immediately by the renal pelvis, which is prefixed in the manner of a head to those tubes. In their course, they receive increments of motion from the peritonæum and the muscles of the abdomen; besides which, they are furnished with motive fibres of their own. That the ureters, like the before-mentioned ligaments, communicate the whole of their motion to the bladder, may be inferred with certainty from the fact of their insertion into it, and from the passage of their extremities between its coats: particularly from the circumstance that all their coats are continuous respectively with the coats of the bladder, and absolutely produce themselves into the latter. With respect to the peritonæum or the external coat, it is evident that it proceeds continuously from one to the other. With respect to the second or muscular coat, the same thing is plain, from the passage of the ureters between the muscular and nervous coats of the bladder. After the ureters have deposited this coat, that is, continued it into the muscular coat of the bladder, they lastly interweave their innermost coat with the innermost coat of the bladder, and make a single coat of the two; as may best be seen from a description of their course or passage. "[The ureters]" says Winslow, "at about a finger-breadth from each other, begin to pierce the coats of the bladder. They run a little way between the muscular and nervous coats. . . . The edge of the orifices [of the ureters in the bladder] is very thin, and appears to be only a membranous fold, formed by the

303. All these causes conspiring, generate the peculiar action of the bladder, which is directed to throwing out the urine, to unlocking the sphincter, and to putting aside any obstacles which stand in the way. In order that these results may be produced according to nature's appointment, effect and end, the uppermost part of the bladder first, must be pushed down, and driven inwards; for the action originates at the very summit of the fundus, and proceeding to the sides, and to the cervix, terminates in the door, and at the lock, which shuts the urethra. That all the parts which are connected on the surface, concur mutually to this object, is shewn by the external and general impulsive forces mentioned above (*l*): by the indrawing of the region round about the umbilicus, and which indrawing is even sensibly evident during micturition (*m*): by the threefold liga-

meeting of the internal coat of the bladder with the internal coat of the ureters" (n. 297). Respecting the length of their course between the coats of the bladder, and the manner of their progress, see the Tabulæ of our authors. This external action, therefore, calls forth an internal action, which seems indeed to arise from internal causes, yet which, without external aid, would never be sufficient to produce the effect in ultimates, as will be explained in the following paragraph.

(*l*) See n. 301. All the forces there mentioned are concentrated towards the fascia or linea alba of the abdominal muscles, and particularly towards the common ligament which enters the bladder from above; consequently these are external forces. But wherever there are external forces, there also there are internal forces: without the concurrence and correspondence of external with internal forces, no action can be produced. The internal forces, then, are the fibres of the two brains, or the nervous fibres, for these act upon the blood-vessels and upon the muscular fibres. The lungs excite these also into operation. Consequently, the forces proceeding from the action of the lungs, are at once internal and external, in short, intermediate. But the actions of the viscera under the middle septum, as well as of the muscles and many other parts specified above, are only external. According to this order, the bladder is excited to its peculiar motion. From a distinct notion of acting forces, we obtain a distinct notion of causes, which are always both internal and external, with intermediate causes conjoining them.

(*m*) Two in a manner opposite actions occur, to prevent the fæces and the urine from being discharged simultaneously. At the time the ment proceeding from the umbilicus, and which falls upon the middle of the vertex of the bladder, and afterwards divaricating, proceeds to the sides (*n*) : by the peritonæum, which is extended over the bladder in the manner of a hood, and descends in ample folds at its sides (*o*) : by the muscular fibres commencing at the very top, and descending straight along the whole circumference, all the way to the cervix, where they deviate and are deflected towards the centre (*p*) : by the sinewy cords surrounding the muscular surface, and serving as so many directors of its constrictions (*q*) : by the blood-vessels themselves, and the

fæces are discharged, the bottom of the belly is expanded, and the active forces determined backwards—*versus posteriora:* at the time the urine is discharged, the belly is contracted, and the action determined from the umbilicus directly upon the superior part or fundus of the bladder.

(*n*) See what we lately said above respecting the ligaments (*h*).

(*o*) The viscera, as we before remarked, act upon the bladder by means of the peritonæum, with the assistance also of the muscles which surround the peritonæum. When therefore the viscera of the abdomen are acting, the action of all is determined, according to the determination of the peritonæum, upon the bladder. "That part of the external or common coat," says Winslow, "which covers the upper, posterior, and lateral portions of the bladder, consists of the true lamina or membrane of the peritonæum: the rest of the viscus is surrounded by cellular tissue" (n. 297).

(*p*) That the external muscular fibres commence from the summit, and from it, as a kind of pole, descend in straight lines, is very conspicuously shewn in the delineations of the authors cited above. "The muscular coat," says Winslow, "is composed of several strata of fleshy fibres; the external fibres are mostly longitudinal, and the next to these are more inclined toward each side. . . . The anterior external [fibres], which are the longest, form a semicircle round the urachus at the top of the bladder, much like that of one of the fleshy portions which surround the upper orifice of the stomach and lower extremity of the œsophagus" (n. 297).

(*q*) Respecting these tendinous cords, see Eustachius, *De Renibus*, cap. xix.; and *Tabul. Anat.*, tab. iii., fig. 18. (See n. 299, above). They seem to serve as tendinous bonds to the muscular fibres, and when the fundus is depressed, to relax those fibres, and to give the nervous fibres full power of acting upon the muscles of the cervix.

vascular fibres, taking a course exactly accommodated to such a modification of action (*r*): by the three striæ of as it were fascia alba, descending rectilinearly and obliquely on the concave surface, and tending jointly to the caput gallinaginis: by the analogous course of the fibres proceeding from the little eminences or tubercles in which the ureters terminate (*s*): espe-

Whether they have any other origin or use, I leave for further investigation. It is very evident that they are also directors of the expansion and constriction of the bladder.

(*r*) See Ruysch, *Thes. Anat.* x., n. 126; where he mentions that "innumerable blood-vessels run over the fundus of the bladder in a serpentine manner, very differently from those which are distributed through the skin, the kidney, the spleen, and the liver;" that is to say, exactly accommodated for influxion into the motive fibres, when the bladder is kept in the state of depression. But we ought to remember, that a small part only of the vessels injected and shewn by Ruysch are true blood-vessels in the natural state of the body; the greater portion consisting of vascular capillaments and fibres which proceed in immense numbers from the sides of the vessels at all points, and draw off the serosity; and which are white in living animals, and so numerous, that they construct whole membranes, looking the while like nervous fibres; as we see in the skin, the kidneys, and other parts. These vascular fibres are never filled with blood, except by violent pressure, or during inflammation. At other times, the blood flows through only the larger branches, and does not turn aside into these lateral vessels: if it does so turn aside, then the whole skin is suffused with intense redness and heat. Meanwhile, as they form membranous webs, we may infer from them and their direction what mode of compression and dilatation prevails.

(*s*) In one of the Tabulæ of Terraneus, these white striæ are represented, one of them descending from the vertex direct to the sphincter; the two others obliquely, from the sides, through the apertures of the ureters, to the former one; and afterwards all proceeding together to the sphincter; answering in a measure to the fluxion of the ligaments to the vertex of the fundus, on the convex part of the bladder. Morgagni speaks of them as follows: "These two fleshy bodies [muscular cords] after a short course, unite, seldom in a straight line, namely, between the two ureters, seldom even in the manner of a curve, but passing obliquely downwards, they incline to form an angle, from which, in males, I have frequently found a kind of indefinable line procially, by the remission or slackening of the nervous fibres during the intropression of the fundus of the bladder, and the consequent relaxation of the sphincter (*t*). That these latter particulars are true, is demonstrated by the drawing up of the fundus of the bladder close to the umbilicus, in the fœtus, whereby the urine, which at this time drips in by very small quantities, is all retained, and the sphincter closely shut (*u*).

304. The terrestrial, gravitating and miry parts of the whole animal microcosm, are banished and thrown into the bladder and rectum; much in the same way, if we may be allowed the comparison, as the earthy and inert masses of the universe or macrocosm, are consigned to the planet or terraqueous globe. The respective matters, in both cases, are circumpressed by the forces of the whole system; and in both cases rest in equilibrium on their centres of motion, although in the last and lowest places (*x*). The outcast fæces likewise, exempted from the

longed downwards, and continued to that prominence of the urethra, termed caput gallinaceum" (n. 298).

(*t*) The influx of nerves affords us the principal means of judging of the state of motion of the bodies into which they flow; for the muscles are excited to motion by the nervous fibres. Wherefore, when the fibres are slackened by the depression of the fundus, the relaxation and spontaneous opening of the sphincter follows as a consequence. But these particulars will be explained more fully in the Part on Muscular Fibre, and on the Organism of the Motion of the Body; where we shall also shew that external and internal forces must necessarily concur to every action.

(*u*) This finishing clause not only corroborates but absolutely demonstrates the proposition of the article. For the embryonic bladder is drawn up close to the umbilicus, and the fundus being raised in this manner, therefore the sphincter at the cervix is closely shut, so that not a drop of urine can escape. "In the fœtus," says Heister, "the bladder is of a longer shape, and extends almost to the umbilicus" (n. 296).

(*x*) There are as many centres of motion or hypomochlia, as there are points of the pelvis; for thither, as to their rest and centre of gravity, the motions and weights of the bones are directed: also the ligaments, the muscles, the skin, and the membranes; whence the most secure and tranquil region of the whole body is situated in the pelvis.

powers of the circumambient world, gravitate by their own weight (*y*). For the muddy urinous humor there collected,

Hither all the heavy liquids tend from their circumferences, as by a kind of centripetency; consequently here is the most general port, that is to say, the port of the whole, for all gravitating substances—for all things which can take no part in the respective circulations. You need not wonder that I speak so often of centres of forces; for such centres there are, both universal and singular, general and particular. Every circumference respects its own centre; all the circumferences together respect theirs; thus there are infinite centres, and these likewise accurately disposed into a form, from which they again tend to their centres. Indeed, if we choose to examine the corporeal machine still more deeply, we shall find in it nothing but mere centres of gravity, which, by their combination, form diameters and circumferences; and this, in such a manner, that the very points of the diameters and circumferences are relatively points of a centre. In such fluxion consists the perfection of the higher forms. But if we examine them with only the lowest form in view, we shall never perceive any centre but what is related to some simply circular periphery, which very possibly we can nowhere find; and thus we shall entirely overlook nature's wonderful gyre in her living universes, where you can never light upon an end, except it be at the same time, as a point in a perpetual circle, also a beginning, and again a respective centre; so as to seize even minds of a geometric cast with dizzy astonishment.

(*y*) It was shewn above, in the Chapter on the Intestines, that the animal microcosm imitates the macrocosm in all its properties; and whatever it receives from the visible and circumambient world, it withdraws from the powers of that world, and subjugates, and appropriates to itself: that consequently, to avoid being misled in investigating the organic animal body, it is necessary to put aside ideas derived through the senses from without, touching the extension of the government of the general sphere to the interiors of the animal world, and to confine ourselves to the consequences of similar causes within (n. 133). The above laws hold universally in the body, all the way to these ultimate sinks or drains; but there a mixed action seems to commence,—an action at once of the macrocosm and the microcosm. For in the bladder the urine itself begins to gravitate, and to weigh down the fundus of the cervix: the earthy and muddy particles, being at rest, also act by their own gravity, and do not tend to perpetual circles. Nevertheless the microcosm, according to its nature, still claims the sometimes to overflowing, by its own volume and proper gravity distends the bladder, dilates the cervix, and converts it into the fundus; and when, from the preponderating quantity of the urine, and from the bladder being altered in shape to an inverted cone, the suspensory ligaments no longer sustain the vertex, there is then a necessity of discharge arising from external causes; and while this is happening, the same effect takes place from internal causes (*z*). And in case the urine be unusually acrid, although small in quantity, it stimulates and irritates the sensile glands and villi (*a*), and carries away the whole bladder into a kind of spasm or convulsive motion. By the deep depression of the fundus, and the strong action of the irritated fibres, the sphincter is not only unlocked, but its lips seem also to be separated so widely, that the neighboring parts of the cervix roll into its open mouth, and thus express and eject the last drop of urine, together with any viscid or gravelly substances (*b*).

principal command over them, as may be inferred from the ejection of the urine, which whether thrown upwards or obliquely, still, by virtue of the force which it has received within, may be considered in one sense as always carried or falling downwards: of the source and cause of this *nisus*, we shall speak presently.

(*z*) They are called external causes relatively to the interior and the very innermost causes. They are not to be termed internal simply because they are excited by the proper muscles of the bladder. Internal, indeed, they are, relatively to our senses, because they are internal with respect to the body, and what is more, internal with respect to the bladder; but relatively to the true interior causes, which are causes depending upon the nervous or cerebral fibres, excited by the very innermost mind, they are only external. Whatever is inferior and posterior, is also exterior; and on the other hand, whatever is superior and prior, is also interior: thus the innermost nature is also the highest and the first, &c.

(*a*) That the internal coat is villous and glandular, consequently sensile, is thus declared by Winslow, "The internal, [third or villous] coat," says he, "is slightly granulated and glandular" (n. 297). This is particularly the case about the neck of the bladder, according to Heister; see above, n. 296.

(*b*) This appears to be pretty clearly deducible from the expression

305. As the viscera, muscles and coverings of the whole body, act and press upon the bladder, not by mass and weight, but by force of motion, so the bladder itself also acts and presses upon the pool of urine, and by this means upon the now relaxed and open sphincter or door of the urethra, not by total closing, but by the power of the motive fibres (*c*). Hence the action of the urine out of the bladder, to the last drop—even to the glairy matter and sediment which comes out the last: from the figure of the sphincter, and the complication of the membranes which there unite: from the state of contraction of the cervix, when the fundus is relaxed and depressed: also from the parallelism of all similar locks and sphincters throughout the body; that is to say, from the pharynx, the anus and the pylorus, respecting which see above.

(*c*) According to our senses, (placed in the world's general sphere, and deriving from it their images and their modes, and insinuating ideas into the mind,) it appears altogether as if the bladder drove the urine out of doors by total compression or contraction, and as if this were the cause of the force whereby that fluid is ejected. But as I have already warned the reader (*y*), this force originates entirely from causes generated and excited within the animal world; and as far as possible, prevents the circumambient world or macrocosm from exercising any control over its actions. That an active force internal to the microcosm, is what impels the urine, and this, without a total compression of the bladder, is plainly shewn by the discharge of the urine, when its quantity is only small, occupying but a part of the cervix; notwithstanding which it is evacuated to the very last drop, grit and grain. The idea of the compression of the bladder into so small a space,—in fact, to the size of a walnut,—is *prima facie* absurd, and must appear particularly so to those who have an anatomical knowledge of its size, attachments, and suspension. We must recur, therefore, to the organism of the bladder, which, indeed, at the time of discharge, is compressed laterally, particularly about the cervix, where it lies in rugæ, like the stomach and intestines. "The nervous coat of the bladder," says Winslow, "is nearly of the same structure as the nervous coat of the stomach. . . . The internal coat appears sometimes altogether uneven on the inner side, being full of little projections and rugæ when empty and in its natural state of contraction. These inequalities disappear when the bladder is full" (n. 297). We must, therefore, admit the same kind of action here as in the stomach and intestines, respecting the action of which, see the analyses of those parts. But in the bladupon the urine is not propagated by continuity, but by series of forces or motions, divided and subdivided, from generic to specific, and from specific to particular. Thus the action, multiplied by degrees in its passage, and concentrated upon the beginning of the urethra, is continued similarly through every point of the entire tube from the head to the bulb. Hence whatever direction the urine may take, it always pushes on with perpetually augmented celerity and powers of flight, consequently in the same duplicate ratio as other falling streams or weights: which is the reason why it comes forth with such rapidity, and is jetted out to so great a distance.

306. Besides the spasmodic motion, during the expression of the urine, the bladder has also the same constrictile and expansile or alternate motion, as the peritonæum and the viscera enclosed therein (*d*): and this, in order that by the alternate contraction and relaxation of its fibres, it may give the ureters the power, not only of correcting and tearing asunder the viscid and conglutinated portions of the urine supplied to them (*e*), but

der, as shewn above, a kind of mixed action occurs; that is to say, the volume of urine gravitates and presses upon the parietes in the ratio of its depth, and expands them in the same ratio.

(*d*) That is to say, with the same alternate movements as the lungs. This is proved by the action of the peritonæum, which sits upon the bladder like a cap, and covers its vertex and a large portion of its sides: also, by the action of the abdominal muscles upon its ligaments, which lie between two laminæ of the peritonæum: likewise, of all the viscera moving together in the same alternations: by the action also of the lungs through means of the diaphragm: and lastly, by the action of the cerebrum and cerebellum by means of the fibres.

(*e*) That the ureters break up the grumous, and separate the viscid parts of the urine, and perform in a manner the same operation upon them, as the intestines perform upon the food they have received,—this is conspicuously evident from their contraction, and from the passage of the contracted tubes for some distance between the muscular and nervous coats of the bladder; and lastly, from their power of expelling concreted, glairy and gravelly substances, and calculi. The same kind of action, only more obtuse, appears to go on in the bladder; which, like the intestines, contracts into rugæ, and likewise decreases in width, terminating in a cone.

also of instilling, eructating and expressing it, by turns, into the cavity of this primary urinal: also, in order that the bladder may gently turn about and work the urinous pool and volume, and prevent the heavier parts from subsiding in the form of a sediment, and sticking to the folds: and in order that a continual vapor may irrigate and soften its walls, and the roof of the concave fundus, and when use requires, summon and incite them to action: and lastly, that the bladder may at all times preserve the fibres of its three coats in a state of motion, in a state of power, and in integrity, and keep them ready and separable for every degree of expansion and contraction.

307. Before we quit the bladder, situated as it is on the extreme verge of the abdomen, and at the boundary of the abdominal viscera, it behoves us to fulfil the intention of our analysis, and to obtain instruction from the greatest things respecting the least,—from those which fill the whole field of sight, respecting those which scarcely influence a single ray thereof (*f*): that is to say, from the vesica or bladder, respect-

(*f*) For the object of analysis, is, to conduct us, under the guidance of reason, from the visible objects of the senses, to a knowledge of those things which are above or beyond the sensual sphere: in short, to cause the eye to deliver the phenomena received by sight, to the rational mind, in order that it may distinguish them, open them into causes, and in a manner digest them. The path of analysis lies through the senses to the understanding, that is to say, from posterior to prior things—*a posterioribus ad priora;* the senses being in the posterior or inferior world, and the rational mind in the prior and superior world: wherefore this which is termed the analytic way, is absolutely natural to man, and in it he must travel, if he wishes to penetrate into the secrets of nature; as we before explained in the Prologue (n. 11, 12, 13). Respecting the state, *modus agendi,* and operation of the minute corpuscules which we term glands, vesicles, follicles, granules, acini, &c., we may derive instruction not from the urinary bladder, ureters and kidneys only, but equally also from the stomach, intestines, and other viscera of this region, which likewise in the ovum were once corpuscules of a similar form, and are so still in the minute bodies of the insect tribes. By those things then which are clearly manifested in the ultimate world and before the senses, we may be informed respecting the structures and modes of action of those others ing the vesiculæ or little bladders of the minutest glands; from the urethra, respecting their little emissary ducts; from the ureters, respecting the capillary vessels which flow into them; and from the kidneys, as the grand strainers, respecting the simplest sieves or incernicula of the glands. For the bladder, or if you please, the whole cavity of the abdomen, was precisely such a follicle or vesicle, at conception, and in its initiament, in the ovum: precisely such also it is in little worms and crysalisses, whose entire bodies form scarcely a single point to our senses. Such, therefore, it is in itself, in principles and causes, but enlarged, strengthened and fortified, so as to suit all kinds of effect and use. The least acinus of the gland flows through a little pervious line into its follicle; much in the same manner as the kidneys flow through the tubular ureters into the bladder. Still following the parallel, this little pervious line, or most simple ureter, dips into and entwines with the little membranes of its follicle; connects and continues its own membranes therewith (*g*); opens into the cavity; shuts on the outside, and marks its little door (*h*), to prevent anything from passing back; collects and concentrates its little fibres thither (*i*); contracts and relaxes them alternately, locking and unlocking its little sphincter; throws its exquisitely fine dew headlong, into either some larger cavity, or else absolutely out of doors; begins the very motion which the little duct continues at all points (*k*); thus

which are most minute: if only, according to the laws of the Doctrine of Order, we substitute delicately motive fibres in the place of muscles, or in the place of the former, nervous fibres, which act the part of motive fibres in the purer sphere, &c.

(*g*) See n. 302 (*k*).

(*h*) That the ureters produce certain little protuberances about their apertures, and in this way mark and seal their terminations, see Morgagni, n. 298, and several of Bianchi's Tabulæ. And not thus only, but also by their prolonged passage between the muscular and nervous coats, they close their doors, and prevent the least drop of urine or vapor from returning.

(*i*) See n. 305.

(*k*) See n. 305. For in the greatest and least things, nature intrudes and extrudes the fluids, and determines them to the spots where uses are intended, with every possible diversity of method: and to expels and ejaculates its liquid, in the same manner as the bladder expels and ejaculates the urine through the urethra. It would be easy to pursue further these analogies and points of comparison, but all the glands, as well as the liquids which they extract and filter, and the uses which they yield, are specifically different from each other.

give effect to her decrees, she generally narrows and contracts the little canals from beginning to end; and thus prevents the humors from being driven down by any force, excepting what is proper and peculiar to the living body. Thus she contracts the ureters; and the bladder towards the cervix; the urethra also towards the bulb; and its aperture or slit is narrower than the cavity, which acts upon the escaping urine and impels it, particularly in the beginning; and likewise in the rest of the course, if the member containing this canal should become tense and rigid. But respecting the urethra, and its action upon the passing urine and semen, we shall treat in Part IV., when we come to speak of the Genital Members.

CHAPTER XVI.

THE PERITONÆUM.

308. Heister. "The peritonæum is immediately subjacent and adherent to the transverse muscles of the abdomen. It is a thin, smooth, and lubricous membrane, investing the whole internal surface of the abdomen, and most of the abdominal viscera, and containing the latter as it were in a bag. It is also connected with the diaphragm, and with all the viscera of this region. It entirely encloses the stomach, intestines, mesentery, omentum, liver, spleen, and pancreas, which are, therefore, said to be lodged in a duplicature of the peritonæum. The kidneys, ureters, receptaculum chyli, and great vessels of the abdomen, &c., are covered only on their anterior part; and the bladder only on the superior part; differently from what is the case in brutes. The peritonæum consists of two laminæ; the umbilical vessels lie between them, and in many places they are covered with a cellular substance. It forms the ligaments of the liver, and of the umbilical vein and arteries, the various ligaments of the intestines, the broad ligaments of the uterus, &c. It sends two processes out of the abdomen, which encase the spermatic vessels and the testicles. In regard to these we are to observe, 1. Their singular passage between the muscles of the abdomen. 2. The hiatus visible in the abdomen of the dog, but wanting in the human subject. 3. The septum separating the testicles on the upper part. Also the involucra of the round ligaments in females; and their diverticula. Nuck has delineated these in his *Adenographia*, fig. 39, 40. The arteries and veins of the peritonæum come from the epigastric, mammary, lumbar and phrenic vessels; to these some add lymphatics. The nerves come from the phrenic, dorsal, lumbar and sacral nerves. Some writers have also mentioned glands in the peritonæum, but they are not observable in the natural state of the mem-

brane. Uses,—1. To enclose the contents of the abdomen; for when the peritonæum is dilated unduly, wounded, or ruptured, they fall out of their proper places, or hernia arises. 2. To give an external coat to almost all the parts contained in the abdomen; which are therefore generally said to have their external membrane from the peritonæum. 3. To form the processes of the peritonæum, and the tunica vaginalis of the testes." *(Comp. Anat.*, n. 206.)

309. WINSLOW. "The xiphoid cartilage of the sternum, the cartilaginous portions of the last pair of true ribs, those of the first four pairs of false ribs, all the fifth pair, the five lumbar vertebræ, the ossa innominata and the sacrum, form the bony sides of the cavity of the abdomen. The diaphragm, the muscles of the abdomen in particular, the quadrati lumborum, psoas, iliaci, the muscles of the coccyx and rectum, form the chief part of the circumference of this cavity, and its internal surface is lined by a membrane termed the peritonæum. As additional or accessory to these we may likewise add some portions of the sacro-lumbales, longissimi dorsi, vertebrales, glutæi, &c. The cavity of the abdomen is of an irregularly oval shape, but still symmetrical. *(Exp. Anat., Tr. du Bas-Vent.*, n. 19, 20, 21.)

310. "Having carefully removed the muscles of the abdomen, the first thing we discover is a very considerable membranous covering, which adheres immediately to the internal surface of the transverse muscles, and of all the rest of this cavity; and involves and invests its viscera as in a kind of bag. This membrane is termed peritonæum. It is of a pretty close texture, and yet flexible, and capable of very great extension, after which it easily regains its ordinary size; as we see in pregnancy, dropsies, corpulency and repletion. It seems to be made up of two portions, one internal, the other external, which have been looked upon by many anatomists as a duplicature of two distinct membranous laminæ. But properly speaking the internal portion alone deserves the name of a membranous lamina, as being the main body of the peritonæum. The external portion is no more than a kind of fibrous or follicular apophysis of the internal; and may properly enough be termed the cellular substance of the peritonæum. The true membranous lamina, commonly called the internal lamina, is very smooth, and polished on that side which is turned to the cavity and viscera of the abdomen; and continually moistened by a serous fluid discharged through almost imperceptible pores. These pores may be seen by spreading a portion of the peritonæum on the end of the finger, and then pulling it tight on all sides; for then the pores are dilated, and small drops may be observed to run from them, even without the microscope. The sources of this fluid are not as yet well understood;

perhaps it comes out by a kind of transudation or transpiration, like that which we observe in animals newly killed. The whitish corpuscules found in diseased subjects are no proof of the glands which some anatomists place there in the natural state. The cellular substance or external portion of the peritonæum adheres very closely to the parts which form the inside of the abdominal cavity; and is not everywhere of equal thickness: in some places it is in very small quantity; and scarcely any appears at the tendinous or aponeurotic portions of the transverse muscles, and on the lower surface of the diaphragm. In all other places it is thicker, and forms cells expanded into fine laminæ or leaves, which in some diseased subjects become so large and thick as to resemble so many distinct membranes. In some places this substance is exactly like membrana adiposa, being filled with fat; as round the kidneys, and along the fleshy portions of the transverse muscles, to which it adheres. It entirely surrounds some parts, as the bladder, ureters, kidneys, spermatic vessels, &c.; in these places it is commonly but improperly termed, the duplicature of the peritonæum. Besides these differences in thickness, the cellular substance has several elongations, which have been called, productions of the peritonæum. Two of these accompany and invest the spermatic cords in males, and the vascular cords, commonly called the round ligaments, in females. There are other two which pass under the ligaments of Fallopius, that is, the tendinous ligaments of the abdominal muscles, with the crural vessels, which they cover, and they are gradually lost in their course downwards. To these four elongations of the cellular substance of the peritonæum, we may add a fifth, which is spread on the neck of the bladder; and perhaps a sixth, which accompanies the rectum. All these prolongations pass out of the abdominal cavity, and may be termed external, to distinguish them from others that remain in the abdomen, and are called internal. The great blood-vessels, that is, the aorta and vena cava, are likewise enclosed in this cellular substance of the peritonæum. In a word, it involves, immediately and separately, all the parts and organs which are commonly said to lie in the duplicature of the peritonæum. The true lamina or membranous portion of the peritonæum, is connected by the cellular substance to the internal surface of the abdominal cavity; but in the human subject, it does not naturally accompany the external elongations of that substance; it only covers the origin or basis of these elongations, without any interruption or alteration of its own surface or level at these places. It has nevertheless elongations of its own, but they are very different from those of the cellular substance; for they run from without inwards; that is, they advance from the convex side of the great bag of the peritonæum, into the cavity of that bag; some more, some less, and also in different manners; as if the sides of a large bladder were pushed inwards in various places into its cavity. Of these internal elongations some are simply folded, like a duplicature; others are expanded like inverted bags or sacculi, to contain some viscus; some begin by a simple duplicature, and end in a divarication or cavity, which likewise contains some organ: some are alternately extended in the form of simple duplicatures and of cavities; and lastly, some form only a slight eminence on the inner surface of the great cavity of the peritonæum. To the first species of these elongations, we may refer the membranous ligaments, such as those of the liver, colon, &c. We see the second species in the external membrane of the liver; the third, in the mesentery; the fourth, in the mesocolon; and the fifth, over the kidneys and ureters. Besides the external elongations of the cellular substance of the peritonæum, it has the same number of internal elongations as the true membranous lamina, which lie between all the duplicatures, and line the insides of all the cavities, or the sides next the viscera. *(Ibid.,* n. 22—38.) I must here observe that three of the umbilical ligaments are invested by a falciform membranous production or duplicature, which the peritonæum sends into the cavity of the abdomen." *(Ibid.,* n. 42.)

311. Verheyen. "The peritonæum is a thin, soft, and dilatable membrane. As extended naturally, its figure is oval, corresponding in length and breadth to the abdomen. It is double throughout, and manifestly so from the umbilicus to the os pubis, in females particularly, in whom it is also thicker than in males. Its external surface is somewhat rough and fibrous, on account of its connexion with the muscles. The internal surface is smooth, and covered with an unctuous humor. Anteriorly, it is connected to the muscles of the abdomen; superiorly, to the diaphragm; inferiorly, to the os pubis and ischium; laterally, to the os ileum; posteriorly, to the sacrum, and the lumbar vertebræ, particularly the first and third. Superiorly, where it is connected to the diaphragm, it is perforated by the gullet, the vena cava, and the nerves of the par vagum. Inferiorly, by the rectum and the vagina; and at any rate one lamina of it by the urethra: anteriorly, in the fœtus, by the umbilical vessels; but as these vessels shrivel after birth, and become very much attenuated, and moreover as the peritonæum is closely connected to them, therefore this perforation is not noticeable in adults. But that the peritonæum is not so strong there as in other parts, is evident from the facility with which it is there relaxed and perforated by the action of preternatural forces, as by the air in tympanitis, by the serous fluid in ascites, &c. The exterior

lamina, at its under part, sends down two processes in the male to the scrotum; these contain the spermatic vessels; in the female they enclose the round ligaments of the uterus. In the scrotum they dilate, and constitute the tunica vaginalis testis. These processes arise at the sides, and slant downwards to the front. . . . In dogs they form simple membranous tubes, opening each by a large aperture into the cavity of the abdomen. In the human subject we find no openings of the kind; because the spermatic veins and arteries, from their origin, lie in a duplicature of the peritonæum, and the processes are made up of the external lamina only, the internal lamina being placed over their orifices, and being very thick and firm in this situation. Moreover, in dogs, these processes are altogether free inside, so that when air is blown into them, it immediately passes through them. In the human subject they contain a number of membranous partitions, which connect the spermatic vessels to each other, and to the sides of the processes; so that scarcely any cavity can be seen, sufficient to hold the air, unless one be formed by the dilatation of the membranes, or their preternatural separation. In the vicinity of this part, dangerous and often fatal herniæ occur; we mean at the place where the iliac veins and arteries pass to the thighs; for if the peritonæum be dilated, or what is a very uncommon occurrence, ruptured, a small portion of the ileum descends, and becomes adherent to the neighboring parts; and by such adhesion I have seen a small portion of the ileum entirely closed up, and the patient miserably dying, having for many days previously had no alvine excretions, but at last vomiting feculent matter from the mouth. I find a case of this kind also recorded by Nuck. The peritonæum receives arteries and veins superiorly from the mammary and phrenic vessels; inferiorly, from the epigastric and sacral vessels; also, a few little branches from the spermatic vessels; laterally, from the intercostal and the lumbar vessels. Its nerves come from the lumbar and sacral portions of the spinal cord, and from the intercostal and phrenic nerves." *(Corp. Hum. Anat.,* tract. ii., cap. vii.)

ANALYSIS.

312. The animal body, which, according to a common mode of speaking, is also a kind of whole, a universe, microcosm or kingdom, is most distinctly divided into integral parts, members, viscera or organs; these again into lesser organic parts; and these ultimately into least parts, simple parts, and unities (*a*). This whole with its members and parts, is divided as it were into regions, hollow chambers, and cavities. The cerebrum inhabits the highest chamber; the heart and lungs inhabit the middle chamber; the stomach, intestines, liver, &c., the lower, and the genital organs, the lowest chamber. Each chamber is surrounded by a common membrane; the highest or cerebral chamber, by two membranes or meninges; the middle or thoracic chamber, by the pleura; the lower or abdominal, by the peritonæum; the lowest or scrotal chamber, by its own peculiar sheath or capsule: all together are surrounded by one most general membrane; that is, by the skin. These common membranes or coverings are distinctly separated from each other, and distinctly connected with each other, and communicate, in the exact proportion in which the organs they enclose are dignified in office and act in society. Each of these four coverings also makes common, special, and particular cause, with the viscera and members which it folds in its embrace. But in order to understand the manner whereby the *Common* or *General* acts

(*a*) The lesser organic parts are the miliary glands, or least glandular acini, in which the viscera of the chest and abdomen for the most part terminate internally. But the simple parts are the vascular and other similar threads of which these glandular termini themselves are constructed.

upon the parts of the whole, it is necessary to understand also the order by which the *Universal* flows into them. First of all, however, the terms, universal and general, must be defined; for an obscure notion of the meaning of the term general, and an indistinct notion of the meaning of the term universal, entangles, involves and confuses the one with the other; although the human rational mind,—mistress of philosophy,—is perfectly acquainted with both their essences, and when left to herself and to her own powers, we being unconscious the while, in ordinary speech and expression, she very rarely substitutes the one for the other (*b*).

313. A UNIVERSAL is that which exists and acts universally in the whole, and in all parts of the whole. Or to make this more apprehensible, instead of the whole, let us assume some limited universe or kingdom, or some particular body (*c*), as for

(*b*) That the human rational mind is in its own nature when engaged in ratiocination or philosophy, is very evident from the circumstance, that all philosophical science is derived from it, as from its proper fountain: for to construct a philosophy is nothing more than to give the best attention to the operations of one's own mind, and from its *modi operandi* to draw forth that which we are to bring into our code, in the form of precepts, laws, and axioms: thus we learn this high science from our very selves. This affords a plain proof that something flows in from above into the thoughts of the rational mind, and gives it the faculty of distributing the objects of the memory into true analytic orders, and of digesting the things below it into rational quantities. A proof also that there is something below the rational mind, which flows in with objects, as instruments and materials for constructing similar analytic edifices: which faculty thus endeavors to be instructed by its superior faculty, by means of a kind of store drawn forth by it, and disposed into the form of science. But on these subjects we must refer the reader to our Psychology.

(*c*) This rule has a most extensive range of applications: it applies not only to the whole circumambient world, which we term properly the universe, and to our organic animal body, which may also be termed a universe, world, or microcosm, or even a kingdom; but also to all empires and forms of government; to particular societies; to every individual man; to all his actions, sensations, laws, sciences, arts, and even speeches and conversations: in a word, to everything which is a subject instance, the human animal body. Here, the *Universal Essence* is that which is universally present, potent, active, conscious and provident, in the whole body, and in all its members, viscera, or organs (*d*), and in their lesser organic parts, and lastly, in the least or simplest particles of these parts. Thus a universal is recognized by its presence, potency, activity, consciousness, and providence (*e*). In the human microcosm, the soul is such a universal; the simplest fibre being the universal radius of its determinations (*f*).

314. Every whole, whether it be a universe, a microcosm, a kingdom, or a body, &c., has its proper superior universal, inferior universal, and ultimate or lowest universal (*g*). That which is superior, is also prior, interior, simpler, and more

at all, or has in it a form that derives its essence, nature, and quality from the order according to which universals flow in, and from the manner according to which the general circumscribes the series. For the present we had better confine ourselves to the subject before us, to wit, the human organic body; we therefore use terms adapted thereto; but which must be immediately transmuted into other terms when these universal rules are applied to other subjects.

(*d*) Members, viscera, and organs are here assumed as synonyms; although, properly speaking, they mean distinct things.

(*e*) By providence we here mean that circumspection and prudence whereby we contrive and provide for ourselves in the present and for the future. In the very operation of the rational mind we have sufficiently plain proofs of this kind of providence.

(*f*) On these subjects we refer the reader to the Part on the Cortical Substance and Medullary Fibre of the Cerebrum; and to the Part on the Soul.

(*g*) Where the degrees of universality are still more numerous, they may perhaps be subdivided differently; namely, into a supreme, a superior, an inferior, a sub-inferior, and an ultimate universal. In the human animal body there is a universal to which we ought to assign a still lower place than the third; we mean the motive muscular fibre, which is constructed by the blood-vessels; but at present we are unwilling to extend our exposition of universals to this point; for the first rudiments of the doctrine are all that can properly be given here, where our purpose is limited to explaining clearly what is meant by the term general, as preparatory to shewing what uses the peritonæum affords as a general membrane.

perfect; in other words, more present, more potent, more active, more intelligent, and more provident. The inferior universal derives its essence and its possibilities from the superior; consequently depends upon it, in the same way as an agent depends upon its power, or a cause, upon its principle. If the superior universal in the human body, is the simplest fibre, which is acted upon by the soul immediately (*h*), then the proximately inferior universal is the fibre immediately derived from this,—the nervous fibre of the body,—which is acted upon by the animal spirit (*i*), and the remotely inferior universal, or the lowest of the three, is the artery and vein, which are acted upon by the blood. In the human body, these are the three universals, which derive, produce, form and generate all things.

315. In the living body, the derivation, production and generation of the inferior and ultimate universal essences from the superior or supreme, are as follows:—from the supreme universal is derived a proximately inferior universal (*k*); by this

(*h*) This will be shewn in the Part on the Fibre and on the Soul.

(*i*) In the same Part we shall have to explain what the nervous fibre is in its primary form, that is, in its principles; taking for our guides microscopic observations, and the experience of effects and phenomena.

(*k*) To wit, the nervous fibre in its simple form, which resembles the purest vessel, and conveys what are termed the animal spirits. This fibre rules universally in the body, and indeed, more universally than the blood-vessel; for the nervous fibre absolutely enters, forms, constitutes, and actuates the blood-vessel; as may be seen from the numberless nervous fibres which accompany the vessels, particularly the arteries, constantly dip into their membranes, and even construct their very canals. Yet still the nervous fibre rules less universally than the simplest or first fibre of all, which latter produces the nervous fibre as the executrix of its commands: for whatever the nervous fibre, regarded in its primary form, possesses, it derives from a prior fibre,—the first of the body,—which we have termed above, the universal radius of the soul's determinations. The nervous fibre cannot *be* from itself, nor yet exist immediately in a pervious state: the little membrane with which it is surrounded must inevitably be derived from a fibre prior to itself. We are obliged to use the term *derived*, because no other expresses our meaning, or squares in all respects with the means, a kind of new principle is formed, which may be termed the corporeal and material principle (*l*), from which a still inferior or ultimate universal is produced (*m*): by this means again the whole body, with its members, and their parts, is formed (*n*).

mode of formation of this fibre, which mode is a sort of modification of the primitive force.

(*l*) Respecting this new principle, or respecting the origin of the new fibre which must be termed the corporeal fibre, we shall speak in the Part on the Organs of the Senses, (when we come to treat of the skin, or of the sensorium of touch,) and also in the Part on the Fibres. The corporeal fibres are what grow or proceed from the most minute glandular forms almost resembling in appearance the cortical glands of the cerebrum, in the extremes of the body particularly; that is, in the skin, and in other parts also; and which are so very numerous and small, that they resemble true nervous fibres or capillary vessels. These most subtly organic forms, constructed by the nervous fibrillæ in the extreme goals of the kingdom of the body, are as it were really new principles formed in the body, and produce little emissary ducts as fibres, which latter therefore ought to be termed corporeal and material fibres. These glands in the ultimates of the kingdom correspond to the cortical glands of the cerebrum and cerebellum. But we shall treat of them more at length in the Part before mentioned.

(*m*) To wit, the blood-vessel. It will be proved in the sequel, that the first or innermost membrane of the arteries is composed entirely of these corporeal fibres; for the innermost coat of the artery is what is termed the nervous coat, next to which is the muscular coat. But that which is the innermost is also the first, and always derives its origin from the fibres formed by the before-mentioned purest glandular forms of the body. This will be proved both from the experience of microscopists, and by phenomena and effects when unfolded to their first causes; also by comparative, and finally by rational, anatomy.

(*n*) That the organic body, with all its members and parts, is woven and constructed by the blood-vessels and their ramifications,—this is a point established beyond all question; inasmuch as myriads of vessels are seen under the microscope in spaces only a few lines in diameter, and may be even injected, particularly by the Ruyschian method. The proximate formation of the body, then, must be ascribed to the blood-vessels. But the blood-vessel itself derives this power and determination from the nervous fibre, which enters into the composition of the blood-vessel; and this fibre obtains its power and determination from the simplest fibre, from which also it is derived. Or if we substitute

When this gyre and course of formation are accomplished, then the ultimate universal passes to the supreme universal, and enters into absolute and intimate union with it (*o*); and thus the inferior universal proceeds from both (*p*).

316. A GENERAL is the summary and complex of parts and singulars; as a number is the complex of several unities, and an equation, of several analytic ratios. Now in order that singulars may be kept in mutual association in their complex, it is necessary that they have a common bond or tie; and in the living body, the coats, membranes and coverings constitute such bonds. Inasmuch as this bond represents a common or general cause, we, therefore, in the present case, term it a general, and contrast it with a universal; in which sense, a *general* is what

the blood, the animal spirit, and the soul, in the place of the fibres, it will amount to the same thing; for the fibre is the instrumental acting cause, and the blood, &c., the principal acting cause, and each two respectively, in the most harmonious manner, make one unanimous cause. This formation respects not only the organic form of the body, but also whatever gives it the power of acting according to this its form.

(*o*) In the Part on the Cerebrum and the Cortical Substance, it will be demonstrated in the plainest manner, that the innermost membrane of the artery, (which, as we said above, is generated by the corporeal fibres,) in the cerebrum, unites with, and absolutely insinuates itself and plunges into, the cortical glands, which are the principles or beginnings of the simplest or superior universal fibres, and the parts from which the proximate or nervous, that is, the inferior universal fibres, are put forth: so that this ultimate or lowest universal fibre—the arterial vessel—passes at last to the first of all the fibres, and enters into the closest union with it. Hence there is a perpetual circle from the last things to the first, and from the first to the last, and everything throughout is held in perpetual connexion,—so long as it is not detached or separated, but forms a part within this gyre.

(*p*) For if the last fibre—the fibre of the third order—unites with the first, and this, in the very beginnings or principles of the fibres, that is to say, in the cortical glands, from which the nervous fibre—the inferior universal fibre—or the animal spirit, is produced, then the consequence is, that this nervous fibre proceeds from both. That such is really the case, will be proved to even sensible demonstration in the whole progress of our analyses.

contains and distinguishes a universe, its integers, and singulars (*q*): or speaking more definitely, and substituting the body and its members for a universe, integers, and singulars,—a general is that which embraces, demarcates, limits and connects (*r*) the body, any region thereof, any members of this region, or any parts of these members. A general, therefore, is recognized from its being the complex, limit and nexus of those things which are associated.

317. The *Universal* gives the essence, and determines it; the *Common Bond* defines and bounds it. Thus from both together exists the finite and limited entity termed a substance, which subsists by itself distinct from other things, and is the subject of accidents; quality being predicated of its form, and nature, of its power and activity.

318. There are, then, as many general limitations and boundings, as there are essential parts, or determinations of essentials (*s*). The most general is that which involves and com-

(*q*) These are mere indeterminates—general, universal, integers, singulars; but this definition, educed, according to the analytic method, from determinate and finite things, is here prefixed solely for the sake of order and intelligibility.

(*r*) That which circumscribes the aggregates of things, not only contains and *embraces* them, but also divides, *demarcates*, and distinguishes them from contiguous things, as well as *limits* them with respect to shape, and *connects* them. For if every quantity be similarly circumscribed and distinguished, that is to say, the lesser quantities which constitute the larger, they must of course be connected with the larger, and this cannot be accomplished excepting by prolongations from the most general covering. This is the case not only with the viscera of the body, but also with the muscles, with all the motive fibres, and even with the nervous fibres, the common fasciculus whereof is covered with a membrane which in like manner detaches or gives off filaments and little membranes, that connect the lesser fasciculi, and even the very fibres, with the entire nerve. It may therefore be regarded as a law, and it is particularly manifest in the brain.

(*s*) The fibres are determinant of the above-mentioned triple order. For the soul, the animal spirit, and the blood, are the veriest essences of the body, and by means of the fibres and vessels determine all things. The coats of the latter are the instrumental causes, which prehends the whole; the less general is that which comprehends a part thereof; and so on. If we apply these limitations and boundings to the coverings of the body, then the skin, which surrounds and encloses the whole body, is the most general: the meninges of the brain, the pleura of the chest, the peritonæum of the abdomen, and the skin and dartos of the scrotum, are the less general: the still less general are the integuments of the respective viscera, organs or members: the still less general are the tunics of their organic parts: and the least general are the little membranes of the singulars of the parts. Hence it appears, that in this sense a general is more universal in proportion as it is less general, because in this proportion it is more simple, more internal, and nearer to all the essentials of its body (*t*).

319. The general coverings, which are many in the body which is one, communicate with each other exactly in proportion as the viscera or organs enclosed by them act in society: or inversely, the members of one region communicate with the members of another by means of their coverings, exactly in proportion to the general or universal character of their object. Thus the cerebrum, by its fibres and by its vessels, by means of

united with the principal causes make one cause. The viscera, organs, members, glands, and muscles, from their greatest mass to their least parts, are the determinations of these essentials. The determination itself proceeds from the nature of the soul,—the *ens universale* of the body. But what the determinations are, it is the province of anatomy and physiology to teach us.

(*t*) Universality may be predicated of all things which increase and decrease in order, and by degrees; consequently of the membranes or coverings. All the membranes, whether they contain fluid or continuous matters, are instrumental causes, and make one cause with their principals. The coats of the vessels, as we before indicated, are the instrumental causes of the fluids permeating them and acting as principal causes. So also the coats of the viscera and organs are instrumental causes, respectively to the agents which they surround and enclose, whether these agents be fibres or vessels, or congeries of such; consequently the coats come under the notion of universality, in a similar manner.

the meninges (*u*), communicates with all the members, both thoracic, abdominal, and genital. The heart, by its blood, by means of the arterial vessels; and the lungs, by their motion, by means of the vertebræ, the ribs, the muscles, the œsophagus, the diaphragm, and the pleura, communicate with both the cranial, the abdominal, and the genital members. The stomach, the intestines, the liver, and the other viscera under the septum, by their chyle, and their blood that is to be, by means of the venous vessels, the thoracic duct, the diaphragm, and the peritonæum, communicate with the members above and below them.

320. The common covering produces itself in order, into all its viscera, and from the viscera into all their parts, and parts of parts, all the way to the very innermost; with a view to keeping them all in connexion; to giving all things distinct limits; to properly distributing among all, the powers and actions of their superiors, according to the nature of their functions; and to reducing the forces and motions of all to one constant standard; and thus repressing impetuosity, quelling disturbances, and like a balance, equalizing the vibrations of the scales. Therefore the peritonæum produces itself, or puts forth from itself fræna and ligaments, whereby it binds down, coerces and suspends the viscera of this region: from these it proceeds further, and by capsules, penetrates to the innermost and deepest things (*x*), and by similar chains and bits,

(*u*) The principles or beginnings of the nerves,—the primitive fasciculi of the fibres of the cerebrum, cerebellum, medulla oblongata, and medulla spinalis,—are covered by the pia mater, at the first threshold of their departure from their natal soil. These fasciculi, again united or compacted into a nerve, are covered by the dura mater on the second threshold, that is, when they reach the cranium or the vertebræ; and thus they proceed into the kingdom of the body. (See Baglivi, *Specimen quatuor Librorum de Fibra Motrice.)* But they unfold and evolve these coats, in an order corresponding to the explication and evolution of the fibres. The venous channels also—the jugular veins of the cerebrum and cerebellum—likewise borrow and bring with them their coats from the dura and pia mater. The respective parts which the fibres and blood-vessels of the two brains play in the body, will be shewn in the proper places.

(*x*) That the peritonæum passes all the way to the minutest glands curbs the very parts or unities of the viscera,—for example, of the liver and spleen,—and subjects and devotes them to its own general government, and to the government of the whole.

321. The members of the abdomen are kept bound and guarded by the general bond of the peritonæum, and by the still more general bond of the diaphragm, closely and thoroughly, in proportion to their priority of place, as determined by their dignity of office, and to their usefulness in the kingdom; and this, according to the cause which they undertake in the unanimous society. On this account, the stomach (*y*), the

or glandular vesicles of the liver, by means of the capsule which proceeds from the very ligaments put forth by the peritonæum and the diaphragm, see the Chapter on the Liver, n. 207 (*o*), and n. 209. That it passes likewise to the glandular vesicles of the spleen, see n. 242; and to those of the suprarenal glands, see n. 275 (*s*) and (*u*). That the peritonæum passes in a different way into every different member enclosed in the cavity of the abdomen, and wherefore, will be shewn presently.

(*y*) Among the viscera of the abdomen, the STOMACH plays the principal part, and indeed is the principal; for it is a kind of wheel or axis to the whole machine, and all the other viscera, (which are either continuous with it, as the intestines, mesentery and omentum, or which stand around it,) are its lesser wheels and pulleys. The offices of all the members of this region commence from, and revolve round, this wheel or axis: which is the reason that the stomach is more immediately excited and disposed by, and dependent upon, not only the peritonæum, but also the superior chamber containing the lungs, and the supreme chamber containing the brain: primarily, by the immediate immersion of the œsophagus in the universal gorge and cavity of the stomach; for whatever the œsophagus, in its passage through the thorax, derives from the pleura, from the mediastinum, and, in fine, from the diaphragm, it communicates unreservedly to the entire stomach. According to n. 82 (in the Chapter on the Œsophagus), that tube "brings together and unites the lowest things and the highest; it connects all things belonging to the tongue and the palate, to the mouth and the lips; consequently to the temples, the cheeks and the forehead; also to the nares and the cranium, and thereby to the cerebrum; in a word, to the head; as well as all things belonging to the chest—in short, all things above the diaphragm—with all things belonging to the abdomen, or with all beneath the diaphragm; in particular, with the stomach, by means of its coats, external and internal;

liver (*z*), and the renal capsules (*a*), are connected to both bonds in the closest manner; but the pancreas, the spleen and the

and by means of its passage down the mediastinum and through the diaphragm. . . . And as it connects the substances of parts, so it connects their particular and general forces." And according to n. 83, "The œsophagus also puts forth certain motions, which pass uninterruptedly through the stomach, and ultimately through all the viscera subjacent and appended to it; in a word, it constantly impresses, not the cardiac, but the alternate respiratory motions of the lungs," &c. The stomach derives its most external coat from the peritonæum, as a result of its connexion with the diaphragm; respecting which, Winslow says, "The first or external coat [of the stomach] is simply membranous, being one of the internal prolongations or continuations of the peritonæum. This appears evidently at the connexion of the superior orifice with the diaphragm, where the external coat of the stomach is really continuous with the membrane that lines the inferior surface of the diaphragm" (n. 88). In order that there may be no interruption of continuity between the superior chamber, the peritonæum, and the stomach, there also runs forward a ligament close under this very membrane, along both the small and great curvatures, which ligament communicates, repeatedly and throughout, with all the muscular fibres, and thus with the interior coats. Respecting this ligament, and its connexion with the common coat, see Winslow above, n. 88. Along this same ligament, from the insertion of the œsophagus on the further side of the diaphragm, pour in and extend the grand nerves of the cerebellum—the sympathetic nerves—that is to say, the par vagum, with the fibres of the intercostal nerve; by which means, the stomach is made to act completely under the command and at the beck of the lungs, the head, and the cerebellum: besides which, the general motion of the lungs, (by means of the diaphragm,) during every act of the respiration, communicates itself to, and associates itself with, the similar general motion of the stomach. The stomach, as we before remarked, like a great wheel sets in motion the lesser wheels of the abdomen, and performs almost the same office internally, in the middle of the abdominal cavity, as the peritonæum performs externally, or at the sides thereof: and by the meeting of the two, the other viscera, as intermediates, are kept steadily in their stated motions and forces. In order to strengthen this association, the stomach has the omentum bound down and connected to it, and closely inserted along its great curvature; and by this means it has all the abdominal viscera that assist it in its functions, attached to it, and tied up with it. For one omentum, not so closely (*b*): most closely also, the receptaculum chyli, the thoracic duct, the aorta and the vena cava, the emul-

border of the omentum is connected to the great curvature of the stomach; the other, to the great curvature and convexity of the colon; and the commissure of the two is fixed on the right side to the common ligament of the duodenum and colon, and to the neighboring parts of these intestines; on the left side, to the longitudinal fissure of the spleen, to the extremity of the pancreas, and to the convexity of the great curvature of the stomach. The omentum is connected also to the membranous ligament which supports the ductus cholidochus, and by the lobulus Spigelii it is continued to the trunk of the vena portæ. See Winslow above, on the omentum, n. 252. To say nothing of the intestines, (which are continuous with the stomach,) in that they are connected in almost their whole extent to the peritonæum, and externally to the stomach also, by the omentum, as we said above.

(*z*) The *liver* comes next to the stomach, in the dignity, necessity, and usefulness of its office; for the liver is the end and complement of the works of all the abdominal viscera, as the stomach is the beginning. All the chyle that is forwarded by the stomach and intestines into the inferior cava, as well as all the venous blood which irrigates this region, is collected in the channels and streams that discharge themselves into the vena portæ. Thus the liver together with the mesentery, has equally as extensive and important duties as the stomach together with the intestines. The liver conveys its chyle into the inferior vena cava; the mesentery conveys its chyle into the superior vena cava. Thus the liver and the mesentery have this province divided between them: for which reason not only the liver, but also the mesentery and the receptaculum chyli, are so closely connected, both to the peritonæum, and, by means of the diaphragm, to the superior chamber: thus the liver is absolutely subject to the motions of both, or, if you will allow the metaphor, it lies bound, like a janitor at a gate, with the bits and chains of both the thorax and abdomen. "The convexity of the liver," says Winslow, "is connected to the diaphragm by three ligaments, which are only continuations of the membranous lamina of the peritonæum. . . . The right ligament sometimes connects the great lobe to the cartilages of the false ribs; and the left ligament, or that of the small lobe, is often double, and advances towards the middle ligament. The middle ligament begins below, in the great fissure of the liver, near the eminences called portæ. . . . It is fixed all along the upper and inner part of the sheath of the right rectus abdominis muscle, in an oblique gent and spermatic vessels (*c*). From this common obligation or bond, that is to say, from the ligaments put forth by the peri-

manner. The liver is likewise connected to the diaphragm, not by ligament, but by a broad and immediate adhesion, without the intervention of the peritonæum, which is only folded round this adhesion, to form the external membrane of the body of the liver" (n. 194). Hence it appears, that the liver is so tied up to, and suspended by, the peritonæum and the diaphragm, that it can scarcely open or raise itself at all, except according to the breath and respiration of both these parts; and thus, should it happen to be carried away by any irregular motion,—should any rush of hostile blood take place into its chambers, then, by means of these bonds, the attack would be checked at the first onset, and reduced to the natural standard of the pulmonic motion; under the auspices of which all things proceed according to the tenor of nature. Moreover, the common or external membrane of the liver, which is a continuation or appendage of the peritonæum, is propagated in the form of capsules, and penetrates to the minutest vesicles or glands, and covers and completes these likewise, as we shewed in the Chapter on the Liver.

(*a*) The *renal capsules*, like the liver, were among the few members that had a kind of command and leadership, and administered a part of the government of the kingdom, in the primeval, natural, or uterine state; consequently they must be numbered with those organs, which if they do not now excel in dignity of office, yet once did excel: and this is the reason why they likewise are bound with both bonds,—with a common bond derived from the peritonæum, and with a still more common bond derived from the diaphragm. They are in fact so closely connected to the peritonæum that lines the inferior surface of the diaphragm, that they may easily be overlooked by enquirers, according to the account of Eustachius. See above, n. 270.

(*b*) The case is otherwise with the *spleen, pancreas and omentum*, not one of which, so far as I am aware, is immediately connected to the peritonæum, still less subnected to the diaphragm; but only mediately, by ligaments, to the intestines, and by the omentum; wherefore these must be counted as members of inferior rank and dignity, and which merely assist and subserve the stomach, intestines, and liver. That this is their character, is sufficiently evident from the fact that they may be excised, and the processes of chylification and sanguification go on in their own way notwithstanding. With respect to the spleen, it has been extirpated in numberless cases. With respect to tonæum and the diaphragm, we may infer and understand, what authority each organ possesses as a member in the empire; and from the influx of blood through the vessels, we may infer what the same organ administers, and what dignity and purple it wears. The organic parts likewise, as the glands, and other similar minute structures in the remote recesses of the several viscera, communicate by capsules, vessels, and nerves, with the common integument, and thereby with the peritonæum and the diaphragm,—each according to the dignity of its cause and office.

322. In proportion as any of the members of the abdomen are loose in their attachments, or more slightly held, confined and bridled by the general bond of the peritonæum, in the same proportion they are more apt and frequent to rush into preternatural modes and motions, and they are more difficult to

the pancreas, see Mangetus, *Theatr. Anat.*, cap. *De Pancreate.* With respect to the omentum, and the result which followed when Galen removed it from a gladiator, see n. 255.

(*c*) Respecting the use and necessity, and consequent dignity of office, of the *receptaculum chyli* and *thoracic duct,* see Chapters VI. and VII. The receptaculum, therefore, is almost doubly connected to the peritonæum, and planted close to the extremity of the diaphragm. The thoracic duct, as it rises along the vertebral column toward the subclavian vein, is covered by the pleura: so that both these parts are absolutely in the common bond. The same may be said of the *great artery,* and *vena cava;* the inferior prolongations of which—the iliac or crural vessels—are likewise covered by the peritonæum. See Winslow, n. 310; and Verheyen, n. 311. The *spermatic vessels,* (which supply the testicles, the ovaries, and the members allotted to generation, and which have one entire province of offices,) in their descent, are entirely covered and defended by a duplicature proceeding from the peritonæum. "The cellular substance," says Winslow, "has several elongations, which have been called, productions of the peritonæum. Two of these accompany and invest the spermatic cords in males. . . . There are other two which pass . . . . with the crural vessels, which they cover" (n. 310). "The exterior lamina," says Verheyen, "at its under part, sends down two processes in the male to the scrotum; these contain the spermatic vessels; in the female they enclose the round ligaments of the uterus. In the scrotum they dilate, and constitute the tunica vaginalis testis" (n. 311).

recal to the standard of nature (*d*). We see this plainly from the case of all those members and organs, the ligaments of

(*d*) This is a second and subordinate reason why some of the abdominal viscera are guarded and covered more carefully than, and in a different manner to, others, by a continuous production of the peritonæum. For the viscera of the body, without exception, besides having natural vicissitudes of motion, that is, general alternations of constriction and expansion, are also subject to extraordinary motions. This is the case with the œsophagus, every time it performs deglutition: with the lungs, in coughing, expectoration, sneezing, yawning, and similar actions: with the stomach and intestines frequently, not only when they and their folds roll back and return the ingesta, but also when they are irritated by medicaments, and food containing sharp particles. The liver also, as it would seem, rises very easily into preternatural motions, and loses its normal shape; that is to say, whenever serous, coagulated, and grumous blood, or not sufficiently digested chyle, is carried into it, perhaps from inactivity of the spleen. So likewise the other viscera. So also the muscles, nearly all the voluntary motions of which are beside or beyond the order of nature, whose motions are most constant, and conspire to the conservation of the state of all things. In order, then, that these preternatural or extraordinary motions may not pervert the states of the viscera, it is the business and office of the peritonæum, as a common bond, to reduce them perpetually to the constant natural motions, that is, to those single general motions, under the auspices of which all things are performed properly. In furtherance of this secondary end, the viscera which are subject or obnoxious to many irregularities of motion,—for instance, the stomach and intestines,—are bound together more tightly by the common bond of the peritonæum: the duodenum also, and the colon, are connected to it, and also to the other members, in very many places and very closely; in order that if they are carried away by any violence, they may be whirled back instantaneously, by a number of powers and forces, to the constancy of nature. For the same end also, the mesentery is wholly surrounded and covered by an expansion of the peritonæum; and this, because it is attached to all the intestines,—parts easily excited to unruly motions, (particularly the ileum, which being under very little restraint, is apt to change its stated motion, and to vary its serpentine inflexions, obediently to every animus and motion of the cerebrum): wherefore the entire mesentery lies in the embrace of the peritonæum, that being thus rendered perfectly safe from all attacks, it may flow back, by an inevitable determination, into the common stream which, either by nature, age, disease, or accident, have become over extended, flaccid, passive, or atonic; or over expanded, inflamed, indurated, inflexible, unyielding, or indolent; or obstructed by callous, tendinous, or cartilaginous matter; or lastly, wasted, or ruptured. Equally so if any of these changes befal the little members or parts of any of the viscera. In proportion then as any part is loose from, and unconfined by, its general—in proportion as its natural tendency to action is impeded; or as it is too little or too much attached to the covering of its particular viscus, and this covering to the peritonæum, and this again to the viscera of the pleura or of the thorax, and these to their cerebra,—in the same proportion its importance is diminished in its society, republic, or kingdom, that is to say, in the unanimous living body. But those parts that are more or less detached—to prevent them from weakening the adjoining parts—are recalled to the general, by the three universal essences which hold the rod of empire, by all sorts of different methods and devices,—sometimes by the imposition of bonds, nooses and fetters (*e*). But if the attempt to effect this prove

of nature. For the same reason, again, the kidneys, the ureters, and the bladder are more or less covered by the peritonæum; namely, that whenever they are forced into unnatural states, they may instantly, as by the pressure of a spring, be restored and glide back to the natural state. Hence, lastly, the spleen is restrained with difficulty, because it is not devoted to the peritonæum immediately, or by any proper ligament.

(*e*) It is perpetually observable in the animal kingdom, that nature has infinite ways of repairing disasters, and for this purpose is always taking the field with new and most ingenious methods and contrivances. Sometimes she inserts glandular congeries in the injured part, and plants in it corpuscules and tubercles of the same nature as itself, either hard or soft,—raising them from a plane surface. Sometimes she inseminates muscular fibres, and carries them round by various paths, and compensates the defect in some measure by means of number. Sometimes she expands the minute lateral ducts of the blood-vessels into considerable vessels, and carries streams of blood through them, and from these vessels again educes other lateral ducts in immense abundance, wherewith she builds membranes, and lays up her new organ within them; this being the origin of steotomata, supernuabortive, it is then all over with such parts, and they are cut off from communion with the rest, and thrown out of the system. Hence it may be seen, that there is the same state, order and form of government in the body, as in a kingdom.

323. All these considerations shew, that the peritonæum is the instrument of union of the members of its society, or the common external bond of the viscera of the abdomen; and the stomach, the similar, but common internal bond of the same viscera; and that these two respect each other mutually, like the circumference and axis of a wheel (*f*). They further shew, that the peritonæum is also the common but internal bond of the muscles, cartilages, and bones of this region. And in order to constitute it such a bond, it requires to be bound down and connected to each member, muscle and part, with express reference to its situation, forces and motion.

324. But as the peritonæum is the [immediate] bond of the members of this region, and the proximate and proper general centre of motion of its little wheels and levers, so it is the proximately remote or mediate general bond between its members and the members of the superior or thoracic region; and the

merary spleens, pancreases, renal capsules, &c., strumous corpuscules, &c. But who can enumerate the methods which the universal essence of the body, that is to say, which the soul brings forth,—not from any faculty either produced or instructed by way of the senses, but from nature, the mistress of all arts and sciences, and the principle of all minds and faculties; and determines without in any way consulting memory, and instantly commits to use; and thus either mends the injured part, or so hides it, and separates it from its fellows in office, that it has no intercourse with them, and no power of infecting the adjoining parts. These statements are supported by an infinity of experience in medico-anatomical records.

(*f*) That the stomach is in a manner the axis of a wheel, see just above, note (*y*). The stomach and intestines extend from the diaphragm, that is, from the superior or cardiac orifice, all the way to the rectum and anus. The other viscera revolve round this axis, as little wheels round the axes of machines. The whole alimentary tube, commencing from the mouth, or aperture of the lips, and extending through the fauces, pharynx, œsophagus, stomach, and intestines, represents a kind of continuous and general axis of the body.

still more remote or ultimate general bond between its members and the organs of the supreme region of the head and cerebrum. And as the peritonæum establishes their connexions as a bond, so it sustains their motions as a centre of motion. It takes up or receives in a general manner, as a uniting medium, the pulmonary and cardiac motions; and as a uniting ultimum, the superior and supreme motions of the head and the cerebrum; it initiates them, distributes them among the viscera of its cavity, collects them again therefrom, absorbs and bounds them. This membrane is, therefore, expansile, contractile, elastic (*g*); variously cellular, constricted and thickened (*h*); tied up and

(*g*) In order that the peritonæum may serve as a bond of the kind described, it must of course be furnished with powers of binding the bodies contained inside its cavity, as well as those extended round it. Elasticity, or extensility and contractility, are of the first importance in the organic body: by means of these properties the peritonæum is rendered accommodable to all the destined uses. In proportion as any membrane ceases to partake of the properties of inelastic or heavy bodies, in the same proportion it is better suited and applicable to all things, even to bodies possessing gravity; inasmuch as it pours forth and communicates fully and entirely all the forces and shocks which it receives. That the peritonæum was exceedingly yielding in its earliest infancy, and from all its individua conspired most distinctly to the general action of the whole, both of the body and of itself,—this is very conspicuous from its state during that time. And that it still continues to have this property, even after adult age, may be seen by examining it anatomically. "It [the peritonæum]," says Winslow, "is flexible, and capable of very great extension, after which it easily regains its ordinary size; as we see in pregnancy, dropsies, corpulency, and repletion" (n. 310).

(*h*) The cellular tissue whereby the peritonæum is applied to the muscles, the vertebræ, and the other external levers, greatly augments and strengthens its power of serving as a common bond; for by this means all assailant external motions are intercepted, stayed, and extinguished in their first effort and onset, and prevented from penetrating far into the internal membrane, and disturbing the motions of the internal viscera. All the external muscles, and the ribs and vertebræ to which they are attached, are subject during the day-time to the dominion of the will; but the motions of the internal or abdominal viscera are exempt therefrom, and devoted to nature alone. Hence, to connected to the last ribs, the vertebræ, the muscles, and the diaphragm (*i*); continued to the œsophagus; produced into

prevent the voluntary impulses from infringing and destroying the order of the natural motions,—albeit in the most general and external manner,—the interior membrane is divided by numberless sections into fine membranous films, which form a cellular down or tissue, and are applied to either the muscles or the bones,—all so many levers which are being momently drawn in new and different directions; by which means, the total action does not communicate, excepting feebly and very generally, with the interior membrane which puts forth ligaments to the viscera of the abdomen. But to go into details,—to describe, for instance, how the peritonæum is inserted into one muscle, and how into another, and why it is adapted to each in the manner it is,—this would be to enter the field of specialities and particulars, whither were we to expatiate while we are dwelling on generals and universals, as in the present Chapter, we should have to write a complete treatise on the subject. "Properly speaking," says Winslow, "the internal portion alone deserves the name of a membranous lamina, as being the main body of the peritonæum. The external portion is no more than a kind of fibrous or follicular apophysis of the internal. . . . [It] is not everywhere of equal thickness," &c., &c. (n. 310). In a word, in every possible situation it is suited to its use,—being constricted, attenuated, thickened, or dilated, accordingly; so that from its mere attenuation, condensation, and attachments, we are empowered to conclude respecting the nature of its function, collection of forces, and accommodation to assailant motions; for not the minutest thread in any cell, still less the entire fabric of a cell, is connected to any fibre, external or internal, without a necessity derived from use. Hence, in the female sex the peritonæum is generally thicker, to meet the case of pregnancy, and of the expansion of the uterine mass. The membrane of the peritonæum, according to Verheyen, "is double throughout, and manifestly so from the umbilicus to the os pubis, in females particularly, in whom it is also thicker than in males" (n. 311).

(*i*) In order that the peritonæum may serve as a common bond, it requires to be most closely connected to the ribs, vertebræ and muscles, in this situation; for in proportion as its connexion to these parts is close, and suitable to their mode of operation, in the same proportion the membrane makes a better general instrument. Thus the more closely the dura mater cerebri is connected to the cranium, the better it acts as an internal periosteum, being proportionably more bound to its office; as during infancy. Throughout old age, however, these internal, and elongated into external processes; saluted by numberless vessels from the thorax (*k*), and by numberless fibres from the brains and spinal cord (*l*); and thus invited and compelled to all manner of obedience. Such are the general statements and results derived from experience in all its spheres, anatomical, physical, and philosophical,—from the threefold experience of art or science, of nature, and of the human mind.

* * * * * *

325. I had intended to treat of the circulation of the humor and serosity through the cellular tissue of the peritonæum, and of its sources and places of discharge; but inasmuch as the foregoing experience has hitherto afforded me no clear and distinct ideas upon the subject, therefore I dare not venture to determine anything with certainty. However, I see plainly enough, that the fluid which irrigates and sometimes inundates the cellular tissue of the peritonæum, comes originally from no other source than the viscera enclosed in the cavity of the abdomen; and that the stream is constantly circulating through the whole of the cellular tissue; and never escapes from its

bonds are usually relaxed by degrees; first those which are relatively simple; afterwards, as life advances, those which are more general, and which gradually adapt themselves to their singulars. Hence when the peritonæum begins to be slackened or remiss, all the operations of the viscera are neglected proportionably. Respecting the mode of connexion of the peritonæum to the muscles, ribs, and vertebræ, see Winslow, n. 309;—its connexion, we mean, to the muscles of the abdomen, the quadratus lumborum, the psoas, iliacus, the muscles of the coccyx and rectum, the longissimi dorsi, sacro-lumbales, and vertebral muscles; also to the cartilaginous portions of the last pair of true ribs, those of the first four pairs of false ribs, the ossa innominata, and the sacrum. See also Verheyen, n. 311.

(*k*) The vessels and nerves afford the best indications of the cause which every member makes in the society, that is to say, in the body. For the vessels and nerves are the essential determinations or emissaries of the universals—the *entia universalia*. What we are to conclude therefrom, will be shewn in the Part on the Organism of Animal cells or follicles into the cavity itself, unless they are ruptured by over distention, or by the wearing out of their membranous partitions; but rather that the peritonæum absorbs the fatty effluvial vapor with which the cavity of the abdomen abounds, and carries it in a determinate channel, with the rest of its serosity, to some place of discharge, which place I have (in Chap. XIV.) maintained to be situated at the kidneys. FIRSTLY; *That the fluid which irrigates and sometimes inundates the cellular tissue of the peritonæum, comes originally from the abdominal viscera themselves:* that is to say, from the stomach, intestines, liver, and other viscera, according to the nature of their communication and operation. This is proved by the cellular tissue of these viscera—by their sometimes having two cellular tissues, one above and one below their muscular membranes; and by their continuity—the continuity of the external cellular tissue particularly,—by means of foramina, with the innermost membrane of the viscus; and by its immediate communication with the cellular tissue of the peritonæum: which latter circumstance is proved by the abundance of serous liquid which comes to the surface [of the viscus], where all but the purer portion, (which is absorbed by the numerous lymphatics,) is committed into circulation through the peritonæum. The free continuation of this cellular tissue from the liver and intestines to the peritonæum, is pretty evident, not only from ordinary anatomy, but also from artificial anatomy, as pursued by injections; for when the cellular tissue of the one is distended by inflation, the adjoining part of the tissue of the other swells. This is likewise abundantly shewn in tympanitis, ascites, dropsy, and other

Motion. The cause which the peritonæum makes is clearly shewn by the vessels and nerves. On account of this cause, the peritonæum is supplied not only by vessels of its own region, but also by great numbers from the superior or thoracic region. According to Heister, "The arteries and veins of the peritonæum come from the epigastric, mammary, lumbar and phrenic vessels" (n. 308): and according to Verheyen, "The peritonæum receives . . . . a few little branches from the spermatic vessels, [and] laterally, from the intercostals" (n. 311).

(*l*) Its nerves come from almost the same quarters as its blood-vessels; that is to say, from the phrenic, dorsal, lumbar, and sacral nerves. See Heister n. 308, and Verheyen, 311.

inundations: and we also see clearly, that the quantity of wind and water in these diseases cannot have any other source than the viscera themselves, where the air is disengaged in great quantities in consequence of the disruption of the solid parts; and the viscera—as the stomach, the small and large intestines, and the liver—are inundated with a constant torrent of water. SECONDLY; *That this stream circulates through the whole of the cellular tissue.* This proposition, being a consequence [of the preceding proposition], admits of no doubt whatever. For granting the continuation of the cellular tissue through the peritonæum, and granting the constant agitation of it by the muscles, consequently the expansion and constriction, the inevitable result must be, that the fluid circulates whither the tissue is permeable, and is determined to certain places of discharge; it cannot stagnate, without producing the destruction of the whole. THIRDLY; *That it never escapes from the cells or follicles into the cavity of the abdomen, unless the membranous partitions are ruptured by over distension.* This is fully proved by the above-mentioned morbid swellings of the peritonæum, as tympanitis, in which the abdomen is turgid with air; ascites, when the cells are distended with water till they resemble hydatids; and particularly dropsy of the peritonæum. For if the internal coat of the peritonæum were pervious into its general cavity, or perforated by little foramina leading thither, then no dropsical stoppage of the kind could thus augment the peritonæum with respect to mass and thickness, but the wind or water would escape immediately through the porous pellicle, and fill the cavity. The same thing may be shewn artificially by the injection of air or water, which distends the cellular tissue, but is not found to escape elsewhere: exactly as in the corresponding coat of the omentum, which like the peritonæum has numberless pores, (so some anatomists have observed,) and yet no evaporation takes place from its little cellular cavities. "When we handle these membranes [of the omentum]," says Winslow, "with dry fingers, the membranes stick to them so closely as hardly to be separated without being torn; as we see by the reticular holes that appear in those portions of the membranes that have been thus handled. In that case it is to no purpose to blow through the orifice already mentioned [the foramen of

Winslow]; and it is owing to these small accidental holes that the membranes of the omentum have been supposed to be naturally reticular" (n. 252). FOURTHLY; *That the peritonæum rather absorbs the fatty effluvial vapor with which the cavity of the abdomen abounds.* For wherever any vapor or humor is present, and indeed supplied in a continual stream, there must necessarily be places of discharge provided; consequently from the cavity of the abdomen, as from all the other cavities. Whether the discharge takes place from this cavity by any external prolongations, as for instance, towards the scrotum, or along the ischiadic vessels, or elsewhere, is a subject for enquiry; also, whither the humors go next after leaving those places. Experience clearly shews, that the peritonæum throughout is full of perforations, and that the foramina afford no passage either for air or liquid into the cavity from the cells outwards. Granting, then, the existence of little foramina, (like those of the omentum, according to the observation just quoted,) and the possibility of their being bibulous and attractile, and the converse follows as a matter of course,—namely, that this membrane is permeable from the cavity into the cells of the cellular tissue. The surface of the internal membrane of the peritonæum, according to Winslow, "is continually moistened by a serous fluid discharged through almost imperceptible pores. These pores may be seen by spreading a portion of the peritonæum on the end of the finger, and then pulling it tight on all sides; for then the pores are dilated, and small drops may be observed to run from them, even without the microscope. The sources of this fluid are not as yet well understood" (n. 310). FIFTHLY; *That the peritonæum conveys this serosity in a determinate channel, to some place of discharge;* which place I have maintained to be situated in the kidneys; and during the uterine state, in the renal capsules. See the Chapters on the Kidneys and Succenturiate Kidneys. The greatest care seems to be taken, to prevent the intestines from communicating with the urinary bladder, through the cellular tissue of the peritonæum, and the more pure and chyliferous portion of the serum from thus entirely escaping by a short cut once allowed to be opened, along with the worthless and aqueous portion. Hence the passage intervening between them, as for instance, about the concentration of the abdominal muscles, is comparatively narrow and constricted. But if we attentively consider the structure of the bladder, it will appear very probable, that the bladder itself does imbibe some part of the thin vapor, although relatively to the whole, a very small part, and like the rest of the viscera, throws it out into its cellular tissue, as into the surrounding peritonæum, and perhaps around the ureters.

EPILOGUE.

326. At the end of each Part of my Analyses, I intend to subjoin an Epilogue, for the purpose of collecting the several uses of the viscera treated of, and the several amounts of their offices, into one sum, and drawing a line underneath it; that is to say, of reducing and connecting the scattered ideas of particulars, and the vague and faint notions, into a single and general idea, just in the same manner as the peritonæum reduces and connects the functions of the viscera of the abdomen (*a*). The whole of the viscera which inhabit or constitute this earth, or subthoracic region of the body, are altogether devoted to preparing and refining the chyle and the blood. Chylification, Sanguification, and Purification are the sum of their offices. These offices form a circle and everlasting chain; that is to say, *chylification must take place; what is chylified, must undergo sanguification; what is sanguified, must undergo purification; and what is purified, must undergo chylification; and so on perpetually.* But let us examine the links of this chain, one by one.

327. *Firstly,* chylification must take place; in other words, food must be taken; when taken, comminuted; and when comminuted, digested into chyle (*b*). *Secondly,* what is chylified, must undergo sanguification; that is to say, the chyle must be introduced into the veins, inaugurated into the blood, copulated with it, and thus itself reduced into blood (*c*). *Thirdly,* what is

(*a*) Respecting the office of the Peritonæum, see the whole of the preceding Chapter.

(*b*) See the Chapters on the Tongue, the Pharynx and Œsophagus, and particularly those on the Stomach and Intestines.

(*c*) See the Chapters on the Liver and Thoracic Duct.

sanguified must undergo purification; that is to say, the purer portion of both the blood and the serum, must be separated from the less pure portion, or that of a middle quality, and from the stale portion, or that of the lowest quality (*d*). *Fourthly*, what is purified, must undergo chylification; the purest blood and its corresponding serum must be remitted into circulation; and the vilest blood and its corresponding serum must be excreted and eliminated; but the blood of a middle quality must be corrected, reduced into salivary juices, and restored to the chylopoietic organs, to serve as a menstruum both for preparing the new chyle, and for introducing it into the blood (*e*). Thus this last link is succeeded by the first of the chain.

328. The *first* department of all, or that of *Chylification*, is

(*d*) The blood, and particularly the serum thereof, as we observed in our Analyses, may be divided into three kinds or classes: to the first class belongs the pure, virgin, newly-conceived or regenerated blood, perfectly accommodated to every use, and perfectly obedient to the cerebrum, the sensoria, and the muscles,—the true universal essence of the third order (n. 314); thus the vicegerent of the soul in the ultimates of the kingdom, that is to say, in the body. To this class also belongs a corresponding serum—a serum kindred and proximate to such blood. To the second class belong the blood and serum of a middle quality,—the blood which is comparatively old, has circulated many times, is less obedient, more obdurate, and acts rather from gravity than elastic force, and being numbered among antiquated things, requires, therefore, to be recruited and regenerated: likewise the corresponding serum, which has in it many elements of the genuine blood, but which are connected and fettered one with another; and which serum, therefore, requires to be corrected and rectified, that is, remitted into the intestines. To the third class belongs the absolutely outworn, lifeless, worthless, rigid, concreted, and inert blood; and the corresponding serum. Respecting the blood and serum of the second order, see Chapters I. and II., where they treat of the Salivary Glands; also the Chapters on the Pancreas, the Spleen, and the Liver; which organs reduce and break up this middle blood, and send it away into the salivary juices,—into the succus pancreaticus, or the hepatic bile. Respecting the blood of the last kind or sort, see the Chapter on the Liver and Gall-bladder; and respecting the serum of the same kind, see the Chapters on the Kidneys and the Urinary Bladder.

(*e*) That only the blood and serum of a middle quality is expended assigned to the stomach, and to the small and large intestines; and that of lustrating and refining the chyle, to the liver. The *second*, or that of *Sanguification*, is assigned to the mesentery, the thoracic duct, and again to the liver. These members introduce the chyle as a bride into the bed of her husband, that is, into the veins, and to the blood; in order that it may be inaugurated, copulated, and itself made into blood. The *third* department, or that of *Purification*, is committed to the pancreas, the spleen, and here again, to the liver, which correct the blood of a middle quality, reduce it into salivary juices, and restore it to the chylopoietic organs, that is, to the intestines. But the worthless serum is committed to the kidneys and urinary bladder; and the corresponding blood, to the gall-bladder, and to the intestines; and they excrete and eliminate it. Thus each member of the abdomen contributes in some respect to the generation and regeneration of the blood, and each in its place, and by its office, takes part in the everlasting circle. Hence all their regards are directed to the blood, which is the common object of the abdominal viscera.

329. The case is different in EMBRYOS, which pass an innocent, golden, and scarcely personal life, in the womb, and neither generate nor produce blood from new chyle, but elect and suck it,—ready filtered and lustrated in the placenta, the umbilical veins, and the liver,—immediately from the maternal store (*f*); and simply transfer it into the nascent viscera; whose incernicula or strainers sift the blood from the serum, and separate both into their component parts; which, again reuniting, they remit, through a short passage, into a simple gyre. This office is enjoined upon the liver and the renal capsules. But in case anything spurious, or of no use to the genuine blood, should be intermingled in the current, with the serum, and not be intercepted in the strainers or pipes by the veins, nor transude or be driven through the parietal foramina; and should it consequently not be raised and sublimated through the lanuginous or downy tissue, to the surface of the viscus,—whether the

in preparing the chyle, see the Chapters on the Liver, the Pancreas, and the Spleen.

(*f*) These subjects will be explained more fully in Part V.

stomach, the liver, the gall-bladder, the pancreas, the spleen, or the kidneys,—and so not reach the cells of the peritonæum (*g*),—in this case it descends to the excretory outlets:

(*g*) Throughout our expositions hitherto it has been stated, and, as I believe, proved from experience, that the innermost membranes of all the abdominal viscera, are not only covered with an infinity of little venous orifices, but are also perforated with immense numbers of other foramina, (this is conspicuously the case in the stomach and intestines); and that these little foramina communicate with the outermost coat, where they meet a cellular tissue which is continuous with the cellular tissue of the peritonæum. The consequence of this structure is, *firstly*, that the most refined humors are partly absorbed by the little lips of the veins, and partly expressed through the little foramina into this cellular tissue, whence they are sent forth to describe a certain field and circle through the peritonæum. And *secondly*, that all the humors secreted from the blood, and passed through the little tubes, are likewise either reabsorbed by the veins in these exceedingly narrow passages, or transuded and expressed in the form of vapor through the cribrous membranes of the parietes: the minute tubes having thus the same nature imprinted upon them as the large cavities. This is very conspicuous in the liver and spleen, and particularly in the kidneys. Surrounding these tubes there is an uninterrupted cellular down, which is continued from the minutest interstices all the way to the surface; so that wherever the humor escapes or is expressed from the tubuli, it rises as on wings towards the surface of the viscus, where it is partly taken up by the lymphatics, and partly effused into the continuous cellular passages of the peritonæum. From these facts, founded upon observations, it necessarily follows, that in the primeval or embryonic state, whatever, either in the great cavities or canals, or in the small cavities and tubuli, is not intercepted by the little veins, and sent through the parietal foramina, pursues its way naturally along the course of the cavity, canal, or tubulus,—as for instance, from the stomach towards the intestines; likewise, from the liver, the gall-bladder, and the pancreas, through their ducts, and again through the ductus cholidochus into the duodenum, and thence into the ileum, the cœcum, and the colon; that is, to the places of excretion: likewise, from the kidneys, through the ureters, into the urinary bladder; and so in all other parts. Such is the origin of the glutinous humor, the meconium, and the urinous fluid in the fœtus. In order therefore that such meconium and urine may not so increase in quantity as actually to open hence the glutinous humor in the fœtal stomach, the meconium in the intestines, in the large intestines particularly, and in the vermiform appendix; the scanty portion of urinous fluid in the bladder, &c.

330. The case is different again in different living creatures, —both in those which walk, or creep, and in those which fly, or swim. In each, the nature of the soul determines the nature of the blood, which acts as the soul's vicegerent in the ultimates of the kingdom, that is to say, in the body: and the nature of the blood that is about to be formed, determines a strong desire for food of a corresponding nature: and the nature of the blood and the food together, produces a corresponding structure in the viscera which prepare the chyle from the food, and the blood from the chyle. In minute reptile and winged creatures, this structure is most simple, merely a web woven in the loom of interior nature, so naked as to be perfectly transparent, and with no swathings or coverings,—no manifold composition of fibres,—to entangle or intercept the sight. Let us then compare these small things with large, for the sake both of confirmation, and of informing ourselves respecting the variety of nature's ways.

331. The PEDICULUS or louse, as appears from its anatomy, lives for scarcely any other object than the indulgence of the gullet and the stomach; for the latter, together with the intestines, fills and distends the greater part of its molecular frame (*h*). From these creatures it is perfectly manifest, that a vermicular

or burst the locks at the ends of the canals, none but select and well-filtered blood is allowed to be emulged by the embryo from the womb, nor is the blood thus purified suffered to be infested and defiled during uterine life by any emotions of the body, the senses, the animus, or the mind.

(*h*) See Swammerdam's *Biblia Naturæ*, p. 74—80, and tab. ii., fig. 3, 6; also our Chapter on the Stomach, n. 91. To understand the remarks which follow, the reader ought to have our author's *Biblia Naturæ* at hand; otherwise the meaning will be obscure. In the work alluded to, Swammerdam shews that this ignoble animalcule leads a most low and earthy life,—that it lives for the body and the belly only; as indeed is the case with all the worm tribe, before they assume the form of nymphs or crysallises, and before they are furnished with and peristaltic motion passes continuously from the gullet to the stomach, and from the stomach to the small and large intestines, but interrupted according to the articulations and intervening spaces of the canal (*i*) : and that this their motion coincides exactly with the respiratory motion of the lungs; and both, with the animatory motion of the brain, and with the systaltic motion of the spinal marrow; for a complete system of pulmonary pipes permeates and almost constitutes the tunics of these canals, and even performs the motive offices of a muscle; the gullet also lies close to the brain, and the stomach, to the spine (*k*). Further, that all these guts or swallows are

wings: when this happens, they rise from the earth, and are carried into the atmosphere as their celestial aura.

(*i*) For the louse has a gullet, a stomach, a small intestine, a colon, and a rectum; also a pylorus, and articulations similar to those of more perfect creatures. "The œsophagus," says Swammerdam, "is a very small canal, which terminates in the stomach. * * * At the lower region of the stomach is seen the pylorus, and immediately after this the small intestine; which is dilated here and there, and formed like the stomach: after this appears the colon, and at the end of the colon there is a manifest dilatation, which is the cloaca: below the cloaca is the rectum." (*Bib. Nat.*, p. 75. 77.) The motion also of these parts was perfectly apparent, for as our esteemed author says, "The motion of the stomach is truly wonderful; insomuch that, by reason of its strong agitations, contractions, dilatations, corrugations, and expansions, (which are plainly seen through the body, and strike one with amazement,) one might suppose it an animal within an animal. It is sometimes observed, that the remainder of the old aliment is mixed with the new food, and shaken and agitated up and down, and on every side, in the stomach." (*Ibid.*, p. 77.) That the same is the case in the stomachs of larger animals, and in the human stomach, may be seen in our Chapter on the Stomach. But in the instance before us, these actions take place still more readily, both because the fabric is simpler, and because the stomach sends two processes into the chest, (see *ibid.*, tab. ii., fig. 3, *k*, *k*,) by means of which it obtains a power of expansion, and of adaptation to every space; for these processes are in a manner general respondents, which swell up as the stomach contracts, and *vice versa*.

(*k*) That the pulmonary pipes, that is, the ramifications of the trachea, so completely pervade the coats of the stomach and intestines, as apparently to constitute the very texture of the external coat, is thus pliable, contractile and expansile, into every possible form of motion, and can make use of any form, with strength, ease and freedom (*l*). Also, that the stomach is furnished with a salivary spring of its own, placed in its own cavity, and which irrigates and washes the materials of the food, suitably to all

stated by our author. "The external coat of the stomach," says he, "is furnished with so great a number of pulmonary pipes, as can hardly be expressed in words. . . . The small intestine is also provided with a great many pulmonary pipes." (*Ibid.*, p. 76, 77.) If then this coat be endowed with these pulmonary ramifications and tracheal pipes, and almost entirely composed of them; and if they are constantly being filled with the air of the lungs, it follows as a matter of course that the stomach and lungs cannot possibly be carried away by contrary torrents of motion. See our declaration respecting the similar coincidence between the motions of the stomach and lungs in more perfect animals, in the Chapters on the Stomach and Intestines. In fact, if we consider duly the numberless ramifications of the pulmonary pipes in insects, it would appear that they perform the office of muscular fibres; for they expand their little tubes in the same manner as muscle expands during the action of its motive fibres; thus they similarly extrude and draw back the coat lying under them, or constructed of them, precisely as the veriest butterflies' wings: respecting which subject we shall speak in our next Part. For in those insects in which, according to our author, "the blood is like cow's milk, and consists of pellucid globules" (*Bib. Nat.*, p. 69), and consequently is not of such a nature that it can flow in and out of the fibres with a rapidity proportioned to the rapid momenta of the nervous fibre,—in those insects, the motive fibres appear to be supplanted by pulmonary pipes; particularly since, according to Swammerdam, no muscular coat of the stomach has been discovered in them: but of these subjects we shall speak at greater length elsewhere. That this motion coincides also with the animatory motion of the brain, may in some measure be inferred from the contiguity of the gullet and the brain, and from the contiguity of the stomach and the spinal marrow. But solitary proofs are never sufficient,—numbers and combination are always requisite. Respecting the contiguity of the gullet to the brain, our author says, "The gullet is a very small canal, situated a little behind the eyes, where it seems to be carried up above the brain." (*Ibid.*, p. 75.)

(*l*) See note (*i*), just above, and the delineations by our author of the forms of the stomach when in motion, *Bib. Nat.*, tab. ii., fig. 6.

the digestive motions (*m*). That prefixed to the large intestines, just at their place of articulation with the small intestines, a number of cœcal or vermiform appendages are inserted—the vasa varicosa—which sprinkle in a liquid menstruum, for mace-

(*m*) There is a little glandular body concealed in the stomach, which our author regards as supplying the place of the pancreas, but Hooke considers that it performs the office of the liver. That in its way it resembles both, appears from a comparison of the organs of chylification, sanguification, and purification, in simple animals, with the organs of the same functions in more perfect animals. For the liver, besides purifying the blood, and bestowing the purified portion upon the chyle, in order that it may be again inaugurated into the blood, also pours the part that is still susceptible of correction, into its salivary juice, that is, into the hepatic bile, in order that it may serve as a salivary menstruum for the intestines. The same is the case with the pancreas: it remits the thickened blood and serum, similarly corrected, into the pancreatic juice, and sends it forth into the duodenum by a duct similar to that by which the liver sends forth the hepatic bile. Inasmuch, then, as both the hepatic bile and the pancreatic juice are salivary menstrua, therefore this single organ here enclosed in the stomach, may serve for both liver and pancreas; particularly since the blood of this animalcule does not require to be elaborated by any very operose process, or in any great quantity. Thus both these viscera—the liver and the pancreas—seem to be here combined in one; that is to say, in the before-mentioned glandular organ, which perhaps also supplies the place of the salivary glands of both the mouth and œsophagus. Hence this organ has its place in the stomach itself, in which all the salivary streams are also concentrated in more perfect animals; and consequently it supplies the place at once of the liver and pancreas, in order that a similar salivary juice may flow down into the small intestine, which is perfectly continuous with the stomach; first, however, being all poured out over the alimentary mass. "Nearly in the middle of the stomach of the louse," says Swammerdam, "there is a certain corpuscule, which Hooke apprehends may be the liver; but I should rather take it to be the pancreas. Its color is somewhat inclining to citrine yellow. It is strongly connected to the stomach. If it be placed under the microscope, it may easily be divided into many little grains, like glands, but these are not very transparent. Pulmonary pipes also appear in it," &c. (*Ibid.*, p. 76, and tab. ii., fig. 3, *m*; also the Chapter on the Liver in this Work, n. 198.) Thus this conglomerate rating the exhausted food in the colon (*n*). Lastly, that the chylified essences do not pass through any elaborate filters or lustratory organs, but immediately through glands planted in the coat of the stomach, and thus through a brief venous course, into the comparatively ignoble blood (*o*). Thus what is made into chyle, is instantly made into blood, and what is made into blood, is purified by the glandular corpuscule placed inside the stomach itself,—and which may be considered as either the liver or the pancreas (*p*),—and is thus a second time made into chyle. Such appears to be the gyre of the generation and regeneration of the blood, in the pediculus.

332. But in no species of living creatures have we a clearer or more distinct manifestation, or better means of judging, of the general use of the viscera, than in the testacea, where the

gland is excited to motion by the stomach, in the same manner as the parotid and other glands are excited to motion by the tongue and palate, according to our explanation in Chapters I. and II.

(*n*) Respecting these vessels, of which insects generally have four, see Swammerdam, *Bib. Nat.*, tab. ii., fig. 3, *p p p p*. "At the end of this small gut," says he, "are discovered four little vessels, called by Malpighi, in the silk-worm, vasa varicosa. They are of considerable length, and of the same structure as the intestines. These are the four intestina cœca, which are found in all insects. They open into the intestine from whence they arise: after these comes the colon." (*Ibid.*, p. 77.) They are very similar to the vermiform process, which is double in fowls, and multiple in some fishes. That they supply a liquid for macerating the refuse of the food, see Chapter V., on the Intestines, n. 138.

(*o*) Respecting these glands, our author says, "The coats consist of a vast quantity of globular granules, which are, however, irregular in form; the greatest part of them, when often touched, comes off from the stomach." (*Bib. Nat.*, p. 76.) That these glandular congeries adhere very slightly to the coats, and nevertheless are abundantly furnished with fibres and absorbents, see Leeuwenhoek above, n. 120: particularly in worms, in which similar glandular and villous coats are generally cleared away about the time the worms change into nymphs and butterflies; and this, in order to prepare the way for a new regeneration of the blood, and a new kind of corporeal life.

(*p*) See note (*m*), above.

external covering or shell is twisted and drawn out into spires corresponding to the very gyres of the governing nature,—as in verticilli or turbines, pinnæ, nautili, cornua ammonis, penicilli, conchæ, and the various species of volutæ and snails. For the common covering, formed to the measure and rule of nature, whether such covering be a skin or a shell, gives to the parts or members which it circumscribes, a more suitable and regular situation, connexion, mutual relation, field of action, and consequently motion; which considered together, call forth and represent the idea of use in a comparatively simple and perfect manner. But it will be sufficient to examine the interiors of one species only. In the COVERED SNAIL—that species which lives in vineyards (*q*)—the intestines are observed to be circumvoluted into a perfectly spiral gyre, for they follow the helix or spiral of the shell (*r*). The mesentery is identical with the liver, and even with the pancreas, and placed in the very top or centre of the whole, that is to say, in the apex of the cone (*s*). Both

(*q*) Respecting the covered snail, see Swammerdam, *Bib. Nat.*, p. 123—125; and tab. v., fig. 6, 7, 8: also, our Chapter on the Stomach, n. 91, and on the Liver, n. 198.

(*r*) That in man, animals, and birds, the intestines are convoluted into a perpetual-circular or spiral gyre, and that the principles of the causes of digestion, and the proportions and rationale of the digestive motions, must be referred to the properties of this form, see our Chapter on the Intestines. We have a particular confirmation of the correctness of these views, in the gyration of the intestines in snails, where the spire is perfectly regular, and follows the spiral revolutions of the shell. "The liver of the covered snail," says Swammerdam, "with the intestines, completes the spiral of the body, * * * making convolutions corresponding to two or three of its turnings." (*Bib. Nat.*, p. 123, 124.)

(*s*) This organic and glandular body which we name the liver, is surrounded and intercepted by the intestines, and enclosed in the same manner as the mesentery in more perfect animals. "The liver of the covered snail," says Swammerdam, "with the intestines, completes the spiral of the body. It is divided into divers lobes, according to the different course of the intestines, which make as many depressions on it as they have turnings and windings." (*Ibid.*, p. 123; and n. 198 above.) That the mesentery, and particularly the cisterna mesenterii the stomach, the intestines, the liver, and all the other organs, here act, constrict, expand and digest, manifestly under the auspices of the pulmonic or respiratory motion, and by virtue of the assistance derived therefrom, express the contents of the belly, as in large animals they express the fæces: for the air is taken in under the shell, round the whole circumference of the body, near the orifice for the discharge of the fæces (*t*). A gland, made up of two parts, and divided into lobules and fringes, supplies both the gullet and stomach with an abundance

or receptaculum chyli, resides in the centre of the whole body, (like the corpuscule here spoken of,) see above, n. 148 and 154. And inasmuch as the mesentery and the liver perform parallel and social offices; that is to say, inasmuch as they both transport the chyle of the stomach and intestines into the veins, the former, by way of the lymphatics, into the vena cava superior, the latter, by way of the veins, into the vena cava inferior, therefore in this comparatively simple reptile, a single organ discharges the offices of both; just as in birds, in which all the chyle passes by way of the blood, and afterwards by way of the liver, into the veins. This organ also assumes the office of the pancreas,—it eructates a salivary juice through a number of ducts into the intestines.

(*t*) This is better seen in the snail than in any other animal: for in this creature, the air is introduced by an orifice not far from the orifice by which the alvine fæces are discharged; and thus introduced, is conveyed all over the circumference of the body, even towards the interiors; so that the animal, by making use of the enclosed air, is enabled to extrude itself entirely from the shell. "The verge or lip," says Swammerdam, "which surrounds the whole body of the snail, is connected very closely to the extremity of the shell, both externally and internally. Underneath the right side of the belly it has a remarkable aperture, which serves to take in air; besides which, it has another, to discharge the fæces. As the snail rolls its body out of the shell, so in proportion it drives the air into the cavity of the verge." (*Bib. Nat.*, p. 111, 112.) Thus it is manifest, that the air in the lungs makes common cause in constricting and dilating both the stomach and intestines, and the other viscera, abdominal and thoracic; and that it concurs to all the general actions of the body, (exactly according to our declaration in the Analyses of the viscera,) particularly to the discharge of the fæces and urine. See our Chapter on the Urinary Bladder.

of saliva (*u*). The liver itself supplies the intestines through a sort of ductus cholidochi (*x*). Further, a glutinous humor is deposited in a certain bladder, termed the saccus calcarius, and with this humor the last fæces are moistened: perhaps also the slime on the depressed and hollow surface of the snail is derived from this source (*y*). From these particulars we may readily

(*u*) Our author thus describes this gland. "Two beautiful vessels," says he, "run along the stomach and the gullet, and discharge themselves into the upper part of the palate. On the inside they are hollow, and contain a clear liquor, which they discharge by two small apertures into the mouth; they are, therefore, salivary ducts. They arise from two small, clear and snowy parts, joined together in the middle, and divided into various lobuli or fringes. These little parts are laterally connected with the stomach by several vessels, which look like so many delicate filaments. They are not fatty, for they are not inflammable, or melted by heat." (*Bib. Nat.*, p. 124, 125.)

(*x*) "The liver," says our author, "is very full of vessels; and it has its particular ductus cholidochi, which discharge themselves into the intestines without any intermediate gall-bladder; in the same manner as in horses, pigeons, &c. The bile of the snail is not remarkably bitter." (*Ibid.*, p. 123.)

(*y*) The most likely conclusion from comparative anatomy respecting this sacculus, appears to be, that it is at once the renal gland and the bladder, and secretes the urinous and glutinous serum, and discharges it by way of the rectum along with the fæces. For according to our author, this animal is polyphagous, and a quantity of undigested chyle enters its blood, not passing through pulmonary vesicles, but only through a little heart of very simple construction; whence the blood in its tardigrade, cold, and slimy body, is to the highest degree phlegmatic, and a part of it requires to be excreted to make the rest suitable and serviceable to the sensory and motory organs. "The blood of the snail," says our author, "is whitish, somewhat inclining to blue." (*Ibid.*, p. 119, 120.) But as there is a perpetual chain of uses in the animal kingdom, so even the excrementitious part of this serum appears to be expended on the ultimate use of moistening the fæces, and perhaps of yielding the slimy matter which constantly covers the concave portion of the animal, and by the assistance of which it adheres to the various branches and stalks in its way: although its very skin is full of glands from which a similar viscid slime exudes. Respecting this little vesicle or bladder, our author says, "The saccus calcarius understand the circle, order, and mode, wherein chylification, sanguification, and purification are performed in the simple fabric of this little creature:—we may understand that the chyle, prepared in the stomach and intestines, with the assistance of the saliva, by gentle motion and friction (*z*), is submitted to a general organ, which may be considered as the liver; and through this gate, as in more perfect animals, passes into the blood, and from it, as a common centre, migrates and ramifies over the whole circumference; also, that the antiquated and mediastinal or worthless blood is recalled to the liver and the salivary gland (*a*), in order to be purified, and in order,—after it has served as a menstruum and hymeneal medium to the chyle,—to be regenerated.

333. In the COSSUS or VERMIS SCARABÆI—the worm of the rhinoceros beetle (*b*)—which lives upon hard wood and sumach,

discharges itself into a pretty large duct, which runs near the rectum, and, in my opinion, opens into it: and accordingly we find a calcareous matter sometimes mixed with the excrements. It exhibits a glandular structure, and as it abounds with grey calcareous humor, hence it always renders water turbid." (*Ibid.*, p. 122.) Similar bladders are also found in some insects, and the fluid which they contain serves to agglutinate the insects themselves, or their eggs, to walls. But this bladder, inasmuch as it is of a glandular structure, seems to resemble both bladder and kidney. These points, however, had better be committed to the exploration of a future age.

(*z*) Inasmuch as the structure of the intestines has a perfectly spiral circumvolution, and the intestines themselves are excited so immediately by the air of the lungs; also, inasmuch as the stomach is provided besides with muscular fibres.

(*a*) That a great part of the blood flows to this common organ, as to the liver and pancreas in more perfect animals, is evident from our author's description. "The liver," says he, "is very full of vessels. It seems to consist of small equidistant granules. It is of a dark brown color, a little approaching to green." (*Ibid.*, p. 123.) Respecting the glands which are the common sources of the saliva, our author says, "A beautiful little vessel runs over their whole surface, and gives a great many branches to each of them." (*Ibid.*, p. 124.)

(*b*) Respecting this worm, see Swammerdam, *ibid.*, p. 313, 314, 315, tab. xxvii., fig. 11, 12; and our Chapter on the Stomach, n. 91.

we find a more remarkable structure in the organs that reduce the aliment into chyle, the chyle into blood, and the blood again into chyle. The ventricular cavities are both larger and more numerous, united together by short and narrow tubuli, and crowned and beset with ducts opening inwards in various directions and eructating a salivary humor (*c*): besides which, at the lower part of the stomach, near the pylorus, there are certain little cœcal intestines, or fistulæ varicosæ (*d*). All these chylopoietic organs are so constructed, placed in such successive order, and irrigated by such numerous salivary streams and

But the structure of this stomach can scarcely be comprehended from any mere description: it ought to be seen in the figure.

(*c*) On these subjects, our author says, "The stomach [of the cossus] is very narrow where it forms the gullet; a little after it expands, until it makes its upper orifice, answering to the œsophagus. In that part, the stomach is armed with about seventy little tooth-like processes. These are divided into six rows; of which the two upper ones point inwards, but the other four point partly forwards, partly backwards. All these tooth-like tubuli open into the stomach, in the same manner as the appendages, or the pancreas in fishes, opens into the intestines: each opens separately into an ecphysis, as in the salmon. A little lower down, the stomach is rendered conspicuous by twenty-two whitish, glandular tubes, with their ends directed backwards. Finally, about the lower part of the stomach, a little above the beginning of the pylorus, we again observe thirty such tubes, which open into the stomach in this situation, with their ends pointing forwards," &c. (*Bib. Nat.*, p. 313, 314.) But a still better idea may be obtained from Swammerdam's Figures and his descriptions thereof; where we shall find that the gullet is small at first, and afterwards dilates like a bell: that it is succeeded by the narrow superior orifice: and this immediately, by the upper part of the stomach, of an interrupted cylindrical form; this again, by a narrow orifice: that after this, the stomach is of considerable length; being followed by the small intestine, which ends in another of large size, that may be termed the colon: this ends in the rectum, which is incurvated and reflected towards the stomach.

(*d*) Respecting these little cœcal intestines, which are four in number, see our author, *ibid.*, p. 319. And that in the cœca of the worm of the bee, he found coagulated excrements, *ibid.*, p. 415. Hence it follows, that these cœca, like the liver and gall-bladder, are designed for excreting the scoriæ of the blood.

rivers, in order to enable them to scrape down, shave, levigate, corrode, reduce and dissolve, the barren, juiceless, and crude aliment upon which the creature lives (*e*); and this, by continual agitation, elixation, and maceration; that is to say, by repeated acts of kneading, stirring and cleansing, and by repeated changes of place, and supplies of more active menstrua. *Chylification* is performed in this manner. But what is chylified, seems to be immediately absorbed by the venous orifices, and by the glands; and this constitutes *Sanguification*. The sanguified product again, is refunded into the numberless glandular ducts, and this constitutes *Purification*, or *new Chylification*. The inference deducible from these anatomical phenomena, is, that the numerous tubular appendages and intestinula cœca, sustain the office of both liver and pancreas, and also of salivary glands. Circumstances, however, are altered when the worm rises out of this obscure and simply vegetative and corporeal life, and is intoduced into the freer and less limited life which it leads as a scarabæus (*f*).

334. But it is useless to prosecute the subject into further details; for the variety is infinite; there is no conceivable difference in any particular whatever, nor any intermediate distinction between differences, which does not actually exist. The nature of the principle, capable of assuming all possible varieties or changes of state, represents this immense diversity of images in ultimates, and thereby bewilders human minds,—which scarcely know anything beyond what the eye can see, or transcend the common boundaries of vision,—and astounds them with apparent prodigies (*g*). Nevertheless, in all these infinite varieties,

(*e*) Namely, the wood and sumach. "The stomach," says Swammerdam, "is found distended and filled with chewed wood and sumach." (*Ibid.*, p. 313.)

(*f*) That this worm leads only a ventricular and nutritive life, is evident from our author's description. "The cossus," says he, "is almost filled by the stomach." (*Ibid.*, p. 313.) Respecting the stomach and intestines of the scarabæus or beetle into which the cossus changes, see *ibid.*, p. 319, tab. xxviii., fig. 5.

(*g*) In minute reptiles, for instance, in little worms, whose whole bodies scarcely occupy a single ray of our vision, we meet with more varieties than in all the genera and species of larger animals: for these there is one thing intended, or one universal end; which end is, that a certain deputy and vicegerent something may exist and act in the body, and be the immediate agent of the soul's

little creatures are the play of nature in her simpler sphere, and in that midway where she acts with perfect freedom: where no impediment arises from gravity, (which increases gradually in proportion to the descent to ultimates and composites;) for a force the most active—the principle of gravity—here rules in all the parts and connexions of parts, as in its own peculiar sphere: where also no obstacle is presented, of forms made up of gross muscular fibres, filaments, cartilages, bones, and the like, put together in all sorts of ways, and which naturally involve rest, torpor, or vis inertiæ: where consequently there is no hindrance from bulk or mass,—nor from the general product of all these circumstances,—the forcible separation of the last things from the first. Now inasmuch as in these most simple animalcules, nature takes her own course with absolute freedom, therefore of consequence infinite variety is possible in them; and that such variety actually exists, is evident from their anatomy, as pursued by men of the greatest experimental genius. Not to repeat what we have brought forward already, there are some creatures whose stomachs are provided with little teeth and saws, with which they comminute and grind their hard food—even mud, gravel, and the like—as in a mill; abrading its angles by friction and delay, and reducing all irregular pieces and refuse globules with the most suitable menstruum, which their organs prepare in the most skilful manner. There are other creatures again, as the working bees, and other nimble honey-suckers, which in their stomachs not only refine the chyle, and convert it instantaneously into blood; but also prepare a store of future chyle or pabulum. There are others which consume and ruminate the rude and primitive chyle, and mingle it occasionally with new chyle: others, which frequently recal the food itself, and transmit it from one stomach to another, according to every state of requirement and digestion. In a word, there are infinite varieties, all arising, as we before pointed out, from the state of the essences of prior nature. By an infinity of phenomena we are presented with the following Truth,—That to the highest and simplest Nature, as she descends by means provided, through many degrees of composition, into the ultimate world, all things accommodate themselves, so that nothing can prevent her from producing a plenary image of herself in ultimates. Thus it is an everlasting truth, that the Spiritual World holds the physical and material world at its beck and nod, subject and perfectly ready to yield it all manner of obedience.

behests therein, instead of the soul itself, and yet act under the auspices of the soul: this something is the blood. But the universal medium for the creation of the blood, is the food and aliment, from which, as a beginning, an entire series of means is described, which series ends in the blood, and commences from the blood. The instrumental causes, which in the living body are termed organs, are in a similar series; and they are the viscera, those of the abdomen particularly, by whose ministrations, this illustrious offspring, the blood, with respect to its corporeal part, is generated, formed, and born. These viscera constitute the lowest region of their body or world, as it were, an earth or ground, which, to pursue the analogy, produces from the food certain noble progeny, as it were, trees and shrubs. On the other hand, the viscera of the thorax, which constitute the middle region,—as the lungs, with their air and respiration,—press and actuate the inferior viscera, as the atmospheres press and actuate the earth. But we have sufficiently cultivated the present ground: we shall now therefore pass on to the higher or thoracic sphere.

END OF PART I.

SCIENTIFIC WORKS BY THE SAME AUTHOR.

Parts II. and III. of the ANIMAL KINGDOM are already translated, and will be put to press as soon as the requisite number of Subscribers is procured.

CONTENTS:

Part II.—Chapter 1. The nose and the uvula. 2. The larynx and the epiglottis. 3. The trachea. 4. The lungs. 5. The pleura, mediastinum, and pericardium. 6. The thymus gland. 7. The diaphragm.—Epilogue.

Part III.—Prologue.—Chapter 1. The skin and the sense of touch. Organic forms generally. The sense and sensorium of touch in particular. The use of touch. 2. The sense of taste.

IN THE PRESS,

(Translated from the Latin by the Rev. Augustus Clissold, A.M.)

THE PRINCIPIA.

CONTENTS:

Part I.—Chapter 1. The means of attaining to a true philosophy, and the character of the true philosopher. 2. The first natural point, and its existence from the Infinite. 3. The first, or simple finite. 4. The second finite; also generically on the active of the first finite. 5. Specifically on the active of the first finite. 6. The first element of the mundane system. 7. The actives of the second and third finite. 8. The third finite, or substantial. 9. The second, or magnetic element of the mundane system. 10. The existence of the sun, and the formation of a solar vortex.

Part II.—Chapter 1. The causes and mechanism of the magnetic forces. 2. The attractive forces of magnets. 3. The same when the poles are alternated. 4. The attractive forces at the magnetic equators. 5. The repulsive forces of magnets. 6 The forces of attraction between the magnet and iron. 7. The influence of the magnet upon ignited iron. 8. The penetration of effluvia through metals. 9. Chemical experiments upon the magnet. 10. The friction of iron upon the magnet. 11. The conjunctive force between the magnet and pieces of iron. 12. The motion of the magnetic needle. 13. The modes of rendering iron magnetic. 14. The declinations of the magnet reduced to calculation upon the foregoing principles. 15. The causes of the declination of the magnet. 16. Calculations of the declination of the magnet.

Part III.—Chapter 1. Comparison of the sidereal heavens with the magnetic sphere. 2. The diversities of worlds. 3. Resumption of the argument on the fourth finite. 4. The solar and planetary chaos. 5. Ether. 6. The fifth finite. 7. Air. 8. Fire. 9. Water. 10. Aqueous vapor. 11. The progression of the earth from the sun to its final orbit. 12. Paradise and the first created human being.

PREPARING FOR IMMEDIATE PUBLICATION,

(Translated from the Latin by the Rev. Augustus Clissold, A.M.)

THE ECONOMY OF THE ANIMAL KINGDOM.

CONTENTS:

Treatise I.—Chapter 1. The composition and genuine essence of the blood. 2. The arteries, veins, their coats, and the circulation of the blood. 3. The formation of the chick in ovo, and the rudiments of the arteries, veins, and heart. 4. The circulation of the blood in the fœtus: the foramen ovale, and the ductus arteriosus of the embryonic and infantile heart. 5. The heart of the marine tortoise. 6. The arteries, veins, and coronary vessels of the heart. 7. The motion of the heart of the adult. 8. Introduction to a ratiocinative psychology.

Treatise II.—Chapter 1. The motion of the brain,—shewing that it possesses an animatory motion which coincides with the respiratory motion of the lungs. 2. The specific nature of the cortical substance of the brain. 3. The human soul.

These Works will be printed uniform with Part I. of the ANIMAL KINGDOM.

LONDON: WALTON AND MITCHELL, PRINTERS, WARDOUR STREET.

Printed in Dunstable, United Kingdom

91220863R00308